MONOGRAPHS ON
PHYSICAL BIOCHEMISTRY

EDITORS

A.R. PEACOCKE, W.F. HARRINGTON

ELECTROPHORESIS

Theory, Techniques and Biochemical and Clinical Applications

SECOND EDITION

ANTHONY T. ANDREWS

Food Research Institute, Reading

CLARENDON PRESS · OXFORD

Oxford University Press, Walton Street, Oxford OX2 6DP
Oxford New York Toronto
Delhi Bombay Calcutta Madras Karachi
Petaling Jaya Singapore Hong Kong Tokyo
Nairobi Dar es Salaam Cape Town
Melbourne Auckland
and associated companies in
Berlin Ibadan

Oxford is a trade mark of Oxford University Press

Published in the United States
by Oxford University Press, New York

First published 1981
Reprinted (with corrections) 1983
Second edition 1986
Reprinted 1987 (with corrections), 1988 (twice)

British Library Cataloguing in Publication Data
Andrews, Anthony T.
Electrophoresis: theory, techniques, and
biochemical and clinical applications.—2nd ed.
—(Monographs on physical biochemistry)
1. Electrophoresis
I. Title II. Series
541.3'7   QD79.E44
ISBN 0–19–854633–5
ISBN 0–19–854632–7 Pbk

Library of Congress Cataloging in Publication Data
Andrews, Anthony T.
Electrophoresis: theory, techniques, and biochemical
and clinical applications.
(Monographs on physical biochemistry)
Bibliography: p.
Includes indexes.
1. Electrophoresis. I. Title. II. Series.
QD79.E44A53 1985 543'.0871 85–8847
ISBN 0–19–854633–5
ISBN 0–19–854632–7 (pbk.)

Printed in Great Britain
at the University Printing House, Oxford

PREFACE
TO THE
SECOND EDITION

ALTHOUGH it is only just over three years since the appearance of the first edition of this book, it covers a field that is extremely active in terms of both exploitation of the methodology described and development of techniques. It is not surprising therefore that in spite of this relatively short interlude there have been a number of significant advances which merit inclusion. The considerable success of the first edition, which has accelerated the advent of this revised edition, has given me the opportunity to incorporate many of the most recent advances and also to take due note of the many helpful comments from colleagues and reviewers, to whom I am most grateful.

The popular layout of the first edition, in which for most chapters a particular method is introduced with a brief theoretical discussion followed by experimental procedures and the chapter closed with a section on applications, is retained. Apart from a general updating and the inclusion of over 200 additional new references there are a number of more substantial improvements. In Chapter 2, which remains what may be termed the basic methodology chapter, improved silver staining methods are described and represent a substantial advance over the earlier methods which had just been introduced at the time of the first edition. More importantly, Chapter 2 now also includes a substantial section on transfer blotting techniques, whereby macromolecules are eluted from a gel matrix onto an immobilizing membrane; a procedure which has become very widely used for the mapping and isolation of DNA restriction fragments as well as for RNA and protein work. There has been a dramatic upsurge in genetic manipulation studies and this has been recognized by an updating and expansion on the sections dealing with RNA and DNA sequencing in Chapter 6. In the first edition isoelectric focusing (IEF) and isotachophoresis (ITP) were treated together in a single chapter but with their increasing importance they have now been allocated separate chapters. Chapter 9 on IEF techniques has been very much extended with far more emphasis than previously on methods in gel media, including the measurement of protein titration curves, techniques for modifying the shape of pH gradients and the recently developed Immobiline systems for IEF in immobilized pH gradients, which appear to have considerable future potential. While a number of less widely used methods remain grouped into a single chapter (12) of miscellaneous techniques, during the last few years there has been a very great increase in two-dimensional procedures, particularly for protein mapping, in both purely research applications and in those with potential for the clinical diagnosis of disease. It is clear therefore that two-dimensional procedures now merit more extensive discussion than could be given in one section of a chapter, so they are now covered in much more detail in a Chapter (11) of their own.

Most of the material in the first edition has stood the test of the relatively short time that has elapsed quite well, so very little of it has been omitted for this second edition, which has therefore grown substantially due to the inclusion of new material. I believe that this new material not only brings the book thoroughly up to date but should further broaden its appeal to students of biochemistry at all stages of their careers by opening up new vistas in experimentation, data handling and interpretation that will extend electrophoretic methods into a still wider range of applications in the future and confirm their dominant role as analytical techniques of prime importance.

Shinfield   A.T.A.
March 1985

PREFACE
TO THE
FIRST EDITION

THE key step in the great majority of studies on molecules of chemical, biochemical, or biological interest is a separation process. This is true whether the objective is merely to analyse a complex mixture, to describe its composition in terms of what is or is not present, or to separate and isolate a particular constituent for further examination. When the mixture involved is a natural product, including such materials as plant and animal tissues and fluids, the great majority of separation regimes involve at least one of the various forms of electrophoresis. Within the last 20 years or so electrophoresis has rapidly evolved from a generally low resolution method of relatively limited application into a wide variety of analytical and small-scale preparative techniques of unrivalled resolving power and exceptional versatility. These qualities have resulted in a virtual explosion in their use and it is probably no exaggeration to say that well over half and perhaps as many as three-quarters of all research papers in the whole field of biochemistry make some use of electrophoresis.

This tremendous expansion means that methods are being constantly improved and modified, new variations introduced, new equipment built, and yet new areas of exploitation opened up. During the past few years a number of colleagues and students have come to our laboratory either in search of rather general information on the capabilities of electrophoretic methods or for comment on a specific separation problem. It became apparent that there was a gap between a small number of rather elementary treatments of the subject and a rather larger number of highly specialized works. The latter include symposium proceedings or books with individual chapters contributed by leading experts. Excellent although these are, they generally assume some measure of basic knowledge or expertise on the part of the reader. This book is intended to fill the gap.

If I have succeeded, I hope that the student or scientist who has no previous knowledge of electrophoresis may come to understand the principles and capabilities of the various methods which fall within the general definition of electrophoresis and then go on to apply them to specific problems. I have included a wide range of methods in some detail and many of the most recent developments in technique and application which I hope will be of interest and value to the research worker who already uses electrophoretic methods. Inevitably in a book by a single author the choice of what to include and what to leave out, as well as the views expressed, must to some extent reflect his own personal prejudices. I must apologize therefore to those who feel that their own contributions to the field should have been included or more favourably commented upon, but I hope I have achieved a reasonable balance.

The book places strong emphasis on the various forms of gel electrophoresis, particularly on the methods based on polyacrylamide gel which in recent years have largely superseded earlier techniques. The basic approach to this is described in Chapter 2, which includes a great deal of the methodology common to most electrophoretic measurements. I hope this provides a sound basis from which the reader can go forward with confidence to consider the various aspects covered in later chapters. Each chapter covers the equipment required as well as the actual performance of experiments and ends with a section on biochemical applications which includes reference to many of the most recent uses to which the particular method has been put. Some of these will be to areas of clinical importance but a more general review of clinical applications is reserved until the last chapter. As implied above, I have included extensive experimental details and I hope therefore that this book will find a place as a useful handbook at the laboratory bench and not merely as a reference book sitting on a library shelf.

Shinfield   A.T.A.
June 1980

ACKNOWLEDGEMENTS

I SHOULD like to thank all the authors and publishers who gave permission for the reproduction of their material. Mr R. A. Young kindly prepared a number of the photographic plates and Mrs J.M. Geens drew some of the figures, and Mrs B. Boel helped with many of the last minute corrections and modifications. My sincere thanks to all three of them. I should like also to thank a number of colleagues for their encouraging remarks during the preparation of the manuscript and particularly Dr G. C. Cheeseman for reading it through. However, pride of place among the acknowledgements must go to my wife, Caroline, who not only endured many solitary evenings while I worked but then demonstrated extreme heroism by typing the entire manuscript and helping me to check through it—a truly magnificent effort.

TO MY WIFE

whose long-suffering patience and support has made this book possible

CONTENTS

1
INTRODUCTION

THE scientist beginning a study of a material of biological origin or of a biological process is often faced with the problem of having to separate and examine the properties of molecules of high molecular weight such as proteins and enzymes, nucleic acids, and complex lipids and carbohydrates. In almost all cases it is also necessary to cause as little damage as possible to the molecules so that their properties are not changed significantly. Thus current separation methods usually lean heavily on physical processes which cause the minimum disturbance to both the physical and chemical properties of the molecules and which result in the maximum retention of any biological activity which the molecule may possess.

Within the fields of protein and nucleic acid chemistry the vast majority of separation methods fall into three broad categories, namely those based on size differences, those based on differences in the electrical charge carried by the molecules, and those based on some specific biological or chemical property of the molecule under investigation. It must be stressed at the outset that it is not always possible or necessary that these categories should be completely distinguished from one another. For example some separations based primarily on charge differences are influenced to some extent by size differences, as will be seen later. However, separation methods which are based mainly on size differences, such as gel filtration, ultrafiltration, and the use of ultracentrifuges, and also methods based on some specific property of the molecules, such as affinity chromatography, immuno-adsorption, precipitation procedures, etc., fall outside the scope of this book and will not be considered further.

The two general areas covered by separations based completely or largely on charge differences between molecules include those using ion-exchange media and those using the various forms of electrophoresis. The former can be used for both analytical and preparative work and may be comparatively readily scaled up to the industrial scale. Although the different methods of electrophoretic separation have their own particular advantages and disadvantages, at present they are all confined to the analytical or small-scale preparative range. At the analytical level electrophoretic methods can be unrivalled for resolution and sensitivity.

1.1. Basic principles of electrophoresis

The term electrophoresis is used to describe the migration of a charged particle under the influence of an electric field. Under conditions of constant velocity the driving force on the particle is the product of the effective charge on the particle Q and the potential gradient E, and this is balanced by the frictional resistance f of the medium. In free solution this obeys Stokes' law so that

$$f = 6\pi r v \eta,$$

where r is the radius of the particle moving with velocity v through a medium of viscosity η. However, Stokes' law is not obeyed strictly in gels, and f then depends on a number of factors which include gel density and particle size (Maurer 1971).

The electrophoretic mobility m is defined as the distance d travelled in time t by the particle under the influence of the potential gradient E so that

$$m = d/tE \quad \text{or} \quad m = v/E.$$

Thus measured migration distances are proportional to electrophoretic mobilities but direct comparisons between different experiments can only be made if the products tE are equal in all cases. It also follows that, ideally, if all other conditions are equal a second experiment run at double the potential gradient (voltage) for half the time would result in the particle migrating the same distance d as in the first experiment. However, this relation is only approximately true and it is influenced by a number of factors including particularly the effects of the extra heat generated by increasing current. Nevertheless, a quick calculation of volts multiplied by time can be a useful rough practical guide when seeking an idea of how altering either of these two factors will influence the course of an electrophoretic separation.

The potential gradient E also corresponds to the ratio of the current density J to the specific conductivity κ, so the velocity of the charged particles can also be expressed as

$$v = Em = mJ/\kappa.$$

Most of the large molecules with which we shall be concerned possess both anionic and cationic groupings as part of their structure and hence are termed zwitterions. Since the dissociation constants (pK values) of these groups will differ widely, the net charge on such a molecule will depend upon the pH of its environment so that pH will also influence the mobility of the molecule. The ionic strength determines the electrokinetic potential which reduces the net charge to the effective charge and it is found that the mobility of the charged particle is approximately inversely proportional to the square root of the ionic strength. Low ionic strengths permit high rates of migration, while high ionic strengths give slower rates but in practice sharper zones of separation than low ionic strength buffers (Maurer 1971).

Unfortunately, the higher the ionic strength of the buffer the greater the conductivity and the greater the amount of heat generated. Increasing temperature causes an increase in the rates of diffusion of the ions and also an increase in the ionic mobility amounting to about 2·4 per cent per degree Celsius rise in temperature. At the same time the viscosity of the medium falls with rising temperature. Thus the electrical resistance decreases and at constant voltage the current will rise increasing the heat output still further. The choice of buffer strength then may be seen to be crucial since it effectively determines the amount of electrical power which can be applied to the system. Too high a power input results in excessive heat generation which may lead to an unacceptable rate of evaporation of solvent from the medium, and in free-solution systems can result in convection currents and a mixing of separated zones. In cases which are rather sensitive to heat there may even be a denaturation of proteins or a loss of enzymic activity. In contrast, too low a power input may overcome any heating problem but can also lead to poor separations as a result of the increased amount of diffusion that may occur if the running time is too long.

The removal of heat generated by the passage of the electrical curent is one of the major problems for most forms of electrophoresis since cooling inevitably results in the formation of a temperature gradient between those parts of the medium that are better cooled than others. Because of the factors discussed above, any temperature gradient or temperature difference will lead to distortions in the bands of the separated molecules due to variations in the rates of migration through the medium. In the forms of electrophoresis carried out in cylindrical tubes or in slabs cooling is more effective at the outer edge of the medium than in the middle and curved bands result.

Heating therefore causes variations in both the current and voltage, and in order to minimize these fluctuations it is usual to carry out electrophoresis with power supplies which can be regulated to provide an output at constant voltage or constant current. Although neither of these can provide complete control of heat generation, they are relatively inexpensive and adequate for most purposes. Recently power supplies delivering a constant power output, independent of changes in the electrical resistance, have become commercially available and are probably to be recommended for analytical work of the highest accuracy. These of course only ensure that heat generation is constant throughout the run and in no way overcome the problems associated with heat production in the first place or of attempts to remove it, so it seems likely that the extra expense of such units is not justifiable in most cases. It is clearly best if electrophoresis experiments could be carried out at constant temperature, and as an aid to this when preparative columns are used they can be jacketed and in other cases the experiments can be conducted in a cold room.

Electrophoresis refers simply to the movement of ions through a medium, so that the factors discussed so far which affect this are applicable to all forms of electrophoresis, whether in free solution as in moving-boundary electrophoresis or when a supporting medium such as starch gel, polyacrylamide gel, cellulose acetate, or paper is employed. However, when a supporting medium is used additional factors may also influence mobility and the sharpness of separation. These include adsorption effects on to the support, inhomogeneities within the matrix of the supporting material, ion exchange with charged groups of support molecules, and electro-endosmosis. The first three of these are largely self-explanatory, but electro-endosmosis generally occurs by virtue of charged groups on the supporting medium. For example, paper has a small content of carboxyl groups and agarose possesses sulphonic acid groups. In neutral or basic buffers these will be ionized and will be attracted towards the anode during electrophoresis. In a solid medium movement towards the electrodes is not possible of course, so the effect is compensated for by a migration of H^+ ions (as hydrated protons H_3O^+) towards the cathode which effectively results in a movement of solvent relative to the support medium. Uncharged molecules are then carried towards the cathode in spite of having no ionized groups. The extent of this electro-endosmosis can be measured by studying the movement of suitable uncharged molecules like urea, dextran, sucrose, deoxyribose, or blue dextran (commercially available from Pharmacia Fine Chemicals AB). Blue dextran is especially suitable since being strongly coloured it can readily be observed without further analysis.

We can see then that the basic concept of electrophoresis is a very simple one but that the progress of the charged particle or ion through the medium is influenced by an almost bewilderingly large number of factors. However, it is precisely because of these various influences, that the principle can be turned to good use by the investigator. For example, if size was not a factor there would be no separation between a large particle with a particular charge and a small one with a similar charge. Likewise, a non-homogeneous supporting matrix is deliberately introduced in the technique of gradient gel electrophoresis to aid separation, while an equalization of the charges of cationic and anionic groups to give molecules with no net charge is the basic requirement for isoelectric focusing. The various methods that will be described in subsequent chapters therefore have a fundamental unity, and their apparent diversity is the result of their development to exploit to maximum advantage one or more of these influencing factors in order to achieve the desired goal.

2

POLYACRYLAMIDE GEL
ELECTROPHORESIS (PAGE).
HOMOGENEOUS GEL AND BUFFER
SYSTEMS

Moving-boundary electrophoresis performed with the components in free solution was historically the first form of electrophoresis to be used widely, but has now been largely superceded by methods employing a supporting medium and especially by electrophoresis on polyacrylamide gels (PAGE). The purpose of a supporting medium is to cut down convection currents and diffusion so that the separated components remain as sharp zones with maximum resolution between them. In most cases it is desirable that this supporting medium should be chemically inert during the separation process. It clearly should be uniform in its properties and be able to be readily and reproducibly prepared. Polyacrylamide gel has been adopted widely for this purpose because it fulfils these requirements very well. It also has the advantage that the composition of the gel can be modified in a controlled way to achieve the best conditions for the problem in hand.

In any discussions of the properties of polyacrylamide gel the nomenclature introduced by Hjertén (1962) to describe gel composition is very useful. It has been widely adopted and will be used throughout this book. In this nomenclature T represents the total concentration of monomer (acrylamide + Bis) expressed in grams per 100 ml (i.e. weight per volume per cent) and the term C is the percentage (by weight) of total monomer T which is due to the cross-linking agent (Bis).

2.1.   The formation and structure of polyacrylamide gel

The gel is formed by the vinyl polymerization of acrylamide monomers $CH_2=CH–CO–NH_2$, into long polyacrylamide chains and cross-linking the chains by the inclusion of an appropriate bifunctional co-monomer, usually N,N'-methylene-bis-acrylamide (or Bis for short) $CH_2=CH–CO$ $–NH–CH_2–NH–CO–CH=CH_2$. The polymerization reaction therefore produces random chains of polyacrylamide incorporating a small proportion of Bis molecules, and these can then react with groups in other chains forming cross-links resulting in a three-dimensional network with the general structure shown in Fig. 2.1.

The concentration of acrylamide used determines the average polymer chain length while the Bis concentration determines the extent of cross-link formation. Thus both are important in determining such physical

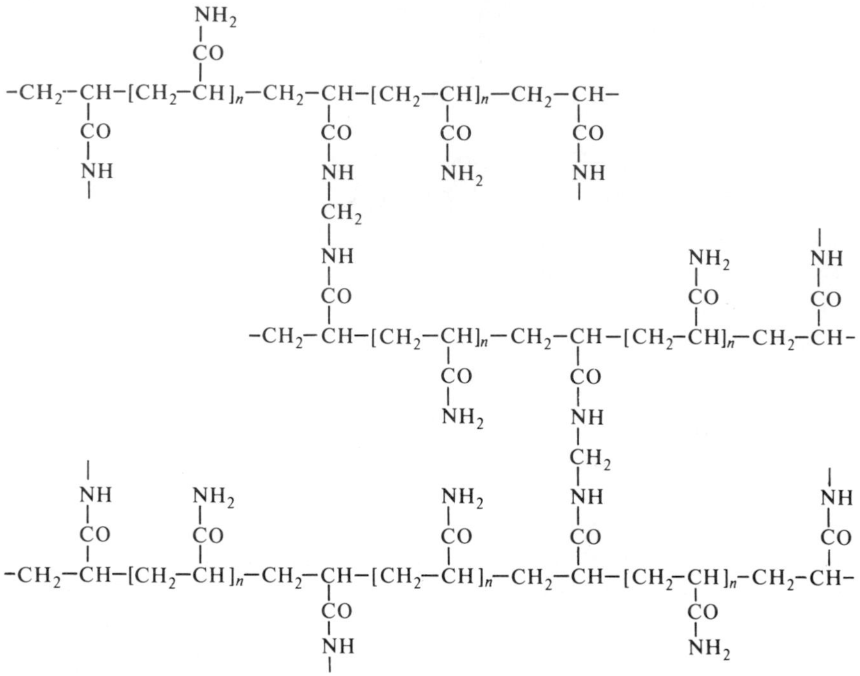

$$
\begin{array}{c}
NH_2 \\
| \\
CO \\
| \\
-CH_2-CH-[CH_2-CH]_n-CH_2-CH-[CH_2-CH]_n-CH_2-CH-
\end{array}
$$

FIG. 2.1. Structure of the polyacrylamide gel matrix formed by copolymerization of acrylamide monomer and N,N′-methylenebisacrylamide cross-linking agent.

properties of the gel as density, elasticity, mechanical strength, and pore size.

The polymerization kinetics of polyacrylamide gels have been investigated by Gelfi and Righetti (1981*a*, *b*), and Righetti, Gelfi and Bosisio (1981), their first paper examining the role of different bifunctional cross-linkers, the names, abbreviations and structures of which are shown in Fig. 2.14. (p.54). From this it is apparent that the order of reactivity in the polymerization reaction with acrylamide is: Bis ≈ DHEBA > EDIA ≈ BAC ≫ DATD. Apart from Bis all other reagents were introduced in order to give gels which after the separation could be readily solubilized to aid the recovery of separated materials or to improve detection (e.g. by counting radiolabelled zones). Gelfi and Righetti (1981*a*) found however that only DHEBA was nearly as good a cross-linker as Bis, and that EDIA and BAC tended to give gels with poorer mechanical properties and rather variable sieving properties (see below), but they could undoubtedly be useful for certain applications. They recommend that DATD should never be used because it is actually an inhibitor of the polymerization reaction and gives few cross links. This results in gels still containing large amounts of unpolymerized monomer, which is a neurotoxin even by skin absorption, so that the gels are also hazardous to handle. In all cases the rate of the polymerization reaction was also dependent on the proportion of cross-linker used (percentage C), so that for $T = 5\%$ gels if C was 3–5% using Bis or DHEBA gels were effectively completely polymerized (defined as at least 95 per cent polymerization efficiency) within about 15 min, while if C was 10 per cent 30–60 min was needed. When EDIA or BAC were used in $T = 5$ per cent gels with C in the range 3–5 per cent at least 3 h was needed for complete polymerization and with all cross linkers when C was more than 10 per cent an overnight reaction time was recommended before the gels were used.

2.2. Pore size effects

In gel media the passage of any particle is hindered by the structure of the matrix, the extent depending upon the relative sizes of the particle and the pores in the gel network. Therefore both molecular size and charge play a role in the separation process. In agar gels the pore sizes are large, so the influence of size (molecular sieving as it is often termed) is less apparent for most proteins than it is in starch or polyacrylamide gels. The properties of starch gels differ considerably with the type of starch used and from batch to batch so that reproducibility can be a problem, but this difficulty does not exist in the case of polyacrylamide which is made up entirely of synthetic monomers. Polyacrylamide also possesses another major advantage in the relative ease with which the pore size, and hence the degree of molecular sieving, can be altered by simply changing the concentration of acrylamide and/or the proportion of cross-linker.

From gel chromatography experiments Fawcett and Morris (1966) found that with a fixed proportion of Bis the pore size varied inversely with and approximately linearly to the total monomer concentration T. They also found that at any given value of T the pore diameter showed a minimum when C was about 5 per cent. More recent work (Margolis and Wrigley 1975; Gelfi and Righetti 1981; Campbell, Wrigley and Margolis 1983) has shown however that for high values of T, above about 15 per cent, the value of C required for minimum pore size is influenced by T. The pore size for a typical $T = 5$ per cent, $C = 5$ per cent gel with Bis or DHEBA as cross-linker is about 20 nm. As the proportion of cross-linker is increased Gelfi and Righetti (1981a) have found that the reaction kinetics slow down and they suggest that the gel matrix becomes progressively less homogeneous, with clumps of bundles of polymerized fibres forming with comparatively much less dense areas in between. In this way highly cross-linked (e.g. $C = 30$–50 per cent) Bis or DHEBA gels can attain effective pore sizes as high as 500–600 nm (Righetti, Brost, and Snyder 1981). In spite of the fact that the average diameter of a linear polyacrylamide chain is less than 0.5 nm, clumps of the size found in these gels can be large enough to scatter light so it is found that porosity is directly correlated to the opacity of the gel; the more turbid it is, the larger the pore size.

Gelfi and Righetti (1981*b*) also found that the temperature at which gels were polymerized was a very important factor. In spite of the fact that many workers recommend that gel mixtures are cooled and polymerization performed at 0–4 °C, it was quite clear that the polymerization kinetics were markedly inferior at such temperatures, leading to non-homogeneous and non-reproducible gels which were much more turbid than if polymerization with the same gel mixtures was performed at higher temperatures. High temperatures (> 50 °C) are also unsuitable and lead to short polymer chains and inelastic matrices, so it was concluded that the optimum temperature for gel formation was 25–30 °C. Difficulties associated with gels polymerized at 0–4 °C were attributed to the formation of hydrogen bonds between molecules of cross-linker and completely homogeneous, non-turbid gels were obtained even at 1 °C if hydrogen-bond disrupting agents such as 3–8 M urea or formamide were present.

Experimentally it is found that during PAGE both the absolute and relative mobilities of molecules are influenced by many factors which affect the average chain length of the polyacrylamide molecules. These include not only the acrylamide and Bis concentrations and temperature but also the concentration of the catalysts used to initiate the polymerization reaction and the time elapsing between addition of the catalysts and gelation (Kingsbury and Masters 1970).

2.3. Some chemical properties of polyacrylamide gel constituents

The polymerization of acrylamide monomers can be initiated by natural light so both acrylamide and Bis should be stored in brown bottles away from light. A low storage temperature (e.g. 4 °C) may also be beneficial. Solid monomer is quite stable under these conditions, but solutions are less stable and should not be kept for more than 1–2 months. This is especially relevant since it is common practice to prepare concentrated stock solutions. During storage of acrylamide solutions some hydrolysis to acrylic acid occurs. This results in a decrease in pH and may cause reduced electrophoretic migrations. To avoid possible difficulties arising from the presence of acrylic acid, stock acrylamide solutions should be stored with a little basic ion-exchange resin (e.g. Dowex 1 or 2, Amberlite IRA-400 etc.) added to 'mop up' acrylic acid as it is formed. N,N,N′,N′-tetramethylethylenediamine (TEMED) is also slightly light sensitive and should be stored in the dark. Ammonium persulphate solid is stable, but solutions should not be stored for more than a few days and it is usually preferable to prepare fresh solutions of this and TEMED shortly before use.

The purity of acrylamide and Bis may vary somewhat but they are generally satisfactory for use without further purification. Both monomers are highly toxic and for this reason it must be emphasized that further purification should not be attempted without good reason. A number of recrystallization procedures have been published (e.g. Maurer 1971) and in a limited number of cases this may be necessary. Monomers of very high purity expressly intended for electrophoresis are commercially available (e.g. British Drug Houses Ltd., BioRad Laboratories, etc.) and as they are relatively inexpensive they are to be recommended.

The monomers exhibit toxicity via skin absorption or by inhalation of particles of acrylamide dust. Reported symptoms caused by even very dilute solutions include skin irritation and disturbances of the central nervous system. The lethal dose (LD_{50}) in mice is 170 mg kg^{-1}. Once polymerization has occurred the resulting gels are relatively non-toxic and can be handled safely, although it is advisable to wear gloves and avoid excessive contact.

2.4. Choosing a suitable gel and buffer system

A mixture of proteins or nucleic acids to be analysed by PAGE almost always consists of molecules differing both in size and in net charge, and by virtue of such differences they can be separated into a number of distinct zones or bands. Although many factors play a role in the separation most of them are not critical, so that some degree of separation can be achieved under widely different conditions of pH, concentrations of acrylamide and Bis, ionic strength, potential gradient, running time, etc. However, selection of conditions for the optimum separation (Rodbard, Chrambach, and Weiss 1974) is more difficult and often requires a number of preliminary runs in which the influence of changes in these factors is examined.

The nature of the sample itself usually provides some guide to the conditions which are likely to give good separations. For example, if one is examining a mixture of high molecular weight virus proteins or nucleic acids one would choose a large-pore gel with an acrylamide concentration T of less than 5 per cent to avoid excessive molecular sieving. Likewise 15–20 per cent or even more would be a logical choice for small molecular weight polypeptides or histones since such concentrations will accentuate small size differences within this molecular weight range. For separations based mainly on charge differences, large pore gels should be used even for comparatively small molecules.

An example of how pore size or gel concentration can affect the progress of a separation is illustrated in Fig. 2.2. Consider the two proteins A and B, where A is smaller than B but B has the higher molecular charge. It can be seen that in a 5 per cent gel B runs ahead of A and is well separated from it. As the gel concentration is raised the effect of molecular sieving becomes greater and B is retarded to a greater extent than A until in gels of about 9 per cent these two proteins fail to separate. At still higher concentrations they again separate but this time A has the higher mobility. Considering the case of C and D (Fig. 2.2), when the smaller component C has the greater charge and hence higher mobility in large pore gels, increasing the gel concentration leads to a greater separation. There are two possible consequences of this type of behaviour. Firstly, when a mixture of proteins is to be separated it is unlikely that any one set of gel conditions can be chosen which will give the very best degree of separation of each component from all the others, so that all separations must to some extent be compromises. Secondly, if a single band is obtained in a gel this does not

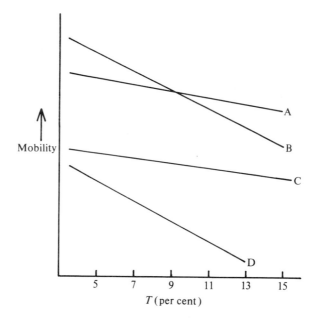

FIG. 2.2. The influence of gel concentration T on the mobilities of sample zones. Considering two different sample constituents A and B, if the larger one (B) also has the higher net molecular charge it will move rapidly at low gel strengths, but as T is increased it will be slowed more than the smaller consituent A by molecular sieving. Considering a different case represented by C and D, if the more highly charged constituent C is also the smaller one then increasing T progressively improves separation owing to the effect of sieving.

in itself constitute proof that only a single homogeneous component is present. To prove homogeneity a second or even third experiment should be run with gels of a different pore size to give different molecular sieving effects or with a gel system of different pH which will alter the contributions due to molecular charge.

If some components of the mixture are unstable or precipitate at a particular pH value, then this should be avoided during the separation. While this seems obvious, it may be difficult to determine precisely what pH conditions arise during the actual electrophoresis, especially when using multiphasic buffers (Chapter 3). In free solution the mobility of a protein increases the further away the pH of the buffer is from the isoelectric point of the protein, so a pH value should be chosen that will maximize differences between sample components. When examining an unknown mixture by PAGE it is therefore sometimes desirable to start with an extreme pH value to ensure that all components move initially in the same direction and in subsequent runs to adjust the pH to maximize the separation. PAGE can be performed at any pH within the range 2·0–11·0 but protein de-amidation and hydrolytic reactions may be severe below about pH 3 or above about pH 10.

As a generalization, basic proteins such as histones are best separated at acidic pH values where their mobilities are greatest, while the great majority of proteins which have isoelectric points within the pH range 4–7·5 and hence are weakly acidic tend to be separated best in alkaline gels in the pH range 8–9. There are of course many exceptions to such a sweeping statement, and in principle two proteins of the same size with differing isoelectric points should be separable at most pH values.

The ionic strength of the buffer system is another parameter that is open to some degree of choice by the experimenter. The ionic strength must be maintained at a level sufficient to keep the sample in solution and also to provide sufficient buffering capacity. The more concentrated the buffer the lower is the electrical resistance, and so the higher the current carried at a given applied voltage and the greater the heat generated. If this extra heat can be adequately dissipated the higher concentration of buffer should result in rather sharper bands being formed since apparent diffusion coefficients are lower at high ionic strength. More usually, however, too high an ionic strength necessitates the use of low voltage gradients to avoid excessive heating. This leads to long electrophoresis times with risks of increased denaturation and band spreading due to diffusion.

In most systems buffer strengths within the range 0·05–0·1 M are typical but concentrations as low as 0·025 M or as high as 1·0 M may be employed. High buffer strengths of over 0·5 M are most usual with strongly acidic buffer systems, since at low pH values many of the weak acids which are commonly used are poorly ionized and high buffer strengths are required to give adequate buffering capacity. Being poorly ionized these buffers do not give rise to excessive current or heating effects. This emphasizes the fact that, although it is almost universal practice to refer to buffers of a particular molarity, the significant parameter is of course ionic strength and not molarity.

2.5. Slabs or cylindrical gels?

When first introduced by Raymond and Wang (1960) PAGE was performed using slabs of gel, but this was shortly followed by the introduction of cylindrical gels in narrow glass tubes (Ornstein 1964; Davis 1964), the so-called disc electrophoresis method. There are currently three different geometrical forms in which PAGE is carried out, namely on horizontal slabs, vertical slabs, or vertical cylinders (rods) of gel. Each has its own particular advantages and disadvantages.

With horizontal gel slabs it is usual for samples to be applied by cutting a slot in the gel into which is inserted a small piece of filter paper impregnated with a known volume of sample solution, as is often done for starch gel electrophoresis. This has the advantage that the sample can be placed anywhere in the slab, so that by positioning it in the middle the migration of components towards either or both electrodes can be detected on a single gel. A further advantage of horizontal slabs is that they can be run on simple inexpensive apparatus which can also be used for other flat-bed electrophoresis systems (e.g. starch and agar gels, paper, cellulose acetate, etc.) or for isoelectric focusing and isotachophoresis.

Vertical gels, either slabs or rods, are technically easier to use with multiphasic buffer systems. With this configuration only components moving towards one elctrode are seen, so either pH conditions must be chosen so that all components migrate in the same direction or alternatively two separate runs must be performed, the second with the electrodes reversed. Resolution in vertical slabs and rods is similar, but more material can usually be applied to slab gels which is useful for preparative work.

A major advantage of slab gels is that a number of sample slots can be formed within a single slab so that several samples can be run side by side under identical conditions and compared without ambiguity. This is particularly useful in screening procedures where a large number of samples must be examined and compared, as in some clinical investigations or when multiple applications of a single sample are made to different slots and the gel subsequently sliced and examined in different ways (e.g. one slice stained for protein, another for enzyme activity, a third for autoradiography, etc.). When a gel rod apparatus is used each sample is run on a separate rod, so that for an accurate comparison of different samples conditions must be identical in all rods throughout the experiment. This can be difficult to ensure.

2.6. The choice of apparatus and of setting-up procedures

Many designs of equipment are available commercially and many more 'home-made' designs are built in individual laboratories. With due care the construction of suitable equipment is a relatively straightforward and simple matter.

(a) *Cylindrical gel rods*

In this method the acrylamide mixture is polymerized in glass tubes usually with an internal diameter of about 4–5 mm and 7–11 cm long. These are then mounted vertically between two buffer chambers fitted with electrodes (Fig. 2.3). Increasing the tube length beyond about 12–15 cm seldom enables one to obtain any worthwhile improvement in resolution, and removal of the gels from such long tubes after electrophoresis becomes much more difficult.

The prime requirements for good and reproducible separations are that all the tubes must have identical dimensions, must be precisely vertical, and should contain the same length of gel, and that the upper surfaces of the gels should be accurately flat so that identical running conditions occur in all gels. The setting up procedure is as follows.

The bottom ends of the tubes are temporarily sealed before introducing the gel solutions, usually by squeezing tight-fitting caps over the ends of the tubes (the polythene caps from small sample vials are often satifactory) or by tightly wrapping the ends with a paraffin wax film such as Parafilm. The tubes are then put vertically into a rack and placed in a water bath at about the temperature at which it is planned to perform the subsequent electrophoresis so that the heat generated during the polymerization reaction is rapidly dissipated. Separating gel solution is then introduced into each tube with a syringe or pasteur pipette (*never* mouth-pipette acrylamide solutions as they are highly toxic) to a level of about 10–15 mm from the tops of the tubes and remove any trapped air bubbles by tapping the tubes. A layer 2–5 mm thick of water or 5 per cent ethanol or isobutanol is then very carefully layered above the gel solution using a syringe with a bent needle or a pasteur pipette with a drawn out and bent tip so that no mixing with the gel solution occurs. No more than 5 min should be taken to cover the surfaces of all the gels in this way. This water layer reduces the surface tension forces at the top of the gel solution and gives a flat surface to the gel. It also protects the surface of the gel solution from atmospheric oxygen which may inhibit polymerization and give rise to an uneven surface.

After the water layer has been added there will be an apparent blurring of the gel solution–water interface due to changes in the refractive index of the gel solution, but once gelation occurs a sharp boundary will be re-established. Once gelation has become clearly apparent the liquid remaining above the gel should be removed and

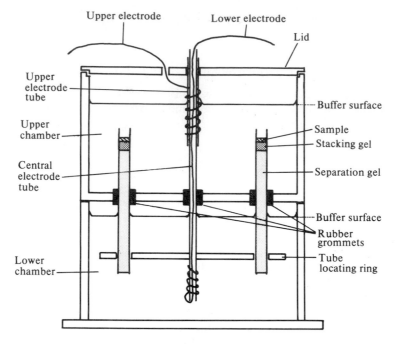

FIG. 2.3. Typical simple apparatus for PAGE using vertical gel rods. The apparatus consists of two circular chambers, one above the other. The gel tubes (usually 10 or 12) are mounted in a circle around the central electrode tube and equidistant from it. They pass through rubber grommets in the base of the upper chamber and then through holes in a locating ring cemented to the central elecrode tube to ensure that they are kept precisely vertical. The lid bearing the upper electrode is placed in position after filling the buffer chambers and loading the samples.

the gel surface rinsed with the buffer used in making up the gel solution (Ames 1974). This is recommended because there will inevitably have been some diffusion and mixing between the gel solution and the water layer, and this more dilute region of gel solution will gradually polymerize over the next few minutes to give gels with an acrylamide concentration gradient in this region rather than gels with a really sharp boundary. The stages involved in preparing gels in cylindrical tubes are shown diagramatically in Fig. 2.4. The gel tubes are then mounted in the apparatus (Fig. 2.3) and the electrode chambers filled with buffer ready for electrophoresis.

Samples can be applied and electrophoresis begun within as little as about 15 min after polymerization or if necessary gels can be stored for up to several days before use. Polyacrylamide gel shrinks and swells depending upon the osmolarity of the surrounding solution so it is best to keep gels under a layer of the gel buffer until use to avoid the development of a concave or convex surface. Stored gels often have slightly different

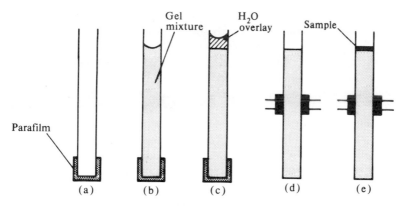

FIG. 2.4. Preparation of gel rods. (a) Glass tubes (internal diameter 4–5 mm) are closed with Parafilm and (b) filled with gel mixture to within 10–15 mm of the top. (c) The gel solution is carefully overlaid with about 5 mm of water or 5 per cent ethanol. (d) When polymerization has occurred the tubes are pushed through the rubber grommets in the apparatus, the Parafilm is removed, and the two buffer chambers are filled with buffer. (e) The tops of the gels are flushed with buffer to remove the water or ethanol overlaying layer and the samples are applied to the gels with a micro-syringe after the addition of sucrose, glycerol, urea etc. to increase the sample density since otherwise they may float upwards and disperse.

electrophoretic characteristics to gels from the same batch run immediately after polymerization. This is because the polymerization reaction follows an approximately sigmoidal course, with an induction period followed by a rapid polymerization stage corresponding to visible gelation and a final 'aging' period in which most of the residual active groups remaining after the gelation stage also gradually react (Richards and Lecanidou 1974a).

Normally such gel rod systems are used analytically with samples of about 10–50 µg applied to each tube of diameter 4–5 mm, but separations of rather larger scale can be performed for preparative work if tubes of diameters up to about 1 cm are used. Within tubes larger than this there will be a marked temperature gradient between the centre of the gel rod and the outer layers next to the glass walls owing to the limited cooling capability. This will result in curved electrophoretic bands.

(b) *Vertical gel slabs*

A very simple 'home-made' design for this type of apparatus has been described by Kerckaert (1978), but there are also many designs of suitable equipment available commercially. Earlier designs of apparatus making gel slabs 3–6 mm thick may be valuable for small scale preparative work, as relatively large sample loads can be applied to them, but are steadily losing favour compared to designs for gels 0.5–1.5 mm thick. With thinner gels staining and destaining steps are faster and more efficient and cooling during the electrophoresis run itself is easier and more even. This in turn permits higher running voltages and shorter running times with consequent improvement in resolution. A widely used design of apparatus that is well suited to this work originates from Hoefer Scientific Instruments (Box 77387, San Francisco, California 94107, U.S.A.) and is also marketed by LKB Produkter AB. In this design (Fig. 2.5) the gels are made in separate moulds which are then inserted into rubber gaskets in the apparatus and immersed in the surrounding buffer, which provides the cooling medium itself being kept cool via water jackets. The use of separate gel moulds means that it is possible to prepare a set of gels while earlier ones are running, and since in the standard apparatus two can be run at a time, it is possible to examine the suitability of many different separation conditions (e.g. with or without a dissociating agent such as urea, various values of T or C etc.) quite quickly.

Another versatile design of apparatus particularly popular for nucleic acid work where efficient cooling is often not necessary (see p.169) is that described by Studier (1973). The apparatus (Fig. 2.6) is particularly simple in concept and employs a gel slab sandwiched between two glass plates. Since glass has a much higher thermal conductivity than Plexiglass only air cooling is required. It is thus easy to modify the apparatus further for gels

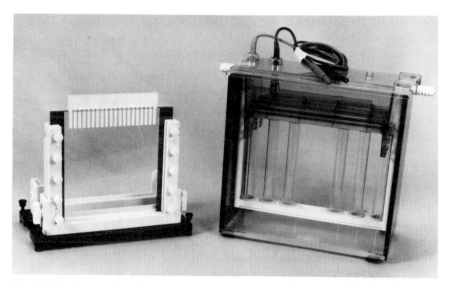

FIG 2.5.Typical vertical slab gel electrophoresis apparatus (LKB Produkter AB).

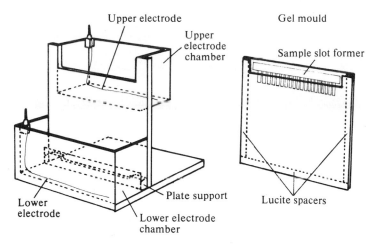

Fig. 2.6. The vertical slab gel apparatus of Studier (1973). The gel is polymerized between two glass plates (140 mm x 160 mm x 3 mm) in a separate mould. The front plate has a slot 20 mm deep cut in it and the two plates are held apart with Lucite spacers 6 mm wide and 2 mm thick. The plates are clamped together and sealed round the edges with melted 1·5 per cent agar solution to ensure a leak-free seal. The acrylamide solution is poured in and the sample slot former added. After polymerization this is removed to leave a series of sample wells. The lower Lucite spacer is also removed, and the mould is then fitted into the main apparatus with the lower end resting on the plate support and clamped into position with the slot in the glass plate aligned with the slot in the side of the upper electrode chamber. A liquid-tight seal is again made with the aid of melted 1·5 per cent agar. Enough buffer is then added to the upper chamber to flow across the slots and cover the top of the gel. The lower chamber is filled with buffer. The samples are applied, and the apparatus is then ready for electrophoresis. (Taken from Studier, F.W. (1973) *J. molec. Biol.* **79**, 237. Copyright by Academic Press Inc. (London) Ltd.)

of different width, thickness, and/or length (e.g. Ames 1974). The standard equipment is commercially available from Aquebogue Machine and Repair Shop, Box 205, Main Road, Aquebogue, N.Y. 11931, U.S.A.

(c) *Horizontal gel slabs*

A simple laboratory-constructed gel mould for horizontal PAGE is shown in Fig. 2.7. This gives gels 3 mm thick, but gels of any thickness can be made in a similar way. The mould consists of three pieces which can all be cut from a single sheet of Plexiglass 6 mm thick. A step 3 mm deep is cut around the edge of the base plate A into which the rim piece B fits closely. To avoid leaks the step is smeared with petroleum jelly or silicone grease before the rim piece is pressed home. The top face of the rim piece is then also smeared with petroleum jelly and the cover plate C placed in position and pressed firmly down onto the greased face of the rim piece.

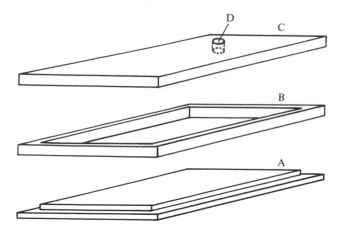

FIG. 2.7. Gel mould for preparation of gel slabs for horizontal PAGE apparatus.

The whole assembly is then fixed tightly together with the help of two or more clamps or spring clips and gel solution is poured into the mould through the hole D, care being taken that no air bubbles are trapped inside. After polymerization is complete the cover plate C is carefully removed. Samples are then applied by cutting small slots in the gel slab and inserting pieces of filter paper (Whatman 3 MM) 5–10 mm wide either presoaked in the sample solution or impregnated with a known volume of it by microsyringe. Alternatively a number of short lengths (5–10 mm long) of plastic rod (1–2 mm square) are glued onto the inside of the cover plate C to produce a series of sample wells in the surface of the gel when the cover plate is removed. The sample solution (5–10 μl) can then be applied directly to the wells with a microsyringe.

The slab is then mounted in a simple electrophoresis tank (Fig. 2.8) of the type also used for cellulose acetate or starch gel electrophoresis. Suitable designs are also available commercially (e.g. Shandon Model 600). Electrical contact is made by lint or filter paper wicks between the gel slab and the buffer compartments. When in place the surface of the gel slab must be covered to prevent drying out during the electrophoresis. If the filter paper sample addition technique has been used, the gel surface can be covered directly with a thin sheet of polythene; otherwise a rigid cover supported just clear of the gel must be used. Typical running conditions with simple uncooled slabs of this kind are overnight at 4 °C with low applied voltages (e.g. 40–80 V for a gel 17 cm long).

More sophisticated, but also more expensive, commercially available apparatus which can be used for horizontal PAGE as well as other forms of horizontal electrophoresis include the Multiphor equipment marketed by LKB-Produkter AB, the Desaphor and Mediphor from Desaga GmbH,

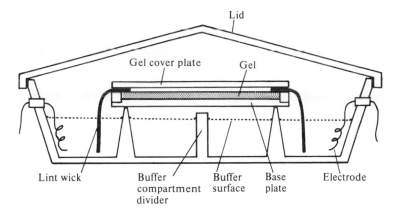

FIG. 2.8. Apparatus for horizontal PAGE. The apparatus is divided into two buffer chambers with one electrode in each and the gel slab is supported 1–3 cm above the buffer surface. The buffer level should be the same in both chambers and the apparatus should be accurately horizontal to avoid capillary flow of buffer through the gel due to differences in hydrostatic pressure.

and the Model FBE-3000 from Pharmacia AB. In general terms the use of this equipment is very similar to that described above and it employs thin (about 2 mm) gel slabs mounted on a cool platen and in which a number of 5 μl or 10 μl sample wells are made with the aid of a slot former. Further details are available in a series of *Application Notes* available from LKB on request and in other literature from the manufacturers.

2.7. Typical gel formulations and pre-electrophoresis treatments

From what we have said so far it will be apparent that as well as differences in the shape and orientations of gels for PAGE (i.e. rods or slabs, vertical or horizontal) there are also a number of different systems which can be used in preparing the gel itself. The most obvious choice which must first be made is whether to opt for a uniform gel phase and homogeneous buffer system or whether to employ a method involving discontinuities in the buffer or gel. In this section we shall confine our attention to the former, while the latter will be discussed in Chapter 3. The only disadvantage of a uniform gel phase and a single buffer throughout is that the resolution is not as good as with non-homogeneous systems. Since there is no concentrating effect in the initial stages, dilute sample solutions are difficult to handle because large volumes of sample give correspondingly broad bands in the gel. If the highest resolution is not necessary for the purpose in hand, this may be outweighed by the advantages of rapid and simple gel preparation. There is also the advantage that in homogeneous systems the precise composition of the buffer is known and the pH value is constant and uniform throughout the separation. This can be helpful when substances with limited pH stability are being separated.

In recent years PAGE has been very widely used for a variety of separation problems, and many different gel and buffer formulations have been employed. Thus a lengthy recitation of precise recipes is impractical but some general guidelines will now be discussed.

Since the buffer system is homogeneous there are no considerations of differing mobilites between various ions to be taken into account, as is the case with discontinuous systems so that any one of a very large number of different acids and bases can be used. Among the most commonly used buffers are those shown in Table 2.1, but this list is by no means exhaustive.

TABLE 2.1

Buffers often used for PAGE in homogeneous buffer systems

Approximate pH range	Primary buffer constituent	pH adjusted to the desired value with
2·4– 6.0	0·1 M citric acid	1 M NaOH
2·8– 3·8	0.05 M formic acid	1 M NaOH
4·0– 5·5	0·05 M acetic acid	1 M NaOH or tris
5·2– 7·0	0·05 M maleic acid	1 M NaOH or tris
6·0– 8·0	0·05 M KH_2PO_4 or NaH_2PO_4	1 M NaOH
7·0– 8·5	0·05 M Na diethyl-barbiturate (veronal)	1 M HCl
7·2– 9.0	0·05 M tris	1 M HCl or glycine
8·5–10·0	0·015 M $Na_2B_4O_7$	1 M HCl or NaOH
9·0–10·5	0·05 M glycine	1 M NaOH
9·0–11·0	0·025 M $NaHCO_3$	1 M NaOH

With regard to gel composition easily manageable gels are obtained when T varies between about 3 and 25 per cent, and gelation does not occur much below $T = 2·5$ per cent even if C is increased to 20 per cent or more. The ratio of acrylamide and Bis concentrations is important in determining the physical properties of the gel, and if C is much greater than 10 per cent gels tend to become brittle and opaque. When C is less than 1 per cent gels become glue-like and are very difficult to handle. Davis (1964) found that as the concentration of acrylamide is increased the Bis concentration should be decreased to give gels with optimum handling characteristics. Richards, Coll, and Gratzer (1965) suggested that within the range $T = 5$–20 per cent the proportion of C of Bis can conveniently be calculated by the formula

$$C = 6·5 - 0·3T.$$

However, this calculated value of C is not very critical, and a commercially available mixture of acrylamide and Bis (Cyanogum 41) which contains 5 per cent Bis is quite widely used for gels within the range $T \approx 4$–15 per cent.

Molecular sieving as well as charge plays a role in the separation of molecules by PAGE, so the acrylamide concentration required to give optimum results depends on the size of the molecules to be separated. As a rough guide the gel concentrations T for the molecular weight ranges shown in Table 2.2 are generally suitable.

TABLE 2.2

Separation ranges of acrylamide
gels of various concentrations T

T (per cent)	Optimum molecular weight range
3– 5	Above 100 000
5–12	20 000–150 000
10–15	10 000–80 000
15 +	Below 15 000

The polymerization of monomers proceeds via a free-radical mechanism which is initiated by the addition of catalysts. A variety of these have been used, but the most common system is ammonium persulphate which produces free-oxygen radicals by a base-catalysed mechanism. The bases most often used to catalyse this reaction are the tertiary aliphatic amines N,N,N′,N′-tetramethylethylenediamine (TEMED) or 3-dimethylamino-propionitrile (DMAPN). Alternatively, free radicals can be generated photochemically using very small amounts of riboflavin, again with the addition of a base such as TEMED. This has the advantage that the onset of gelation can be controlled to some extent by varying the illumination. For both chemical and photochemical production of free radicals the bases TEMED and DMAPN must be in the free-base form, so at acid pH values higher levels of catalysts must be used to avoid undesirably long delays in the gelation reaction or alternative catalytic systems must be employed (see below).

The photochemical polymerization reaction involves the photodecomposition of riboflavin to leucoflavin and in the absence of oxygen no free radicals are formed. However, when traces of oxygen are present re-oxidation of leucoflavin occurs with free-radical generation. Thus, in contrast to the ammonium persulphate reaction, traces of oxygen are needed for gelation to occur, but with both catalytic systems excess oxygen subsequently inhibits the polymerization of acrylamide and may limit chain growth. For this reason acrylamide solutions which have been de-aerated to avoid problems with bubble formation during gelation polymerize much more rapidly after addition of the catalysts than do non-deaerated solutions.

Bubbles, small splits, and areas of separation between the gel and the walls of the glass tubes or slabs are often the result of dissolved air coming out of solution under the influence of the heat which is generated during the polymerization reaction itself. Problems of this type can be minimized either by deaeration of gel solutions or by cooling them before adding the catalysts and by then keeping the gel tubes in a constant temperature bath until after gelation has occurred.

Cooling in this way has the effect of prolonging the time between catalyst addition and gelation but deaeration reduces it. This time interval should be between about 10 and 30 min. Gels which have polymerized in less than 10 min or solutions which have not gelled after 60 min should be discarded because the polymerization will not proceed evenly under these conditions and non-homogeneous gels will result which will give poor separations. At acid pH values slightly longer times may be required but should not exceed about 90 min.

The necessary adjustments which may be required to achieve optimal gelation times can usually be made by changing the amounts of the catalysts added. However, excess catalyst can inhibit gelation by leading to very short polymer chains which cannot form an appropriate gel structure (Richards and Lecanidou 1974a). The optimum concentration of the catalysts at a particular pH depends on the monomer concentration T, the presence or absence of inhibitors, and the desired length of time before gelation occurs. The most suitable amounts are generally within the range 1–10 mM, and equimolar quantities of TEMED and ammonium persulphate are usually found to be best. The persulphate is best added in the form of a small volume of concentrated freshly made up aqueous solution.

The amounts of reagents needed to make up typical gels of various values of T and $C = 5$ per cent are shown in Table 2.3. For gels of differing cross-linkage C the amounts of Bis should be modified accordingly. If photopolymerization with riboflavin instead of ammonium persulphate is desired a concentration of 0·5 mg riboflavin per 100 ml of solution is usual. Photochemical polymerization is less efficient than chemical polymerization and though many workers rely on 1 h illumination this may only result in 50–60 per cent polymerization efficiency, so Righetti et al. (1981) recommend that, although gelation should occur in less than 1 h, illumination should be continued for 8 h or more before gels are used. The amounts of TEMED and catalyst shown in Table 2.3 are only an approximate guide and as stated above should be reduced or increased to give the correct gelation times. In particular, smaller quantities of

TABLE 2.3

The amounts of reagents required to prepare 100 ml of typical poly-acrylamide gel solution

Constituent	Amounts required for gels with		
	$T = 5$ per cent	$T = 7 \cdot 5$ per cent	$T = 10$ per cent
Acrylamide	4·75 g	7·125 g	9·50 g
Bisacrylamide (for $C = 5$ per cent)	0·25 g	0·375 g	0·50 g
TEMED or DMAPN	0·05 ml	0·05 ml	0·05 ml
Ammonium persulphate	0·05 g	0·05 g	0·05 g

ammonium persulphate will be needed if acrylamide solutions are deaerated before use.

When markedly acidic buffers are employed modified catalytic systems may be needed. Stegemann (1967) reported that the addition of 3 mM sodium sulphite as well as the usual TEMED or DMAPN and ammonium persulphate reduced gelation times very considerably. With their highly acidic gels made up in 1 M propionic acid Choules and Zimm (1965) used 6 mg ammonium persulphate and 9 mg sodium sulphite per 100 ml of final gel solution mixture and no TEMED. In similar very acidic gels Jordan and Raymond (1969) used a completely different catalytic system in which 250 μl of 10 wt per cent ascorbic acid and 250 μl of 0·25 wt per cent $FeSO_4$ were mixed with 100 ml of monomer solution and 100 μl of 3 vol. per cent H_2O_2 added immediately before casting the gels.

If sample molecules are insoluble, aggregated, or liable to become aggregated under the conditions that occur during the electrophoresis, it is usual to add urea (up to 8 M or more), glycerol (up to about 70 per cent), or non-ionic detergents (e.g. 5 per cent Triton X-100 or Brij 35). Urea addition is particularly commonly used and usually satisfactory, but it should be remembered that urea is in a slow base-catalysed equilibrium with cyanate (at equilibrium 8 M urea contains 20 mM cyanate which is a large excess over the number of reactive protein amino groups usually present) which can react with some proteins and give rise to spurious bands (e.g. Kim, Yaguchi, and Rose 1969). Cyanate can be removed by pretreatment of urea solutions with a basic ion-exchange resin or cleared out of urea-containing gels by a pre-electrophoresis step in which the gels are run for 30–40 min under conditions similar to those for the subsequent separation before the sample is applied. It is also helpful to buffer urea solutions with tris buffers as this reacts with the cyanate ions and effectively protects proteins in the sample. Treatment of samples with ionic detergents such as sodium dodecyl sulphate, after which individual proteins are separated according to molecular size and independently of charge, is discussed in greater detail in Chapter 5.

If it is necessary for gels to be run under reducing conditions then 0·1–1·0 mM thioglycolate (Brewer 1967; Woods 1967) or 10 mM 3-mercaptopropionic acid (Lane 1978) can be added to the upper (cathode) buffer chamber. The reducing agent will migrate rapidly into the gel as soon as the electrophoresis starts and produce reducing conditions throughout the gel.

Other authors (Arai and Watanabe, 1968; Petropakis, Angelmeir, and Montgomery 1972) have incorporated 1 mM dithiothreitol into gels and buffers to overcome possible oxidation artefacts or loss of enzyme activity caused by residual persulphate. However, the presence of reducing agents during gel preparation may inhibit gelation (Woods 1967) and then electrophoretic introduction of the reducing agent should be used. At the concentrations required to be an effective reducing agent (eg 10 mM or more) 2-mercaptoethanol inhibits gelation and cannot be used.

There are certain other occasions when a pre-electrophoresis step may be beneficial. Gels always contain a small residual fraction of unpolymerized acrylamide monomer and this can react with components in the sample. Likewise the ammonium persulphate catalyst is an oxidizing agent and photopolymerization produces traces of hydrogen peroxide so that both of these can cause oxidation and a loss of enzymic or other biological activity. Such problems can be largely overcome by carrying out a pre-electrophoresis run which will remove any charged ions or alternatively gels can be soaked for a period of days in several changes of buffer.

Of course if the added step of dismantling gel moulds and soaking gels is acceptable, buffers used for preparing gel solutions before polymerization become irrelevant to the separation, because gels can be polymerized at an ideal slightly alkaline pH (with conventional catalyst systems) and the pH changed after polymerization by soaking in several changes of any desired buffer. The same approach can be used to incorporate additives (such as 2-mercaptoethanol) which might interfere with the gelation process, ligands, enzyme substrates, dissociating agents, etc.

2.8. Sample application and the electrophoresis

When homogeneous gels and buffers are used there is usually no concentration of sample components into narrow zones in the early stages of the electrophoresis as there is when multiphasic buffers are used (see Chapter 3). Thus it is important to use comparatively small volumes of sample, e.g. 5–25 μl for a typical 5 mm diameter gel rod. Above about 20–25 μl the thickness of the sample layer begins to have a noticeable effect on the sharpness of the bands and the resolution. If however the sample has a markedly lower conductivity than the surrounding buffer (i.e. is very dilute or has been dialysed versus buffer diluted five-fold with H_2O) then there may be a noticeable zone sharpening (Hjertén et al. 1965), resolution may approach that of multiphasic buffer systems (see p.80) and relatively large sample volumes may be tolerated. With most vertical apparatus (slab and rod type the sample is applied through the buffer in the upper electrode chamber using a microsyringe or disposable micropipette and so the sample must be of a higher density than the buffer. This is achieved by adding small amounts (2–10 per cent) of sucrose, urea or 5–10 per cent glycerol to the sample. It is also common practice at this stage to add a very small amount of tracking dye to the sample. A suitable dye for use in alkaline buffer systems is Bromophenol Blue, while methylene green, methylene blue or Pyronine Y is appropriate in acid systems. The purpose of a tracking dye is to migrate at a higher rate than any of the macromolecular components in the sample and to indicate when electro-phoresis should be terminated (usually when the dye reaches a point about 5 mm from the bottom of the gel). The position of the tracking dye is also useful as a reference point and the mobilities of components relative to it are often measured.

Once the samples have been added the electrophoresis should be started immediately to avoid diffusion of sample constituents. The best electro-phoretic conditions will be determined by all the factors discussed earlier that affect mobility and so differ from one experiment to the next. However, as a very rough guide voltage gradients of the order of 10–30 V cm^{-1} of gel length are often best, and in 5 mm gel rods these correspond to curents of about 1–3 mA per tube. With gels 7 cm long typical separation times are 40–120 min. With gel slabs 20 cm long it may be more convenient to reduce the voltage gradient to 3–5 V cm^{-1} and to leave the sample running overnight.

Provided that temperature rises within the gel are kept small, the rate of migration is proportional to the voltage gradient and separations should be carried out under constant voltage conditions. However, if substantial ohmic heating occurs the rise in temperature reduces viscosity and increases both ionic mobility and diffusion so that if a constant voltage is applied the sample components gradually accelerate during the electro-phoresis run. It is thus quite common to use constant-current conditions. Power packs for electrophoresis should preferably be capable of giving both constant-voltage and constant-current outputs. For apparatus with vertical glass tubes small power packs giving up to about 300 V or up to 50 mA are suitable (e.g. the Vokam power supply from Shandon Southern Products Ltd.). Apparatus of the slab gel type generally requires power supplies with a greater output such as the Vokam 500/150 or Vokam 500/500 (Shandon), Model 500/200 (BioRad), Model 2002 (LKB), or EPS 500/400 (Pharmacia).

2.9.   Gel staining methods

Once the tracking dye has reached the desired position the electric power is switched off and the gel is removed from the apparatus for localization of the separated components. With slab systems it is usually a simple task to dismantle the apparatus and slide the gel into a suitable staining bath. with the gel-rod type of equipment the tube containing the gel is removed from the apparatus and the gel freed from the glass tube with the aid of a syringe filled with water or glycerol solution and fitted with a long (e.g. 6–8 cm) thin needle. This is slowly pushed down between the gel and the wall of the tube while at the same time the tube is rotated with the other hand and the syringe is gradually emptied so that the injected water serves as a lubricant. This process may have to be repeated, if necessary from both ends of the tube, but generally the gel rod either slips free as the needle is withdrawn or can be gently blown out with air pressure from a rubber bulb slipped over one end of the tube. To avoid damaging the gels many researchers use needles with rounded or sawn-off ends. We have found that a very effective procedure for freeing gels is to use a solid wire rod (the plunger from a 10 μl or 25 μl microsyringe is ideal), especially if this is inserted from the top end of the gel where a small volume of buffer solution usually remains and can act as a lubricant. If severe difficulties are encountered, Ghadge et al. (1983) suggest that a piece of rolled Mylar sheet slightly longer than the glass tube is placed in the tube before pouring in the gel mixture. After the separation the Mylar is pulled out bringing the gel out with it. Without this very simple and neat trick, gels of $T = 15$ per cent or more are extremely difficult to remove successfully from the glass tubes, but using it Ghadge et al. (1983) were able to extract gels of up to $T = 40$ per cent.

For staining, gel rods are often placed in individual test tubes, but a staining tray such as that shown in Fig. 2.9 is particularly useful. This consists of a nylon mesh supported in a Perspex frame onto which the gel rods are laid in sequence (to avoid the very difficult problem of how to mark them). The upper part of the tray, which also consists of a nylon mesh in a Perspex frame, is then dropped into place on top of the gels and secured with the aid of two rubber bands around the whole tray. The spacing between the two layers of mesh is about 5 mm so that the gels are free to move slightly but are sufficiently restrained to avoid them getting out of sequence by moving over or past each other. The whole gel tray is immersed in a bath of dye for the required staining time and can then be transferred to the destaining solvent until the gels are adequately destained. All the gels in the run are therefore subjected to identical staining and destaining conditions, a particular advantage for quantitative work, which is difficult to ensure when gels are stained in individual test tubes.

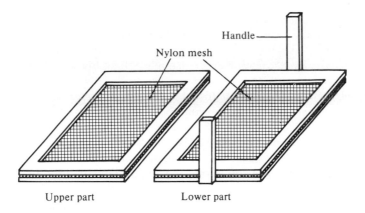

FIG. 2.9. Gel staining tray.

A very large number of staining mixtures are described in the literature, ranging from very general stains to highly specific ones for the identification of a single particular constituent. Comments on these are confined to a few general remarks followed by a more detailed discussion of some more widely used techniques of fairly general applicability.

It is worth pointing out that among the stains for specific constituents there are many for the detection of various types of enzyme activity. Since any form of fixing will often be accompanied by inactivation it is quite common for enzyme staining to be performed with the gels immersed in aqueous buffers, perhaps for several hours, and under these conditions resolution will be greatly impaired by diffusion if the products of the enzymic reaction on which detection depends are soluble and of low molecular weight. In many cases the best sensitivity and localization of activity can be achieved by using methods similar in many ways to histochemical techniques where substrates are converted to sparingly soluble products which can then be coupled chemically to azo-dyes. For example the localization of phosphatase activity (acid or alkaline) with *p*-nitrophenylphosphate or phenolphthalein phosphate which relies on the yellow colour of *p*-nitrophenol or pink phenolphthalein is not nearly so satisfactory as using one of the complex naphthol substrates (e.g. naphthol–ASCl phosphate) and coupling the dephosphorylated product with an azo-dye, such as Fast Garnet GBC (Andrews and Pallavicini 1973).

A particularly mild method is the contact print method which has the merit that the enzyme is not modified, inactivated, or extracted, and so is readily available for further analysis. For this, substrate is impregnated into a sheet of paper (or included in a layer of agarose or polyacrylamide gel on a glass plate) which is then placed in contact with the enzyme-containing gel and incubated at a suitable temperature for an appropriate length of time. The paper or plate is then separated and examined for enzyme activity by whatever method is applicable to the particular enzyme. Alternatively enzymes can often be located by simply cutting up the gel, homogenizing the pieces in a suitable buffer solution (e.g. by extruding them through a syringe), and assaying the mixture or eluates for activity in the usual way.

When the retention of enzymic or biological activity is not an important factor it is normal for the separated components to be fixed in the gel to avoid diffusion and losses during staining and destaining. This fixing can be performed either as a separate step before staining or the fixing agent (e.g. trichloroacetic acid (TCA)) can be included in the dye mixture itself so that fixing and staining take place together. Some general methods for the localization of the major classes of macromolecules which can be separated by PAGE are as follows.

(a) *Proteins*

Historically the earliest protein stain is Amido Black 10 B (sometimes also called Napthalene Black 10B, Naphthol Blue Black B, or Buffalo Black) and it is still used extensively, particularly for qualitative work. The relative merits of Amido Black, Coomassie Blue R250, and Fast Green FCF as protein stains have been discussed by Wilson (1979). A typical staining method is shown in Table 2.4. Gels are placed in the staining mixture for up to 2 h and then transferred to the destaining mixture for at least 24 h during which several changes of solution may be required. The dye solution itself can be stored and reused several times. The lower limit of detection corresponds to about 1 μg per band. One possible difficulty is that the solvent in which the dye is made up is only weakly denaturing and, while this is not usually a problem, in some cases where particularly soluble constituents or proteins of low molecular weight are involved there may be inadequate fixing of proteins within the gel. Difficulties arising from this cause can generally be rectified by including a preliminary fixing stage in which the gels are soaked in 12·5 per cent TCA for 1–3 h before staining in the usual way.

As an alternative to the usual fixation by precipitation with TCA or sulphosalicyclic acid the separated zones can be immobilized chemically by binding to the gel matrix with formaldehyde (Steck, Leuthard, and Bürk, 1980). The method may be particularly valuable for peptides, glycopeptides, strongly basic proteins etc., which are not efficiently fixed by the usual acid precipitants. Staining is performed in a fume cupboard (hood) by immersing the gel for 1 h in a mixture consisting of 0·8 g Coomassie Blue R 250 in 180 ml ethanol, 420 ml H_2O, and 100 ml of 35 per cent formaldehyde. For SDS-containing gels this is followed by a further 1 h in a staining mixture with a reduced concentration of formaldehyde to avoid gel shrinkage, consisting of 1·2 g Coomassie Blue R 250 in 250 ml ethanol, 750 ml H_2O, and 10 ml of 35 per cent formaldehyde. The destaining solvent mixture consisted of 250 ml ethanol, 750 ml H_2O, and 10 ml of 35 per cent formaldehyde without dye.

Coomassie Brilliant Blue R250 is about three times more sensitive than Amido Black 10B and forms electrostatic bonds with NH_3^+ groups and non-covalent bonds

TABLE 2.4
Summary of staining procedures for proteins

Staining mixture	Staining time (h)	Destaining mixture	Comments
Amido Black 10B 1–10 g dissolved in 250 ml methanol; then add 100 ml acetic acid + 650 ml H_2O	0·5–2·0	10 per cent acetic acid	Very widely used, simple; prefixing in 12·5 per cent TCA may be advantageous
Coomassie Blue R250 1 g dissolved in 500 ml methanol; then add 100 ml acetic acid + 400 ml H_2O	1·0	500 ml methanol + 100 ml acetic acid + 400 ml H_2O (store in 7 per cent acetic acid	Very sensitive
or 0·5 g in 50 ml H_2O mixed with 950 ml 12·5 per cent TCA	0·5–1·0	10 per cent TCA for 20–30 min (store in 7 per cent acetic acid)	Before staining, gel can be prefixed for 0·5 h in 12·5 per cent TCA (or overnight if desired); very sensitive
Xylene Cyanine Brilliant G 2·5 g dissolved in 500 ml methanol; then add 500 ml 25 per cent TCA	0·5–1·0	5 per cent TCA for 10–60 min (store in 7 per cent acetic acid	Easier to use than Coomassie Blue R250 and nearly as sensitive; Good for densitometry
or 0·2 g dissolved in 100 ml H_2O; add 100 ml 2N H_2SO_4; stand for at least 3 h and filter; add 22·2 ml 10 N KOH to filtrate; add 28.7 g TCA	5–8	Store in staining mixture, or better, H_2O	Complicated to make up but no destaining required; rapid and sensitive
Anilinonaphthalene sulphonate 30 mg in 1l of 0·1 N sodium phosphate pH 6·8	0.05	–	Fluorescent labelling; less sensitive than staining methods but faster and preserves biological/enzymic activity
o-Phthaldialdehyde (OPA) (1) Soak gels for 10 min in solution of 2-mercaptoethanol (70–80 µl per 100 ml) in 0·1 M barbiturate, phosphate or hydrogen carbonate	25 min	–	Sensitive fluorescent staining; very rapid; no destaining step; sensitivity varies with different proteins; enzymic and biological activities usually retained
(2) Add OPA (20mg) in 2 ml methanol; mix in and allow to stand in dark for 15 min (3) Examine under u.v. illumination at 365 nm			

with non-polar regions in the proteins. In this case also staining is uniform, stable, and suitable for densitometry, although the binding begins to deviate from Beer's Law at higher protein levels and it is therefore not so suitable for quantitative measurements in gels having widely differing amounts of protein in the various stained bands. Coomassie Blue R250 can be used in an acetic acid-methanol solution (Table 2.4) but a better procedure is probably that described by Chrambach et al. (1967) in which TCA is used (Table 2.4). After destaining it is found that the bands often intensify during the first day or two of storage. Bands containing as little as $0 \cdot 1$ μg of protein can be detected. If band patterns are very faint they may also sometimes be intensified by adding a few drops of dye mixture to the 7 per cent acetic acid storage solution and leaving overnight.

Another general protein stain (Fig. 2.10) of similar sensitivity to Coomassie Blue R250 is Xylene Cyanine Brilliant G (sometimes rather confusingly referred to as Coomassie Blue G 250). It is also rather easier to use since it does not precipitate so readily. The methanolic TCA solvent in the staining mixture (Table 2.4) is strongly denaturing, and fixing and staining take place simultaneously. In our own work we have concluded that this particular stain represents the optimum at the present time for localizing proteins and that it is also very suitable for quantitative densitometry (e.g. Anderson and Andrews 1977; Hillier 1976). If a very rapid localization of a band is needed gels can be briefly stained for 5–10 min (or even less if the dye mixture is warmed to about 60 °C) and the major protein bands will be seen either immediately or after only a few minutes in 5 per cent TCA. This can be useful if gels are being used to monitor the progress of a separation, and the gels can then be returned to the staining mixture to be stained more thoroughly to localize minor bands or for subsequent quantitative measurements. The stain can be used repeatedly, but under the influence of light the dye causes a gradual photodecomposition of the TCA which becomes manifest by the development of a strong smell of chloroform. Fresh batches of dye should therefore be made up every two weeks or so.

In the method of Blakesley and Boezi (1977) (Table 2.4) the Xylene Cyanine Brilliant G dye is made up into a solution the pH of which is such that the dye is in the almost colourless leuco form, but on binding to a protein the pK of the dye is shifted and the usual blue colour develops. Thus when gels are immersed in this stain the background remains almost colourless and coloured protein bands gradually become apparent. Since no destaining step is involved this method can give a very rapid answer, and in our experience strong bands begin to be apparent after a few minutes while bands containing 5–10 μg of protein are visible within 30 min. Maximum sensitivity is reached in 5–8 h but after transferring the gels into water the bands intensify further. The background remains very free of colour so the gels are particularly suitable for densitometry.

Although much less sensitive (down to about 20 μg) than the above staining procedures, fluorescent labelling with anilinonaphthalene sulphonate (ANS) is useful as a simple rapid method for staining proteins while still retaining enzymic or antigenic activity since there can be minimal denaturation (Hartmann and Udenfriend, 1969). The gels are immersed in the staining mixture for about 3 min (Table 2.4) and examined with a u.v. lamp. If permissible, the intensity of the staining is greatly enhanced by denaturing the proteins by washing the gels for up to 2 min in 3 N HCl before placing them in the ANS solution.

In our own laboratories we have found fluorescent labelling with o-phthaldialdehyde (OPA) useful (Table 2.4), and a band containing less than 0·25 μg can be detected by u.v. illumination at 340–360 nm (Fig. 2.10). It is likely to prove particularly useful in preparative work for the rapid location of bands for

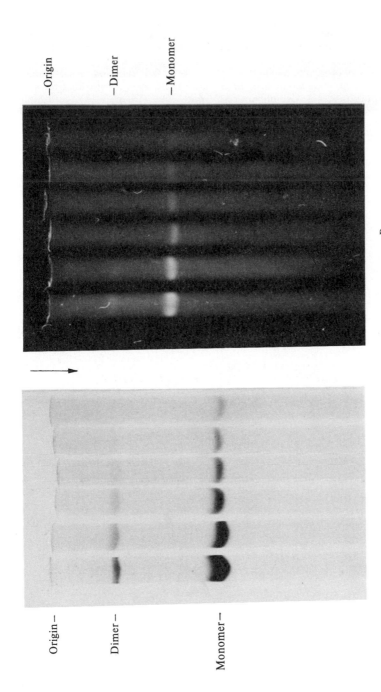

B

A

FIG. 2.10. A comparison of the detection of separated protein zones in polyacrylamide gels by (a) staining with Xylene Cyanine Brilliant G and (b) fluorescent detection using *o*-phthaldialdehyde. Gels were of $T = 7.3$ per cent and $C = 5$ per cent and were run in a homogeneous buffer system of 0·025 M NaHCO$_3$ (pH 8·5). The samples were bovine serum albumin and the concentrations from left to right were 8·0, 4·0, 2·0, 1·0, 0·5 and 0·25 μg per gel. The staining method is slightly more sensitive but the fluorescent method is more rapid.

subsequent extraction since the reaction does not require denaturing conditions, unlike dye-staining procedures. To avoid diffusion, if gels need to be preserved for a longer time, a prefixing step is required and gels then equilibrated at pH 7 before OPA labelling. Alternatively for permanent records gels should be run, labelled, and immediately photographed with u.v. illumination. The only major limitations are that buffers containing amine groupings and the use of urea must be avoided.

For some purposes, e.g. two-dimensional maps, etc. (Chapter 11), very high sensitivity is required for adequate localization of the separated proteins and this has nearly always been achieved to date by autoradiographic procedures with radioactively labelled proteins. These may not always be convenient or applicable, especially for *in vivo* work for example, but now the introduction of silver staining (Switzer, Merril, and Shifrin 1979), which is claimed to be 100 times more sensitive than Coomassie Blue R250 extends staining methods down to the sensitivity levels of radioactive procedures. The method was developed (Oakley, Kirsch, and Morris 1980) into a simplified procedure with a less costly use of reagents and this version is summarized in Table 2.5. Glass-distilled water should be used for making up all solutions and gloves worn whenever gels are handled, in order to avoid marking them with fingerprints. In step 4, use sufficient solution to allow the gel to float freely and ensure even staining. After use the ammoniacal silver solution should be disposed of immediately because it can become explosive upon drying. Oakley *et al.* (1980) recommend precipitating the silver as AgCl by acidification with HCl and if sufficient quantities are used it may become economic to sell the AgCl recovered. In steps 5 and 6 it may be advantageous to transfer the gel(s) into clean containers rather than just changing the solutions, to avoid the gel sticking to the staining vessel.

TABLE 2.5
High-sensitivity silver staining method for proteins

Staining procedure (successive steps)	Time
(1) Soak gel(s) in 10 per cent aqueous glutaraldehyde	30 min
(2) Rinse three times in 500–1000 ml distilled water	10 min
(3) Soak in water	> 2 h or overnight
(4) Drain off water, add fresh ammoniacal silver solution. Stir continuously to prevent deposition of silver on the gel surface. (Ammoniacal silver solution prepared by adding 1·4 ml of conc. NH_4OH to 21 ml of 0·36 per cent NaOH. Stir vigorously and slowly add 4 ml of 19·4 per cent $AgNO_3$. When the transient brown precipitate has cleared make up to 100 ml with H_2O)	≤ 15 min
(5) After no more than 15 min wash gel in H_2O	2 min
(6) Transfer into a fresh solution of 0·005 per cent citric acid and 0·019 per cent formaldehyde. Protein zones become visible at this stage. Remove gel from solution when background begins to darken	5–30 min
(7) Wash gels thoroughly in distilled H_2O (at least 3 changes)	> 1 h

Very faint bands may not become visible in step 6 until substantial background coloration has developed. Overstained gels can be destained with photographic reducers such as Kodak rapid fixer, Farmer's reducer (Kodak) or the reducing mixture described by Switzer *et al.* (1979) and prepared by mixing 100 ml of absolute ethanol, 5 ml of 1 per cent citric acid, and 5 ml of 3.7 per cent formaldehyde made up to 1 litre with H_2O. Destaining can be stopped by placing the gel in photographic hypo solution (about 5 per cent aqueous sodium thiosulphate). After such destaining treatment the gel(s) should be thoroughly washed in water.

According to Oakley *et al.* (1980), with polyacrylamide gels of $T > 10$ per cent the background may be unacceptably high and such gels should be prefixed in methanol, glacial acetic acid, water 5:1:4 (by vol.) for at least 30 min followed by overnight washing in 5 per cent methanol containing 7 per cent acetic acid before beginning the staining procedure given in Table 2.5.

Since 1980 there has been widespread use of silver staining for many types of gels and a proliferation of staining methods. Six of the earlier variations have been compared by Ochs, McConkey, and Sammons (1981) who concluded that the most sensitive was that described by Sammons, Adams, and Nishizawa (1981). This procedure which was at least 10-fold more sensitive than that of Oakley *et al.* (1980) is summarized in Table 2.6 and has the additional advantage of requiring even less silver nitrate. It is important to agitate the gels on a platform shaker throughout steps 4–9. In this method there is a further bonus in that different proteins appear to show up as different colours and Sammons *et al.* (1981) claim that this can be an aid in identifying individual proteins and in resolving overlaping protein bands or protein spots on two-dimensional gels. While undoubtedly useful the significance of colour differences obtained in this procedure (most other silver staining methods are more or less monochromatic) has been disputed however (Dunn and Burghes, 1983). The precise mechanism of silver staining is not known but it appears that initially silver ions are bound to protein zones and then act as nucleation sites for the deposition of further amounts of silver. It is possible that the colour may be more a reflection of silver grain size and rate of silver ion reduction rather than the identity of the protein concerned (Dunn and Burghes, 1983). The colour may be influenced also by the amount of protein in a zone as well as by its identity.

The first variants of silver staining were derived from histochemical methods (e.g. Switzer *et al.* 1979; Oakley *et al.* 1981) and can be accompanied by formation of deposits on the surface of the gels which causes them to stick to one another or the containers. Hence gels should be stained in separate containers. There is also some doubt about the use of glutaraldehyde as a fixative for proteins in the gels because it can give rise to a yellow background staining due to reaction with glycine in the electrophoresis buffer present in the gel, although it may be the most appropriate fixative for small or basic peptides (Steck *et al.* 1980). In spite of these drawbacks the method of Oakley *et al.* (1980) has probably been more widely used than any other to date.

The other principal group of silver staining methods are photochemically based (e.g. Sammons *et al.* 1981; Morrissey, 1981; Merril, Goldman, and Van Keuren, 1981) and are in general terms somewhat more simple to perform, are usually more rapid, give less deposit formation on surfaces, and are less costly in terms of expensive silver reagents. The differences are small however and the proliferation of methods of both groups points to the conclusion that there is no clearly superior method at present. Most of this second group of methods employ ethanol or methanol and acetic acid as fixatives but there seems to be no objection to the use of trichloroacetic acid (TCA) and this is a better fixative for proteins so should probably be recommended. A major purpose of the fixatives is also to remove interfering substances such as glycine and detergents before staining. It is often claimed that the photochemically based stains loose sensitivity when used with slab gels thicker than 1 mm, but the method of Sammons *et al.* (1981) was actually optimized for 1.5 mm thick gels and it seems that with thorough prefixing and prewashing there should be little difficulty with even thicker gels. There are many similarities between the most popular methods of this group. In the particularly simple and rapid method of Morrissey (1981), shown in Table 2.6, fixing is performed in methanol-acetic acid followed by glutaraldehyde treatment (which is not particularly effective at acid pH values), while Sammons *et al.* (1981) use ethanol and acetic acid only and Merril *et al.* (1981) fix with methanol-acetic acid, or better, for 20 min in 20 per cent TCA followed by ethanol-acetic acid. One might conclude from this that as long as it is done thoroughly the precise conditions for fixation are not very critical. This can be followed by treatment with a reducing agent such as dithiothreitol (Morrissey, 1981) or an oxidising agent such as 3.4 mM potassium dichromate containing 3.2 mM nitric acid (Merril *et al.* 1982). This step does not seem to be necessary (e.g. Sammons *et al.* 1981) but it may increase the sensitivity a little. In all methods the gel is then impregnated with silver nitrate and the stain developed with an alkaline solution of formaldehyde. The colour may then be further enhanced (Sammons *et al.* 1981) with sodium carbonate alone, but more often is first stopped by acid treatment, usually dilute citric or acetic. At all stages the gels should be gently agitated and washed. Plastic gloves should be worn if gels are handled to avoid fingerprints on the gel.

If greater sensitivity is required it is possible to recycle the stained gels through the procedure again from the silver nitrate treatment stage (Merril *et al.* 1982). Berson (1983) has described a blue toning procedure claimed to be fast, simple and more effective than this recycling through silver nitrate, and resulting in a 3 to 7-fold further increase in sensitivity. For this, gels were first silver stained and then washed thoroughly in tap water for at least 30 min. The toning solution was made up just before use by mixing 10 ml of 5 per cent aqueous ferric chloride, 10 ml of 3 per cent aqueous oxalic acid, 10 ml of 3·5 per cent aqueous potassium hexacyanoferrate III, and 70 ml H_2O. (Stock solutions of these reagents are stable for months at room temperature.) The toning solution should be brown and is discarded if blue-green. When gels are placed in it the protein bands turn blue and increase in intensity, but the background gradually increases also and when the desired degree of enhancement has occurred, typically 1–2 min, the treatment is stopped by washing for 10 min in tap water. Finally the gels are stabilized by soaking for 10 min in 20 per cent aqueous methanol containing 5 per cent acetic acid.

Most of these silver staining procedures are of the order of 100 times more sensitive for protein detection than Coomassie Blue staining (Fig. 2.11) and in fact silver staining can be applied perfectly satisfactorily to gels that have already been stained with Coomassie Blue (Hallinan 1983). This can be useful when sample constituents vary widely in concentration, silver staining being used to detect minor constituents after the principal ones have previously been revealed with dye staining. The vast majority of applications of silver staining have been to two-dimensional gels (see Chapter 11), and nearly always with the detergent SDS present in the gel and buffers. Under these conditions there may be some differences in staining efficiency between different proteins, as with dye staining methods, but this is seldom a serious problem, although Merril *et al.* (1982) did observe an acidic protein which failed to stain initially and only showed up on recycling the gel through the silver nitrate stage. Hallinan (1983) has reported that with gels and samples containing no detergent or other denaturing agents this differential staining of proteins is potentially far more serious, many zones staining with variable efficiency and some not at all well. Thorough prefixing did not correct

TABLE 2.6
Photochemically based silver staining methods for proteins

Staining procedure (successive steps)	Time
Method of Sammons *et al.* (1981)	
(1) Fix gels in 50 per cent aqueous ethanol containing 10 per cent acetic acid	2 h or more
(2) First wash in 50 per cent aqueous ethanol containing 10 per cent acetic acid	2 h
(3) Two washes in 25 per cent aqueous ethanol containing 10 per cent acetic acid	2 x 1 h
(4) Two washes in 10 per cent aqueous ethanol containing 0.5 per cent acetic acid	2 x 1 h
(5) Equilibrate gel in a degassed aqueous solution of silver nitrate $(1.9 \, \text{g} \, \text{l}^{-1})$, the volume of solution being about 3 times the volume of gel	2 h or more
(6) Rinse briefly in degassed H_2O	10–20 s
(7) Place in a reducing bath consisting of 0.75 M NaOH, 87.5 mg l^{-1} of NaBH and 7.5 ml l^{-1} of 37 per cent formaldehyde which is added to the solution immediately before use. The volume used should be 5.5 times the gel volume	10 min
(8) Place in colour enhancing solution (5.5 times gel volume) consisting of 7.5 g l^{-1} Na_2CO_3 in H_2O	1 h
(9) Transfer into fresh Na_2CO_3 (7.5 g l^{-1} H_2O)	1 h
(10) Store in Na_2CO_3 (7.5 g $l^{-2}H_2O$)	
Method of Morrissey (1981)	
(1) Prefix gel in 50 per cent aqueous methanol containing 10 per cent acetic acid	30 min
(2) Wash in 5 per cent aqueous methanol containing 7 per cent acetic acid	7 min
(3) Fix in 10 per cent aqueous glutaraldehyde using 6 ml per ml of gel	30 min
(4) Soak in distilled H_2O overnight and then rinse in fresh H_2O for for 30 min, or rinse in several changes of H_2O for at least 2 h	overnight + 30 min
(5) Soak in dithiothreitol (5 μg ml^{-1} in H_2O) using 12 ml per ml of gel	30 min
(6) Pour off dithiothreitol solution and without rinsing add 0.1 per cent aqueous silver nitrate (12 ml ml^{-1} gel). After 30 min,	30 min
(7) rinse rapidly with a small volume of H_2O	10–20 s
(8) Rinse rapidly twice with a small volume of developer solution consisting of 50 μl of 37 per cent formaldehyde in 100 ml 3 per cent sodium carbonate	30–40 s
(9) Soak in developer solution (12 ml ml^{-1} gel) until desired level of staining is reached	5–15 min
(10) Stop staining by adding 2.3 M citric acid directly to the developer, using precisely 5 ml for every 100 ml of developer solution and agitate for 10 min. (Note, the amount of citric acid should balance the amount of carbonate in the developer giving a neutral pH).	10 min
(11) Wash gel several times in H_2O	30 min
(12) Soak gel in 0.03 per cent sodium carbonate	10 min
(13) Wrap in cellophane or seal in plastic bags for storage	

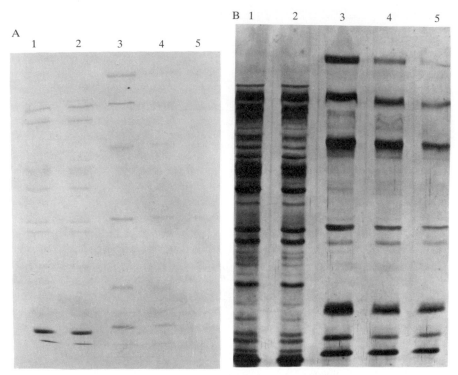

FIG 2.11. Comparison of (A) Coomassie Brilliant Blue R250 staining with (B) silver staining using the Bio Rad silver staining kit which is based on the method of Merril, Switzer and Van Keuren (1979). The 0·75 mm thick SDS–PAGE gel was of $T = 12$ per cent. Samples 1 and 2 were a soluble fraction of bovine brain homogenate and samples 3, 4 and 5 were of Bio Rad low molecular weight standards for SDS–PAGE diluted 40-fold, 80-fold and 160-fold respectively. (Photography kindly supplied by Bio Rad Laboratories.)

this, but nearly all the variability was eliminated and sensitivity enhanced if gels were prestained with Coomassie Blue, destained in the usual way and then silver stained. It appears (Nielsen and Brown 1984) that ionic amino acid side chains may play an important role in the silver staining of proteins, so it seems likely that when proteins have a low content of these and are therefore poorly stained a preliminary Coomassie Blue staining can make good this deficiency by virtue of the sulphonic acid groupings on the dye molecules. This greater reliability and sensitivity of silver staining after a preliminary staining with Coommassie Blue has been observed by several groups of workers using various different silver staining techniques, and it appears that the provision of extra silver ion binding sites by dye molecules bound to protein in the gel may be generally beneficial. In our own laboratory we have found that a very good procedure is to stain with Xylene Cyanine Brilliant G, destain and then silver stain by Morrissey's procedure (Table 2.6) starting at step 4, because of course the proteins will have already been fixed during the Xylene

Cyanine Brilliant G staining. This also has the advantage of avoiding the rather noxious glutaraldehyde fixing treatment step.

(b) *Phosphoproteins*

Many proteins are coloured red by the nucleic acid stain 1-ethyl-2-3-(1-ethylnaptho-{1,2d}-thiazolin-2-ylidene)-2-methyl-propenyl-naphth-{1,2d}-thiazolium bromide, fortunately abbreviated to Stains-All, while phosphoproteins are dyed blue, so this can be used for differentiating between them (Table 2.7). The sensitivity varies but is generally in the range 1–5 μg for phosphoproteins. Some non-phosphorylated proteins do not stain at all (Green, Pastewka, and Peacock 1973).

A slightly less sensitive but more specific stain for phosphoproteins using methyl green and molybdate has been described by Cutting and Roth (1973) and this also is shown in Table 2.7. The method can detect as little as 1 nmol of phosphate, which in a protein band containing 30 μg of protein would correspond to approximately one phosphate group for every 300 amino acid residues.

(c) *Lipoproteins*

Being partly protein in character lipoproteins can be localized and quantified with all the above protein stains, but in addition they can be prestained before electrophoresis in order to distinguish them from other proteins. Sudan Black B is most commonly used for this (Ressler, Springgate, and Kaufman 1961). One volume of a saturated solution of Sudan Black B in ethylene glycol is mixed with 2 volumes of the lipoprotein solution and the mixture kept at 4 °C for at least 24 h. The mixture is applied directly to the gel for the electrophoresis in which the excess dye remains at the origin. A 1 per cent solution of Lipid Crimson in ethylene glycol can be used in the same way.

Alternatively, after electrophoresis lipoproteins can be localized (Prat, Lamy, and Weill 1969) with a stain made up by dissolving 500 mg of Sudan Black B in 20 ml acetone and adding a mixture of 15 ml acetic acid and 80 ml H_2O. The mixture is stirred for 30 min and centrifuged. Gels are stained in this overnight and then destained with three successive washings in a mixture of 150 ml acetic acid, 200 ml acetone, and 650 ml H_2O. According to Tsai and Frasch (1982) lipopolysaccharides can also be located on gels by silver staining. They developed a procedure specifically intended for this but also successfully used the method of Sammons *et al.* (1981), which is shown in Table 2.6.

(d) *Glycoproteins*

Like phosphoproteins and lipoproteins, glycoproteins can be stained by all the methods described above for ordinary proteins but the presence of carbohydrate means that other procedures can also be used both for their localization and in order to distinguish which components of a mixture are glycoproteins. At the present time the most widely used specific glycoprotein staining method (Table 2.8) is the very sensitive periodic acid-Schiff (PAS) procedure of Zacharius, Zell, Morrison, and Woodlock (1969). The fuchsin–sulphite (Schiff's reagent) staining solution for this is prepared as follows: 2 g basic fuchsin (caution–carcinogenic) are dissolved in 400 ml H_2O with warming and then cooled and filtered. Add 10 ml 2N

TABLE 2.7

Staining procedures suitable for the identification of phosphoproteins

Staining procedure (successive steps)	Time (h)	Comments
Methyl green		
(1) Prefix gels in 10 per cent sulphosalicylic acid	12+	Use several changes over at least 12 h; also removes any residual traces of phosphate in samples and gels
(2) Transfer into 10 per cent sulphosalicylic acid + 0·5 M CaCl$_2$	1+	Ca^{2+} precipitates phosphate released in step (4)
(3) Rinse briefly in H$_2$O	—	Removes loosely adhering CaCl$_2$
(4) Heat in 0·5 N NaOH at 60 °C	0·5	Hydrolyses much of the phosphate off phosphoserine or phosphothreonine residues; stronger hydrolysis would release a little more but also cause excessive gel shrinkage
(5) Rinse twice at 10 min intervals with 1 per cent ammonium molybdate	—	Mo^{2+} ions diffuse into gel
(6) Immerse in 1 per cent ammonium molybdate in 1 N HNO$_3$	0·5	Further diffusion of Mo^{2+} ions which in the acid solution form an insoluble nitrophosphomolybdate complex
(7) Transfer into 0·5 per cent methyl green in 7 per cent acetic acid	0·5	Basic dye enhances the sensitivity of the ammonium molybdate reagent by forming an insoluble lake
(8) Destain in 10 per cent sulpho-salicylic acid		
(9) Store in 7 per cent acetic acid		
Stains-All		
(1) Warm gels at 50 °C in 30 vol 25 per cent isopropanol	0·25	Serves to fix the gels and to remove SDS from gels containing this detergent; not necessary for gels run at acid pH values
(2) Rinse briefly	—	
(3) Stain in mixture containing 10 mg Stains-All in 20 ml formamide mixed with 50 ml isopropanol + 1 ml 3·0 M tris-HCl buffer pH 8·8 + H$_2$O to 200 ml	Overnight	Keep in the dark
(4) Rinse and destain in H$_2$O		Keep in the dark or at least away from direct light

HCl and 4 g K$_2$S$_2$O$_5$, place in a stoppered bottle, and keep cool and dark overnight. Stir in 1 g activated charcoal, filter, and add sufficient 2 N HCl (10 ml or more) until a drop dried on a glass slide does not turn red. Store stoppered in a cool dark place and discard if it turns pink. After following the procedure outlined in Table 2.8 for localization of the glycoproteins, if a small amount of either the Coomassie Blue R250 or the Xylene Cyanine Brilliant G staining mixtures (Table 2.4) is added to the 5 per cent acetic acid storing solution (e.g. 1–3 ml per 100 ml) then on standing for a further 24–48 h other protein (i.e. non-glycoprotein) bands will become coloured blue.

Two related methods developed from the PAS procedure have now gained wide acceptance. In both, the fixation, periodate oxidation, and bisulphite treatments are performed essentially as described in steps (1)–(6) of Table 2.8 except that in both methods step (5), the fuchsin staining is omitted. After step (6) the gels are washed with H_2O and in the procedure of Wardi and Allen (1972) the gels are then placed in a 0·5 per cent solution of Alcian Blue in 3 per cent acetic acid for 4 h. Destaining, electrolytically or by diffusion, is in 7 per cent acetic acid. Like the PAS procedure, the stain detects a lower limit of about 2–3 μg of protein-bound carbohydrate. The second method (Eckhardt, Hayes, and Goldstein 1976) is much more sensitive (down to about 40 ng of bound carbohydrate). For this, after bisulphite treatment the gels are washed with H_2O and then stained for 2 h at 60 °C in sealed tubes with a mixture of equal volumes of acidified dimethyl sulphoxide (DMSO) containing 0·6 ml conc. HCl per litre of DMSO and a freshly prepared solution of dansyl hydrazine (2 mg ml^{-1}) in DMSO. The staining solution is then decanted off and the gels treated with a solution of NaBH$_4$ (0·2 mg ml^{-1}) in DMSO for 30 min at room temperature. The gels are rinsed with H_2O and destained with 1 per cent acetic acid until the background no longer fluoresces under u.v. illumination.

The most sensitive staining method at present for the detection of glycoprotein and polysaccharide zones in polyacrylamide gels is a version of the silver staining procedure (Dubray and Bezard 1982), claimed to be about 64 times more sensitive than the PAS procedure and capable of detecting 0.4 ng of bound carbohydrate making it comparable to autoradiography of radiolabelled zones. The method utilizes periodic acid for the cleavage of 1,2-diol groups in saccharide units generating aldehyde groups which are then detected with ammoniacal silver nitrate. Dubray and Bezard (1982) first soaked gels overnight in 25 per cent

TABLE 2.8

Glycoprotein staining method of Zacharius et al. (1969)

Staining procedure (successive steps)	Time (min)
(1) Immerse gels in 12·5 per cent TCA (25–50 ml per gel rod of diameter 5 mm	30
(2) Rinse in H_2O	0·25
(3) Immerse in 1 per cent periodic acid in 3 per cent aqueous acetic acid	50
(4) Wash thoroughly with H_2O (at least six changes over 1–2 h with stirring or overnight)	Overnight
(5) Transfer into fuchsin–sulphite stain in dark[a]	50
(6) Wash with three changes (10 min each) of 0·5 per cent sodium metabisulphite (25–50 ml per gel rod)	30
(7) Wash in frequent changes of H_2O until excess stain is	Overnight
(8) Store in 5 per cent acetic acid	

[a] See text for preparation

isopropanol containing 10 per cent acetic acid and then for 30 min in 7.5 per cent acetic acid. The gel slab was then placed in 0.2 per cent aqueous periodic acid and kept at 4 °C for 60 min. This was then followed by the silver staining procedure (Table 2.5) of Oakley et al. (1980) but omitting the glutaraldehyde treatment (step 1 of Table 2.5). No doubt fixing with TCA, methanol/acetic acid or ethanol/acetic acid could be used instead of isopropanol and after the periodic acid treatment other silver staining methods may also be applicable (e.g. those in Table 2.6).

Acid glycoproteins can also be stained (Caldwell and Pigman 1965) by immersing the gels in 0·2 per cent Alcian Blue, followed by electrolytic destaining in 7·5 per cent acetic acid. For acid mucolysaccharides, staining for 1 h in a 0·1 per cent solution of Toluidine Blue O in 1 per cent acetic acid (Maurer 1971) or 30 per cent (v/v) methanol (Funderburgh and Chandler 1978) can be used, with washing in 1 per cent acetic acid or H_2O for the destaining step. Using the method of Dahlberg, Dingman, and Peacock (1969) (see below), Bader, Ray, and Steck (1972) found that Stains-All was suitable for the location of a number of glycosaminoglycans. Heparin stained a rust-brown colour, hyaluronic acid blue, and the chondroitin sulphates purple. The method gave a linear quantitative response over at least the range 0·5–5 µg for hyaluronate.

A convenient, specific and mild method for the detection of glycoproteins utilizes the interactions between lectins and carbohydrates. The method (Furlan, Perret, and Beck 1979) consists of washing the acid-fixed gels in four changes of 0·05 M tris-HCl buffer pH 7·0 containing 0·1 M NaCl, 1 mM $CaCl_2$, and 1 mM $MnCl_2$ until they attain pH 7·0. Fluorescein isothiocyanate (FITC)-labelled lectins are dialysed against this buffer, adjusted to 1 mg ml^{-1}, and then used for staining the gels at room temperature for 12 h. The same buffer is used for destaining (2 days). The FITC-lectin solution can be re-used three times for economy, and the fluorescent bands which are visible under u.v. illumination are stable for at least 1 month at 4 °C. Sensitivity is of the order of 100 ng or better of bound hexose, but will of course depend upon the carbohydrate binding specificity of the lectin. Some FITC-lectins are available commercially but others can be prepared by the method of The and Feltkamp (1970) for example (see Section 8.4.(b)). Cotrufo et al. (1983) claim that the sensitivity of the method can be increased 20-fold if laser light is used for excitation rather than conventional u.v. illumination. Koch and Smith (1982) used [125] I-labelled lectins followed by autoradiography for still greater sensitivity.

(e) Ribonucleic acids (RNA)

There are a number of procedures which can be applied to the staining of RNA. Among the most widely used is Stains-All which has been described already (Table 2.7) for phosphoproteins. It stains RNA bluish-purple. For RNA staining Dahlberg et al. (1969) used a solution of 1 vol. 0·1 per cent Stains-All in formamide which is diluted with 19 vol. aqueous 47 per cent formamide so that the final formamide concentration is 50 per cent. Gels are soaked in this stain overnight in a tray protected from light by wrapping in aluminium foil and then washed free of excess stain with running tap water. Owing to the photosensitivity of the stain, bands fade if exposed to bright light. Proteins (e.g. from polyribosomes) are also stained by this procedure, but are coloured red.

Alternatively one can immerse the gels overnight in a staining bath composed of 0·5 per cent pyronine Y or G and 1 per cent lanthanum acetate in 10 per cent acetic acid containing 10 per cent methanol (Marinka 1972) and then destain the gels in a similar solvent mixture (i.e. acetic acid/methanol/H_2O). Gels are subsequently stored in about 5 per cent acetic acid. The method has been stated to be specific for RNA, especially low-molecular-weight forms such as t-RNA.

Unquestionably the most popular method recently has been the use of the highly sensitive fluorescent reagent ethidium bromide which stains both RNA and DNA. The conditions for staining do not appear to be very critical. Sharp, Sugden, and Sambrook (1973) dissolved ethidium bromide at a level of 0.5 μg ml^{-1} in a buffer of about pH 7.8 containing 40 mM tris, 5 mM sodium acetate, and 1 mM EDTA and immersed the gels in this for 30 min. Prunell, Kopecka, Strauss, and Bernardi (1977) used 2 μg ml^{-1} in a very similar buffer with a 2 h staining time, while Bailey and Davidson (1976) and Lehrach, Diamond, Wozney, and Boedtker (1977) used a lower level of 1 μg ml^{-1} in 0.01M ammonium acetate solution but with overnight staining. Dolja, Negruk, Novikov, and Atabekov (1977) used a pH 7.2 buffer of similar composition to that used by Sharp et $al.$ (1973) but containing in addition 0.05 per cent sodium dodecyl sulphate (SDS). They used this buffer both for the electrophoresis and for subsequent staining after adding 10 μg ml^{-1} of ethidium bromide (15 min staining time) and it appears that it is often possible to use the same buffer both for electrophoresis and staining in this way. No destaining is required. Stained bands containing as little as 0.05 μg RNA or DNA are visualized by illuminating the gels with a short-wavelength u.v. lamp. Most of the above groups of workers then photographed the illuminated gels with for example a polaroid camera using type 55 P/N, 57, or 105 film and a red (e.g. Kodak no. 23A) or yellow filter.

Silver staining as used for proteins (e.g. Table 2.6) can also be applied without modification to the detection of RNA, although it may be possible to abbreviate the fixing and washing stages reducing the overall time taken because RNA gels generally do not contain glycine in the buffers or detergents which interfere with staining (Igoi, 1983). Sensitivity was even better than ethidium bromide staining (Goldman and Merril, 1982) and similar to methods with radiolabelled RNA (e.g. Berry and Samuel, 1982; Beidler, Hilliard, and Hill, 1982). In addition the reagents used are far less hazardous than ethidium bromide which is a potent mutagen. A number of silver staining procedures have also been specifically developed for nucleic acids but that of Kolodny (1984) is claimed to give both higher sensitivity and sharper stained zones than earlier versions. The method (Table 2.9) employs preliminary staining with methylene blue followed by silver staining. In this regard it resembles the double-staining of protein containing gels with Coomassie Blue and silver (see p.36). It is important to thoroughly clean the glass plates or tubes of the gel apparatus with chromic acid followed by copious rinsing with H_2O before use. Gels must be handled throughout using gloves to avoid contamination. Steps 4–6 should be performed away from direct light and with continuous agitation of the gels. The developer solution (step 6) should be made up 1 h before use, otherwise the colour appears too rapidly and the gel can easily be over-developed. In any case gels should not be left in the developer for more than 5 min because noticeable further darkening continues during the first 15 min in the fixative. The background should be a homogeneous light yellow-tan colour and the sensitivity was such that several bands could be distinguished when a 100 pg of crude t-RNA was separated. Peats (1984) has also published a similar sensitive silver stain for DNA separated in agarose gel slabs.

(f) Deoxyribonucleic acids (DNA)

Ethidium bromide used as described above is unquestionably a sensitive and widely

TABLE 2.9
Silver staining of nucleic acids

Staining step	duration
(1) Fix gel in 1 M acetic acid	15 min
(2) Stain in 0·2 per cent methylene blue in 0·4 M sodium acetate containing 0·4 M acetic acid (pH 4·7)	45 min
(3) Destain in several changes of H_2O	3–4 h
(4) Silver stain (in a solution made up by mixing 40 ml 0·1 M NaOH and 3 ml 15 M NH_4OH and adding 1·6 g $AgNO_3$ dissolved in 10 ml H_2O followed by H_2O to 200 ml)	20 min
(5) Wash continuously in H_2O	
(6) Treat with a developer solution (5 ml 1 per cent citric acid and 0·5 ml 37 per cent formaldehyde diluted with H_2O to 500 ml).	< 5 min
(7) Place in a fixative solution of 45 per cent methanol and 10 per cent acetic acid in H_2O	

used method for DNA and probably the method of choice in most cases. An example of a gel containing DNA bands stained with ethidium bromide is shown in Fig. 6.1.

Among earlier staining procedures one of the most convenient for the identification of native DNA is to immerse the gels in a 0·25 per cent solution of methyl green in 0·2 M sodium acetate buffer pH 4·1 for 1 h (Kurnick 1950). Before use the stain should be extracted with chloroform until the organic phase is colourless. Destaining is achieved by repeated washing for 1–3 days with the acetate buffer. The stain should be stored in a brown bottle and it can be used several times provided the pH does not exceed 4·5.

Denatured DNA can be localized (Boyd and Mitchell 1965) by pre-soaking the gels for 1–2 h in 0·1 M sodium acetate buffer pH 4·5 and then staining them for 24 h in a pyronine B solution. This is prepared by making a 2 per cent aqueous stock solution of the dye, extracting it exhaustively with chloroform, and then diluting a portion with 9 vol 0·2 M sodium acetate buffer pH 4·5 before use. The dilute stain should only be used once but the aqueous stock solution can be stored in the dark at 4 °C. Destaining of the gels is achieved by washing with acetate buffer.

Alternatively, Stains-All when used as described above for RNA stains DNA bands blue (Dahlberg et al. 1969).

DNA can also be detected using silver staining procedures such as those in Table 2.6, the sensitivity being at least as great as with ethidium bromide, and as little as 1 ng can be detected (Goldman and Merril 1982). According to Merril et al. (1982) it may not be necessary to fix the gel before staining, and it is sufficient to wash the gels with H_2O only, in order to remove buffer salts. Ethidium bromide stained gels can subsequently be silver stained if desired.

A useful way of locating nucleic acid zones (both DNA and RNA) that is particularly appropriate in preparative work where high sensitivity detection is not usually necessary is u.v. imaging (Clarke, Lin, and Wilcox 1982). It is a development of earlier u.v. shadowing methods and consists essentially of placing the gel on a sheet of plastic wrapping material taped to a piece of wood. This was taken to the dark room and a sheet of photographic printing paper (e.g. Agfa No.

5, Ilford Ilfospeed No. 5, or Ilfomat No. 3, etc.) inserted under the plastic wrapping. The gel was then illuminated with an ordinary hand-held u.v. lamp and the printing paper subsequently developed in the conventional manner. Ideally the u.v. lamp should emit light at about 260 nm but this is not essential. Exposure will depend upon the lamp used, lamp to gel distance, gel thickness, photographic paper emulsion, etc., but with their lamp at a distance of 1 m from the gel Clarke *et al.* (1982) found exposure times were inconveniently short for easy reproducibility and so reduced the lamp output very substantially by taping over the face of the lamp a piece of aluminium foil with a small (6 mm) hole cut in it. The method was about 100-fold less sensitive than ethidium bromide staining but bands with less than 1 μg of DNA restriction fragments of differing sizes could be readily detected in polyacrylamide or agarose gels. The photographic image obtained was 'life-size' and it was very easy to use it to locate bands and cut them out for preparative extractions. It also provided a convenient permanent record of the gel.

(g) *Radiolabelling*

As an alternative to staining methods labelling with radioactive isotopes is widely used. Compared with staining there are considerable advantages in terms of sensitivity and conditions can be non-denaturing so that the separated components can subsequently be extracted in biologically active form. Radiolabelling is nearly always carried out before the sample is examined by electrophoresis (see Appendix 1), but can be performed *in situ*, for example by radioiodinating the separated protein bands in a polyacrylamide gel using a Chloramine T procedure (Christopher *et al.* 1978).

2.10. Destaining

Most staining procedures colour the whole gel and not just zones containing the bands of the substance being examined. This non-specific background absorption must be reduced by a destaining step so that the bands can be seen clearly and photographed and quantified if required. With many stains the bonds formed between the dye molecules and the separated material are comparatively weak so care must be taken not to reduce the intensity of the bands or even to decolorize them completely by excessive destaining.

There are two ways to destain gels, by washing or by further electrophoresis, the former being much the most widely used. The washing method consists simply of leaching out the excess dye by immersing the gels in a large volume of a suitable solvent (Section 2.9 and Tables 2.4, 2.7 and 2.8) for a comparatively long period, often 24 h or more. If the gels have been stained in a staining tray, such as that shown in Fig. 2.8, after staining the tray is simply transferred into a bath of the destaining solvent. The rate of diffusion of dye molecules out of the gel is accelerated by agitation (stirring) and by frequent changes of the solvent mixture to maintain the greatest possible concentration gradient of dye between the gel and the solvent. Destaining is also accelerated by suspending a bag containing activated charcoal or ion-exchange resin (e.g. Dowex 1 for acidic dyes or Dowex 50 for basic ones) in the destaining bath to absorb the dye molecules as they diffuse out of the gels.

In electrophoretic destaining the essential requirement is to position the gel rods or slab between two electrodes so that the electric potential is applied *across* the gels. This reduces the time required for migration of unbound dye out of the gel. Solvents used are usually the same or similar to those used for diffusion destaining. Compared with the initial electrophoresis, destaining requires a relatively low voltage but high current, typically up to about 50 V and with currents up to 1 A. Thus home-made designs of destainer have been described which rely on the power supplied by a 12 V car battery charger (Margolis and Kenrick 1968), but commercial power supplies specifically intended for destainers are available (e.g. Pharmacia DPS; Bio Rad Model 250/2.5, etc.). destaining times are usually of the order of 20–60 min.

Electrophoretic destaining is applicable when the separated substances can be easily and completely fixed during the staining stage. It has the advantage that destaining is rapid and complete, but there can be a loss of material if it is inadequately fixed and it is also relatively easy to destain excessively. For these reasons diffusion destaining is preferred in most quantitative work.

2.11. Quantitative determination of separated components

Quantitative measurements fall into two classes: firstly those where only relative measurements are needed, as for example when monitoring the progress of an isolation procedure or of a chemical reaction or in many screening procedures, and secondly those where the results must be expressed in absolute terms, i.e. concentration, units of activity, etc. Quantitative methods usually employ densitometry. This can be applied directly by scanning the stained or unstained gels themselves or indirectly by scanning negatives of photographs of the gels. The measurement of isotopically labelled bands is considered in Section 2.12.

(a) *Densitometry of stained gel patterns*

This is much the most widely used method for quantifying the results of PAGE separations. Comparatively low-cost and low-resolution densitometers such as those marketed for scanning paper or cellulose acetate strips can be successfully applied to the densitometry of PAGE slab gels providing bands are well separated and the best resolution is not required. They are generally less accurate when applied to cylindrical gel rods and are not suitable when a fine band structure is to be resolved. High-resolution densitometers are relatively expensive, and a good approach is to purchase a good quality spectrophotometer such as the Zeiss PMQ-II, Gilford Model 250 or Perkin-Elmer No. 5 or 7 which can also be used for other work but to which suitable gel-scanning attachments can be added. The Gilford Model 2410S and 2520 Linear Transport Attachments (for scanning gels up to 10 cm and up to 20 cm long respectively) enable gels to be passed at a constant rate through the light beam and can be used in conjunction with modified Unicam SP-500 or Beckman DU spectrophotometers if one of these should be available. There are also further attachments which enable gels to be scanned for fluorescent bands. The latest development in scanning densitometers is the use of a laser light source (e.g. LKB Model 2202, Ultro Scan Laser Densitometer) and these are claimed to give superior performance when scanning gels with the extremely high levels of resolution which are now attainable. For example narrow bands in PAGIF gels (Chapter 9) may be less than 100 nm apart. The scanner output can be computerized (e.g. the LKB Gel Scan software package for running on an Apple II computer) to enable the data to be recorded, displayed, manipulated, evaluated, stored on disk or printed, and two different gels can be compared if required. For many routine PAGE separations however the resolution, and hence scanning profile, is determined by the separation on the gel and not by limitations in the optics of the densitometer, so cheaper and less sophisticated equipment may be adequate.

The quality of quantitative densitometric measurements depends on a number of factors (Fig. 2.12). Inadequate fixing can result in unfixed (and unstained) material continuing to diffuse through the gel. When this comes into contact with fixative or dye molecules in the outer regions of the gel it becomes fixed and stained, and this results in broad diffuse bands or in extreme cases a 'halo' effect. In this an apparently sharp band of stained material forms a ring around the surface of a gel rod which is partly obscured by an inner spherical region representing the front formed between slowly penetrating dye molecules and unfixed material diffusing out of the inner regions of the electrophoretic band (Fig. 2.12). This type of error is more likely to occur at high sample loadings and with thick gels, so for optimum quantitative measurement thin gel slabs (1 mm or less) or narrow-diameter gel rods (< 3 mm) are often recommended.

Any difficulties due to aberration, such as can occur from a 'lensing' effect with cylindrical gel rods, are generally overcome by scanning the gels in a cuvette filled with 7 per cent acetic acid which has a similar refractive index to polyacrylamide gel.

The densitometer output is often fed to a chart recorder and quantified by measuring the areas under individual peaks. Areas are calculated from

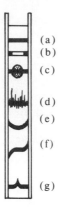

FIG. 2.12. Some common band patterns. (a) Ideal pattern: gel stained well into centre; outermost portions may become partly decolorized during prolonged destaining. (b) Centre not stained; staining time too short. (c) 'Halo effect' due to combination of inadequate fixing and high loading. (d) 'Streaky' bands due to particulate material in sample. (e) Curved bands due to uneven gel polymerization, or more usually to too high a current leading to a substantial temperature gradient (and hence mobility differences) between the centre and outer parts of the gel rod. (f) Distorted bands due to an uneven gel surface (can be produced by 'bombing' the gel mixture with an overlying water layer before polymerization). (g) Distorted band produced by an air bubble rising during polymerization.

measurements of peak height times the width at half height, by tracing the recorder plot and cutting out appropriate areas and weighing them, or by planimetry. A magnifying lens incorporating a measuring scale is a useful aid. Alternatively, electronic integration or data-logging equipment can be used so that peak areas can be calculated by computer, and this is certainly the method of choice if a considerable number of gels have to be scanned. Yakin *et al.* (1982) have published a particularly versatile approach for smoothing data, background and baseline corrections, peak fitting, and peak area integration all implemented as BASIC programmes for laboratory minicomputers. The method was readily adaptable to a wide variety of electrophoretic separations for research and clinical applications. Kitazoe *et al.* (1983) have described a computerized Gaussian fitting of data from a conventional Gilford Model 250 gel scanner using the least squares method which could resolve over 20 overlapped components simultaneously. Although developed for SDS-PAGE gels it should be equally applicable to other single dimension gel separations (e.g. PAGE, PAGIF, etc.)

Less accurate than peak areas is quantitative measurement from peak height only. Peak height depends not only on the amount of material present but also on the distance travelled through the gel. This can be seen from Fig. 2.13 in which a densitometer scan of a gel to which repeated 10 μl samples containing 12 μg of ß-lactoglobulin A were applied to a single gel at about 10 min intervals is shown. The areas under all the peaks are identical, but owing to spreading caused by diffusion the peak height and width both depend on the distance from the origin. Thus it is essential to run a series of standards at the same time to which the peak heights in the sample can be referred. Another important source of error lies in the accuracy with which the sample can be measured out and applied to the gel. It is difficult to dispense a 10 μl volume of solution with an accuracy to better than 1–2 per cent, even using a microsyringe, and with smaller volumes the errors will be much larger.

With virtually all staining methods which rely on dye binding it must be remembered that each separated component is likely to take up a different amount of the dye. Thus when a stained gel is scanned in a densitometer, *if the area under the peak given by a particular band is expressed as a percentage of the total area under all the peaks added together, a highly*

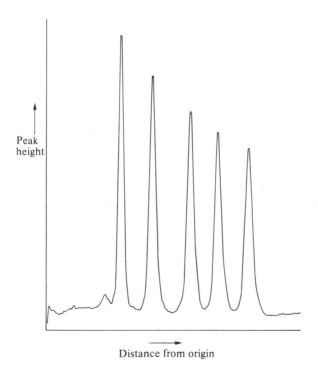

Distance from origin

FIG 2.13. Variation of peak height and width with migration distance. All peaks contain the same protein load (12 μg of ß-lactoglobulin A) and are of the same area. (Provided by Dr. R.M. Hillier.)

inaccurate value for the amount of material in that band may result. An example of this has been given by Anderson and Andrews (1977) who showed that in the milk of cows infected with mastitis, in which there are elevated levels of bovine serum albumin and immunoglobulin, failure to correct for the differential binding of Xylene Cyanine Brilliant G results in a 25 per cent overestimate of bovine serum albumin while immunoglobulin is underestimated by a similar amount. Accurate quantitative measurement in absolute terms therefore requires a knowledge of the dye binding of each constituent of interest and standard curves should be plotted by running known amounts of the substances on a series of gels. It is then only necessary in each subsequent experiment to run a single standard sample or standard mixture to which the unknowns can be related using the standard curves.

For the greatest quantitative accuracy it is a good idea to include if possible an internal standard with each unknown sample. This can be done by mixing with the sample a known amount of a standard substance having a mobility different from any components in the sample. Alternatively, a known amount of one of the sample components (e.g. the fastest moving) can be applied and the gels run for a few minutes before switching off and applying the sample proper. The amount added should be such that the standard band on the gel is of a similar intensity after staining to the bands of interest in the sample.

The choice of a suitable standard will depend upon the particular experiment but some proteins useful in this context are given in Table. 2.10.

TABLE 2.10

Some proteins which may be suitable for use as internal standards in quantitative PAGE

Protein	Isoelectric point (pI at 25 °C)	Molecular weight
Lysozyme	10·0	14 000
Cytochrome C (horse)	9·3	12 256
Chymotrypsinogen A (ox)	9·0	23 600
Ribonuclease A	8·9	13 500
Myoglobin (sperm whale)	8·2	17 500
Myoglobin (horse)	7·3	17 500
Erythroagglutinin (red kiney bean)	6·5	130 000
Insulin (beef)	5·7	11 466
ß-Lactoglobulin B	5·3	36 552
ß-Lactoglobulin A	5·1	36 724
Bovine serum albumin	5·1	67 000
Ovalbumin	4·7	45 000
Alkaline phosphatase (calf intestine)	4·4	140 000
α-Lactalbumin	4·3	14 146

The extent of staining of a gel depends upon the length of time of staining and of destaining, both of which should be kept constant for reproducible results, and upon the shape, thickness, and composition of the gel, which determine the rates of diffusion of fixing agents and dye molecules. As well as correcting for any small differences in these factors, constructing a standard curve enables one to verify that the plot is linear within the concentration range being investigated and obeys Beers Law. The precise range over which a linear plot is obtained depends partly upon the characteristics of the densitometer and recorder but is also affected by migration distance (Fishbein 1972). At short distances bands are more compact and the linear response range of the densitometer is exceeded sooner than with the broader more diffuse bands that have migrated further. Kruski and Narayan (1974) have shown that the limitations are in the densitometer and that the uptake of dye can be proportional to the amount of material in a band over a much wider range. These authors also reported that with Amido Black staining deviations from a linear response began when the protein band exceeded a maximum optical density of about 1·5. This is rather similar to the value we have found using Xylene Cyanine Brilliant G staining, although we would prefer to set this limit rather lower, at 1·2–1·3.

(b) *Densitometry of unstained gels*

Quantitative determination of unstained gels by u.v. scanning is one of the occasions when it may be advantageous to prepare the gels from highly purified monomers in order to give gels with low absorption in the region of 250–300 nm. Below about 250 nm the background absorption of the gel becomes excessive even with pure reagents and there should never be much absorption above about 300 nm with any reagents unless high proportions of Bis are used. One of the major factors influencing the opacity of gels is the proportion of the cross-linking agent. Owing to light scattering the absorption of gels in the u.v. varies approximately exponentially with Bis concentration, so that $C = 3$ per cent or less is helpful. In some buffer systems large u.v. absorbing peaks, probably due to unpolymerized acrylamide, migrate through the gel, and in these cases a pre-electrophoresis step may be beneficial to remove both this and traces of catalysts. Cylindrical gel rods are generally scanned in the u.v. immediately after electrophoresis in quartz tubes without removing the gels from the tubes (Fries and Hjertén 1975). Scanning patterns of unstained gels may not be identical to patterns obtained from the same gels after staining and relative peak areas may be different. This is due to differences in the dye-binding capacities of different components, and with proteins scanned at 280 nm to differing contents of tyrosine and tryptophan which are the amino acids determining absorption at this wavelength. The extinction coefficients for most proteins at 280 nm are only about 3–5 per cent of those for nucleic acids at 260 nm, and consequently the method is much less sensitive for proteins than staining procedures.

Quantitative measurement of unstained gels containing protein can also be achieved by utilizing the intrinsic fluorescence of tryptophan residues (Easton, Lipner, Hines, and Lief 1971). Gels made with highly purified monomers are prepared in quartz tubes, and after the electrophoresis they are scanned in a densitometer with a fluorescence detection attachment using excitation at 280 nm and detection of emission at 340 nm. For blood serum proteins the sensitivity appeared to be similar to that achieved by Amido Black staining. Peak areas showed a curved relationship with the amount of material present, owing to self-absorption by the protein within the gel, so that accurate quantitative measurement required the use of standards and a standard plot.

(c) *Densitometry of photographic negatives*

This method is most often used for the quantification of fluorescent bands, especially if the fluorescence is relatively short-lived or if a fluorescence attachment for the densitometer is not available. The equipment required is generally the same as that used for stained gels with a film holder in place of the normal cuvette.

Photographic conditions must be kept constant, and once a particular combination of lighting, camera, lens, filters, film, etc. has been decided upon a set of standards should be photographed so that a standard curve can be constructed. Subsequent photographs on the same film can then be directly related to this standard curve. The last frame on a film should be the same subject as the first frame on the next film so that slight differences in the film emulsion and in processing conditions can be corrected for. The evenness of illumination and the distance of all parts of the gel from the camera lens must also be considered. Problems due to the latter, which may mean that bands towards the ends of the gel are viewed slightly obliquely, can be overcome by bending the gel slightly towards the lens.

Because it involves additional factors which can influence both resolution and quantitative accuracy, the scanning of photographic negatives can never be expected to equal in these respects direct scanning of stained or fluorescent bands, but the results may be adequate for many purposes and there is the advantage of making a permanent record at the same time. A good example of the method in its most sophisticated form has been given by Prunell, Strauss, and Leblanc (1977) who were able to apply it to fluorescent bands containing as little as 0·2 μg of DNA labelled with ethidium bromide. A design for a computer controlled two-dimensional scanner for obtaining quantitative results from the photographic negatives of ethidium bromide stained DNA-agarose gels has been described (Sutherland *et al.* 1984).

(d) *Spectrophotmetry of eluted dye*

With Amido Black staining and destaining the bands of interest are cut out from the gel with a scalpel or razor blade and chopped into very small pieces. The gel pieces are then mixed with 1 ml 0·5 N NaOH in a solution of methanol:water (1:1 by volume) for each 20–40 μg of protein in the gel band and heated at 55 °C for at least 1·5 h (Kruski and Narayan 1974). After cooling, the supernatant is decanted off from the clear gel pieces and the optical density measured at 625 nm. Optical density is proportional to dye concentration up to a value of 1·0, with a 1 cm path length cell, but then begins to deviate from linearity. At the same time as the sample band is cut from the gel, a blank piece of gel of the same size is cut out and passed through the procedure so that a correction for background staining can be made. Quantitative measurement is achieved by reference to a suitable standard plot.

When protein bands are stained with Coomassie Blue R 250 the gel pieces are extracted with 25 vol. per cent dioxane or 25 vol. per cent pyridine in water (Fenner, Traut, Mason, and Coffelt-Wickman 1975). Among the advantages given is the ease of quantifying proteins with bands of very different widths and staining intensities owing to the wide range within which protein concentration obeys Beer's Law. The method is also particularly suitable for two-dimensional gels which are technically difficult to scan by conventional densitometry.

2.12. Localization and quantitative measurement of radioactively labelled components

Most macromolecules, if not all, can be labelled with radioactive isotopes such as ^{3}H, ^{14}C, ^{32}P, ^{35}S, ^{131}I etc. either by adding a labelled precurser to the appropriate biological system and examining the formation of labelled product or by a direct chemical reaction, e.g. alkylation with ^{14}C-iodoacetate, iodination with ^{131}I, etc. Some procedures for radiolabelling of proteins and nucleic acids are given in Appendix 1. A major advantage of the method is that a suitable isotope can often be incorporated without altering biological or enzymic properties and, of particular relevance to electrophoretic separations, without changing the net molecular charge.

Once separated, the radioactively labelled bands can be localized in gels by procedures involving liquid scintillation counting, autoradiography, or fluorography. The methods are highly sensitive and can be very accurate.

An electronic device has been described for the rapid quantitative measurement of radiolabelled bands in gels (Goulianos, Smith and White 1980).

(a) *Liquid scintillation counting*

The first requirement for liquid scintillation counting is to slice up the gel accurately so that individual slices can be placed in separate vials for measurement. Because of the highly elastic nature of polyacrylamide accurate manual slicing is almost impossible so a number of gel-cutting devices have been designed. In the simplest form (e.g. Aronson and Borris 1967) a suitable device consists of a number of razor blades fixed in a block or bolted together with regular spacing between blades (e.g. 1 mm apart). Because gels distort very readily it is advisable to place the gels on a sheet of Parafilm or aluminium foil and freeze them at about –20 °C before cutting. The typical slice thickness required is usually about 0·5–1·0 mm. Tissue-chopping devices (McIlwain and Buddle 1953; Loening 1967) which are commercially available (e.g. Mickle Engineering Co. Ltd., Gomshall, Surrey, U.K.) are generally more suitable for gel slicing than microtomes. In this case also gels should be frozen by placing in a tray over solid CO_2 before being sliced. Chao and Chao (1981) found that gels, particularly thin polyacrylamide gels, were more easily sliced if first dried down on a backing sheet of filter paper.

For ^{14}C or isotopes with higher energies the gel slices can simply be placed on small pieces of filter paper and dried. They are then transferred, complete with the paper, into counting vials, covered with an appropriate volume of scintillation mixture, and the radioactivity is measured in a liquid scintillation counter. Alternatively, Weber and Osborn (1975) recommend that the wet slices are placed in vials and shaken for 6–12 h at 37 °C with 0·5 ml of 0·1 per cent sodium dodecyl sulphate. After addition of 3 ml of Aquasol (New England Nuclear Corporation) the samples are ready for counting. Rice and Means (1971) soaked the gel slices in 70 μl of 2M piperidine for 30 min, added 0·5 ml of NCS reagent (Nuclear Chicago Corporation), and after allowing the slices to swell for 4 h at 65 °C added 10 ml of 0·5 per cent Omnifluor (New England Nuclear) in toluene before counting.

For separated substances of low energy (e.g. ^3H) or where the amounts of activity are close to the lower limit of detection, Maurer (1968) favours a microcombustion method in which the gel slice is dried on a small piece of optical lens paper and folded into a small wide-meshed platinum basket. The protruding tip of paper is blackened with ink and the wire coil, basket, and sample are placed in a counting vial which is then flushed with oxygen. The vial is capped and the paper tip ignited from outside by powerful illumination with a projector lamp. The water formed by the combustion is condensed by cooling the vial in liquid nitrogen or a solid CO_2-acetone bath and scintillation fluid is added. When this method is used to measure ^{14}C or ^{35}S, a fibreglass disc soaked in 2-phenylethylamine is placed in the bottom of the vial before flushing with oxygen to absorb the $^{14}CO_2$ and $^{35}SO_2$ formed. A rather similar combustion method has been described by McEwan (1968) and such methods give high counting efficiencies.

A completely different approach to the preparation of samples for liquid scintillation counting is to solubilize the gel. With conventional Bis cross-linked gels this is most usually achieved by treatment with 30 per cent hydrogen peroxide (Young and Fulhorst 1965). In a typical method (Tishler and Epstein 1968) the gel slices are thoroughly dried, then incubated in 0·1 ml 30 per cent H_2O_2 at 50 °C for

1–6 h until solubilized. 1·0 ml NCS solubilizer reagent is added followed by 10 ml scintillation liquid (3·0 g 2,5-diphenyloxazole (PPO) + 0·1 g of 1,3-bis-2-(5-phenyloxazole) (POPOP) per litre of toluene) or 10 ml of the commercially available fluid Biofluor (New England Nuclear Corp.). Counting efficiencies are about 80 per cent for ^{14}C and 23 per cent for ^{3}H. However, there may be some loss of ^{14}C as ^{14}CO$_2$ and some quenching during the counting. These difficulties are less severe with gels which have cross-links susceptible to chemical cleavage and can therefore be solubilized more readily.

In a study of the polymerization kinetics of various cross-linkers such as those discussed below, Gelfi and Righetti (1981a) found very great differences, the order of reactivity being Bis ≈ DHEBA > EDIA ≈ BAC ≫ DATD. In fact DATD was even found to be an inhibitor of the polymerization process, leading to gels with high proportions of unreacted monomer present. Since the monomers are neurotoxins, it is positively dangerous to handle such gels and the use of DATD should almost certainly now be abandoned.

Ethylenediacrylate (Fig. 2.14) contains base-cleavable ester groupings (Choules and Zimm 1965) instead of the amide bonds which are present when the usual Bis is used as cross-linker. The zones of interest in the gel can then be cut out and dissolved in 1 M NH$_4$OH using 1 ml for every 400 mm^3 of gel (Cain and Pitney 1968). Complete solution takes a few hours. the NH$_4$OH solution is then evaporated to dryness at 65 °C and the residue resuspended on 0·2 ml H$_2$O followed by 1·5 ml of NCS solubilizer solution. After a few hours at room temperature 10 ml of toluene-based scintillation liquid (4 g PPO + 0·3 g POPOP per litre) is added. Counting efficiency for ^{3}H is 26–30 per cent.

Unfortunately ethylenediacrylate can only be used in acidic and neutral gel systems. Since most proteins and nucleic acids are neutral or acidic they are often best separated at alkaline pH so this is quite a serious limitation.

An alternative cross-linking agent (Fig. 2.14) is N,N′-diallyltartardiamide (DATD) which is commerically available from Pierce Chemical Co. If this is substituted for Bis on a mole for mole basis (Anker 1970) the resulting gel can be dissolved in 2 per cent periodic acid in about 20–30 min at room temperature or 10 min at 37 °C. The gels take rather longer to set (e.g. 1 h) and alkaline gels must be neutralized by immersion in acetic acid before the periodic acid treatment. Periodic acid causes no additional quenching so the solubilized gel solution can be mixed with scintillation fluid in the usual way, remembering that 0·2 ml is about the limiting amount of any aqueous solution that can be mixed with 1·5 ml NCS reagent and 10 ml scintillation fluid. Because of the relatively poor mechanical properties of the gels due to inefficient polymerization kinetics and the consequent neurotoxic properties of gels containing large proportions of remaining monomer (see p.6), the use of DATD as cross linker should be discouraged.

Hansen (1976) used N,N′-bis-acrylylcystamine (Fig. 2.14) (Pierce Chemical Co.) which when substituted for Bis on a molar basis gave gels solubilizable with 2-mercaptoethanol or dithiothreitol. The method works best with gels of low T, but with modifications to the catalyst system it can be used to prepare soluble gels up to $T = 20$ per cent (Hansen, Pheiffer, and Boehnert 1980; Faulkner, Carraway, and Bhatnagar, 1982).

O'Connell and Brady (1976) describe the synthesis and use of a number of allyl and acrylamido cross-linkers and conclude that the latter give gels with better separation characteristics. One of these, N,N′-(1,2-dihydroxyethylene) bisacrylamide (DHEBA) (Fig. 2.14) appeared to be especially suitable and is substituted for Bis on a mole for mole basis. The cross-link possesses both a periodate-sensitive 1,2-diol structure and two amido methylol bonds which are base cleavable but more

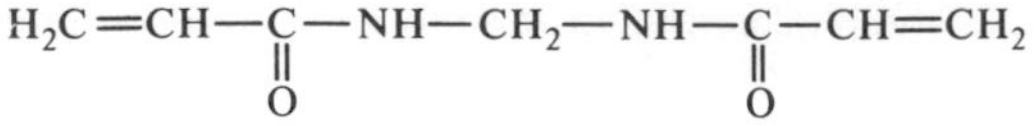

$$H_2C=CH-\underset{\underset{O}{\|}}{C}-NH-CH_2-NH-\underset{\underset{O}{\|}}{C}-CH=CH_2$$

N,N'-methylenebisacrylamide (Bis)

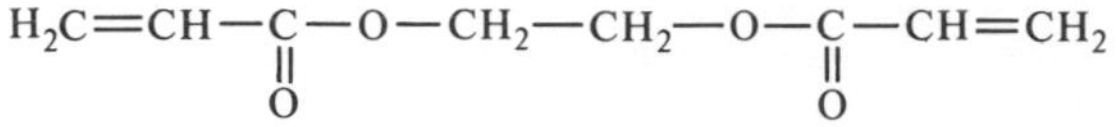

$$H_2C=CH-\underset{\underset{O}{\|}}{C}-O-CH_2-CH_2-O-\underset{\underset{O}{\|}}{C}-CH=CH_2$$

ethylenediacrylate (EDIA)

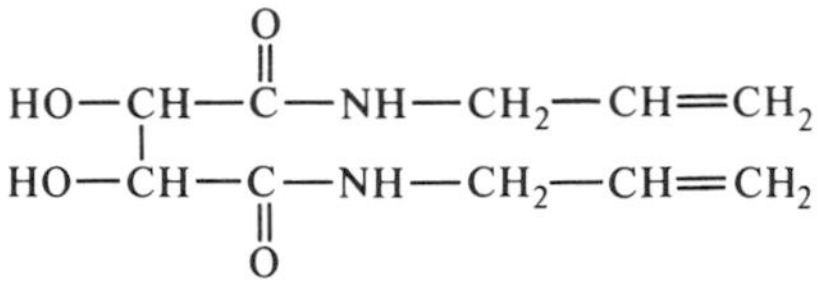

$$HO-CH-\overset{\overset{O}{\|}}{C}-NH-CH_2-CH=CH_2$$
$$HO-CH-\underset{\underset{O}{\|}}{C}-NH-CH_2-CH=CH_2$$

N,N'-diallyltartardiamide (DATD)

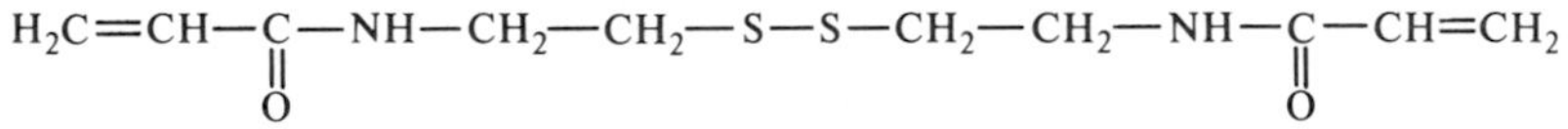

$$H_2C=CH-\underset{\underset{O}{\|}}{C}-NH-CH_2-CH_2-S-S-CH_2-CH_2-NH-\underset{\underset{O}{\|}}{C}-CH=CH_2$$

N,N' - bis-acrylylcystamine (BAC)

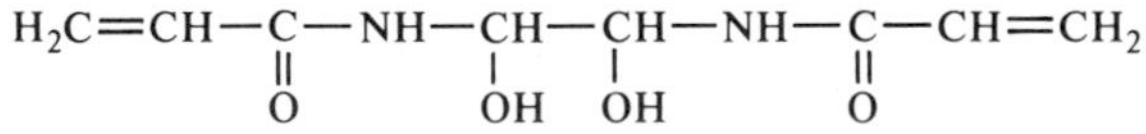

$$H_2C=CH-\underset{\underset{O}{\|}}{C}-NH-\underset{\underset{OH}{|}}{CH}-\underset{\underset{OH}{|}}{CH}-NH-\underset{\underset{O}{\|}}{C}-CH=CH_2$$

N,N'-(1,2-dihydroxyethylene) bisacrylamide (DHEBA)

FIG 2.14 Some cross-linking agents used in polyacrylamide gel formation.

stable to base than the ester bonds of ethylenediacrylate so that the common alkaline buffer systems can be used. Borate buffers should be avoided since they form negatively charged complexes with cis-1,2-diol groupings which might lead to a charged gel matrix. Gels were stained with Xylene Cyanine Brilliant G, destained, and sliced according to the band pattern. Each slice was then placed in a counting vial and 1·0 ml 0·025 M periodic acid added. The vials were tightly capped, incubated at 50 °C for 48 h until the slices were liquified and the dye decolorized, and 10 ml scintillation fluid (e.g. Mix tT21) (Patterson and Green 1965) was added. If solubilization with weak bases (e.g. NH_4OH) is used the dye is not decolorized and a separate bleaching step is required for efficient scintillation counting (Cain and Pitney 1968). Mechanical, separation, and solubilization characteristics over a wide range of gel concentration were good, so that it would appear that DHEBA is probably the best cleavable cross-linking agent available at the present time for most applications.

(b) *Autoradiography*

Autoradiography is a highly sensitive technique for the localization of substances labelled with isotopes such as ^{14}C, ^{32}P, ^{35}S, ^{59}Fe, ^{125}I, or ^{131}I. The power of the method is well illustrated by the very elegant work of O'Farrell (1975). In a two-dimensional protein mapping procedure which gave very compact spots in the gel, he found that staining with Coomassie Blue R250 was capable of revealing as little as 0.01 μg of protein, but autoradiography was 100-fold more sensitive and in favourable cases 1000-fold more so. Autoradiography is not suitable for ^{3}H-labelled material.

Autoradiography can be applied to slabs of gel or to longitudinal slices of gel rods. Simple devices for slicing gel rods have been described by Fairbanks, Levinthal, and Reeder (1965), by Ho (1979) and by Watts, King, and Sakai (1977). Watts *et al.* describe a device consisting of a stack of thin brass plates which they claim is easy to use with gels of low T. The gel rod is placed in a long slot cut in the stack which is then cooled in dry ice until the gel is frozen. The side of the gel protruding out of the slot is then sliced off with a razor blade. One plate is removed from the stack and another slice cut, and so on.

For autoradiography of high-energy isotopes (e.g. ^{32}P, ^{59}Fe, ^{64}Cu, ^{131}I, etc.) the gel slice can merely be covered with a thin sheet of polythene and placed in contact with a sensitive X-ray film (e.g. Kodak XAR, Kodak SB-5, Kodak X-OMAT, Kodak Direct exposure medical film, Ilfex X-ray film, etc.). For increased sensitivity X-ray intensifying screens such as the Kodak X-OMATIC or LANEX and the Dupont Cronex Lighting Plus or Cronex Quanta II can be used (Swanstrom and Shank 1978). These gave up to a 30–40-fold increase in the detection of ^{125}I and up to an 8–10-fold increase for ^{32}P.

It is very common practice, and necessary for the localization of ^{14}C, first to dry the gel slices (e.g. Fairbanks *et al.* 1965). This can be achieved (Reid and Bieleski 1968) by placing the thin gel slabs or slices onto a clean glass plate big enough for there to be a margin of at least 2 cm all around the gel. This is placed in a dish and melted 2 per cent agar is poured over the gel, covering it with a uniform layer 4 mm thick. After the agar has set, the plate is dried slowly and evenly, at first using an i.r. lamp until nearly dry and then just at room temperature.

Popescu (1983) used a rather similar approach for drying slab gels of various concentrations (7·5–15 per cent T). After the destaining stages the gels were placed in 7·5 per cent acetic acid solution containing 1·5 per cent glycerol for 1 h, placed on a plexiglass plate, and air bubbles carefully removed from between the gel and plate. The gel is then covered with a 5 per cent aqueous solution of gelatin (prepared by heating on a water bath and filtering if necessary while hot) containing 0·02 per cent NaN$_3$, using 30 μl of solution for every cm^2 of gel surface area. This is spread evenly over the gel surface without spilling any over the edges of the gel. When it has become more viscous (after 1–2 h) it is then spread over the edges of the gel with the aid of a spatula to give a border round the gel 1·5–2·0 cm wide. The gel is then air dried at room temperature for 24 h or more and when completely dry the edges are freed with a scalpel or razor blade, and the gel can simply be peeled off the plexiglass plate. Drying time depends upon composition and thickness of the gels but once dry they can be easily handled, stored or used for autoradiography. Dry gels can be reconstituted to their original size by heating at 60 °C in H$_2$O for 2–3 h.

Juang *et al.* (1984) describe an even more simple method which is suitable for drying all types of polyacrylamide gel. After final destaining (during which some swelling often occurs) gels are shrunk back to their original size in 65 per cent methanol containing 0·5 per cent glycerol. A 25 x 25 cm square of wet cellophane is then spread over a 20 cm square glass plate, the gel placed in the centre, avoiding trapping air bubbles between the gel and the cellophane, and another square of wet cellophane placed over it. This is most easily done by first rolling the wet cellophane around a clean glass tube and then unrolling it over the surface of the gel leaving the cellophane behind. The cellophane borders are then folded back under the glass plate and the edges clamped with spring clips. The gel 'sandwich' is then oven-dried at 70 °C (30–120 min is usually sufficient), although for thick (≥ 1.5 mm) or concentrated gels ($T \geq 15$ per cent) a longer time at a lower temperature is preferable. If the dry gel is to be used for autoradiography of ^{3}H or ^{14}C-labelled zones, it is better to siliconize the glass plate and place the gel directly on it without an intervening cellophane layer, so that after drying the uncovered gel face is placed directly onto the x-ray film. This avoids reductions in sensitivity due to absorption of the low energy emissions from these isotopes by the cellophane .

Other methods of drying gels have been briefly reviewed by Maurer (1971) and simple designs of drying apparatus which can be built relatively easily have also been described by Maizel (1971) and Studier (1972). Commercial equipment is available (e.g. BioRad Model 224 Vacuum Gel Dryer, LKB 2003, etc.) but quite expensive.

For autoradiography the drying step is important to the quality of the final result and dried gels should not be cracked or curled. The dried gels are clamped or taped to the X-ray film, which after exposure is developed in the usual way. As a rough guideline, according to Weber and Osborn (1975) a band containing 50 000 dpm may require a 20 h exposure, but this clearly depends upon the area over which it is spread and the moisture content of the gel which greatly influences the degree of quenching. Albanese and Goodman (1977) state that a band 1 mm wide in a 5 mm diameter gel rod containing 1 nCi of ^{14}C after slicing gives an autoradiogram suitable for scanning in 7–10 days. With the higher-energy isotope ^{32}P only 24–48 h is needed with 0·1 nCi. O'Farrell (1975) found that very compact spots containing only 1 dpm ^{14}C could be detected after a 20 day exposure. In order to detect minor components it may be necessary to over-exposure major ones, and this can lead to difficulties caused by autoradiographic spreading, due to which there is a loss in resolution as exposure is prolonged. This amounts to about a 2-fold increase in spot size for a 10-fold increase in exposure time (O'Farrell 1975), which while it may not seem very important can result in minor components adjacent to major zones becoming obscured.

A number of procedures have been developed for dealing with doubly-labelled material, usually relying on differences in energy of the two types of emission. For example Cooper and Burgess (1982) detected ^{35}S- and ^{32}P-labelled proteins in a single gel simultaneously using two films, the first recording directly the ß-emissions of ^{35}S and the other (shielded by 0.015 mm thick aluminium foil) recorded scintillation photons from an intensifying screen excited by ^{32}P emissions. When autoradiography was performed at –80 °C the istopic discrimination was virtually complete.

Quantitative measurement of autoradiograms is achieved by densitometric scanning of the film after development. Film response is only linear up to an optical density of 0·5 (O'Farrell 1975) so several exposure times are needed to quantify a number of components accurately in a single gel if they differ widely in amount or activity. To measure the total radioactivity the integrated optical density (peak area) of the band on the film is measured and related to radioactivity with a standard curve. Calibration is achieved (O'Farrell 1975) with the help of standardization wedges made by incorporating known amounts of labelled material into a series of gels, each gel containing half of the amount of the previous gel. These gels are fixed in 7 per cent acetic acid and cut into small pieces (e.g. 2·5 x 4 mm). One piece of each gel in turn is arranged into a sequence on a piece of filter paper and dried. Each of these sequences forms a standardization wedge which is exposed on the same film alongside each unknown gel. Scanning the wedge at the same time as the unknown also acts as an internal control correcting for any variations that may occur during autoradiography, film processing, or from one batch of film to another. The standardization wedges can be stored indefinitely, so their preparation is a once and for all undertaking.

As an alternative to densitometry, Suissa (1983) has suggested that individual bands can be cut out of the film, soaked in 1 ml portions of 1 M NaOH for 2 h (or until all the silver grains have been eluted) and glycerol added to 30 per cent (to retard settling of the grains). The solution is then mixed on a Vortex or similar mixer and the absorbance at 500 nm of the colloidal suspension measured immediately. If dust on the film is carefully removed beforehand with an antistatic cloth and the NaOH is filtered before use the response should be linear up to an optical density of 0.5 or more.

(c) *Fluorography*

Tritium (^{3}H) labels cannot be detected by autoradiography because of the low energy of the ß-particle emission which is almost entirely absorbed within the sample itself. Fluorography overcomes this difficulty by impregnating the gel with a scintillator before exposing the gel to X-ray film. The film is thus not exposed directly by the ß-particle radiation but indirectly by light generated by interaction of the ß-particles with the scintillator molecules. The limits of detection of ^{3}H present in a band 1 cm x 0·1 cm are about 3000 dpm using a 24 h exposure or 500 dpm with a one week exposure (Bonner and Laskey 1974). Compared with conventional autoradiography the method also gives a 10-fold increase in the sensitivity of localizing ^{14}C and ^{35}S. The sensitivity of fluorography can be still further enhanced by hypersensitizing the film before the main exposure (Laskey and Mills 1975). With this modification 300 dpm of ^{3}H or 30 dpm of ^{14}C in a gel band can be detected in 24 h. This pre-exposure step also corrects the non-linear relationship between radioactivity of the sample and absorbance of the film image, allowing quantitative measurements to be made (Laskey and Mills 1975).

The optimal procedure for fluorography has been described (Bonner and Laskey 1974) as follows.

(1) Directly after electrophoresis or after staining soak the gel in about 20 times its volume of dimethyl sulphoxide (DMSO) for 30 min, followed by a second 30 min immersion in fresh DMSO.

(2) Immerse the gel in four volumes of 20 wt per cent PPO in DMSO for 3 h. The PPO concentration should always be within the range 14–19 wt per cent after equilibration with the gel so solutions should only be used once or the concentration will fall below the optimum. PPO can be recovered from used solutions for re-use by precipitation with water.

(3) Immerse the gel in 20 volumes of water for 1 h to replace the bulk of the DMSO with water before drying. Precipitation of PPO within the gel does not reduce fluorographic efficiency.

(4) Dry the gel as for autoradiography (Section 2.12(b)). If the dried gel is sticky the DMSO has not been adequately removed, so soak the gel in H_2O again and re-dry.

(5) Place the dried gel in contact with Kodak R P Royal X-OMAT or equivalent medical X-ray film between two glass plates to form a 'sandwich' which can then be taped together. Exposure takes place preferably at –70 °C or failing this the 'sandwich' can be buried in solid CO_2. The influence of film type and exposure temperature have been discussed by Bonner and Laskey (1974)

who concluded that this was the best film and that greater sensitivity is achieved at lower temperatures.

This somewhat laborious process for exchanging the aqueous solvent with organic solvent in order to enable efficient infusion into the gel of the water-insoluble and rather expensive scintillant PPO has now been largely replaced by use of the commercially-available fluor EN^3HANCE (New England Nuclear Corpn.), although this sometimes gives counting efficiencies that are slightly inferior to those obtained with PPO (McConkey and Anderson 1984a). Kodak X-OMAT film has now been largely replaced by Kodak XAR-5 or SB-5 film for autoradiography and fluorography but the principles and procedures are otherwise largely unchanged.

Hypersensitizing the film (Laskey and Mills 1975) is achieved by exposing it to a single flash from a photographic electronic flash unit (e.g. Metz mecablitz 181 or Panagor 77) fitted with three filters taped over the 'window' to reduce and diffuse the light output. The first filter (nearest to the window) is an i.r.-absorbing heat filter (this filter is optional), the second is a coloured filter to reduce light output (e.g. Kodak Wratten No. 22 Deep Orange or for weaker flash units Kodak Wratten No. 21 Orange), and the third is a piece of porous paper (e.g. Whatman No. 1 filter paper) to diffuse the light. The film was backed by yellow paper during the exposure and the surface nearest the light source was applied to the gel for the fluorography. For the hypersensitizing flash the film was usually placed about 70 cm from the flash unit.

After the fluorographic exposure the film is developed for 6 min at 20 °C in Ilford PQX-1 or Kodak DX-80 developer and then scanned with a densitometer as in conventional autoradiography (Section 2.12(b)) using a standardization wedge in the same way for accurate quantitation.

This fluorographic technique has been very widely used in the last few years but it does suffer from the disadvantage that the scintillator PPO is insoluble in aqueous solvents so gels must first be transferred into DMSO. In addition PPO is very expensive and the excess reagent is therefore often recovered in a separate crystallization step. Both of these difficulties are avoided in a method (Chamberlain 1979) in which the inexpensive sodium salicylate is used as a scintillant. Acid-fixed or stained gels are soaked in 20 vol water for 30 min before treatment to avoid precipitation of salicylic acid and in the standard method are then treated as follows.

(1) Soak the gel slice in 10 vol. 1 M sodium salicylate at a pH between 5 and 7 for 30 min at room temperature.

(2) Lay on moist Whatman 3 MM paper and dry under vacuum for 2 h at 80 °C under an acetate sheet.

(3) Remove the acetate sheet and expose to preflashed X-ray film at –70 °C as above.

Chamberlain (1979) used a Bio-Rad drier in step (2) and found that the gel slices adhered to the Mylar sheet usually used with this drier but they did not stick to acetate sheets once the gel was completely dry and cool. Compared with the PPO–DMSO procedure the fluorographic bands were very slightly more diffuse, but for most purposes this is not likely to be a serious drawback and the method is much more rapid and inexpensive. Heegaard, Hebsgaard, and Bjerrum (1984) have described a variation of this method which is very suitable for the fluorographic measurement of immunoprecipitated proteins (or presumably any other proteins) in agarose gels.

Many aspects of the technique of fluorography have been reviewed by Laskey (1980).

2.13. Blotting techniques

In the present context, blotting refers to techniques for transferring separated zones of protein, RNA or DNA from gels to thin sheets of a derivatized paper or membrane matrix such as nitrocellulose to which they bind and are immobilized. While addition of this process after the electrophoretic separation increases the time and complexity of the whole experiment, there are a number of very substantial advantages to be gained by including a blotting step. Many of the advantages are related to the much greater accessibility of macromolecules absorbed or bound to the surface of a thin sheet compared to those buried within the matrix of a gel. This is particularly important in experiments where gradient gels (see Chapter 4) are used since different parts of the gel have different porosities and separated zones will vary in accessibility to reagents. When transferred to the blotting membrane all immobilized macromolecules are equally accessible and usually much smaller amounts of reagents are needed than would be required for a relatively voluminous gel. Wet blotting membranes are pliable and often easier to handle than gels. Among other advantages pointed out by Gershoni and Palade (1983) are that since the immobilized macromolecules are on the surface of a thin membrane, processing times for staining and destaining reactions, incubations, washings, etc. can be much shorter than for gels, transferred zone patterns can be dried easily and stored for months before analysis if required, multiple replicas of a single gel can be made, often a single protein transfer can be subjected to several successive analyses and some analyses that are difficult or impossible to apply to gels can be applied to blot patterns without difficulty.

(a) *Immobilization matrices*

There are several different types of immobilizing matrix available but without doubt the most widely used is nitrocellulose (NC). A good example of the transfer of proteins from polyacrylamide gel to NC is shown in Fig. 2.15. NC is commercially available from several suppliers (e.g. BioRad, Schleicher & Schuell, Millipore, etc.) and material with 0.45 μm pore size is most commonly employed, although other porosities may sometimes be preferable (e.g. 0.22 μm may be better for low molecular weight proteins with low binding affinities for NC). The mechanism of protein binding to NC is complex and not at present entirely understood, but hydrophobic interactions are probably important since the addition of non-ionic detergents such as Triton X-100 aids subsequent elution of protein from the matrix. At pH 8, which is typical of many protein transfer conditions, both the NC filter and most proteins carry the same (negative) net charge, which argues against a key role for ionic bonding.

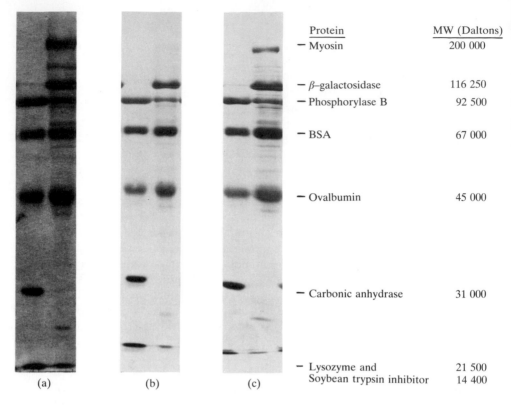

Protein	MW (Daltons)
— Myosin	200 000
— β–galactosidase	116 250
— Phosphorylase B	92 500
— BSA	67 000
— Ovalbumin	45 000
— Carbonic anhydrase	31 000
— Lysozyme and	21 500
Soybean trypsin inhibitor	14 400

(a)   (b)   (c)

FIG 2.15. Protein blotting. Protein molecular weight standards, varying in size from 14 000 to 200 000 daltons, separated on a $T = 10$ per cent SDS–PAGE gel were transferred to a nitrocellulose membrane by electroblotting for 3 h in a 0·025 M Tris, 0·192 M glycine buffer pH 8·3 containing 20 per cent methanol. (a) Gel stained with Coomassie Blue R250; (b) transfer blot obtained by running at 60 V (7·5 V/cm) and 0·15 A at room temperature; (c) 'high-intensity' transfer run at 220 V (55·5 V/cm) and 1·0 A with a final temperature of 35 °C. (Photograph kindly supplied by Bio Rad Laboratories.)

NC is suitable for the transfer of both proteins and nucleic acids (DNA and RNA) from polyacrylamide or agarose gels but some components with relatively weak affinity may be lost during transfer or during subsequent processing of the membrane. In these cases, as well as switching to less porous membranes, losses can be minimized by covalently cross-linking proteins to the filter after transfer. This is done by soaking the filter containing the pattern of transferred protein bands for 15 min in 0·5 per cent glutaraldehyde in phosphate-buffered saline. Alternatively a simple acid treatment of soaking for 1 h in 25 per cent isopropanol containing 10 per cent acetic acid may also suffice (Gershoni and Palade 1982). Some NC membrane filters (e.g. Millipore) are mixed ester matrices and contain cellulose acete in addition to NC and this appears to reduce their capacity to bind protein (Gershoni and Palade, 1983), so care should be taken to choose pure NC filters. This is probably particularly important in electroblotting procedures.

In order to overcome some of these disadvantages diazobenzyloxymethyl paper (DBM) was introduced (Alwine, Kemp, and Stark 1977). With this, negatively charged proteins are thought to be bound initially to the positively charged diazonium groups by electrostatic bonds but this is followed by a slow reaction via azo derivatives leading to covalent bond formation between the protein and the paper. This results in stable transfer patterns although the resolution may not be as good as with NC membranes because of the coarse grain of the DBM paper. Another diazo-paper, diazophenylthioether (DPT) paper has also been used for transfer of proteins and nucleic acids and has been found to be just as efficient as DBM paper but rather more stable (Reiser and Wardale 1981). However neither of these diazo-papers are particularly stable so they are best used in conjunction with electrophoretic transfer (electroblotting) rather than with diffusion or convection blotting both of which are much slower processes. The papers are prepared as the aminobenzyloxymethyl (ABM) derivative (Alwine et al. 1977) or the aminophenyl-thioether (APT) derivatives (Reiser and Wardale 1981), which are quite stable, and they are then diazotized immediately before use by immersing for 30 min at 4 °C in a solution made up of 40 ml H_2O, 80 ml 1·8 M HCl, and 3·2 ml of freshly prepared $NaNO_2$ solution (10 mg ml^{-1}). The resulting DBM or DPT papers are then washed five times for 5 min each with cold H_2O and then twice for 10 min each with cold 50mM sodium borate buffer pH 8·0–9·2 and placed on the gel immediately. Transfers must begin within less than 15 min. The preparation of the ABM and APT papers involves the use of toxic or carcinogenic reagents and must be performed with great care in a fume cupboard (hood) but they can be purchased commercially (e.g. Bio Rad Laboratories Inc.) if desired.

Paper or cellulose acetate can be activated with cyanogen bromide and gives stable transfers in much the same way as DBM paper (Clarke, Hitzeman and Carbon 1979), but this approach has not been so widely employed. Conventional ion-exchange papers, such as diethylaminoethyl (DEAE) paper have been used in many transfer blotting applications and are particularly useful for preparative work since the recovery of the transferred material is particularly simple and efficient. Unfortunately the transfer patterns themselves are less stable than with the other types of immobilization matrix, in that material transferred from the gel may be bound initially but then lost later during the actual transfer operation because the binding is only electrostatic and therefore relatively easily reversible. Thus care and preliminary experimentation are needed in order to optimize the conditions for the most efficient transfer and to maximize recoveries.

In addition to the cellulosic matrices described above two types of nylon-based membrane have been introduced for blotting procedures and possess a number of advantages, being as thin and fine-grained as NC but stronger. These membranes are Gene Screen available from New England Nuclear (Boston, Mass, USA) and Zetabind (ZB) from AMF/CUNO (Meriden, Conn., USA) but also marketed as Zeta Probe by Bio Rad Laboratories. ZB is derived from nylon 66 and incorporates many tertiary amino groupings capable of electrostatic binding to biopolymers such as proteins and nucleic acids. It is mechanically strong and has about a 6-fold higher protein binding capacity than NC. This means that greater amounts of protein can be analyzed and in addition the highly charged cationic nature of the ZB matrix results in better transfers than NC of proteins from SDS-PAGE gels in which the proteins are present as complexes with the anionic detergent molecules. It is also suitable for nucleic acid transfers.

(b) *Methods of transfer*

There are three ways in which proteins or nucleic acids are transported from the gel onto the immobilizing matrix; simple diffusion, solvent flow or under the influence of an electric field. The concept of transferring macromolecules to such a matrix was first introduced by Southern (1975). It consisted of transferring DNA to NC paper by a solvent flow procedure and it subsequently came to be referred to as 'Southern blotting'. When DBM paper was introduced (Alwine, Kemp, and Stark 1977), still employing solvent flow, the method became known as 'Northern blotting', to be followed almost inevitably by the naming of the electrophoretic transfer of proteins and nucleic acids to NC (Towbin, Staehelin, and Gordon 1979; Burnette 1981) as 'Western blotting', while Reinhard and Malamud (1982) referred to the bidirectional transfer by solvent flow of proteins from isoelectric focusing gels to NC as 'Eastern blotting'. The coining of these terms represents one of the little academic jokes in the field of macromolecular research, but like all jokes it should not be carried too far or we shall end up with a highly confusing jargon of terms such as 'South-South-West blotting'! As pointed out by Gershoni and Palade (1983) it is preferable to use much more descriptive and unambiguous terms such as DNA blotting, protein blotting etc.

Blotting by diffusion is the simplest but also the slowest transfer method. In this the gel is placed between two sheets of wet immobilizing membrane or paper material taking care not to trap any air bubbles. Two pieces of foam sponge material (one on each side) or several thicknesses of blotting paper are then added and the whole assembly clamped between two pieces of nylon or stainless steel mesh. The resulting sandwich is then submerged in buffer for 24–48 h, giving two transfers of course, one from each side. Amines should not be present in the buffer if DBM or DPT paper is used, but otherwise almost any buffer can be used.

Transfer by means of a solvent flow (e.g. Southern 1975), sometimes termed capillary blotting, can be achieved by placing the gel onto two or three thicknesses of Whatman 3MM paper or other thick absorbent paper soaked with transfer buffer and supported on a stage or piece of sponge wetted with buffer. A piece of transfer membrane material is placed on top of the gel and then several pieces of dry absorbent paper and paper towel are stacked above the transfer paper and usually held in place with a light weight (Fig. 2.16). Thus buffer flows gradually from the sponge through the wet papers and gel carrying the macromolecules from the gel onto the transfer membrane, through which the buffer passes and is absorbed by the paper towels. Blotting is generally allowed to continue overnight. Buffer composition is not critical and most of the usual buffers can be used, except when transferring to DBM or DPT paper when amines should be avoided. The method is more rapid and more efficient than diffusion blotting and can be made even more rapid by applying a low vacuum (Peferoen, Huybrechts, and De Loof 1982).

A somewhat similar 'protein printing' technique has been described by Johnson *et al.* (1982) in which rod gels were sliced longitudinally (see p.55) and the flat surface of the cut gel applied to a sheet of NC previously soaked in phosphate-buffered saline. A number of gel slices could be applied to a single sheet and the gels are then covered with a glass plate and a 1 kg weight applied. After 15–30 min

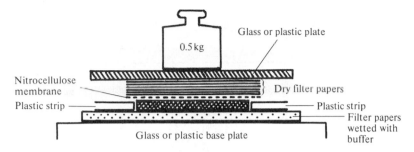

FIG 2.16. 'Southern blotting', the transfer by solvent flow of macromolecules from a gel slab to an immobilising membrane. Diagrammatic representation of a typical gel and membrane 'sandwich' arrangement.

the surface proteins have been transferred and if desired residual proteins in the gel can be revealed by staining in the usual way (Section 2.9).

The most widely used transfer method, however, employs eletrophoretic elution of the proteins or nucleic acids and is therefore sometimes referred to as electro-blotting. Many designs of apparatus have been described for electroblotting, those of Towbin, Staehelin, and Gordon (1979) and of Bittner, Kupferer, and Morris (1980) being particularly widely copied, and several designs are also available commercially (e.g. LKB Produkter AB; Bio Rad Laboratories etc.). Most consist essentially of wetting a sheet of transfer membrane and placing it on the gel to be blotted which in turn is supported by a porous pad (such as 'Scotch Brite'[TM] scouring pads, foam sponge material or layers of wet blotting paper) mounted on a stiff plastic grid (Fig. 2.17). It is important that no air bubbles are trapped between the transfer membrane and the gel, as these greatly reduce transfer efficiency and also distort band patterns. A second porous pad and plastic grid are then added above the transfer membrane and rubber bands placed around the whole 'sandwich' so that the gel and membrane are held tightly in close contact, which helps to prevent skewing of the bands or band spreading during transfer. The 'sandwich' is then placed in a tank of transfer buffer between two electrodes. The electrodes may be attached to the sides of the tank but must cover a large area and be capable of generating a homogeneous electric field over the whole area of the gel. Some designs employ stainless steel mesh electrodes but the most common approach is to use a zig-zag arrangement of platinum wire (e.g. Bittner *et al.* 1980). The transfer buffers used are of low ionic strength and buffers of many different compositions and pH have been used, e.g. 0·7 per cent acetic acid; 25 mM sodium phosphate pH 6·5; 20 mM tris-150 mM glycine pH 8·3; 7·5 mM tris–1·2 mM boric acid pH 8·9 etc, and can contain 20 volume per cent of methanol. Buffers at pH 3·0 may be optimal for transfer of nucleic acids to NC (see Section 2.13d). Glycine and other constituents containing amino groups react with DBM and DPT paper and so must not be present in transfer buffers with these papers. Glycine-containing buffers are commonly used in the electrophoresis gels themselves and if DBM or DPT paper is to be used the glycine must be washed out of the gels with phosphate buffer before transfer. The incorporation of methanol (Towbin *et al.* 1979) tends to stabilize the geometry of the gel during transfer and also increases the binding capacity of NC for protein but it does reduce the efficiency of protein elution from the gel, particularly from SDS-PAGE gels, and long electroelution times (e.g. 12 h

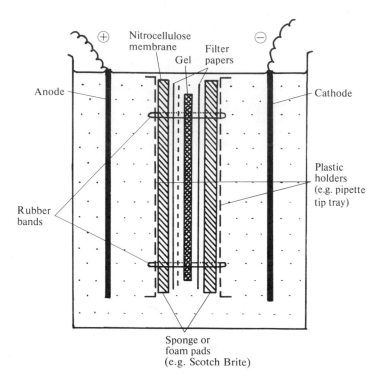

FIG 2.17. 'Western blotting' or electrophoretic transfer of macromolecules from a gel slab to an immobilizing membrane. Diagram showing a typical transfer assembly.

or more) may be required for efficient transfer of large proteins (Burnette, 1981). If methanol is omitted gels may swell a little in low ionic strength buffers which can lead to band distortion on the blot, but this can be overcome by presoaking the gel in transfer buffer for about 1 h before transfer.

The conditions required for electro-blotting depend upon the type of gel (PAGE, SDS–PAGE, PAGE with urea-containing gels, PAGIF, agarose, 2-dimensional slabs, etc. have all been used successfully), the apparatus, the immobilizing membrane or paper and the proteins or nucleic acids to be transferred (Gershoni and Palade 1983). Perhaps rather obviously, the transfer membrane must be on the correct side of the gel, so it is important to know the net charge on the molecules to be transferred under the conditions present. For example when Towbin *et al.* (1979) transferred proteins to NC using 0·7 per cent acetic acid as the buffer in the apparatus by applying a voltage gradient of 6 V cm^{-1} for 1 h the proteins behaved as cations so the NC should be on the cathode side of the gel, but when SDS is present as in SDS–PAGE gels approximately neutral or weakly basic buffers are used, the proteins migrate as anions and the NC must be on the anode side. Relatively low voltage, high current power supplies, such as those used for electrolytic destaining (section 2.10), are suitable for electroblotting (e.g. the BioRad Model 250/2·5) and simple 12 V battery chargers can also be used (Gibson 1981). Most power supplies used for electrophoresis are limited to about 200–500 mA and if these are used Gershoni and Palade (1982) recommend that they are set at the maximum constant current, the voltage gradually falling during the transfer as buffer constituents from the gel are eluted and contribute to the conductivity of the transfer buffer. Relatively steep voltage gradients of 5–10 V cm^{-1} enable transfers to be essentially complete in 1–3 h, but lower gradients naturally require longer times. The completeness also depends on the molecular size of the macromolecules being transferred and Burnette (1981) recommends that for 1.5-mm-thick gels an overnight transfer time (20–22 h) at 6–8 V cm^{-1} is required for > 90 per cent transfer to NC of all proteins with molecular weights up to 100 000.

(c) *Detection of transferred components on blots*

NC membranes can be stained for proteins with all the usual protein stains, with Amido Black being the most commonly used. With proteins immobilized on the surface of a thin membrane, staining and destaining times are much shorter than with a gel slab, and a few minutes in 0.1 per cent Amido Black or 0.2 per cent Coomassie Brilliant Blue R250 in 45 per cent methanol containing 10 per cent acetic acid followed by 5 min destaining in 90 per cent methanol containing 2 per cent acetic acid is adequate. Burnette (1981) cautions that NC membranes disintegrate if left in acidic methanol for more than about 5 min, so destaining should be rapid. Gershoni and Palade (1982) recommend 1 min staining of protein on NC filters with 0·1 per cent Amido Black 10B in 25 per cent aqueous isopropanol containing 10 per cent acetic acid and destaining for 30 min in the same solvent, NC being more stable in isopropanol than in methanol. According to these authors background staining was lower than when Coomassie Blue R250 was used, and indeed Amino Black staining is far more widely employed. In all cases the destained membranes are washed in water for 1–2 h and blotted dry between sheets of filter paper. Drying in a stream of hot air should be avoided. Proteins and nucleic acids radiolabelled with ^{32}P, ^{35}S, ^{125}I, or to high specificity with ^{14}C are readily and efficiently detected by autoradiography. The thinness of the membrane allows much closer contact with the X-ray film than is possible with gels themselves, which both increases sensitivity and aids resolution since there is less scatter of the radiation, and hence zone broadening, when the source of the radiation is close to the film. ^{14}C and ^{3}H-labelled proteins and nucleic acids can be detected by fluorography. The use of transfers blots avoids tedious slicing or drying of gels for autoradiography and the potentially hazardous soaking of gels for fluorography in diphenyloxazole (PPO), only a brief soaking of membranes in 10 per cent PPO in ether being needed (Towbin *et al.* 1979). Nucleic acids can also be detected by ethidium bromide staining as described on p.41, but again much shorter staining times are required compared to gels.

Other transfer matrices (e.g. DBM, DPT, ZB, DEAE-paper, etc.) either react chemically with the common anionic protein stains or bind them so tenaciously that they cannot be removed, so radiolabelling or immunological detection are usually used. Nucleic acid detection with ethidium bromide is suitable however, at least with DEAE paper. Silver staining methods for proteins were reported (Gershoni and Palade 1982) to give rather variable results with ZB membranes and usually high background staining, but that of Sammons *et al.* (1981) appeared to be the most promising. A variation of silver staining specifically designed for use with proteins on NC transfer membranes has been published (Yuen *et al.* 1982) and appears useful for the detection of just a few nanograms of proteins, being several hundred-fold more sensitive than Amido Black staining. Unlike immunochemical methods, which reveal only proteins recognised by the particular antibodies employed, the method should be generally useful for all proteins. It can probably be used for proteins on other transfer matrices but to date it does not seem to have been applied to them. Unfortunately it does appear to be limited to proteins isolated from gels containing the detergent sodium dodecyl sulphate (SDS–PAGE, see Chapter 5) as it is the protein–SDS complex that is detected and uncomplexed proteins are difficult to distinguish from the background. Most proteins retained antigenicity and silver staining could be followed by detection with specific antibodies if desired. The method (Table 2.10) is derived from those of Oakley et al. (1980) and Merril, Dunau, and Goldman (1981) and as with other silver staining methods all steps should be performed on a shaker to ensure rapid and thorough mixing. Protein zones show up as light bands on a darker yellowish-brown background.

Immunological detection of proteins is suitable with all types of immobilizing transfer matrix. After the transfer or blotting step, all additional binding sites on the matrix must be blocked with excess protein; then a specific antibody is bound and finally, a second antibody directed against the first antibody. This second antibody can be fluorescent labelled (e.g. with fluorescein isothiocyanate – see p. 227), radiolabelled or conjugated to an enzyme such as horseradish peroxidase or alkaline phosphatase for location by u.v. illumination, autoradiography, or enzyme activity respectively. According to Towbin et al. (1979) as little as 100 pg of protein can be clearly detected by the enzymatic procedure. With DMB or DPT paper, residual diazonium groupings must be inactivated before immunochemical detection

TABLE 2.10

Detection of proteins on NC membranes by silver staining

Staining procedure	Time
1. Wash NC in 0·01 M tris-HCl, pH 7·4	Overnight
2. Rinse twice in H_2O	2 x 5 min
3. Soak in 100 ml 0·5 per cent potassium ferricyanide while shaking over a light box	5 min
4. Wash briefly in H_2O	10–20 s
5. Place in 100 ml solution containing 0·2 g $AgNO_3$, 0·2 g NH_4NO_3, and 0·5 ml 37 per cent formaldehyde and shake over a light box	20 min
6. Rinse briefly in H_2O	2 x 10 s
7. Wash in H_2O	3 min
8. Soak in 100 ml 3 per cent Na_2CO_3 containing 0·25 ml of 37 per cent formaldehyde	Overnight
9. Wash twice in H_2O	15 min
10. Shake in 100 ml solution containing 1·4 ml NH_4OH, 0·07 g NaOH, and 0·2 g $AgNO_3$	15 min
11. Wash briefly twice in H_2O	2 x 10 s
12. Place in 100 ml solution of 0·005 per cent citric acid containing 0·019 per cent formaldehyde until protein zones show up	
13. Store in H_2O	

by incubation for 2 h at 37 °C in 10 per cent ethanolamine at pH 9 (Renart, Reiser, and Stark 1979).

For the immunological detection of protein on NC filters (Towbin *et al.* 1979), the NC protein blots were soaked for 1 h at 40 °C in 3 per cent bovine serum albumin in 10 mM tris–HCl buffer pH 7·4 containing 0·9 per cent NaCl (TBS) to saturate excess protein binding sites (termed quenching). Saravis (1984) reported much improved quenching when fish skin gelatin (Hipure Liquid Gelatin, Product Number 50–005 from Norland Products Inc., 695 Joyce Kilmer Ave, New Brunswick, New Jersey, USA) was used in place of bovine serum albumin or animal sera. The NC filters were then rinsed in TBS and incubated with the appropriate dilution of antiserum to the proteins on the blot made up in 3 per cent bovine serum albumin in TBS containing 10 per cent carrier serum of a different species to that of the antiserum. The NC filters were then washed in TBS for 30 min, with five changes of buffer, and then incubated with the second (indicator) antibody directed against the immunoglobulins of the first antiserum. As indicator antibodies Towbin *et al.* (1979) used ^{125}I-labelled sheep anti-mouse IgG diluted to 106 cpm ml^{-1} in TBS containing 3 per cent bovine serum albumin and 10 per cent goat carrier serum. An NC sheet of 100 cm^2 area was incubated for 6 h at room temperature in 3 ml of this mixture, to which 0·01 per cent NaN$_3$ was added as preservative. The NC filters were then washed five times in TBS, dried in a stream of hot air, and subjected to autoradiography by exposing to Kodak X-Omat R film for 6 days. They also used commercial preparations (Nordic Laboratories, Tilburg, Netherlands) of fluorescein-and horseradish peroxidase-conjugated rabbit anti-goat IgG diluted into 3 per cent bovine serum albumin containing 10 per cent rabbit carrier serum in TBS in a similar way. The fluorescein-labelled antibodies were diluted 1:50 and NC filters incubated in the mixture for 30 min, washed as above, and inspected (or photographed) under u.v. light. The horseradish-labelled antibodies were diluted 1:2000 in the same way and a 2 h incubation period used for the NC filters. After washing the filter, peroxidase activity is revealed (Hawkes, Niday, and Gordon 1982) by immersing in a freshly prepared solution of 4-chloro-1-naphthol (0·5 mg ml^{-1}) in 50 mM tris-HCl buffer pH 7·4 containing 0·2 M NaCl and 0·01 per cent H$_2$O$_2$. This was prepared by diluting a stock solution of 3 mg ml^{-1} 4-chloro-1-naphthol in methanol with 5 volumes of buffer just before use and adding the necessary amount of H$_2$O$_2$. Positive bands give a blue colour in 2–15 min. Wash the filter with H$_2$O, allow to dry and store in the dark. This procedure was found to be as sensitive as methods employing *o*-dianisidine and 3,3′-diaminobenzidine both of which are suspected carcinogens and it also gave lower background staining than these.

The procedure of Towbin *et al.* (1979) is given in detail here because since its publication this form of electro-blotting has become unquestionably the most widely used of all methods for transferring proteins from gels to various transfer matrices and their detection method also has been the most widely used. More recently, however, two detection methods, very similar in concept to that of Towbin *et al.* but employing alkaline phosphatase-labelled antibodies have been introduced (Blake *et al.* 1984; Knecht and Dimond 1984), the former being particularly sensitive and capable of showing up as little as 30 pg of protein. Blake *et al.* (1984) replaced the bovine serum albumin or gelatin used for blocking residual binding sites on NC filters before antibody treatment with two successive 30 min washes using 0·5 per cent Tween 20 in phosphate buffered saline (PBS) since Tween 20 was found to block these sites perfectly adequately. Subsequent incubation with the appropriate antiserum for 3 h, three washing steps and 2 h incubation with alkaline phosphatase-labelled anti-antibody were also performed with 0·5 per cent Tween

20 present in the PBS buffers. After a final wash in 0·15 M veronal acetate pH 9·6 they revealed the presence of alkaline phosphatase-labelled zones by incubating the NC membrane at 37 °C in a mixture of 9 ml 0·15 M veronal acetate pH 9·6 containing 1 ml of 0·1 per cent nitro blue tetrazolium (in the same veronal buffer), 20 μl of 2 M MgCl$_2$, and 0·1 ml of 5-bromo-4-chloroindoxyl phosphate (5 mg ml^{-1} in dimethylformamide).

Burnette (1981) used a very similar protocol for protein detection on NC membranes except that after incubating for 90 min on a rocking platform with the appropriate antiserum and thorough washing, the membrane was immersed in fresh TBS containing 5 per cent bovine serum albumin and 2–5 x 10^5 cpm/ml of ^{125}I-labelled *Staphylococcus* protein A (IPA). The mixture was rocked gently for 30 min at room temperature to allow binding of IPA and the NC membrane washed, dried, and subjected to autoradiography. The use of IPA instead of radiolabelled primary or secondary antibodies greatly simplified and enhanced the autoradiographic detection of immune complexes on blots reacted with specific antiera.

As general protein detection methods, these immunological systems leave something to be desired in that although extremely sensitive they are somewhat cumbersome and require a general antiserum to all the proteins on the blot. In many cases these antisera will have to be prepared and perhaps a number will be needed to enable different blots to be examined. Wotjkowiak, Briggs, and Hnilica (1983) attempted to overcome some of these difficulties by soaking the membrane for 10 min in a 0·01 per cent solution of 2,4-dinitrofluorobenzene (DNFB) in 50 per cent dimethyl sulphoxide containing 50 mM NaHCO$_3$ to give dinitrophenyl (DNP)-labelled proteins. After washing several times in this same solvent to remove excess DNFB the transfers were washed several times with PBS and then incubated at 40 °C for 1 h in 3 per cent bovine serum albumin in PBS containing 10 per cent heat inactivated calf serum to block excess reactive sites on the membrane. This was followed by an overnight incubation at 4 °C in a 1:200 dilution of rabbit antiserum to DNP (Miles Laboratories Inc.) in the same bovine serum albumin-calf serum mixture, and then, after four 5 min washes in PBS, by a 30 min incubation at room temperature in peroxidase-labelled goat anti-rabbit IgG. The method was 100-fold more sensitive than Amido Black staining. Kittler *et al.* (1984) reported a general protein detection method very similar in concept to this except that the blots were shaken for 20 min in a 0·3 mM solution of pyridoxal 5′-phosphate in PBS at pH 8·0, followed after washing by a 20 min incubation with sodium borohydride (300 mg in 125 ml) in PBS. This converts proteins on the blot into 5′-phosphopyridoxyl derivatives which, after washing and blocking excess binding sites on the filter in the usual way, could be detected with a mouse monoclonal anti-5′-phosphopyridoxyl antibody followed by horseradish peroxidase-conjugated goat anti-mouse F(ab′)$_2$ fragment of IgG. They applied this to ZB membranes as well as NC, the only difference being that more rigorous blocking of excess binding sites was required with ZB, and they used 10 per cent bovine serum albumin and 2·5 per cent human plasma in PBS pH 7·3 at 45 °C for 12 h (the plasma was dialysed against 5 mM hydroxylamine at pH 7·3 beforehand to remove endogenous pyridoxal 5′-phosphate).

Glycoproteins can be detected specifically (Glass, Briggs, and Hnilica 1981) in a rather similar way. In this case the bovine serum albumin solutions used for blocking excess reactive sites on the transfer membrane were pretreated with 0·01 M periodic acid at pH 4·5 (0·1 M sodium acetate) for 6 h to inactivate lectin-binding activity. The excess periodate was then destroyed by adding 0·01 M glycerol and the solution dialysed. After treatment with this serum albumin, the blot was treated with a 5–20 μg/ml solution of concanavalin A, wheat germ agglutinin or other lectin followed by anti-lectin antiserum and then horseradish peroxidase-conjugated antibody against the anti-lectin antiserum. Clegg (1982) avoided the use of antisera altogether by exploiting the divalent nature of concanvalin A in which after the lectin binding step the NC membranes were washed with buffer containing 0·5 per cent Triton X-100 and then incubated directly with a solution of horseradish peroxidase, followed by further washing cycles and detection of bound enzyme activity. Of course radiolabelled- (Gershoni and Palade 1982) or peroxidase-conjugated (Moroi and Jung 1984) lectins can also be used, again avoiding antisera-containing steps.

Lipopolysaccharides have also been detected immunochemically (Bradbury et al. 1984) by blocking excess binding sites with buffered 0·25% gelatin followed by an incubation in a solution of the appropriate antiserum and then in ^{125}I-labelled Staphylococcus protein A followed by autoradiography. The procedure is thus very similar to that used by Burnette (1981) for protein detection.

Transferred DNA or RNA on immobilizing matrices is bound in a manner that retains its ability to hybridize with radioactive DNA or RNA probes in an adjoining liquid phase (Southern 1975). Bittner, Kupferer, and Morris (1980) demonstrated this by digesting unlabelled cytosine containing T4 DNA with Eco RI endonuclease and subjecting the digest to electrophoresis on a 0·75 per cent agarose gel. The separated pattern of restriction fragments was transferred to an NC membrane, hybridised with sonicated ^{32}P-labelled T4 DNA, and the pattern of fragments revealed by autoradiography. Such hydridization steps are now very widely employed for detection of nucleic acids on transfer blots.

(d) Some advantages and disadvantages of blotting techniques

NC is the most commonly used transfer medium, especially for proteins, and binds macromolecules primarily by absorption. It has a relatively high binding capacity (typically 80–100 $\mu g/cm^2$) but naturally if this is exceeded material can be lost. While binding by absorption to NC is helpful for small scale preparative work, as macromolecules can be recovered relatively easily by eluting slices of NC in detergents (e.g. 1 per cent Triton X-100, SDS etc), denaturing agents (e.g. 8 M urea, 3 M NaSCN) or acid (e.g. pH 2–3), it does mean that the efficiency of binding may not be very high and that material is not firmly attached and can be inadvertently leached off. There are reports (e.g. Gershoni and Palade 1982) that the inclusion of Triton X-100 in buffers to reduce non-specific binding for example can cause unwanted losses.

The limited binding capacity of NC may be helpful on occasions however, in that a number of blots can be prepared from a single gel and used for several different analyses or subjected to different detection techniques (i.e. Amido Black staining, enzyme activity stains, immuno-chemical reactions, etc.). For example, Gershoni and Palade (1982) made use of the fact that excess protein passes through an NC membrane, particularly if methanol is not included in the transfer buffer, to prepare no less than 10 blots on successive layers of NC from a single gel. Another approach to preparing several replicate blots is to change the transfer membrane after a short time and then to continue the transfer process. In both cases however, it is most important to remember that blots are only semiquantitative at best. The rate of transfer out of the gel is greatly influenced by molecular weight with both capillary blotting and electroblotting techniques, smaller molecules being eluted much faster than large ones. Thus with sequential blots collected on a time basis the first will be relatively richer in low molecular weight species than the later ones in which they will be under-represented (Howe and Hershey 1981). With successive layers of transfer membrane the smaller molecules will also tend to saturate the local binding capacity most rapidly and to pass quickly through to the next layer, so large molecules will be overrepresented in membranes close to the gel, and small ones overrepresented in the most distant. Smaller molecules also tend to be less well absorbed into the membranes and less firmly bound, so being lost more readily. The efficiency of transfer is also affected by the gel strength, molecules of a given size being transferred more slowly the higher the per cent T of a polyacrylamide gel. This can be an important factor if gradient gels (Chapter 4) of varying per cent T are used (Gershoni and Palade 1983). Finally, but not least important, the mobility of a macromolecule on transfer depends upon its isoelectric point and the pH of the transfer buffer, since naturally if the two coincide the macromolecule is not going to move out of the gel.

Various strategies have been explored for overcoming problems associated with transfer efficiency. Gibson (1981) when transferring large proteins to NC from SDS–PAGE gels, soaked three sheets of Whatman 3 MM filter paper in the SDS–PAGE electrode buffer 0·025 M tris, 0·0192 M glycine, 0·01 per cent SDS containing 25–75 μg/ml of Pronase and placed these sheets immediately adjacent to the gel on the cathode side. During transfer the Pronase caused a limited digestion of large proteins to peptide fragments that were still large enough to reaction immunologically but were small enough to be transferred with much greater efficiency than the parent protein. Other proteinases were also used, but Pronase was preferred because of its broad specificity and resistance to detergents. At these levels Pronase did not appear to produce significant amounts of peptides too small to bind to NC and losses of low molecular weight proteins were also very low.

The relatively slow and inefficient transfer of high molecular weight DNA compared to that of low molecular weight was observed by Bittner *et al.* (1980) and alleviated in a manner similar in concept to the above proteinase treatment of proteins. In this case the strategy involves introducing single-stranded scissions or nicks into the double-stranded DNA by digestion before electrophoretic separation with bovine pancreatic DNAse I under buffer conditions (Mg^{2+} but not Ca^{2+}) which limit the nuclease action to the introduction of single-stranded nicks (Melgar and Goldthwait 1968). This nicking procedure which reduced the single-stranded length to an average of about 500 bases was found not to alter the electrophoretic mobility during separation of the DNA on agarose gels but greatly enhanced the efficiency of subsequent transfer to NC membranes, which was then almost quantitative.

Other approaches explored in order to improve the efficiency of transfer of large proteins from polyacrylamide gels include the omission of methanol from the transfer buffer of Towbin *et al.* (1979) or the addition of 0·01 per cent SDS to it. Both procedures worked well but some care must be exercised for although the presence of SDS in the transfer buffer greatly enhanced the exit of large proteins from the gel (Nielsen *et al.* 1982; Johnson *et al.* 1983) it can also reduce the amount that binds to NC. This work was done using 6 V cm^{-1} for 1 h however, and the fact that considerable amounts of protein remained in the gel may only reflect the lack of completeness of elution within this time, which naturally could also be overcome by using longer times as suggested by Burnette (1981).

The addition of SDS to buffers, or its presence in the gels during SDS–PAGE, can result in irreversible denaturation of some antigenically reactive sites on the fractionated molecules. This is not usually a problem when polyvalent monospecific antisera are used for detection but can cause difficulties when monoclonal antibodies directed against single binding sites are used, so in these cases it should be avoided. One benefit of transfer blotting techniques in relation to the use of SDS (e.g. in SDS–PAGE) is that in many cases some or all of the enzyme activity can be regained on removal of SDS and of course it is much easier to remove SDS when the enzymes are transferred onto a membrane than from proteins embedded in a gel matrix. The SDS can be removed from membrane-bound enzymes or proteins by simple washing or by a further brief electrophoresis in the transfer apparatus with SDS-free buffer.

NC is a convenient, relatively cheap, readily available and simple immobilization matrix which unlike DBM or DPT papers requires no pretreatment before use. While the matrix of choice for most applications, small fragments of DNA or RNA do not bind well to NC under standard transfer conditions, particularly when electrophoretic transfer is employed, DBM and DPT papers were introduced specifically to overcome this difficulty but their binding capacities are lower than NC (about 20–40 μg cm^{-2}) and they are not stable so must be used very shortly after preparation from the precursor ABM or APT papers. Smith *et al.* (1984) subsequently reported a modification of transfer to NC membranes which was claimed to overcome the poor binding previously observed with low molecular weight DNA. Their procedure made the electrophoretic transfer (which is quicker and generally more efficient than capillary blotting) of DNA fragments as small as 40 bp from polyacrylamide gels to NC membranes a practical proposition. The improvement they introduced was based on the observation that binding of DNA fragments to NC in low ionic strength buffers (a requirement for electrophoretic transfer to avoid excessive heat generation) is markedly influenced by pH. Traditionally transfer buffers with pH values of 6·0–8·0 are usually employed and binding of DNA to NC is not good under these conditions but becomes better as the pH is lowered and is almost quantitative below pH 3·0. Smith *et al.* (1984) therefore made up a solution of stock buffer by mixing 200 ml 1 M NaH_2PO_4 with 55 ml 1 M citric acid. Gels were washed for 20 min in this buffer diluted 10-fold, 20 min in 20-fold diluted buffer, and then three times for 20 min each in 40-fold diluted buffer. The same buffer diluted 40-fold was used as the transfer buffer, the composition of which was therefore 19.6 mM phosphate, 5·4 mM citrate and the pH 3·0.

DEAE ion-exchange paper will bind any acidic or amphoteric molecule above its pI and can therefore be applied to RNA, DNA, and protein transfers, but the binding capacity is quite low (e.g. about 15 μg cm^{-2}) and it is mechanically weak which can cause difficulties in subsequent washing and overlaying steps. It is thus not well suited for analytical work but can be very useful for preparative experiments, such as the recovery of purified DNA fragments from gels (Danner 1982), for which it is probably the method of choice at the present time.

It is often very convenient to make multiple use of transfer blots, something which cannot usually be done with gels themselves. This is achieved by removing one probe while retaining the original protein pattern on the membrane for subsequent analysis with another probe (e.g. Geysen, de Loof, and Vandesande 1984). This can be done for example by lowering the pH to dissociate antigen-antibody complexes or by denaturing or removing the probe with denaturing agents such as 5 per cent SDS, 3 M NaSCN or 8M urea while radiolabelled lectins can be removed by competition with an appropriate monosaccharide. If acid treatment or denaturing agents are applied to NC memberanes or DEAE papers the proteins on the blot are nearly always eluted as well as the probe so multiple use is not possible. With DBM and DPT papers the transferred macromolecules are covalently bound and cannot be lost in this way so these papers are to be recommended if multiple detection probes are to be employed. ZB membranes like NC rely on non-covalent bonding but this is much stronger than with NC and the binding capacity of ZB membranes is also much higher than with other membranes (about 480 μg cm^{-2}). The inclusion of methanol improves binding of proteins to NC but reduces the efficiency of elution of proteins, particularly large ones, from the gel, but ZB has a high binding capacity in the absence of methanol (Gershoni and Palade 1982) and is well suited to any transfer application in which transfer efficiency is important or if losses of components (e.g. very small proteins and peptides) from NC membranes are severe due to poor binding. Unlike NC, it is often possible to use ZB in multiple probing regimes, since the high retention of transferred macromolecules helps to prevent loss of the blot pattern.

Blots on various types of membranes have been used for many different purposes but are particularly well suited to the study or detection of protein-ligand interactions, such as protein-DNA or protein-RNA interactions (Bowen *et al.* 1980; Hoch 1982), protein–heparin binding (Cardin, Witt, and Jackson 1984), protein–hormone receptor identification (Fernandez-Pol 1982) and protein–protein interactions (Gorelik *et al.* 1982; Bell and Engvall 1982). Enzyme modifications of transferred macromolecules on blots can be performed more easily than with components still in gels, for example the treatment of glycoproteins with glycosidases such as neuraminidase before lectin overlaying (Gershoni and Palade 1982). Overlaying of protein blots on DPT paper with serum containing polyclonal antibodies followed by washing, excising single bands containing antigen–antibody complexes and then dissociating the complexes by incubating in an acidic (pH 2·8) buffer provided Olmsted (1981) with a small scale method for the preparation of monospecific antibodies. It appears that the vast majority of proteins transferred from media containing denaturants such as urea and SDS retain sufficient structure through the whole procedure (or regain it after initial denaturation) to be recognised immunochemically on transfer membranes. Anderson *et al.* (1982) found that more than 96 per cent of proteins transferred from a two-dimensional gel separation (see Chapter 11) of human plasma proteins could be recognized with appropriate specific rabbit antisera even after storing the dry NC membranes for 5 months. The method thus provided a useful 'third dimension' to two-dimensional separations, namely that of immunological specificity. None of a limited number of monoclonal antibodies also tried in this work bound to their appropriate antigens on the transfer membranes however, and it appears that in these cases the antigenic determinants were lost during the denaturing separation, so that the application of monoclonal antibodies for staining blots transferred under denaturing conditions or from gels containing denaturants will be very limited. Jackson and Thompson (1984) showed that gels that have been stained with Coomassie Blue R250 and destained in the usual way for the general detection of proteins can then be used for electroblotting to NC perfectly satisfactorily, providing that beforehand they are incubated with gentle shaking at 4 °C for at least 1 h in 0·025 M tris, 0·192 M glycine buffer containing 1 per cent SDS. Under these conditions proteins also retained sufficient antigenicity to be revealed on the blot by immuno-chemical methods, but in this case the Coomassie Blue dye staining zones in the gel was also transferred and bound by the NC membrane. Thus all protein zones visible in the gel were in effect 'prestained' on the resulting blot. Immunochemical detection could then be used to locate specific components. If peroxidase-conjugated antibodies are employed and detected with *o*-dianisidine or 3,3'-diaminobenzidine these components show up as brown zones easily distinguishable from the other (blue) protein zones, especially if viewed or photographed through a Wratten 47b blue filter. The method was particularly valuable for identifying proteins on two-dimensional maps (see Chapter 11), because there are often small differences between two gels (maps) of the same sample, one of which has been stained and the other used for immunochemical detection, which makes precise identification by superimposing patterns difficult. Such difficulties are of course avoided if a single gel (or a gel and a blot) can be used for both general and specific detection methods.

2.14. Biochemical applications

PAGE in homogeneous gel and buffer systems is an increasingly widespread method for the analysis of charged macromolecules, requiring only microgram quantities of material, and it is almost universally applicable to mixtures of proteins, lipoproteins, glycoproteins, mucopolysaccharides, peptides, nucleic acids, etc. In this context it is rivalled only by PAGE in multiplastic buffer systems (Chapter 3) and IEF (Chapter 9) as a method of high resolution. While the use of multiphasic buffers is therefore generally preferable for the analysis of complex mixtures, it is slightly more complicated in terms of technique and there are a number of situations in which the more simple procedure of homogeneous gel and buffer phases has advantages. These situations include the following: (a) separations during which the pH must be maintained within known closely defined limits (e.g. to retain labile enzymic or other biological activity); (b) experiments in which polymerization catalysts (oxidizing agents) must be removed by pre-running to avoid sample inactivation; (c) separations of simple mixtures or when a single component of interest is well separated from other components so that high resolution is not necessary; (d) preparative scale experiments, particularly with some designs of apparatus in which the formation of two or more gel phases is technically difficult.

When using PAGE to prove the identity of two unknowns A and B or to demonstrate that an unknown (A) is the same as a standard substance B a good experimental procedure is to apply A to one gel (or sample slot), B to a second, and A + B to a third. If the substances are identical a single band of the same mobility will be obtained in all cases. This approach is necessary, particularly when the vertical gel rod type of apparatus is being used, because it is unfortunately common for there to be small differences in mobility between individual rods so that it is unwise to rely solely on a comparison of the migration distances of two bands in two separate gels. The finding of a single band at a single gel concentration and in one buffer system is not sufficient to prove the identity of A and B, and at least three different pH conditions and a number (e.g. three or more) of different gel concentrations should be used to establish identity rigorously.

Related to this use in identifying unknown substances is the use of PAGE in demonstrating the purity of a particular preparation. Ideally a single symmetrical homogeneous band free of contaminating minor bands should be obtained, again under several different conditions of pH and gel concentration. While these comments generally hold, in a minority of cases this definition may be unecessarily rigorous. For example, when a complex protein composed of two different types of subunit is run under denaturing conditions which break the molecule into subunits, two distinct bands should be obtained even though the protein may be perfectly 'pure'. Even under conditions which are not denaturing it is as well to remember that there can be a number of other situations which can cause electrophoretic heterogeneity without reflecting on the purity of the sample within the generally accepted sense of the word. Among such situations the following should be mentioned:

(a) aggregation or dissociation reactions (e.g. bovine serum albumin monomer, dimer, trimer, etc.);
(b) variations in carbohydrate composition, especially in sialic acid content, of the oligosaccharide side chain(s) of glycoproteins (e.g. κ-casein);
(c) binding of charged ligands (e.g. fatty acids binding to serum albumin);
(d) the occurrence of the molecule in more than one conformational state;
(e) incomplete cleavage of disulphide bonds or partial oxidation of sulphydryl groupings;
(f) partial deamidation (of asparagine or glutamine) or decarboxylation (of aspartic or glutamic acids);
(g) variations in sulphate or phosphate content (e.g. in phosphoproteins such as the caseins);
(h) genetic variations resulting in deletions, insertions, or substitutions of one amino acid for another within a polypeptide chain (e.g. the many variants of haemoglobin).

This list is by no means exhaustive and any mechanism, either chemical or enzymatic, which can cause a change in size, charge, or shape will give rise to electrophoretic heterogeneity. For some purposes heterogeneity from such causes will not be important and it may be possible to regard the substance as essentially 'pure' even though the usual criteria of electrophoretic homogeneity are not met. As a corollary to this, the appearance of electrophoretic heterogeneity can of course be used qualitatively and quantitatively in studying all these factors which give rise to multiple bands on PAGE. Electrophoretic methods are now often the method of choice for studying these factors. A good example of this is the study (Ward and Winzor 1981) of even relatively weak interactions between proteins and charged ligands, such as that between phosphate ions and ovalbumin. The oligosaccharide structure of glycopeptides produced by proteolysis of glycoproteins has been investigated (Poretz and Pieczenik 1981) by examining PAGE mobility changes caused by treatment with various glycosidases.

An interesting application of PAGE has been to the separation of glycopeptides (Weitzman, Scott and Keegstra 1979). The binding of borate ions to the glycopeptides depends upon the composition, sequence, and linkages of the carbohydrate portion. Thus using tris-borate buffers they found that comparisons of the glycopeptide patterns after PAGE separation gave useful structural information, and following treatment with various glycosidases they were able partially to evaluate the monosaccharide sequence of the oligosaccharide side chains.

Separation of peptides and glycopeptides by PAGE is much less straightforward than that of proteins largely because of problems of detection. Thin-layer or paper electrophoresis and thin-layer chromatography have been much more widely used in the past, while newer methods such as HPLC and isotachophoresis are now often the methods of choice. Many of the problems with PAGE analysis of peptides arise from poor fixation during staining and destaining steps, so losses can be both extensive and variable. Radiolabelling of samples before electrophoresis and autoradiography overcomes these problems but unfixed low molecular weight materials may spread extensively by diffusion during the comparatively long times needed for autoradiography. Prestaining with dyes or fluorescent tags has been used but in general changes the net charge and alters mobility so that non-total reaction leads to charge heterogeneity and band spreading. In such cases labelled and unlabelled species do not coincide so this method cannot be used for small scale preparative work where bands are excised and material eluted. Prelabelling is thus confined to situations (e.g. Chapter 5) were differences in molecular charge are not involved in the separation process. A method has been developed for peptide separations by PAGE however (Tsugita *et al.* 1982) and appears to hold considerable promise. It consists of reacting all available amino groups with a large excess of 1,3,6-trisulphonylpyrene 8-isothiocyanate, which imparts a strong negative charge to all the peptides, followed by separation in a volatile triethylamine-formic acid buffer at pH 11·7. The label is fluorescent so peptide zones are readily detected by u.v. illumination without fixation or staining, and since all peptides are negatively charged separation is predominantly due to molecular sieving by the gel matrix and hence largely proportional to molecular weight. In concept the method is thus very similar to SDS-PAGE as described in Chapter 5. The use of volatile buffers throughout facilitated its use for preparative work as they could be removed under vacuum but the labelling conditions required heating of samples to 100 °C, so samples are denatured (as in SDS-PAGE) and biological activity usually lost. Detection sensitivity was rather higher for proteins than Coomassie Blue staining and the method was suitable for peptides and proteins with molecular weights between 200 and 100 000, although for proteins resolution was not as good as with SDS-PAGE.

PAGE has been very successfully applied to the sequence analysis of DNA (see Section 6.6) in which oligonucleotides differing in size by only a single nucleotide can be separated easily, giving characteristic 'ladders' of DNA fragments on the gels. Using similar concepts to study enzymically-produced oligosaccharide fragments from glycosaminoglycans Hampson and Gallagher (1984) and Cowman et al. (1984) have used PAGE to investigate the structure of these macromolecules. Fragments varying in size by a single repeating disaccharide unit from disaccharides up to a molecular weight of several thousand could be separated on a single gel. Hampson and Gallagher (1984) started with radiolabelled glycosamino-glycans and used 40 cm long, $T = 15$ per cent slab gels (as often used in DNA sequence analysis) with fluorographic detection, while Cowman et al. used shorter gels (15 cm) of $T = 10$ per cent, digests of unlabelled glycosaminoglycans and Alcian Blue staining after electrophoretic separation. Both groups however reported linear plots of mobility versus log molecular weight up to at least a molecular weight of 7000 with only the smallest fragments (di- and tetra-saccharides) deviating from the plot. Thus like oligo-nucleotides and protein-SDS complexes (Chapter 5) these highly acidic oligosaccharide fragments also have an essentially identical charge to mass ratio, so that separation is then determined by the sieving properties of the gel matrix and proceeds according to oligosaccharide chain length, although configurational (isomeric) and conformational differences may also play a minor role.

Hjertén (1983a) has described an approach to electrophoretic separations in thin-walled glass capillary tubes using a homogeneous tris-acetic acid buffer system with high separation voltages (e.g. 100 V cm^{-1}) and which has the features of high resolution, short run times, high sensitivity and direct u.v. monitoring (avoiding time-consuming staining procedures). The method is referred to as high-performance electrophoresis (HPE) and is claimed to be the electrophoretic equivalent of HPLC in chromatographic methods. The resolution of low molecular weight substances such as aromatic carboxylic acids in only 8 min and of proteins in 40–45 min was very good with sample loads of the order of 0·01–1·0 μg protein. The method, which therefore has some features in common with various micromethods discussed in Chapter 12, could be performed in agarose gels or in free solution in buffers with no supporting gel medium as well as in polyacrylamide gels. These experiments (Hjertén 1983a) were performed using 'very provisional equipment' capable of extensive development and improvement but the work has clearly shown considerable potential and it appears likely that within the next few years this approach will gain wide acceptance.

3

POLYACRYLAMIDE GEL ELECTROPHORESIS USING MULTIPHASIC BUFFER SYSTEMS

First introduced by Ornstein (1964) and Davis (1964), systems of this type represent the classical 'disc electrophoresis' in which discontinuities in both the voltage and pH gradients are introduced by using buffers of different composition and pH in the different parts of the gel column. The term 'disc-electrophoresis' is widely misused in the literature and it should be noted that 'disc' was originally used as an abbreviation for 'discontinuous' referring to the buffers employed, and does *not* have anything to do with the shape of the separated zones. It does *not* therefore mean PAGE run in small glass tubes in which proteins of course separate as circular zones. 'Disc electrophoresis' can thus be performed in any system of vertical or horizontal slabs, tubes, columns, etc. To avoid any possible misconceptions however it may be preferable to use more descriptive terms such as 'multiphasic buffers'. In concept, the actual separation takes place in a separation gel (sometimes called a 'small-pore' or running gel) in the usual way and is determined by both charge effects and molecular size differences. However, above this is added a stacking (or spacer) gel layer in which the sample components are stacked into very thin, and hence concentrated, starting zones before the separation proper. It is this formation of very sharp zones produced by the gel and buffer discontinuities which determines the subsequent sharpness of the separations and is responsible for the very high resolving power of the method.

The production of thin starting zones also makes this method very suitable for use with dilute sample solutions, and indeed samples of 0·1 ml or more can be applied to a single gel rod of 5 mm diameter and still give sharp electrophoretic bands. While this can be very useful and may sometimes render a sample concentration step unnecessary, there is the possible danger of concentration-produced artefacts, such as irreversible protein–protein interactions, if the concentration in the zone becomes too great. This is particularly likely to occur during the stacking phase. Also, as mentioned previously the pH conditions which occur during the separation should not be such that the chemical, physical, or biological properties of sample constituents are adversely affected.

Once sample components have been stacked into thin starting zones and the Kohlrausch boundary (see below) has migrated past the sample the separation process (and hence resolving power) of the method is no different to that in continuous gels with homogeneous buffers. If similarly sharp sample zones could be formed at the start of the electrophoresis separation, then homogeneous buffers could give the same degree of resolution as multiphasic systems. Hjertén *et al.* (1965) showed that this could be achieved by ensuring that the sample had a lower conductivity than the buffer used in the gel and the apparatus. Since the current is the same across the whole apparatus a zone of lower conductivity will also be a zone of steeper voltage gradient which therefore leads to a sharpening of the sample bands. By dialysing the sample against buffer diluted 1:5 with H_2O (or by diluting the sample solution itself) they achieved resolution as good as that obtained with multiphasic systems using a simple single phase gel and a homogeneous buffer composition. As well as being more rapid and simple to prepare, homogeneous systems are of course more flexible in that there are no restrictions on the choice of buffer composition, pH, etc. Thus when considering the use of multiphasic gel and buffer systems it is as well to remember that more simple continuous systems may not necessarily give less good resolution and may in some circumstances have significant advantages, especially in some preparative work, if it is desirable to define the pH of separation at all stages or to run at pH values difficult to attain with multiphasic buffers.

3.1. The concentration and separation processes in multiphasic buffers

At the start of the experiment, the ionic constituents making up the buffers are the same in both the stacking gel and separation gel phases but the pH is different, while the buffer used in the electrode compartments has a different composition. To illustrate the processes involved let us consider a typical example in which there are tris–HCl buffers in both the separation gel and in the stacking gel while in the electrode compartments a tris–glycine buffer is used. The two ionic species (Cl^- and glycinate ions) which migrate in the same direction as the sample molecules are called leading ions and trailing ions respectively, and at the start of the experiment the leading ions (Cl^-) are in both gel phases. By selecting suitable pH values the effective mobilities of all ionic species are arranged so that the molecules in the sample have mobilities intermediate between those of the leading Cl^- and the trailing glycinate ions. When the electrical current is switched on the leading ions attempt to migrate rapidly away from the other ionic species, leaving behind a zone of lower conductivity. Since the specific conductivity is inversely proportional to the field strength a steeper voltage gradient is formed in this region which accelerates the trailing ions so that they migrate immediately behind the leading ions and at the same velocity. This effectively forms a boundary, often termed a Kohlrausch boundary since it is determined at least in part by the Kohlrausch regulating function (Kohlrausch 1897), between the leading ions and the trailing ions which corresponds to a front between regions of high and low voltage gradients. This boundary then sweeps rapidly through the sample itself (which may be applied either in free solution or in a separate sample gel made up in the same buffer as the stacking gel) and through the stacking gel phase. Because the mobilities of the sample components are intermediate between those of the leading and trailing ions, they collect in the boundary region stacked into layers only a few microns thick in the order of their mobilities and between the regions occupied by the leading and the trailing ions. The final concentrations of sample components in such layers are independent of the initial concentrations but are proportional to the concentration of the leading ion.

When this moving-boundary region reaches the interface between the stacking and separation gels both the pH and the pore size change abruptly. At the pH of the buffer in the separation gel the ionization of the trailing ions is markedly increased and their mobility increases to a value nearly as great as that of the leading ions. The trailing ions then overtake the sample component forming a boundary between themselves and the leading ions. This boundary then accelerates away from the stacked layers of sample molecules. This process is termed 'unstacking' and is accompanied by changes in the voltage gradient so that the sample molecules subsequently move in a region of uniform voltage gradient at constant pH in a similar manner to PAGE performed with homogeneous gel and buffer systems. In the stacking gel it is important that there is no molecular sieving effect and that sample components are selectively stacked by virtue solely of their electrical properties. After passage of the Kohlrausch boundary in the separation gel, the sample components are separated according to molecular weight and shape as well as charge in the usual way.

The above represents a somewhat simplistic and purely qualitative discussion of the processes involved and for a comprehensive examination of the theory the reader is referred to the elegant work of Jovin (1973). The phenomenon of stacking components in this way is in fact identical to the separation process exploited in isotachophoresis and is discussed further in Section 9.5.

3.2. The design of multiphasic buffer systems

Based on the general principles outlined above, Williams and Reisfeld (1964) drew up a practical approach to design of multiphasic systems. The approach they described for an anionic system is as follows.

(1) Choose a pH for the separation gel. The actual running pH will be approximately half a pH unit higher than the initial pH of the separation gel and this should be borne in mind when considering the resolution, stability, and solubility of the sample constituents.

All sample constituents should have the same sign of net charge at the pH chosen.

(2) Choose the leading ion. This should have a high electrophoretic mobility at all the pH values within the system and should have the same sign of charge as the sample molecules. Chloride is a good choice as it has a high mobility over a wide pH range and in most cases has no damaging effect on sample constituents such as proteins.

(3) Choose as the trailing ion a weak acid or amino acid with a pK_a up to one pH unit higher than that of the separation gel. This ensures that about half of the acid molecules will be charged at the actual running pH and will therefore have a higher mobility at this pH than any of the sample constituents.

(4) Choose as the pH of the spacer gel a value 2–3 pH units less than the pK_a of the trailing ion. If the sample is applied in the form of sample gel this should be the same as the spacer gel. Under these pH conditions only $0 \cdot 1$–$1 \cdot 0$ per cent of the molecules of the weak acid will have a net charge and the ions will have a very low mobility, lower than that of any of the sample constituents. If this is not so a higher pH for the separation gel should be chosen in step (1), and steps (2) and (3) repeated.

(5) Choose as the buffering counter ion a weak base with a pK_a up to 1 pH unit less than that of the separation gel. This results in a high buffering capacity in the separation gel, but there is little buffering capacity in the spacer gel so care must be taken not to change the pH by any contamination.

(6) Make the pH in the electrode vessels equal to the pK_a of the buffering counter ion. The pH is not critical but this step minimizes any pH changes caused by the products of electrolysis.

Cationic systems are designed in the same way except for the following changes. In step (1) the actual running pH will be half a pH unit *less* than the initial pH of the separation gel. In step (2) K^+ is a good choice of leading ion in place of Cl^-, while in step (3) a *weak base* or amino acid with pK_a up to 1 pH unit *less* than the pH of the separation gel should be chosen as the trailing ion. The pH of the spacer (and sample) gel (step (4)) is 2–3 pH units *more* than the pK_a of the trailing ion, and in step (5) the pK_a of the buffer is up to 1 pH unit *higher* than that of the separation gel.

The concentrations of the leading and trailing ions and of the buffering counter-ions in the various buffer solutions can be calculated from the equations given by Richards *et al.* (1965), the concentrations being those required to give the desired pH values once a weak acid and a weak base of known pK value have been selected. The concentration of the strong acid used in the stacking and separation gels (i.e. the leading ion, often Cl^-)

may be chosen arbitrarily. It is determined primarily by the electrical resistance of the gels, as too high a concentration leads to too high a current and excessive heating. In practice a concentration in the region of 0·05 M is often best.

These relatively simple guidelines can be useful for the design of gel and buffer systems suitable for a particular separation problem, and if followed should enable a successful separation to be made in most cases. However, the behaviour and properties of multiphasic buffer systems can now be predicted in accurate quantitative terms (Jovin 1973). Several thousand buffer systems have been designed by computer and these can be obtained by using the available retrieval program (Jovin, Dante, and Chrambach 1970).

3.3. Apparatus and setting-up procedures

The equipment needed is identical to that employed for PAGE in homogeneous gel and buffer systems (Sections 2.6. and 2.8). However, it is often technically difficult to arrange discontinuous systems using horizontal gel slabs, and it is more common for vertical apparatus of either the rod or slab type to be used.

With the gel rod method the approach is similar to that described earlier (Section 2.6(a)) except that the glass tubes are filled with the separation gel mixture to a height of only about 15 mm below the top of the tube. This solution is overlaid in the usual way with a layer of water or 5–10 per cent aqueous ethanol to ensure a flat upper surface to the gel. After polymerization the supernatant liquid is removed and the stacking gel mixture added. The depth of the stacking gel layer above the separation gel should be about 5–10 mm, leaving a further 5–10 mm to accommodate the sample. The stacking gel solution is likewise overlaid with water or alcohol solution and polymerized.

It is conventional for the separation gel to be chemically polymerized and for the stacking gel to be photopolymerized since the latter gives more homogeneous pore sizes with gels of low acrylamide concentration and high levels of Bis. Another advantage is that when the sample is applied above the stacking gel it is physically well removed from the separation gel and any residual ammonium persulphate. This is beneficial if samples sensitive to oxidizing agents are being separated, particularly since it is not possible to pre-run gels employing discontinuous buffer systems. The zone-sharpening effect depends on the migration of the Kohlrausch boundary and clearly this effect will be lost if the boundary has already passed through the stacking gel in a pre-running stage (Petropakis *et al.* 1972).

If oxidation caused by traces of persulphate remains a problem the only way in which it can be eliminated from non-homogeneous gels is to pre-run the separation gel phase alone, using separation gel buffer in the electrode compartments, before the addition of the stacking gel. After pre-running the apparatus is switched off and drained of buffer; the surface of the separation gel is then dried and the stacking gel solution is applied, overlaid with water, and photopolymerized as described above. This of course does not overcome any difficulties which may be caused by the catalysts used for photopolymerization of the stacking gel layer, but in some cases it may be possible to omit this without serious loss of resolution (see Section 3.7.).

Aasted (1980) has suggested that for PAGE separations performed in the presence of the detergent SDS (see Chapter 5) the conventional polyacrylamide stacking gel used in discontinuous systems can be replaced with an agarose stacking gel. For this sufficient agarose to give a 1·5 per cent solution is added to stacking gel buffer, which is then heated with constant stirring to near boiling point until the agarose has dissolved. The mixture is then cooled to 50–70 °C and added above the polyacrylamide separation gel (once this has polymerized and after removing the overlaying water layer of course). Once the solution has cooled and the agarose gelled electrophoresis is performed in the usual way. Since agarose gels require no polymerization catalysts the sample is well separated from catalysts present in the separation gel phase and during the electrophoresis these will migrate well ahead of sample components. On the assumption that agarose stacking gels could be used in all systems, with or without detergent, it seems reasonable to conclude that this provides an ideal way to preserve sensitive samples from problems caused by interaction with gelation catalysts.

With vertical gel slab apparatus the sample slots are formed in the stacking gel, so the amount of this layer should be such that sample components migrate through about 10 mm of stacking gel before entering the separation gel phase. With slab apparatus it may be preferable to use the chemical polymerization for both separation and stacking gels since it is difficult to apply illumination evenly to the gel mixture when the top of the slab is partially shaded by water-cooling channels, slot formers, etc.

If gels are to be kept for more than a short time before use the water layer overlaying the stacking gel during polymerization should be replaced with stacking gel buffer. This avoids the formation of a convex curved surface due to the expansion of the top of the gel caused by osmotic pressure differences and also avoids changes in the buffer concentration within the top few millimetres of the stacking gel itself.

3.4. Some typical gel and multiphasic buffer formulations

Although several thousand different multiphasic buffer systems have now been designed (see Section 3.2), the vast majority of them have never been used experimentally and probably never will be. Equally certainly, nearly all separation problems amenable to gel electrophoresis can be resolved with at least one of a comparatively small number of gel and buffer formulations.

Zwitterions such as proteins may run either as anions or as cations depending upon the pH of the surrounding buffer medium. In any given experiment the best pH to use will depend upon the objective since the pH giving the optimum resolution of all components in the sample may be different from the pH which will give the best separation of one particular component from the others. Thus the ideal pH conditions are often determined empirically by one or more preliminary trial experiments.

The general comments made earlier (Section 2.4) concerning factors influencing the separation on the basis of molecular charge differences and size differences are equally applicable to homogeneous and non-homogeneous gel and buffer systems, since apart from the zone-sharpening effect the separation process is virtually identical. Thus the comments made in Section 2.7 and Table 2.2 in relation to the choice of acrylamide and Bis concentrations are again applicable here. The two approaches differ only in the choice of buffers and in the use of a stacking gel. Typical among multiphasic buffer systems which have been widely used are those shown in Table 3.1.

System A
  Separation gel catalysts: 70 mg ammonium persulphate + 30 μl TEMED per 100
          ml
  Stacking gel catalyst: 0·5 mg riboflavin per 100 ml.
  Buffer system A (Davis 1964) with various concentrations of acrylamide and Bis (e.g. T = 3–30 per cent) is widely applicable to many proteins, for example proteins within the molecular weight range < 10^4 to > 10^6.
System B
  Separation gel catalysts: 70 mg ammonium persulphate + 60 μl TEMED per 100
          ml
  Stacking gel catalyst: 0·5 mg riboflavin per 100 ml
  This system (Williams and Reisfeld 1964) has proved particularly useful for proteins (especially enzymes) which show optimum separation at pH 8·0 or are unstable or inactivated at pH values more alkaline than 8·0.
System C
  Separation gel catalysts: 140 mg ammonium persulphate + 0·5 ml TEMED per
          100 ml
  Stacking gel catalyst: 0·5 mg riboflavin per 100 ml
  System C (Reisfeld, Lewis, and Williams 1962) is good for basic proteins such as lysozyme, trypsin, and histones. For histones of molecular weight about 20 000 this buffer system is used in gels with T = 15 per cent.
System D
  Separation gel catalysts: 0·7 g ammonium persulphate + 0·6 ml TEMED per 100
          ml
  Stacking gel catalyst: 0·5 mg riboflavin per 100 ml

TABLE 3.1
Some useful multiphasic buffer systems

pH during stacking	pH during separation	Stacking gel	Buffer constituents per litre Separation gel	Electrode chambers
System A 8·3	9·5	32 ml 1M H_3PO_4 + 7·13 g Tris (pH 6·9)	60 ml 1M HCl + 45·75 g Tris (pH 8·9)	2·88 g glycine + 0·6 g Tris (pH 8·3)
System B 7·0	8·0	48·75 ml 1 M H_3PO_4 + 6·19 g Tris (pH 5·5)	60 ml 1M HCl + 8·56 g Tris (pH 7·5)	5·52 g diethyl- barbituric acid + 1·0 g Tris (pH 7·0)
System C 5·0	3·8	60 ml 1M KOH + 3·59 ml CH_3COOH (pH 6·7)	60 ml 1M KOH + 21·5 ml CH_3COOH (pH 4·3)	3·12 g ß-alanine + 0·8 ml CH_3COOH (pH 4·5)
System D 4·0	2·3	60 ml 1M KOH + 3·7 ml CH_3COOH (pH 5·9)	15 ml 1M KOH + 66·6 ml CH_3COOH (pH 2·9)	2·8 g glycine + 0·3 ml CH_3COOH (pH 4·0)

This is applicable to very basic proteins such as histones, tryptic enzymes, ribonucleases and snake venoms. A very similar system has been used in gels with $T = 20$ per cent for histone separations (Shepherd and Gurley (1966).

As with homogeneous systems (Chapter 2) the desired amounts of acrylamide and Bis are dissolved in the appropriate buffer and TEMED can be added to this. The ammonium persulphate or riboflavin are only added (as small volumes of concentrated solution) immediately before use. It is usual for the stacking gel to have relatively low T (e.g. 3–6 per cent) and for C to be quite high (e.g. 5–20 per cent) because as stated on p. 7, for any given value of T the gel pore size is a minimum when $C = 5$ per cent. Thus gels with low T and high C are highly porous and the gel matrix then shows little sieving effect which would retard sample molecules and may adversely affect the stacking process. The above systems utilize chemical polymerization of the separation gel and photopolymerization of the stacking gel but chemical polymerization can be used throughout if desired (Section 3.3). In acid systems (i.e. C and D) where base catalysis of free-radical formation is rather inefficient large amounts of persulphate and TEMED are used and may give rise to artefacts. It can then be beneficial to use much lower amounts (e.g. 70 mg ammonium persulphate per 100 ml) but to include a small quantity (10–30 mg per 100 ml) of sodium sulphite or bisulphite (see Section 2.7).

3.5. Sample application, electrophoresis and analysis

In the original method Davis (1964) applied the sample in the form of a sample gel. This is a large-pore gel similar or identical in composition to the stacking gel. After the stacking gel has polymerized the overlaying water is removed and the sample is dissolved in stacking gel mixture, applied, and photopolymerized in the usual way. This method of sample application avoids possible thermal convection currents within the sample layer and permits a more orderly movement of sample constituents into the stacking gel resulting in a slightly improved resolution. However, in some cases the presence of the sample can interfere with the polymerization reaction, and it has been argued that the chemical processes (i.e. free-radical formation) that occur during polymerization can be harmful to sample constituents and may cause some denaturation. For these reasons and to save time the addition of sucrose to increase the sample density followed by direct application of sample solution is much more widely used and is generally perfectly satisfactory. The electrophoretic conditions, use of tracking dyes, staining, and quantitative measurement are all similar to those already described in the relevant sections of Chapter 2, except that in non-homogeneous systems the electrophoresis is usually run under constant current conditions. When the various ion boundaries migrate through the gel the electrical resistance changes markedly, so that the application of a constant voltage would cause considerable fluctuations in the current with consequent variations in the amounts of heat generated within the gel. Constant-power-producing power supplies have recently become available (e.g. BioRad 3000/300, Pharmacia model ECPS 2000/300 or ECPS 3000/150; Desaga Desatronic 200/300; LKB model 2197; etc.), and these can be a useful safeguard against overheating, as can the voltage-limiting or current-limiting types of power pack described earlier (Section 2.8). However, this more sophisticated equipment is of course more expensive than simpler designs.

When multiphasic buffer systems are used the tracking dye (e.g. 2 ml l^{-1} of 0·001 per cent bromophenol blue or 0·005 per cent methylene green, methylene blue, or Pyronine Y) can be added to the upper electrode buffer rather than to each individual sample, because it too is stacked at the Kohlrausch boundary and marks this front in all the samples.

3.6. Dissociating agents and disulphide bond cleaving reagents

Sample solutions containing non-ionic detergents or other uncharged substances such as urea can be examined directly with any of the above systems. If sample molecules are liable to aggregate or precipitate during the separation, non-ionic detergents or urea can also be added to the gels themselves. Urea is particularly widely used in this context and is usually employed at concentrations of 6–8 M. The solid urea should be incorporated into both the separation gel and stacking gel buffers and is added to the other constituents before the buffers are made up to volume.

It is not necessary to add urea to the electrode chamber buffer. It is often adequate to add high levels of urea (e.g. 8 M or more) to the sample solutions to bring about the dissociation and to have rather lower levels (e.g. 4–5 M) within the gels merely to maintain the molecules in a dissociated state and prevent reaggregation.

Uncharged disulphide-bond-reducing reagents such as 0·1 M 2-mercaptoethanol or 0·02 M dithiothreitol (either alone or together with urea) can also be added to the sample but inhibit gelation if added to gel-making solutions. Disulphide bonds often play a role in the formation of aggregates and the use of cleaving reagents is important if one wishes to disrupt aggregates or to dissociate into individual subunits a molecule with a complex subunit structure (e.g. many enzymes, immunoglobulins, etc.). However, it is possible to include lower concentrations of dithiothreitol (e.g. 1 mmol) in gels and buffers or to incorporate 1 mmol of thioglycollate into the electrode buffers (see Section 2.7).

High proportions (30–40 per cent or more) of glycerol, ethylene glycol, or sucrose can be added to gel buffers in the same way as urea. They also act as mild protein unfolding agents or dissociating agents, but they are not so effective as urea and are much less widely used. Gel-forming ability is not impaired by the presence of these substances but the induction period after addition of persulphate and before gelation may be changed owing to the increased viscosity of the solvent.

3.7. Omission of the stacking gel

Except when the very highest degree of resolution between components of the sample mixture is required it will often be found possible to omit the stacking gel phase altogether particularly if the sample is applied in a relatively small volume (e.g. 10–20 μl). If bromophenol blue is added to the sample, it can be seen that as soon as the electrical current is connected a sharp boundary is formed and migrates rapidly through the sample, forming a very thin zone in the same way as it would through a stacking gel. The original theories on the behaviour of electrolytes described by the Kohlrausch regulating function refer to the behaviour of ions in free solution and the purpose of the gel matrix within the stacking gel phase is merely to prevent a loss of resolution due to convection currents. Normally the stacking gel buffer provides a pH that will ensure the correct degree of ionization of the leading and trailing ions (Section 3.1), but if the pH and ionic strength of the sample solution itself are within the correct general range then this zone-stacking effect and formation of a Kohlrausch boundary will occur within the column of sample solution in the same way. The migration of the boundary through a thin layer of sample solution takes only a few seconds and convection effects will be negligible, so that the final resolution may be almost indistinguishable from that obtained when a stacking gel is used. However, it is always advisable to perform a preliminary trial experiment with the sample and the chosen gel and buffer system to confirm this before routinely omitting the stacking gel. Clearly if the stacking gel can be dispensed with, a considerable saving in time will be achieved and the use of PAGE with multiphasic buffers then becomes no more complicted than the use of homogeneous gel and buffer phases while still retaining the advantage of superior resolution.

Our own experience suggests that omission of a stacking gel is usually more acceptable when using small gel rods than when vertical slab apparatus is used.

3.8. Selective stacking and selective unstacking

If required, the position of the Kohlrausch boundary can be located at any stage of the separation with the aid of a tracking dye which concentrates at the boundary, or conductance measurements, or by analysis for either the leading or trailing ions. The position of sample molecules during the stacking phase can often be seen from refractive index changes since they are present as sharp concentrated zones. If the sample molecule of interest is localized (e.g. by fixing and staining) and found to coincide with the Kohlrausch boundary it is said to be stacked, but it is perfectly possible for it to migrate either faster or slower than the stack if its effective mobility is not intermediate between the mobilities of the leading and trailing ions. This opens the possibility for separations in multiphasic buffer systems by 'selective stacking' (Rodbard *et al.* 1974). Selective stacking can be exploited in two ways, either the component of interest can be stacked and as many contaminants as possible left unstacked or *vice versa*. Confining the component of interest to the stack gives a sharp highly concentrated band which migrates rapidly, a property of considerable advantage in preparative PAGE. However, there are almost always impurities from both the gel matrix and from the sample itself also present in the stack and these cannot be separated readily from the desired component. The alternative of stacking the contaminants and then selectively unstacking the component of interest by reducing its mobility below the lower stacking limit may be preferable. This can be particularly advantageous in preparative PAGE since gels contain a number of non-selective protein absorbing sites which can give rise to low recoveries and these sites would be saturated by the passage of the highly concentrated stack of contaminants moving ahead of the protein of interest.

Selective stacking of a component can be achieved by a systematic variation of pH in a single gel of single buffer composition and gel concentration, but if one wishes to use selective unstacking then two gels in series must be used (Rodbard *et al.* 1974). Unstacking may be induced by changes in gel concentration, buffer pH, or both. If the buffer is kept constant and the pH of the stacking phase is used throughout both the upper and lower gels (equivalent to the stacking and separation gels in conventional 'disc electrophoresis') then the protein of interest may be unstacked in the lower gel purely by the molecular sieving effect in the higher concentration gel. Rodbard *et al.* (1974) quote two references to practical application of such procedures, namely Myerowitz, Chrambach, Rodbard, and Robbins (1972) and Cantz, Chrambach, Bach, and Neufeld (1972).

3.9. Biochemical applications

At the present time PAGE in multiphasic buffer systems is one of the highest resolution techniques available for the analysis of a mixture of biological macromolecules. It is applicable to any charged molecule and requires only microgram amounts of material. Because of these assets it has naturally been applied to a very wide range of problems requiring the separation of proteins, lipoproteins, glycoproteins, mucopolysaccharides, nucleic acids, peptides, etc. Virtually all the situations in which PAGE in homogeneous gel and buffer systems is appropriate can also be examined by PAGE in multiphasic buffer systems, with some increase in technical difficulty but usually with a useful gain in resolution. It is suitable for large- and small-scale preparative experiments (Chapter 7), but is most widely used in a variety of analytical applications. It can thus be applied to monitoring the progress of a purification scheme (e.g. Fig. 3.1) and to assess the purity of the products as well as to identify and characterize them in the same way as PAGE in homogeneous systems.

Since PAGE is a particularly high-resolution method it has found considerable application in studies into the occurrence and causes of molecular heterogeneity, such as (a) aggregation or dissociation reactions, (b) genetic variants of proteins and glycoproteins, (c) compositional variations (e.g. in sialic acid, phosphate, amide groups etc.), (d) post-synthetic chemical or enzymatic modification reactions, (e) conformational isomers, (f) the occurrence of isoenzymes, and (g) ligand binding.

The ability of the method to separate closely similar molecules and the requirement for only very small amounts of material have led to its use in forensic science in a wide variety of applications. The use of the method for the clinical diagnosis of disease is discussed in Chapter 13.

PAGE is particularly suitable for monitoring change, be it on the very long time scale of evolutionary changes in proteins and nucleic acids, or the rapid changes that may have occurred as a result of an industrial manufacturing process for example. In the former context, PAGE of blood

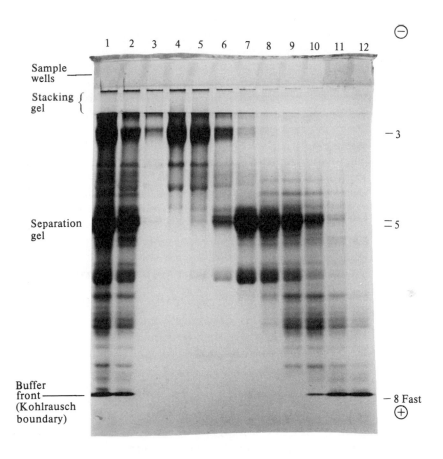

FIG 3.1. The use of PAGE to follow the progress of a purification scheme. Gels had $T = 12.5$ per cent and $C = 4$ per cent in buffer system A (Table 3.1) except that the stacking gel buffer employed HCl in place of the H_3PO_4 and was of pH 7·6. Samples 1 and 2 were two similar crude proteose–peptone fractions prepared from bovine milk and samples 3–12 were successive alternate fractions of sample 1 subjected to gel filtration on a column of Sephadex G-75 (beginning at the void volume). The major components (3, 5, and 8-fast) are well separated from each other but still contaminated by other minor components. An indication of the resolving power of the method is that the two principal bands of component 5 (seen best in samples 6 and 10) represent amino acid residues 1–105 and 1–107 of the ß-casein molecule (Andrews 1978) and hence differ in size by only about 2 per cent and by one unit of net molecular charge (residue 107 is lysine with a positively charged ε-NH_2 group). Major components on this gel slab were deliberately overloaded to reveal minor bands. Xylene Brilliant Cyanine G staining was used.

proteins and milk proteins has played a role in the taxonomy of both extinct and living species and has assisted in the tracing of evolutionary trees.

Considering changes on a shorter time scale, the method is excellent for examining processes of importance in food science and technology. These fall into two main areas. Perhaps most obvious is the quality control aspect since bacteriological breakdown of proteins can be readily detected. This is caused indirectly by proteases produced by the bacteria, and to see a rapid breakdown (within 24–48 h) bacterial counts must usually be very high (e.g. 10^6–10^8 colony-forming units per ml). The precise threshold at which detectable damage occurs depends upon the particular type of bacteria present, but the method is often comparable in sensitivity with other methods. When longer storage of food products is considered, as it often is, much lower levels of enzymatic activity can give rise to marked changes (Law, Andrews, and Sharpe 1977). This is particularly significant, since in a number of cases the bacteria themselves may be destroyed by typical manufacturing heat processes but the bacterial enzymes (proteases and lipases) are sufficiently heat stable to survive and cause deterioration during subsequent storage (e.g. Alichanidis and Andrews 1978; Law *et al.* 1977). Purely chemical changes, such as those resulting from the Maillard reaction (involving interaction between reducing sugars and ε-NH_2 groups of lysine residues), can also be monitored by PAGE (Andrews 1975).

The second area of application in food technology is the use of the method as a testing procedure for adulteration of foodstuffs (e.g. the presence of cows milk in goat-milk or sheep-milk cheeses). Related to this is the identification of varieties of food components, such as hard and soft wheats (Johnson 1967), various potato varieties (Stegemann and Loeschcke 1976; Stegemann 1978), cultivars of legumes, and different types of meat and fish. Quantitative measurement of the intensities of the electrophoretic bands enables a good estimate to be made of the amount of a particular constituent in a mixed foodstuff, e.g. the amount of egg in a cake mix or of soya protein in a processed meat product.

POLYACRYLAMIDE GEL ELECTROPHORESIS. MOLECULAR WEIGHT MEASUREMENT AND THE USE OF GEL CONCENTRATION GRADIENTS

Gel electrophoresis separates molecules on the basis of differences in size or charge or both. However in most cases both factors play a role but if the influence of molecular charge could be eliminated then clearly the method could be used with suitable calibration for measuring molecular weight (MW). This can in fact be done and there are two main ways of overcoming charge effects. In one a relatively large amount of a charged ligand such as sodium dodecyl suphate is bound to the protein (this method has been most widely applied to proteins) effectively swamping the initial charges present on the protein molecules and giving an approximately constant charge to mass ratio. This approach is the subject of the next chapter (Chapter 5). The second method which we shall discuss here relies on a mathematical cancelling of charge effects following measurements of mobility in gels of different concentrations.

4.1. The Ferguson plot

The electrophoretic migration of a molecule in a gel is influenced both by the concentration T of total monomer and by the degree C of cross-linking and Thorun and Maurer (1971) have shown that there is a linear relationship between the logarithm of the mobility and the square root of the percentage of cross-linking agent at a constant T. Therefore for measurements of MW from observed mobilities in gels of different concentrations it is essential that the degree C of cross-linkage be kept constant throughout.

When this is done a straight line is obtained when the logarithim of the mobility m or the relative mobility R_f is plotted *versus* the gel concentration T, so that

$$\log m = \log m_0 - K_R T$$

or

$$\log R_f = \log Y_0 - K_R T$$

where m_0 and Y_0 are the free mobility and relative free mobility respectively which would be obtained in a 'gel' of zero concentration ($T=0$ per cent). The slope of the line is the retardation coefficient K_R. This relationship (Fig. 4.1) was first demonstrated by Ferguson (1964) for starch gels and was later applied to polyacrylamide gels by Hedrick and Smith (1968).

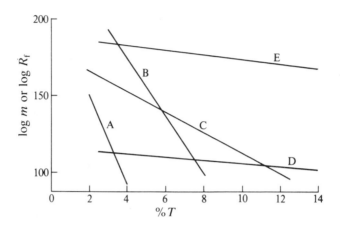

FIG 4.1 The Ferguson plot. The slopes of the lines are the retardation coefficients K_R and are proportional to molecular weight. Hence A is a very large molecule subject to considerable molecular sieving which scarcely penetrates gels with $T > 4$ per cent. B is also large and of higher net charge than A. C is a typical line for a medium-sized molecule. D and E are both small molecules subject to little retardation. D has a low net charge and hence low mobility, whereas the highly charged E moves close to the tracking dye over the whole range of gel concentrations.

If Ferguson plots are constructed (often using $100 \log R_f$ in place of $\log m$ or $\log R_f$ for convenience) for a number of different proteins and the values of K_R calculated, it is found that there is a linear relationship between K_R and MW (Ferguson 1964; Hedrick and Smith 1968). Therefore in order to calculate the MW of an unknown protein all that is necessary is to measure the mobility of the unknown and a number of standard proteins of known size in gels of a number of different concentrations. Rodbard and Chrambach (1971) suggest that at least seven values of T should be taken to give precise values of K_R and Y_0. Ferguson plots are then drawn and the values of K_R for each of the standards and the unknown are calculated. The K_R values of the standards are then plotted against their respective MW and, from the straight line obtained the MW of the unknown can then be read off from its K_R value. All lines drawn on both plots should be least mean squares regression lines.

K_R is dependent upon a number of experimental variables so it is essential that all conditions (e.g. C, polymerization conditions, pH, ionic strength, temperature, buffer compositions, etc.) are kept constant throughout. If any of these parameters are changed a new set of Ferguson plots must be constructed. Thorun and Maurer (1971) point out that the reliability of MW values obtained in this way depends upon the extent to which the different proteins (both unknowns and standards) conform to a common relationship between mobility and molecular weight. They suggest that proteins with partial specific volumes significantly different from a typical average value of 0·73—0·74 ml g^{-1} or having particularly extended and non-spherical shapes would deviate from the K_R versus MW plot. However, the latter is not necessarily true, and fibrinogen for example, behaves in accordance with its correct MW in spite of its extended shape (Rodbard and Chrambach, 1971). This is because K_R is in fact proportional to the effective molecular surface area (or to the radius $\bar{R}$ of a sphere with the same surface area) and not upon MW directly. Rodbard and Chrambach (1971) also state that variations in K_R under different experimental conditions probably reflected changes in surface area as a function of ionic environment. Hence K_R also depends upon the molecular conformation. Differences in conformation in a number of multimeric proteins have indeed been invoked by Bryan (1977) as a possible explanation of non-linearity of deviations in this type of plot. The presence of carbohydrate (e.g. in glycoproteins), lipid (in lipoproteins), or other prosthetic groups may also probably cause deviations (Thorun and Maurer 1971), and with small proteins (below MW = 50 000) some deviation has been reported (Gonenne and Lebowitz 1975).

Nevertheless in spite of these limitations plots of K_R versus MW are usually linear over a MW range of at least 45 000–500 000 (Hedrick and Smith 1968) and the method has been widely applied. In a very thorough investigation Rodbard and Chrambach (1971) have shown that the linear Ferguson relationship itself of log mobility versus T holds over a wide range of experimental conditions (i.e. an MW range of 458–670 000, pH values between 3·6 and 10·2, ionic strengths from 0·0023 to 0·023, at 0 and 25 °C, with either anodic or cathodic migration, with $T = 3$–30 per cent, and with C varied between 1 and 5 per cent).

4.2. Alternative plots for estimation of molecular size

As discussed above, in certain cases there can be difficulties when calculating MW values from plots of K_R versus MW, and Rodbard and Chrambach (1970, 1971, 1974) showed that a straight line relationship is also obtained when $\sqrt{K_R}$ is plotted against the molecular radius $\bar{R}$. They also derived a sound theoretical basis for this relationship, in contrast to the plot of K_R *versus* MW which is merely an empirical relationship with no rigorous theoretical justification. A modified version of their derivation is reproduced with their permission below.

The relationship between $\sqrt{K_R}$ and $\bar{R}$ is based on an extension of the Ogston model for the mechanism of gel electrophoresis and gel filtration (Rodbard and Chrambach 1970). It makes only minimal assumptions about the gel pore structure and applies to non-re-entrant non-spherical molecules as well as spherical ones. Ogston (1958) derived the fractional volume f available to a spherical molecule in a random suspension of linear fibres or beads, and the assumptions made are equivalent to concluding that the number of contacts between a molecule and a gel fibre follows a Poisson distribution. When any arbitrary shaped object is projected into a random distribution of lines (gel fibres), the probability of an end contact is negligible if the fibres are much longer than the molecule. The probability of no tangential contact is then given by

$$f = \exp\left(-l\,s/4\right) \tag{4.1}$$

where l is the total length of gel fibres per unit volume and s is the molecular surface area. When the fibres are very short and the molecule can be regarded as being projected into a random distribution of points (or gel beads), the probability of a tangential contact becomes negligible and the probability of no end contact is

$$f = \exp\left(-n\mathrm{V}\right) \tag{4.2}$$

where n is the number of points per unit volume of gel and V the volume of the molecule.

Equations (4.1) and (4.2) are identical with Ogston's results for a sphere when s is replaced by $4\pi(R + r)^2$ and V is replaced by $\frac{2}{3}\pi(R + r)^3$ where R is the radius of the sphere and r the radius of the gel fibre. Equations (4.1) and (4.2) apply to all molecules, spherical or not, since for any shape of object there exists an equivalent sphere with the same probability of no contact.

The volume fraction of the pores in a gel stucture is equal to the area fraction of the pores in a random cross-section, so the parameter f is also the fractional surface area available to the molecule (Rodbard and Chrambach 1970). If electrophoretic mobility is proportional to the cross-sectional area of the pores, it should therefore be proportional to f.

As a corollary to the above, when eqn (4.1) holds the relationship between the retardation coefficient and the molecular radius is given by

$$K_R^{1/2} = c_1(\bar{R} + r) \tag{4.3}$$

and when eqn (4.2) holds it is given by

$$K_R^{1/3} = c_2(\bar{R} + r) \tag{4.4}$$

where c_1 and c_2 are constants for any specified electrophoretic system and depend on the buffer, ionic strength, pH, temperature, C, etc. Thus these equations explain why there is a good correlation between MW and K_R. Rodbard and Chrambach (1971) subsequently demonstrated experimentally that eqn (4.3) was valid for a large number of standard proteins under widely different experimental conditions, so it can be concluded that a molecule migrating through a polyacrylamide gel behaves as if it were passing through a random network of fibres. Once a plot of $\sqrt{K_R}$ versus $\bar{R}$ has been established in a particular experiment using a number of standard proteins, the molecualr radii (geometric mean radii) of unknown molecules can be read off when their K_R values have been measured. MW values can then be calculated from the relationship.

$$\bar{R} = \left(\frac{3\bar{v}MW}{4\pi N} \right)^{1/3} \tag{4.5}$$

where N is Avogadro's number and $\bar{v}$ the partial specific volume. It is assumed that the molecules are unhydrated and $\bar{v}$ is taken to be 0·74 for all proteins studied, both standard and unknown (Rodbard and Chrambach 1970).

When compared with the simple Ferguson plot of K_R versus MW the plot of $\sqrt{K_R}$ versus $\bar{R}$ does suffer from the disadvantage that calculations have to be made to obtain the values of $\bar{R}$ for the standard proteins using eqn (4.5) and then the reverse calculations made to determine the molecular weights of unknowns using their measured $\bar{R}$ values from the plot. Although computer programs for data analysis have been worked out (Rodbard and Chrambach 1971), the method remains more cumbersome than the K_R versus MW plot and the assumptions made for $\bar{v}$ and for a lack of hydration may not always be entirely valid. Thus the more simple Ferguson plot appears to be perfectly adequate for most purposes, although the $\sqrt{K_R}$ versus $\bar{R}$ plot may be preferable for the most accurate work, esecially with globular proteins.

Working with proteins in the MW range 18 000–70 000 and using PAGE in the presence of the ionic detergent sodium dodecyl sulphate (see Chapter 5), Neville (1971) demonstrated a linear relationship between log K_R and log MW. Fisher and Dingman (1971) used this relationship for single- and double-stranded polynucleotides in composite acrylamide–agarose gels, and Alder, Purich, and Stadtman (1975) and Bryan (1977) have applied it to conventional PAGE with no detergent present for proteins over a wide MW range. The method suffers from the disadvantage that such log–log plots are rather insensitive to differences in experimental measurements which means that answers have relatively large experimental errors, but data handling is almost non-existent and experimental points can merely be plotted on log–log graph paper.

In a recent study involving a number of multimeric proteins with MW values from 60 000–900 000 Bryan (1977) found that there was some deviation from linearity in the conventional plot of K_R versus MW which led to an average error of 18 per cent in the estimated MW values, individual values being as much as 32 per cent in error. With either plots of $\sqrt{K_R}$ versus $\bar{R}$ or of log K_R versus log MW no individual value was incorrect by more than 10 per cent and average errors were a much more acceptable 5 per cent.

4.3. The use of the Ferguson plot in macromolecular separations

The Ferguson plot can provide further useful information in addition to estimates of molecular size. Consider for example the separation of two unknown components separated by PAGE; in plots of log mobility *versus* T there are four possibilities (Rodbard *et al.* 1974).

In Fig. 4.2(a) the plots given by the two components intersect at $T = 0$ per cent showing that in a 0 per cent gel (free solution) with no molecular sieving they would not separate. A system of molecular polymers or of subunits of a multimeric protein can give rise to a family of curves of this sort. In this type of plot the components clearly are better separated the higher the gel concentration. In practice the concentration chosen will depend upon the slopes of the two lines and will depend largely upon considerations of experimental convenience. Since gels with T of more than about 20 per cent become progressively more difficult to prepare, in many cases a good technical compromise may be to use gels with T about 15–20 per cent and $C = 5$ per cent even if this does not give the greatest separation. Of course if both components are relatively large and slow moving the best compromise may be much less concentrated gels than this in order to avoid inordinately long electrophoresis times. Rodbard *et al.* (1974) discussed the optimization of resolution in theoretical terms at some length, but while this is undoubtedly a useful exercise they indicate that there are cases where actual separations are at variance with the calculations and they suggest that the ultimate test is the experimental result.

Figure 4.2(b) shows the shape of the Ferguson plots obtained when the

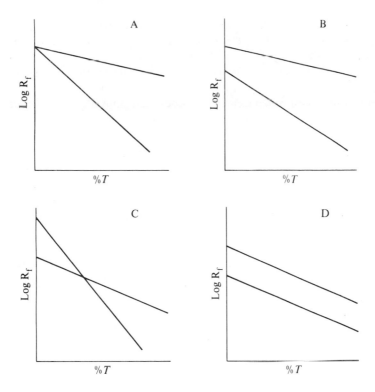

FIG 4.2. Possible relationships between the Ferguson plots of two components. (a) The two components differ in size but not in net molecular charge. (b) The smaller molecule has the higher net molecule charge and free-solution mobility. (c) The larger molecule has the higher net molecular charge. (d) The two components differ in charge but are of the same molecular size. (Reproduced by permission of Walter de Gruyter & Co.).

smaller molecule has the higher free-solution mobility, so that size and charge separation are synergistic. In this case separation is also improved the higher the gel concentration, and the considerations discussed above again apply.

The case where the larger molecule (with the steeper slope) also has the higher free-solution mobility is shown in Fig. 4.2(c). This situation quite often occurs in practice since if the surface charge density remains relatively constant the larger molecules usually have a higher net charge and free-solution mobility. During the PAGE separation the effects of size and charge are antagonistic. By altering the pH and hence the net charges it may be possible to alleviate this antagonism to some extent, but when the components have similar charge to mass ratios a pH change is unlikely to be effective (Rodbard *et al.* 1974). Once again the actual experimental conditions chosen will differ from the ideal of maximum attainable separation and will represent a compromise. In this type of Ferguson plot, the two hypothetical components are best separated either at high values of T or when $T = 0$ per cent. It is not practical to use conventional polyacrylamide gels much below $T = 3$ per cent, and if gels with even lower sieving effects are required the degree C of cross-linking can be increased to 20–50 per cent, which gives large pore sizes for a given value of T. Separations with non-sieving media (e.g. agarose gels, composite agarose–polyacrylamide gels, cellulose acetate) may be better. Alternatively isoelectric focusing in sucrose density gradients may be an appropriate separation method.

The fourth possible appearance of the Ferguson plots is shown in Fig. 4.2(d) and in this case the plots for our two components form two parallel lines. This situation is found when the two components are of similar molecular size but different free mobilities. Since the mobility axis in Fig. 4.2(d) is logarithmic the relative separation of the two components is greatest when $T = 0$ per cent, so that gels of the lowest convenient concentration should be used or alternatively non-sieving media (as above) or isolectric focusing may give better results than PAGE.

4.4. Gels of graded porosity

Since the passage of a macromolecule through a polyacrylamide gel matrix is determined at least in part by its size (or molecular radius), it is not surprising that a useful fractionation can be achieved on gels of graded porosity (gels with acrylamide concentration gradients). In a linear gel gradient of increasing concentration the migration velocity of a band is inversely related to time and varies exponentially with the distance travelled. The band pattern changes continuously during the run and depends upon molecular size, charge, elapsed time, voltage gradient, and upon the gel gradient itself (Rodbard, Kapadia, and Chrambach 1971). Thus there are no true physical constants corresponding to the electro-phoretic mobility m and to R_f as in conventional PAGE.

Rodbard *et al.* (1971) have described the behaviour of macromolecules on gel gradients of any shape (linear, convex, concave, etc.) and from them have predicted the conditions under which a gradient gel will be preferable to non-gradient PAGE. They concluded that when only one homogeneous component was being studied (e.g. for MW determination) it is generally easier to use two or more conventional gels of differing concentration in order to estimate MW values, as described above (Sections 4.1 and 4.2), or to establish homogeneity. For the optimal separation of two components there is usually a particular gel concentration that will give the best results

(Section 4.3) and a non-gradient gel of this concentration will be preferable to the use of a gel gradient. However, with three components a gradient gel may give better resolution than a gel of constant porosity, but it is for the analysis of more complex mixtures (e.g. plasma, cell extracts, urine, etc.) that gradient gels are most appropriate, particularly if the components cover a wide molecular size range. A gel gradient never provides the optimum pore size for the separation of any two of the components present but does provide a compromise in that any pair of components are subjected to optimal separation conditions for part of the time.

Experimentally it is found that not only does the rate of migration of components through a gel gradient vary inversely with time, but also after a sufficient time a stable pattern will develop in which the different components continue to move slowly but their relative positions remain constant. This has led to the concept of the pore limit, which is defined by Slater (1969) as the distance migrated from the origin in a specific gradient after which further migration occurs at a slow rate directly proportional to time. The technique of gel concentration gradient electrophoresis is in fact sometimes referred to as pore-limit electrophoresis. Although widely used, the choice of the term pore limit is perhaps an unfortunate one since it has led to the common misconception that once the pore limit is reached migration stops altogether. And this is not so (e.g. Rodbard et al. 1971), it seems preferable to refer to the method as gel gradient electrophoresis.

For a complex mixture there seems little doubt that under appropriate conditions a gradient gel can give a resolution superior to a gel of single concentration and it is not necessary for this that all the components should have migrated as far as their pore limits. Part of this high resolution results from the fact that throughout the run the leading edge of any particular band is moving through more concentrated gel than the trailing edge and hence encounters greater resistance so that there is a band-sharpening effect (Margolis and Kenrick 1968). There are two further practical advantages of this method. Firstly, since after the initial 'sorting-out' process a relatively stable band pattern is formed it is not necessary to control the electrophoresis conditions so precisely as in other electrophoretic procedures, at least for qualitative work. Thus simple unstabilized d.c. power supplies (of 50–300 V output) can be used and gels can be run for 4–8 h during the day or overnight as convenient. Secondly, once bands have migrated well into the gels and approached their pore limits diffusion is greatly reduced so that gels can be kept unfixed for some days with little loss of resolution (Margolis 1973). This allows plenty of time to test guide strips cut off the gel, to test duplicate gels, or to process 'contact prints' etc. before deciding on how to proceed with material separated in the main gel.

Gels can be prepared with any shape of concentration gradient to suit the particular requirements of the separation. However, the bulk of published work refers to the use of linear or simple concave gradients. Since the pore size is a function of $1/\sqrt{T}$, in a linear gradient the pore size changes more gradually at high values of T than at low values. Thus small molecules have to migrate through a considerable length of gel with a porosity close to the pore limit before finally reaching that limit. Convex concentration gradients would naturally aggravate this situation, but with concave gradients the pore limit is reached more rapidly than with linear gradients.

4.5. The preparation and running of gradient gels

Gradient gels can be prepared either as cylindrical gel rods in glass tubes or as thin slabs, and the types of apparatus used for running them are identical to those used for conventional forms of PAGE (Chapter 2). For most applications samples are compared with known standards run at the same time or a number of samples are run at the same time and compared with each other. Because of the difficulty of ensuring that a number of gel rods are prepared with identical gradients and run under identical conditions, it is more usual to use gel slabs on which several samples can be examined at the same time.

Virtually any device for preparing solution gradients can be applied to the formation of gradient gels. The simplest device for preparing linear gradients requires only two identical vessels, a stirrer, and some tubing (Fig. 4.3). If they are identical the shape of the vessels is immaterial since as long as they are at the same height when one volume of solution is withdrawn from flask A, half a volume will flow over the siphon bridge from flask B into A to maintain the hydrostatic equilibrium. If A is filled with dilute solution and B with concentrated solution and A is continuously stirred then the concentration of solution withdrawn from A will increase linearly. Many published methods are basically only variants of this simple concept. The siphon bridge is often replaced by a tube and stopcock joining the bottoms of the two vessels. Small corrections to the volumes of solution in A and B are made to allow for the different densities of the two solutions, the volume displaced by the stirring bar, and the differences in level produced by the dynamic forces resulting from the stirring action. If flask A (Fig. 4.3) is firmly stoppered so that when 1 volume is withdrawn from A it is replaced by 1 volume from B then an exponential convex gradient will result. A concave gradient can be produced by filling both A and B with dilute solution (and fitting both with stirrers) and then adding a third vessel C filled with concentrated solution connected via another siphon bridge to B.

A simple device relying on the same principle for making linear gradients is shown in Fig. 4.4. In this the above two identical vessels are replaced by two identical chambers machined out of a Plexiglass block and interconnected by a small-diameter channel which can be closed with an appropriate valve or tap. The tap is closed and chamber A filled with dilute (light) solution and B with an equal volume of concentrated (dense) solution. The tap is then opened and the magnetic stirring started so that as liquid is withdrawn from A half a volume of B solution flows into A to maintain the liquid levels, and the solution taken from A therefore

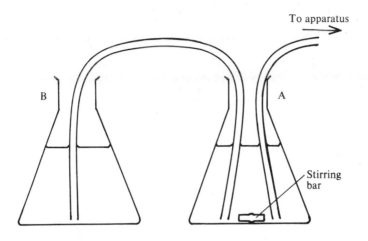

FIG 4.3. Simple arrangement for preparing linear gradients with two flasks.

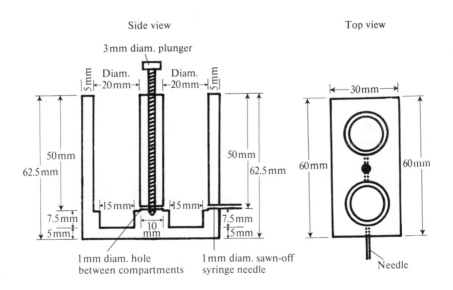

FIG 4.4. Plexiglass gradient making device. If made to the dimensions shown, each chamber will hold up to 17 ml and the device will then be suitable for making gradients up to about 30 ml in total volume. The design can readily be scaled up for larger volumes.

increases linearly in concentration. The magnetic stirring bar in chamber B is merely to correct for the volume of liquid displaced by the bar in A and also to balance the dynamic forces in the two chambers. As with the above two-flask arrangement, here also if chamber A is stoppered (and the pressure equilibrated by briefly inserting a needle between stopper and chamber wall before opening the tap) a convex gradient results, because the volume of liquid in A does not change until that in B is exhausted. Addition of a third chamber could be used to make concave gradients in a manner analogous to that with flasks. In general linear gradients are the most widely used, but concave gradients can be very helpful as they accentuate separations of large molecules where relative percentage size differences between components may be quite small, albeit at the expense of resolution of lower molecular weight components. Convex gradients are much less widely used.

A widely used procedure for gradient gel preparation is that of Margolis and Kenrick (1968) in which the components are fed into the bottom of a rectangular mould in which the gel plates or tubes are supported on a plastic net platform (Fig. 4.5). The mould is first filled with water, which will overlay the gels and give a flat surface, and then the predetermined volume of gradient gel mixture is fed in starting with the most dilute acrylamide solution. Finally 20–30 per cent sucrose solution is added to displace the gel solution out of the tubing and bottom part of the mould. For a defined gradient all of the gel solution must be displaced into the area occupied by the gel plates and the volume of gel mixture used should be such that when the sucrose solution just reaches the bottom of the gel plates the top of the gradient solution is about 0·5–1.0 cm below the top of the plates. The apparatus has the advantage that a number of slabs or gel rods are prepared at the same time and reproducibility of the gel gradients within a batch should be good.

Margolis and Kenrick (1968) stress that the most important precaution is to avoid convection currents, including those generated by the heat produced during the polymerization reaction. This is helped both by the inclusion of a sucrose (or glycerol) density gradient within the gradient gel mixture itself and also by having a gradient in the concentration of the TEMED or DMAPN polymerization catalysts so that the polymerization occurs first in the top (most dilute) part of the gel gradient and proceeds downwards. The composition of gel solutions used by Margolis and Kenrick (1968) to prepare a linear 4–24 per cent gel gradient in their apparatus is shown in Table 4.1. It can be seen that the concentrated solution is actually $T = 26$ per cent and not $T = 24$ per cent; this is because it is impractical to empty both solution reservoirs completely, so 25 ml was left in each of them (50 ml in all) and a correction therefore has to be made for this. They used a buffer of pH 8·28 containing 10·75 g tris, 0·93 disodium EDTA, and 5·04 g boric acid per litre, but this is only one example and the buffer system, catalysts, and gradient shape (both the type of slope and the concentration limits, T and C) can be varied at will.

The above descriptions apply to arrangements such as those shown in Fig. 4.5. where gradient-making solutions are pumped into the bottom of gel moulds, light (dilute) solution first which is gradually displaced upwards as the mould is filled with more dense solution. It is very common with slab gel systems for gradient gel slabs to be made one at a time with filling from the top, the dense (high sucrose and high per cent T) solution being pumped in first and becoming progressively less dense. In this case flask A, or chamber A in the Plexiglass device, is filled with dense solution

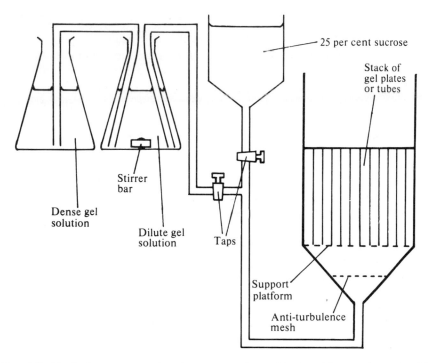

FIG 4.5. Typical apparatus for the preparation of gels with a polyacrylamide concentration gradient.

TABLE 4.1

Solutions required for the preparation of a linear 4–24 per cent poly acrylamide gel gradient

Reagent	Concentrated gel solution (26 per cent)	Dilute gel solution (4 per cent)
Acrylamide	76·57 g	11·78 g
Bis	4·03 g	0·62 g
Sucrose	12·4 g	3·1 g
Buffer	to 302 ml	to 302 ml
DMAPN	0·03 ml	0·09 ml
Ammonium persulphate (10 per cent)	7·75 ml	7·75 ml
Total volume	310 ml	310 ml

From Margolis and Kenrick 1968.

and B with light solution. Stoppering of A as above then gives exponential concave concentration gradients, and for this it is usually appropriate for the initial volume of solution in A to be about one third to one half of that in B. Simple commercially available gradient-making devices have also been used by many workers (e.g. Jeppesen 1974). One such device, with a movable partition to permit easy changes in the gradient shape is the Uniscil Gradient Maker (Universal Scientific Ltd.), but many other manufacturers (e.g. Pharmacia AB, Isolabs Inc., etc.) also market suitable equipment. A rather different approach is to use three channels of a multichannel peristaltic pump: one channel pumps concentrated solution into the stirred vessel of dilute solution, while the other two channels pump solution out of this into the gel moulds or tubes. If the flow rates of the three channels are identical a linear gradient will result, but by changing the relative flow rates concave or convex gradients can be prepared.

A very simple and elegant method for determining the actual concentration gradient in a batch of cylindrical gel rods has been described by Lorentz (1976). This involves the use of a coloured marker added to one of the two component solutions used to prepare the gradient. He used p-nitrophenol added at a level of 1·0–1·2 ml of a 0·1 per cent solution in 10 per cent methanol for every 100 ml of the more concentrated solution, but other markers added to either one of the solutions could also be used. When the gradients have polymerized the whole tube is scanned by densitometry at 405 nm and the proportions of dilute (no marker) and concentrated (positive marker) solutions used at any point in the gel can readily be calculated. Owing to its negative charge, the nitrophenol migrated out of the gel ahead of the bromophenol blue tracking dye during the subsequent electrophoresis and caused no interference.

As an alternative to preparing their own, many researchers use commercially produced gels which are now made to a high standard of reproducibility. Each batch is generally accompanied by analytical data giving the precise specification of that particular batch. Gels with well-defined concave concentration gradients are available from Pharmacia (2–16 per cent and 4–30 per cent gel slabs) and from Gradipore (Gradients Pty. Ltd. and Universal Scientific Ltd.) (4–26 per cent gel slabs). If the buffer system in which the gel has been prepared is not suitable, it can be replaced electrophoretically by a short pre-run (e.g. 15 min at 125 V) in the appropriate buffer before the samples are applied. Discontinuous buffer systems (e.g. a different buffer in the apparatus from that in the gel slab) can be used just as in conventional PAGE.

If desired a stacking gel can be applied above a gradient slab merely by leaving sufficient space between the plates when adding the gradient mixture and then adding a stacking gel mixture in the usual way once the gradient has polymerized. Similarly if it is desired to add a series of sample slots for sample application to a preformed gradient slab this can be done by adding a further portion of gel mixture to the top of the gradient and inserting a slot former. In the method of Margolis and Kenrick (1968) and in the Gradipore and Pharmacia systems the samples are applied into channels in a plastic sampler spacer which is pushed between the glass plates and rests on the actual surface of the gel.

The concentration limits chosen for the gradient depends upon the samples to be separated. Gradients of 4–24, 4–26, 4–30, 2–30 per cent, etc. are generally considered to be suitable for use with most protein samples (e.g. serum, urine, etc.), although if electrophoresis is unduly prolonged it is possible for proteins with MW below about 50 000 to migrate off the end of such gels. Felgenhauer (1979) used 20–50 per cent gradients for proteins of MW 30 000. Gradients of 2·5–12 per cent (Caton and Goldstein 1971) or 2–16 per cent are particularly suited for mixtures of proteins within the MW range 10^5–5 x 10^6 and for nucleic acids up to about 30 S. Jeppesen (1974) used 3·5–7·5 per cent or 2·5–7·5 per cent gradients for fractionation of DNA fragments within the MW range 7 x 10^4–1.4 x 10^7. It is best to maximize the range of pore sizes for any particular gel concentration range, and since this is achieved with 5 per cent cross-linking (Rodbard *et al*. 1971) it is usual to use $C = 5$ per cent in both gradient-making solutions with any value of T. Typical running conditions with an 8 cm x 8 cm slab are 15–24 h at 100–150 V or 3–6 h at 300 V for protein applications and 2–6 h at 50–100 V for experiments with nucleic acids, but these are only very general guidelines and may need to be varied appreciably for any particular experiment.

Campbell, Wrigley, and Margolis (1983) found that while pore size is minimal with about $C = 5$ per cent for gels with T below about 20 per cent, the proportion of cross-linker required for minimum pore size increases with T at higher values of T. They reported that gels with maximum sieving effect were obtained with gels of $T = 40$ per cent and $C = 12·5$ per cent and suggested that these should be the limiting conditions on gradient gels. Using gradient gels (which they termed HX gels) with $T = 3$ per cent, $C = 4$ per cent at the top increasing to $T = 40$ per cent, $C = 12·5$ per cent at the bottom they achieved excellent resolution of proteins covering a very broad size range from 670 000 down to 14 000 and even of peptides produced by trypsin digestion of bovine serum albumin. Gels were run for up to 2400 volt/hours (e.g. 100 V for 24 h or equivalent) but even then no small proteins or peptides had been lost off the end of the gel. Protein zones were very sharp and it is clear that such gels are particularly useful for separating and measuring molecular weights of small proteins and nucleic acids.

Staining and quantitative measurement methods are the same as those used in conventional PAGE (Chapter 2). However, one useful modification to staining methods for gradient gels has been introduced by Manwell (1977). He found that some solvent mixtures distorted gradient slabs and made accurate measurements of migration relative to the original gel size difficult. He therefore used a protein stain comprising 100 mg of Xylene Cynanine Brilliant G in 100 ml of acetic acid ethanol water (1:5:4 by volume), and a few hours before measurement of the positions of the stained bands the gels were placed in water to expand them or in 10 per cent ethanol or 10 per cent glycerol to shrink them to a size as close as possible to the original.

4.6. The estimation of molecular weights on gradient gels

Using a linear 5–30 per cent gradient Slater (1969) reported that once the proteins under study had reached their pore limits the position of a particular protein is such that the logarithm of the distance travelled is directly proportional to the logarithm of MW. Thus if two or more standard proteins of a known size are applied to a gradient gel slab alongside the unknown mixture a simple log–log plot suffices to determine the MW values of components in the unknown. Such plots are actually sigmoidal in shape but for most purposes are essentially linear over quite a wide range (Slater studied the MW range 32 000–480 000). They are more probably a measure of the effective molecular radius $\bar{R}$ rather than of MW directly. In his system Lorentz (1976) also found a linear relationship with MW in the range 21 000–330 000 over the linear portions of his gel gradients, but in this case log MW was plotted directly against the distance travelled. Deviations from linearity were found for the two small proteins cytochrome c and insulin.

This same type of plot (log MW *versus* migration distance) was also used by Andersson, Borg, and Mikaelsson (1972) although they used 4–26 per cent Gradipore gels which have a concave gradient. They obtained a linear plot over the MW range 45 000–270 000 but fibrinogen and α_2-macroglobulin deviated from the plot. They attributed these deviations to the elongated shape of the fibrinogen molecule and to the dissociation of α_2-macroglobulin into subunits, although no evidence was given to support either of these assertions. They also found the linear relation to hold with gels containing 8 M urea. Warren, Naughton, and Fink (1982) also obtained linear plots of log MW versus distance migrated but in this case composite gradient gels of agarose and polyacrylamide were used. This effectively extended the size range of molecules that could be studied using gels that were still mechanically easy to handle. The gels were composed of two opposite gradients and had $T = 0$ per cent polyacrylamide and 1 per cent agarose at the origin and $T = 25$ per cent polyacrylamide and 0 per cent agarose at the other end. Thus very large proteins which would not enter even a dilute (e.g. $T = 3$ per cent) polyacrylamide gel were able to penetrate for a short distance. They used the gels to fractionate proteins and cross-linked proteins varying in size from 10 000 to 650 000 daltons.

Jeppesen (1974) studied DNA fragments on linear 2·5–7·5 per cent gradients and found that as the pore limit was approached an approximately linear relationship held between log MW and log distance travelled. In agreement with Slater's (1969) earlier findings, the plot was in fact sigmoidal but was essentially linear over an MW range of 2×10^5–5×10^6. Smaller DNA fragments also fell on a linear plot if the upper concentration limit of the gel gradient was raised. However, there were some cases where slight deviations from the plot could not readily be explained and these may have been due to factors other than MW (e.g. rigidity etc.) which depended upon particular base sequences. Because of such effects Jeppesen (1974) expressed caution about estimating accurate MW values of nucleic acids and fragments thereof by this method, but nevertheless the method appears to be useful.

The estimation of protein MW using concave (Gradipore) gradient gels has been put on a much more rigorous basis by Manwell (1977). Earlier treatments relied upon the proteins coming virtually to a standstill at their pore limits. Although it was recognized (Rodbard et al. 1971) that the molecules never would stop completely, for most practical purposes this is a close enough approximation and further slow migration alters the slope of the plots a little but does not significantly affect the molecular weight values obtained (Andersson et al. 1972). Manwell's approach differs significantly from this. He realized that an exponential concave gradient should balance the logarithmic relationship between mobility and gel concentration as expressed in the Ferguson equation (Section 4.1) so that the electrophoretic mobility dx/dt is related to a hypothetical limiting distance x_L and the actual distance of migration x by

$$\frac{dx}{dt} = k \, (x_L - x)$$

$$\frac{x_L}{x_L - x} = e^{kt}$$

which approximately equals $1 + kt$ for small values of t ($kt < 1$). This was linearized to

$$\frac{1}{x} - \frac{1}{x_L} = \frac{1}{kx_L} \frac{1}{t}$$

Thus when samples were applied at hourly intervals for 7 h and the Gradipore plates were run at constant 150 V throughout (at 12–14 °C), a linear plot was obtained of the reciprocal of the distance travelled *versus* the reciprocal of the time of electrophoresis. The hypothetical limiting distance x_L is then obtained by extrapolation to $1/t = 0$, i.e. $t = \infty$. Manwell (1977) found that Gradipore gels are in fact made up to two exponential gradients which are described by the equations

$$\log T = 0\cdot091x + 0\cdot785 \qquad \text{for values of } x > 2\cdot6 \text{ cm}$$
$$\log T = 0\cdot133x + 0\cdot670 \qquad \text{for values of } x < 2\cdot6 \text{ cm}$$

where T is the gel concentration (total monomer concentration). Using these equations the hypothetical limiting gel concentration T_L (hypothetical pore limit) is then calculated from x_L and finally the MW is calculated from the equation

$$1/T_L = -3\cdot034 + 0\cdot1732 \, (\text{MW})^{1/3}$$

since there is a linear relationship between a measure of the limiting pore size $1/T_L$ and the effective molecular radius which is directly proportional to the cube root of MW. Such a plot is shown for a number of known standard proteins in Fig. 4.6. These cover the MW range 13 500–900 000 (Manwell 1977). They include a number of glycoproteins which differ considerably in partial specific volume $\bar{v}$ from other proteins, but as these do not appear to deviate consistently from the regression line (Fig. 4.6) it seems that variations in $\bar{v}$ are less important in this method than in other hydrodynamic methods of MW measurement. The method appears to be at least as \ accurate as, and probably more so than, the alternative procedure of measuring MW values from a series of uniform concentration gels (Section 4.1).

Rothe and Purkhanbaba (1982) prepared linear gradient gels and found a linear correlation between the square root of the migration distance $(\sqrt{D})$ and the log of MW or log of Stokes radius. The function $\sqrt{D}$ could in effect also be replaced $\sqrt{T}$. Although the values of the constants (slope and point of intersection) changed during the course of the electrophoresis the linear correlation did not change. The relationship held true over a MW range of 10 000 to 100 000 and the authors present many equations and graphs describing the progress of separations at all stages.

4.7. Transverse gradient gel electrophoresis

This is the technique of applying a single protein or nucleic acid sample across a gel concentration gradient, i.e. compared with the above methods the slab is rotated through 90°. Thus on starting the electrophoresis the

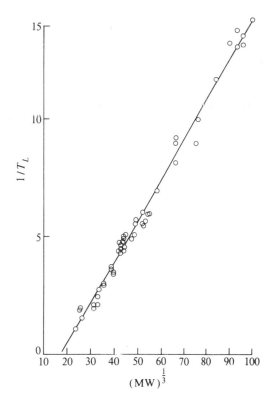

FIG 4.6. The relationship between the limiting pore size T_L in a gradient gel and MW (Manwell 1977). (Reproduced by permission of the *Biochemical Journal*.)

sample will migrate simultaneously into areas of widely differing gel concentration. With 10-cm-square linear 4–24 per cent gel slabs Margolis and Kenrick (1968) used electrophoresis conditions of 220 V for 4 h at a temperature of 12 °C. In our own laboratory we have used 2·5–26 per cent Gradipore gels by carefully unsealing the sides of the gel casettes, removing the side spacers, prising the glass plates apart, and resealing them with the plastic spacers along the other two sides (previously the top and bottom of the slab). Any gaps between the gel and the repositioned spacers should be sealed with 1 per cent agarose or with polyacrylamide made up in the appropriate buffer. With the 8-cm-square slabs we have found 2–3 h at 150–200 V to be the optimal running time. The method is particularly useful to show what concentrations of gel give the best separation for use in subsequent conventional PAGE runs.

When the gels are run in this way, each protein component forms a smooth continuous arc (Fig. 4.7). The relationship between mobility R_f and T is therefore examined directly on a single slab rather than in a series of single concentration gels. Conventional Ferguson plots can in fact be constructed from the data obtained by a single transverse gradient gel run and the MW calculated (Margolis and Kenrick 1968), but they are not likely to be so accurate as those obtained from a series of homogeneous gels. Kapadia, Chrambach, and Rodbard (1974) point out that there remain a number of difficulties in constructing accurate Ferguson plots from transverse gradient gels, the most important being that it is difficult to determine the precise values of T and the migration distance x which are

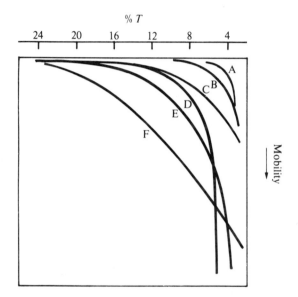

FIG 4.7. Transverse gradient gel electrophoresis. The largest molecular species A scarcely penetrates gels with T greater than 4 per cent. B is slightly smaller and so moves into gels of $T = 4$–7 per cent but it has a lower free-solution mobility (net molecular charge) than A. Thus these components are best separated at T values where A is preferentially retarded (4–7 per cent). C and D are rather similar in size but in this case C has the lower net molecular charge so that they are best separated where charge differences have greatest effect (i.e. at lower values of T). D and E represent a similar situation to A and B and it can be seen that the curves may cross, so that they separate either in the region $T = 6$–16 per cent where D is preferentially retarded or at very low values of T (e.g. 4–6 per cent) where charge differences predominate. F represents a small molecule less influenced by molecular sieving than the other constituents. A quick overall examination shows that all the constituents of this hypothetical mixture can be separated from each other when T is in the region of about 6·5 per cent, although this particular value is not ideal for any one pair of constituents.

required. The measurement problem is exacerbated by differential swelling of different parts of the gel gradient during fixing and staining procedures. Because it is easy to follow the mobility of any component as a function of gel concentration, the method is also useful for the unambiguous identification of bands in the construction of Ferguson plots derived from a series of conventional gels.

4.8. Biochemical applications of PAGE in gels of differing concentrations

The methods for the analysis of macromolecules on a series of gels of different concentrations or on gradient gels described in this chapter can be applied to all samples amenable to analysis by conventional PAGE. The object of this chapter is to highlight the additional information which these methods can give which is not obtainable by the basic method.

The most important parameter that can be obtained is the estimation of MW values from plots derived from the Ferguson equation. For proteins the technique is applicable over a MW range of at least 45 000–900 000. Not only do such plots provide a helpful guide to which techniques are likely to be most fruitful in any further purification scheme, but the relationships have also been applied to studies of ligand binding and subunit structure (Hedrick, Smith, and Bruening 1969). These authors studied the effects of binding a number of ligands to glycogen phosphorylases. They were able to distinguish clearly those ligands (e.g. dextrin, glycogen) which produced a substantial change in the size of the complex formed from those ligands (e.g. AMP) which merely changed the molecular charge. They also used the relationships to demonstrate that phosphorylase a and phosphorylase b differed only in molecular charge and not size. From MW measurements they showed that both enzymes were split from the dimeric form into monomers when gels contained 2 M urea or 0·1 per cent sodium dodecyl sulphate (SDS).

Since the physical constants R_f and K_R which are measured from PAGE runs are characteristic of each molecule and are sensitive to size, net charge, and conformation, they can be used as a criterion of molecular identity and homogeneity. Rodbard and Chrambach (1971, 1974) suggest that a set of three K_R and Y_0 values derived from Ferguson plots in three systems widely differing in pH provides a satisfactory definition of homogeneity and would be sufficient to indicate the identity or otherwise of two molecular species.

The other group of methods discussed in this chapter, namely the use of gels with a concentration gradient, also provides information not obtainable by conventional PAGE methods. Again one of the more important uses is in the estimation of molecular weights. It has been successfully applied to nucleic acids, nucleic acid fragments produced by the action of restriction nucleases, and proteins and glycoproteins with MW values in the range 13 500–900 000. It is important to stress that the methods described here, using either a series of gels or a gradient gel, give molecular weight values of proteins and enzymes in the native state, unlike gel electrophoresis procedures carried out in the presence of detergents such as SDS. Thus these methods are complementary to those described in Chapter 5, and the two approaches together provide a very powerful way of examining molecular structure. A further valuable asset is that all analytical gel electrophoresis techniques require only a few micrograms of material and often only partially purified samples. Even unfractionated crude extracts or tissue homogenates can be employed if coupled with suitable specific detection methods.

Because they are less disruptive to molecular structure than SDS–PAGE, the techniques described in this chapter can be used to investigate the structure in a much more subtle way. For example, rather more mildly dissociating conditions can be produced in a gel by incorporating urea (up to about 8 M if necessary), non-ionic detergents, or other non-charged dissociating agents. When such mild conditions are used alone or in combination with reagents such as 2-mercaptoethanol to disrupt disulphide bonds it may be possible to identify and examine intermediate stages in the dissociation of a multi-subunit molecule (e.g. the dimer stage in the break down of a tetrameric protein) which cannot be detected by SDS–PAGE. On gradient gels it is also possible to examine various aggregation states and to study complex formation between different macromolecules separated purely on a size basis, even if no change in net molecular charge is involved. An extension of this is the use of these methods to study ligand binding and the effects of chemical modification reactions upon either MW or net molecular charge.

As a generalization, gradient gels give a greater resolution of complex mixtures than conventional PAGE in gels of homogeneous concentration and very often with sharper individual bands. Because enzymatic and other biological activities are usually retained, gradient gels may be particularly useful for the study of isoenzymes and have been employed for this in plant, insect, animal, and human material. Since separation is achieved largely on a size basis, gradient gels are often used in two-dimensional procedures, and give particularly good results when used in combination with a method separating predominantly on a charge basis in the other dimension (e.g. conventional PAGE in gels of low T, isoelectric focusing, etc.). This aspect of the use of gradient gels is also discussed in Section 11.5.

Transverse gradient electrophoresis is a technique that has not been widely used in the past but it is likely to be more favoured in the future since the advent of gels of well-defined and highly reproducible gradients.

An important application of this method is as a preliminary screening procedure in the study of an unknown or incompletely characterized mixture. It provides useful information on size and charge relationships between the various constituents and is valuable in indicating the concentrations of gel which are likely to be most beneficial for subsequently separating any particular component or components on conventional PAGE systems. As a test of purity (molecular homogeneity) of for proving the identity of two materials by running them together on the same gel, the appearance of a single arc on a transverse gradient gel is far more powerful evidence than that gained by obtaining a single band with a conventional PAGE run.

An interesting electrophoretic method related to the above was introduced by Creighton (1979). He used slab gels with a transverse gradient of urea from 0 to 8 M to study the unfolding and denaturation of proteins (Fig. 4.8). The method was able not only to show the

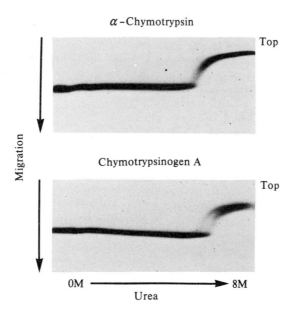

FIG 4.8. Transverse urea-gradient electrophoresis of bovine α-chymotrypsin and chymotrypsinogen A in a gel containing a linear 0–8 M urea gradient, as indicated, superimposed on a slight inverse linear 15–11 per cent acrylamide gradient to compensate for the effects of urea on the electrophoretic properties, unrelated to unfolding of the molecules. The buffer used was 0·05 M tris–acetate pH 4·0, and gels were run at 15–20 °C for 2·5 h at 10 mA per 82 mm wide slab for a few minutes and then at 20 mA. (Reproduced with permission from Creighton (1979). *J. molec. Biol.* **129**, 235. Copyright by Academic Press Inc. (London) Ltd.)

concentrations of urea at which conformational changes causing alterations in mobility occurred but also had sufficient resolution to be able to distinguish subtle differences in the behaviour of various isomers. It has also been applied successfully to the analysis of transferrins (Evans and Williams 1980). The application of transverse gradient gels, and particularly of those with a transverse gradient of urea concentration, in studies of protein and nucleic acid conformation has been reviewed by Goldenberg and Creighton (1984). Feinstein and Moudrianakis (1984) used the same concept of a transverse 0–8 M urea gradient but with the detergent sodium dodecyl sulphate (SDS, see Chapter 5) also present to study the subunit structure of mitochondrial proteins.

An unusual but related technique has been described by Thatcher and Hodson (1981) for studying conformational changes occurring during heat denaturation of proteins and nucleic acids. The arrangement consisted of a conventional polyacrylamide slab gel sandwiched between two thick aluminium plates with hot water circulating through tubes in the plates along one end of the slab and cold water along the other end. After allowing at least 1 h for equilibration there was a stable transverse temperature gradient (typically from about 20 ° to 75 °C or more) across the gel slab. Sample solutions were applied across the top of the slab and after electrophoresis the temperature of transition between the native and slower-moving denatured proteins could be seen. The method could also resolve a complex mixture of double-stranded-DNA restriction digest fragments.

5

POLYACRYLAMIDE GEL
ELECTROPHORESIS IN THE PRESENCE OF
DETERGENTS

One of the first problems that faces the scientist wishing to study the components of tissues, cells, or membranes is to solubilize the material for subsequent investigations. Many different methods have been used to solubilize the constituent lipids and proteins including solvents containing phenol, acetic acid, or high levels of urea, but a large proportion of recent procedures employ detergents. These include non-ionic detergents like Lubrol W, Brij 35, and the Tween and Triton detergents (Triton X-100 is particularly widely used), cationic detergents such as cetyltrimethyl-ammonium bromide (CTAB) and cetylpryridinium chloride (CPC), and anionic detergents such as deoxycholate (DOC) and sodium dodecyl sulphate (SDS). Clearly it is beneficial if the solubilized material can then be utilized directly for further analytical procedures. Not only can this often be done, but when ionic detergents are used most proteins are then separated by PAGE according to molecular size.

Some of the factors to be considered when deciding whether to use detergents, and which ones to use, have been reviewed by Hjelmeland and Chrambach (1981). As a generalization their use is indicated when it is necessary to disrupt protein–lipid or protein–protein interactions which involve hydrophobic bonding. Again as a generalization, detergents are without effect on ionic bonding or polar interactions, which may be influenced by factors such as pH, temperature, ionic strength, ionic composition, etc. Thus detergents by themselves cannot be expected to solubilize for example proteins precipitated at their isoelectric points or heat-denatured insoluble proteins aggregated by virtue of disulphide bond formation. Other steps (e.g. pH modification, addition of reducing agents, etc.) will need to be taken as well as adding detergent to solubilize such materials. Sometimes combinations of detergents will prove to be more effective than any individual detergent, and although this is not often done it is worth considering when working with particularly intractable material.

An important consideration in the use of detergents is often the maintenance of a biological or enzymatic activity. Some detergents (e.g. SDS) are usually regarded as strongly denaturing while others (e.g. Lubrol PX) are often considered to be very mild and non-denaturing. However, as pointed out by Hjelmeland and Chrambach (1981), it is first worth considering what is meant by 'denaturing'; whether the term is being used in its strict physical–chemical context of disruption of tertiary structure to give a random coil configuration or whether it is applied in a broader sense to include loss of a specific binding activity, enzyme activity etc. In most cases (particularly if a reducing agent is present to break disulphide bonds) SDS falls into the category of strongly denaturing detergents and gives rise to essentially random coil configurations. However some 'mild' detergents can also abolish biological activities by various mechanisms, such as interfering with the binding of substrate or essential cofactors, dissociating the subunit structure of multimeric species for which the presence of a multimeric structure is needed for activity, etc. Such changes which may involve the quaternary structure but leave the tertiary structure intact would not usually be thought of as fitting the physical meaning of denaturation but nevertheless are important considerations here. Unfortunately no generalizations can be made regarding the types and structures of detergents which are most or least likely to destroy any given type of activity and this aspect of the application of detergents to electrophoresis (or any other extraction or separation process) has to be determined by trial and error.

5.1. The use of sodium dodecyl sulphate (SDS) for molecular weight measurement

An alternative approach to the methods of molecular weight measurement discussed in Chapter 4 which rely upon a mathematical or physical cancelling of charge effects during electrophoresis separations is to cancel out differences in molecular charge chemically so that all components then migrate solely according to size. On treatment with an ionic detergent such as SDS surprisingly large amounts may be bound. For proteins the figure of about 1·4 g SDS per gram of protein is often quoted as a typical value (Pitt-Rivers and Impiombato 1968; Fish, Reynolds, and Tanford 1970; Reynolds and Tanford 1970a,b). This means that the number of SDS molecules bound is of the order of half the number of amino acid residues in the polypeptide chain. This amount of highly charged detergent molecules is sufficient to overwhelm effectively the intrinsic charges on the polypeptide chain so that the net charge per unit mass becomes approximately constant. Electrophoretic migration is then proportional to the effective molecular radius or approximately to the molecular weight of the polypeptide chain (Shapiro, Viñuela, and Maizel 1967). Although it is an oversimplification to which a number of exceptions are to be found, the relationship does in fact hold true for a very large number of proteins (Weber and Osborn 1969; Dunker and Rueckert 1969) and the method has become one of the most widely used for measurement of protein molecular weights.

The reasons for this rapid acceptance are not difficult to see. The apparatus required is readily available in most laboratories and is inexpensive. Once learnt the procedure is straightforward and highly reproducible, and results can be obtained within a few hours using only a few micrograms of material. As with other PAGE methods, in many cases the sample need not be totally pure. The degree of purity required depends largely upon the sample being studied and the ease with which the component of interest can be identified on the final gel pattern.

The original papers listed above presented the empirical finding that the binding of SDS to proteins was accompanied by the dissociation of oligomeric molecules into subunits which then migrated during electrophoresis in accordance with the size of the individual polypeptide chains. Thus the answers obtained should agree with those given by gel filtration or sedimentation equilibrium experiments when these are performed in strongly dissociating solvents such as 6 M guanidine hydrochloride or 8 M urea.

5.2. Theoretical background of molecular weight measurement by SDS–PAGE

Methods for measurement of molecular size by gel electrophoresis fall into two main categories, namely those making use of a relationship between mobility and gels of various concentrations (see Chapter 4) and those for which a single gel concentration is used. The latter is applicable only to a series of molecules with the same molecular charge density and hence is usually applied to nucleic acids, to a homologous series of proteins, or to molecules in which uniform charge densities have been produced by binding large amounts of a charged ligand such as SDS.

When a series of Ferguson plots of R_F versus T were constructed for a number of SDS-treated proteins (Neville 1971) it was found that the intercept when $T = 0$, which is the apparent relative free mobility Y_0, was almost identical for all the species examined. This demonstrates experimentally that for these materials the ratio of the effective charge to the frictional coefficient is independent of MW. Since Y_0 is virtually a constant, once the average value of log Y_0 has been found for a particular percentage of cross-linking and a particular buffer system, the retardation coefficient can be calculated from a single value of log R_F at any single value of T. Such calculated coefficients K'_R differ from the actual retardation coefficient K_R to the extent that log Y_0 for a particular protein complex differs from the average value of log Y_0 (Neville 1971).

In order to determine molecular weights from K'_R values all that is necessary is to show that there is a uniform dependence of MW on K'_R. Rodbard and Chrambach (1970, 1971) and Chrambach and Rodbard (1971) have shown that K_R is dependent on the effective molecular radius.

Since Reynolds and Tanford (1970b) and Fish et al. (1970) have shown that the hydrodynamic properties of protein–SDS complexes are a unique function of polypeptide chain length, K_R must clearly be dependent on this also. In practice plots of log K_R or log K'_R $versus$ log MW are generally linear with slopes proportional to the relationship between MW and the Stokes radius. However, a number of assumptions are made in this treatment and the linear relationship breaks down when these can no longer be justified, so that changes in slope occur in molecular weight regions below about 15 000 and above 70 000 (Neville 1971).

If the further assumption is made that K_R varies directly with MW, then within the linear range, using two markers of known MW and measured R_f, the MW for an unknown can be determined directly from its $-$log R_f value with a stanard error of $\pm$ 3000. Ferguson plots of log R_f $versus$ T to identify proteins with anomalous behaviour followed by plots of the resulting values of K_R $versus$ MW have been used very successfully to study the subunit composition of cytochrome c oxidase and shown to be markedly superior to log MW $versus$ R_f plots (Tracy and Chan 1979).

Neville (1971) points out that proteins with anomalous behaviour may also be readily observed using plots of log K'_R $versus$ log MW which are superior for this purpose to plots of $-$log R_f $versus$ MW. Both of these plots are preferable to the commonly used plots of log MW $versus$ R_f (Shapiro et al. 1967; Weber and Osborn 1969; Dunker and Rueckert 1969) since these are sigmoidal in shape, although errors introduced by treating them as linear are generally not very great as long as MW is within about 20 000–60 000.

The major sources of error in the above are the assumptions that T_0 is the same for all the standard and unknown proteins and also that there is a constant relationship between the effective molecular radius and the MW. Proteins exhibiting anomalous behaviour for either of these reasons can usually be identified either from plots of log R_f $versus$ K_R or of log K'_R $versus$ log MW, although such diagnostic plots are not necessarily infallible and exceptions do occur (Dunker and Kenyon 1976).

The assumption of a constant value for Y_0 implies that the free mobility of protein–SDS complexes is independent of size and charge. Independence of charge requires the binding of large amounts of SDS so that the intrinsic charge on the polypeptide chain makes an insignificant contribution to the net charge on the complex and also requires that the weight of SDS bound for a given weight of protein is the same for both the standard proteins and the unknowns. In order to achieve complete binding of SDS it is important that polypeptide chains are not conformationally constrained and disulphide bonds should always be reduced. The reagent most commonly used for this is 2-mercaptoethanol. Pitt-Rivers and Impiombato (1968) and Reynolds and Tanford (1970b) found that bovine serum albumin and ribonuclease both bound only 0·9 g of SDS per gram of protein without reduction but bound the usual value of about 1·4 g following reduction of disulphide bonds. As well as conformational effects, in some cases the primary structure of proteins may give rise to anomalous behaviour on electrophoresis. This is caused either by an untypical degree of SDS binding per gram of protein, or because the intrinsic charge of the polypeptide chain makes a significant contribution to the net charge of the protein–SDS complex so that Y_0 deviates markedly from the average value. A case of the latter occurs with histones in which the polypeptide chain carries an unusually high net positive charge.

Apart from proteins with highly unusual amino acid compositions, there are other comparatively readily available proteins which have been reported to behave anomalously and therefore should be avoided when choosing standards for calibration plots. These anomalous proteins can often be detected by comparing the migration rate (measuring K_R and Y_0) of an unknown complex with a group of standard protein–SDS complexes at a number of different gel concentrations (Banker and Cotman 1972).

They include ribonuclease, deoxyribonuclease, chymotrypsinogen haemoglobulin, and proteins with molecular weights below about 15 000. In addition, in our laboratory we have found that the caseins (α_{s1}, ß- and ϰ-casein) are also unsuitable for use as standard proteins. Glycoproteins give satisfactory linear plots of R_f versus K_R (Anderson, Cawston, and Cheeseman 1974), but the plots differ slightly from those obtained with protein standards perhaps because of differences in the extent of SDS binding. Thus glycoproteins should not be employed in constructing calibration plots for proteins. Leach, Collawn, and Fish (1980) in a study with 13 different glycoproteins also obtained abnormally high MW values but the errors diminished as the gel concentration (per cent T) increased. Unfortunately no universal correction factor could be applied however and each glycoprotein behaved differently, so there was no correlation between the error in MW measurement and carbohydrate content.

When using the gel-rod type of apparatus there can be slight differences in migration rates between gels, so a set of standard proteins should be run in each experiment and if possible one standard protein should also be applied to each individual gel as an internal reference point. A minimum of four standard proteins, and preferably more, should be used to construct the calibration plot. Because of these requirements for a number of standards slab gel configurations are preferable to gel rods for MW estimations. Some proteins which have been widely used as MW standards are shown in Table 5.1, and a selection should be made which includes proteins both larger and smaller than the unknown.

Another approach to the choice of standards which can be particularly valuable in the high MW range is to choose a small protein (e.g. lysozyme,

TABLE 5.1

Protein	Molecular weight	Protein	Molecular weight
Immunoglobin M (unreduced)	950 000	Aldolase (muscle)	38 994
α₂-macroglobulin (unreduced)	380 000	Lactate dehydrogenase (muscle)	36 180
Thyroglobulin	355 000	Glyceraldehyde-3-phosphate dehydrogenase (muscle)	35 700
Myosin (heavy chain	220 000		
α₂-macroglobulin (reduced)	190 000	Pepsin	34 700
Immunoglobulin A (unreduced)	160 000	Carboxypeptidase A	34 409
Immunoglobulin G (unreduced)	150 000	Carbonic anhydrase B	28 739
Serum albumin dimer	132 580	Chymotrypsinogen	25 666
ß-galactosidase (*E. coli*)	130 000	Trypsin	23 300
Ceruloplasmin	124 000	Soyabean trypsin inhibitor	20 095
Phosphorylase a (muscle)	100 000		
Lactoperoxidase	93 000	ß-Lactoglobulin	18 363
Plasminogen	81 000	Myoglobin	17 200
Transferin	76 000	Avidin	16 000
Bovine serum albumin	66 290	Haemoglobin	15 500
Catalase (liver)	57 500	Ribonuclease B	14 700
Pyruvate kinase (muscle)	57 000	Lysozyme	14 314
Glutamate dehydrogenase (liver)	53 000	α-Lactalbumin	14 176
Heavy chain of Ig G	50 000	Cytochrome C (muscle)	11 700
Fumarase (muscle)	49 000	Trypsin inhibitor (Lima bean)	9 000
Ovalbumin	43 000		
Enolase (muscle)	41 000	Insulin (reduced)	6 600
Alcohol dehydrogenase (liver)	39 805	Glucagon	3 500
		Bacitracin	1 480

haemoglobin, bovine serum albumin, etc.) and to cross-link the polypeptide chains chemically to form a series of polymers. Payne (1973) used glutaraldehyde in this way to prepare soluble polymers with MW values from 3×10^4 to 2×10^7. Optimum conditions to obtain a good mixture of polymers were to mix rapidly 2–5 μl 5 per cent glutaraldehyde with 200 μl serum albumin or lysozyme (50 mg ml⁻¹) or with 50 μl haemoglobin (200 mg ml⁻¹) dissolved in 0·1 M sodium phosphate buffer pH 7·5. After vigorous stirring for 1 min the solutions were allowed to stand at room temperature for 24 h, suitably diluted, and pretreated with SDS for application to gels in the usual way (see below). Other workers have used diethylpyrocarbonate or dimethylsuberimidate for protein cross-linking.

Instead of preparing standard mixtures or performing cross-linking reactions, a number of MW standard kits are available commercially (e.g.

Sigma Chemical Co.; Bio Rad). These may be either mixtures of different proteins of known size or of individual proteins cross-linked as above to give a series of polymer bands. Several tables listing molecular weight values for over 1200 proteins have been published (Righetti and Caravaggio 1976; Righetti, Tudor, and Ek 1981).

5.3. The standard method

The simple 'standard method' of Weber, Pringle, and Osborn (1972) and Weber and Osborn (1975) is usually adequate to determine the MW of a purified or largely purified protein. It is rapid and easy to perform and is very suitable for use with inexpensive and unsophisticated apparatus. For these reasons this method will be discussed in detail, but it should be remembered that the use of multiphasic buffers gives higher resolution (Section 5.4), and modified methods are required for peptides of MW < 10 000 (Section 5.7). Obviously SDS of high purity should be used, and a product marketed as SDS 'Specially pure' Grade which can be used without further purification is obtainable from BDH Chemicals Ltd. The quality of SDS used may be particularly important if subsequent removal of SDS (see p. 185) and detection of enzymic activity is envisaged (e.g. Blank, Sugiyama and Dekker 1982; Blank et al. 1983), but even with non-enzyme proteins impurities such as the C_{14} and C_{16} homologues of SDS (which is C_{12}) can give rise to the formation of extra bands (e.g. Margulies and Tiffany 1984).

(a) *Pretreatment of samples*

Standard proteins and the unknowns must all be treated before electrophoresis to ensure the maximum and reproducible binding of SDS. The pretreatment conditions should also inactivate any digestive enzymes, especially proteases, which may be present in the sample and which may give rise to spurious bands by producing breakdown products. Although there are exceptional circumstances when undissociated protein may persist (e.g. Bryce, Magnusson, and Crichton 1978), the following pretreatment procedure is generally found satisfactory.

To 1 vol protein solution (0·5 mg ml^{-1} in a stoppered tube or loosely capped vial add 9 vol 0·01 M sodium phosphate buffer (pH 7·0) containing 1 per cent SDS and 1 per cent 2-mercaptoethanol. Heat to 100 °C for 2 min and cool. A small amount (2–10 per cent) of sucrose or glycerol is then added to increase the density of the sample which can usually then be applied directly to gels. However, if the ionic strength of the protein solution is high a preliminary dialysis may be required against 0·01 M sodium phosophate (pH 7·0) buffer, preferably also containing SDS and 2-mercaptoethanol. Bromophenol Blue is a suitable tracking dye and can be added either to the sample or the cathode (upper) buffer chamber.

(b) *The composition of gels and buffers*

Gel concentrations of from 3 to 20 per cent or over can be used in any type of PAGE apparatus (Chapter 2), the choice of gel concentration depending upon the MW range under investigation. Dunker and Rueckert (1969) suggest that a 15 per cent gel is best for MW = 10 000–60 000, a 10 per cent gel for MW = 10 000–100 000 and a 5 per cent gel for 20 000–350 000.

The compositions of solutions for making suitable gels are shown in Table 5.2 (Weber and Osborn 1969). If gels are to be stored before use, cold storage should be avoided because of the poor solubility of SDS at 4 °C. Gels can be kept at room temperature for 1–2 weeks or longer since SDS is powerfully bacteriostatic.

The apparatus buffer used in the standard method is the same as that used in the gels, i.e. 0·1 M sodium phosphate pH 7·2 containing 1 per cent SDS, so that the buffer system is homogeneous throughout.

(c) *Sample size and electrophoretic conditions*

Sample volumes of 10–100 μg of each of the protein components present are generally most suitable. SDS–proteins all have a net negative charge at pH 7·2 and so migrate towards the anode, which should thus be the lower electrode in most types of apparatus. Constant current power supplies are most often used, and with gel rods of 5 mm diameter a current of 4–5mA per gel is appropriate.

TABLE 5.2
The composition of standard–SDS gels of various porosities

Component	$T = 5$	$T = 7·5$	$T = 10$	$T = 15$
Acrylamide (g)	4·85	7·28	9·70	14·55
Bis (g)	0·15	0·22	0·30	0·45
Gel buffer (0·2 M sodium phosphate pH 7·2) (ml)	50	50	50	50
SDS (g)	1·0	1·0	1·0	1·0
H₂O to 95 ml, then add TEMED (ml)	0·15	0·15	0·15	0·15
Ammonium persulphate (1·5 per cent) (ml)	5	5	5	5

5.4. Procedures using multiphasic buffers

As in detergent-free systems the use of multiphasic buffers in SDS–PAGE gives much sharper bands and better resolution. The theory of zone stacking and the other processes involved in SDS-containing gels has been comprehensively discussed by Wyckoff, Rodbard, and Chrambach (1977).

SDS–PAGE is generally performed at neutral or slightly alkaline pH values but for some purposes (e.g. when alkali-labile biological activities must be preserved) low pH runs may be preferable. With homogeneous buffer systems this is of course straightforward, but as with SDS–PAGE at high pH, discontinuous buffers give better resolution and a suitable system of acidic buffers and appropriate gel catalysts operating at about pH 4 has been suggested by Jones, Wilson, and Darley-Usmar (1981). They found lithium dodecyl sulphate to be required as SDS and potassium dodecyl sulphate were insoluble under their conditions, but the results were comparable to SDS–PAGE, so MW values could be calculated and the method used in gradient gels and for 2D procedures (see Chapter 11) in the same way. Nearly all SDS–PAGE experiments employ systems with an alkaline pH however and for this the two multiphasic systems originally introduced by Neville (1971) and by Laemmli (1970) have been particularly widely used, especially the latter which retains its popularity in spite of the introduction of various related or modified procedures (e.g. Baumann, Cao, and Howald 1984).

(a) *The method of Neville (1971)*

Typical compositions of buffers and gels used by Neville (1971) are shown in Table 5.3. The composition shown here for the running gel buffer results in an actual running pH of 9·50 and together with gels with $T = 11·1$ per cent this is suitable for the majority of separations. Neville (1971) also describes alternative compositions, for both the running gel ($T = 7$–15 per cent) and for buffers giving running pHs of 8·64–10·01. Sample preparation is as described for the standard method except that stacking gel buffer is used in place of the 0·1 M sodium phosphate buffer.

Gels of 5 mm diameter are run at 1·5 mA per tube at 25 °C until the bromophenol blue marker dye reaches the bottom of the tube. The exact position of this dye, which migrates at the borate–sulphate ion front, is marked by inserting a short piece of thin stainless steel surgical wire into each gel before staining so that the mobilities of protein bands relative to this can be calculated after staining. Using this procedure Neville and Glossmann (1971) have resolved 35–40 individual protein bands from cell plasma membranes on a single one-dimension gel.

TABLE 5.3

Composition of buffers and gels employed in the method of Neville (1971)

	pH	Composition
Upper reservoir buffer (cathode)	8·64	0·04 M boric acid; 0·041 M tris; 0·1 per cent SDS
Stacking gel buffer	6·10	0·0267 M H_2SO_4; 0·0541 M tris
Running gel buffer	9·18	0·0308 N HCl; 0·4244 M tris
Lower reservoir buffer (anode)	9·18	As running gel buffer
Stacking gel ($T = 3·2$ per cent; $C = 6·25$ per cent)	6·10	3·00 g acrylamide; 0·20 g Bis; Buffer to 100 ml; 0·15 ml TEMED; 0·05 g ammonium persulphate
Running gel ($T = 11·1$ per cent; $C = 0·9$ per cent)	9·18	11·00 g acrylamide; 0·10 g Bis; Buffer to 100 ml; 0·15 ml TEMED; 0·05 g ammonium persulphate

TABLE 5.4

The composition of buffers and gels used in the method of Laemmli (1970)

	pH	Composition
Upper reservoir buffer (cathode)	8·3	0·025 M tris; 0·192 M glycine; 0·1 per cent SDS
Stacking gel buffer	6·8	Tris adjusted to pH 6·8 with HCl and diluted to 0·125 M; 0·1 per cent SDS
Running gel buffer	8·8	Tris adjusted to pH 8·8 with HCl and diluted to 0·375 M; 0·1 per cent SDS
Lower reservoir buffer (anode)	8·3	As upper reservoir buffer
Stacking gel ($T = 3\cdot08$ per cent; $C = 2\cdot6$ per cent)	6·8	3 g acrylamide; 0·080 g Bis; buffer to 100 ml; 0·025 ml TEMED; 0·025 g ammonium persulphate
Running gel ($T = 10\cdot27$ per cent) $C = 2\cdot6$ per cent)	8·8	10 g Acrylamide; 0·267 g Bis; Buffer to 100 ml; 0·025 ml TEMED; 0·025 g ammonium persulphate

(b) *The method of Laemmli (1970)*

Table 5.4 shows the composition of the gels and buffers described by Laemmli (1970). In the original method Laemmli poured 10 cm of running gel into tubes 15 cm long and 6 mm internal diameter. After polymerization this was overlaid with 1 cm of stacking gel. Protein samples 0·2–0·3 ml in volume were made up in a solvent containing 0·0625 M tris–HCl buffer pH 6·8, 2 per cent SDS, 10 per cent glycerol, 5 per cent 2-mercaptoethanol, and 0·001 per cent bromophenol blue tracking dye, and were heated for 1·5 min in a boiling water bath to ensure complete dissociation and optimum SDS binding. The electrophoresis was run at 3 mA per gel until the tracking dye reached the bottom of the tube (about 7 h).

Using slabs of gel 35 cm long and a slightly modified variation of this method, Ames (1974) has separated approximately 60 protein bands in a single run from samples of bacterial cells, the limit of detection being well below 0·2 μg per band. A similar shorter gel (14 cm) is shown in Fig. 5.1.

5.5. Staining and destaining methods

After the electrophoresis the separated constituents are revealed in the great majority of cases by a staining procedure or by examination for the presence of a fluorescent or radioactive label. Two methods for the detection of unlabelled protein bands without a staining step are described in Section 7.2(a) since they are very often used in circumstances where dye or fluorescent molecules would interfere with subsequent analyses.

Membrane   Supernatant

1   2   3   4   5   6   1   2   3   4   5   6

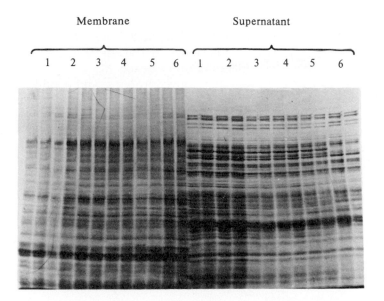

FIG 5.1. Autoradiogram of a dried stained gel ($T = 8$ per cent) of membrane and supernatant fractions of six strains of *Salmonella typhimurium* labelled with ^{14}C-leucine. Each sample (consisting of about 10^5 counts min^{-1} of ^{14}C) was run in duplicate at 15–30 mA per gel slab (about 140 mm x 140 mm x 0·8 mm) for 1·5–4 h using the buffer system of Laemmli (1970). Film exposure was 20 h. (Reproduced from Ames 1974 by permission of the author and publishers).

(a) *Dye staining methods*

The presence of SDS interferes wth Amido Black staining of proteins so the most commonly used stains are those employing Coomassie Blue R250 or Xylene Cyanine Brilliant G (see Table 2.4). Acid Fast Green FCF (1 per cent in 10 per cent acetic acid) has also been used successfully (Bertolini, Tankersley, and Schroeder 1976). Gordon (1975) advocates the use of at least 10 volumes of staining solution per volume of gel in order to dilute out the SDS, and competition of SDS with protein for the dye molecules is not then a serious problem. However, if bands fail to show up after staining or if band patterns are faint, gels can be restained and often show improved results.

For qualitative work and MW measurements the conditions for staining SDS-containing gels are not critical. Short staining and destaining times can be used for localizing a band within 40 min or less providing the band contains adequate material, but for maximum sensitivity and for quantitative measurements a longer staining time ensuring a reproducibly high degree of binding of dye molecules is preferable. For example, using Coomassie Blue R250 in methanol–acetic acid–water (Table 2.4), Weber and Osborn (1975) used staining times of 2–12 h or overnight if more convenient. Either electrophoretic or diffusion destaining can be used (Section 2.10). Some small proteins, glycoproteins, and peptides may not be adequately fixed in the gel in this solvent and can be leached out during staining and destaining. If this is suspected, a prefixing step consisting of immersing the gel for 30 min in 10 per cent sulphosalicylic acid before staining should be included, or alternatively the formaldehyde fixation method of Steck, Leuthard, and Burk (1980) can be used. In this the gel is fixed and stained in a mixture consisting of 180 ml ethanol, 420 ml H_2O, and 100 ml of 35 per cent HCHO, containing 0·8 g Coomassie Blue R250. After 1 h in this the gel is transferred into a mixture of 250 ml ethanol, 750 ml H_2O, and 10 ml of 35 per cent HCHO containing 1·2 g Coomassie Blue R250 for a further 1–3 h. The reduced formaldehyde level used in this second staining bath helps to avoid excessive gel shrinkage. The gel is then destained in the same solvent mixture, but less the Coomassie Blue of course.

A less sensitive but rapid staining and destaining method, useful for MW estimations of proteins has been described by Anderson et al. (1974) and consists of the following steps.

(1) Fix gels for 5 min in 40 wt per cent aqueous trichloroacetic acid.
(2) Wash briefly in water.
(3) Stain for 15 min in a 0·25 wt per cent solution of Xylene Cyanine Brilliant G made up in acetic acid–methanol–water (10:50:40 by volume).
(4) Destain in 5 per cent aqueous acetic acid.

Provided relatively large volumes of staining and destaining solutions compared to the gel volume are used, both procedures employing Xylene Cyanine Brilliant G (Coomassie Blue G 250) described in Table 2.4 can be used, but are less satisfactory than the above methods. Staining times should be the longest shown in Table 2.4 or greater for best results.

Sometimes if SDS has not been adequately removed during staining or a prefixing step this can interfere with dye binding and only very faint (or even non-existent) band patterns will be seen after staining. In these cases the band patterns will usually be revealed if the gel is restained and destained. As with non-SDS containing gels, band patterns will often intensify if stored overnight in 5–7 per cent acetic acid, particularly if a few drops of the dye solution are added to it.

Silver staining procedures (see Section 2.9) can also be used for detecting proteins, glycoproteins and lipoproteins, and many variations abound in the literature but as with other staining methods a key factor for successful application appears to be removal of SDS prior to staining. Various washing regimes with solvent mixtures have been proposed (e.g. Ohsawa and Ebata 1983; Baumann, Cao, and Howald 1984) but as long as it is thorough and virtually all the SDS has been removed, the precise technique is clearly not critical and the acidic methanol or isopropanol mixtures of Glossmann and Neville (1971) and of Dupuis and Doucet (1981) referred to above should be perfectly satisfactory.

Anilinonaphthalene sulphonate (ANS) staining (Table 2.4) has also been used successfully (Daban and Aragay 1984). A particularly simple approach has been suggested by Borejdo and Flynn (1984) who ran the electrophoresis for a few minutes until sample proteins had entered the gel, switched off and then added 0·01 volume of a 2–5 per cent solution of Coomassie Blue R250 to the buffer in the upper electrode chamber. After stirring to distribute the dye evenly, electrophoresis was continued until the dye front reached the bottom of the slab. Destaining was in aqueous 40 per cent ethanol containing 7·5 per cent acetic acid. The method is likely to work best with discontinuous SDS buffer systems (they used the Laemmli system) where the dye migrates rapidly through the gel and builds up at the leading ion–trailing ion front. With the above dye levels staining was said to be quantitative, the main advantages being the evenness of staining and the reduction in time gained by omission of a separate staining step. With lower dye levels staining is no longer quantitative but destaining can also be much shortened or even eliminated.

Glycoproteins can be detected specifically by the periodic acid–Schiff (PAS) procedure (see Section 2.9(d)), but SDS may interfere so the bulk of the SDS should be removed prior to staining by washing overnight in 40 per cent methanol containing 7 per cent acetic acid followed by a further 8 h in fresh solvent (Glossmann and Neville 1971). After location of glycoprotein bands in this way unstained proteins can then be revealed by adding about 1–3 ml of the above Coomassie Blue R250 or Xylene Cyanine Brilliant G staining solutions to each 100 ml of the PAS procedure destaining solution. Proteins will then show up as blue bands in about 24 h.

Glycoproteins can also be localized by ^{125}I-labelled lectin binding provided that the SDS is removed thoroughly from the gel slab first. Dupuis and Doucet (1981) both fixed the gel and removed SDS at the same time with isopropanol:glacial acetic acid:H_2O (5:2:13) and then washed with phosphate-buffered saline (PBS) before incubating the gel for 20 h at room temperature in 50 ml of PBS containing 3 mg of ^{125}I-labelled lectin (specific activity 46 000 dpm μg^{-1}) and finally washing out unbound lectin with 6 or 7 washings in PBS.

(b) *Radioactive labelling*

While staining procedures represent the usual method of detection of proteins after SDS–PAGE, the various alternative detection methods suitable for other forms of PAGE are also applicable here. Thus a radioactive label can be introduced and detected after the electrophoresis either by slicing or solubilization of the gel and liquid scintillation counting, or by autoradiography of gel slices or slabs (Section 2.12).

(c) *Prelabelling with dye or fluorescent markers*

It is in SDS–PAGE that the techniques of prelabelling proteins with a dye or fluorescent marker have found a major application. In most cases the introduction of these markers changes the net molecular charge by reactions involving the terminal α-NH_2 groups and the ε-NH_2 groups of lysine residues. In conventional PAGE mobilities are therefore changed and the labelled proteins cannot be used as markers for the location of the unlabelled molecules. In SDS–PAGE the intrinsic charge on the poly-peptide chain is not important so labelled and unlabelled molecules migrate together according to their size. The small increase in size caused by the introduction of the label is not usually significant.

Although of slightly lower sensitivity than conventional staining methods, Griffith (1972) has advocated the prestaining of proteins (Table 5.5) with a Remazol dye as a method which gives coloured bands that can be watched and photographed during the course of separation. Remazol BBR reacts not only with the −SH groups of proteins but also with primary and secondary amines and alcoholic OH groups so these should be avoided in buffers and sample solutions during the labelling stage. After electro-phoresis gels containing the labelled proteins can be stained with Coomassie Blue in the usual way for greater sensitivity if desired. Variations of this technique have been described by Bosshard and Datyner (1977) using Drimarene Brilliant Blue or Uniblue A which are claimed to give greater sensitivity than Remazol BBR, but at least as good and rather more simple is prelabelling with 4-dimethylamino-azobenzene-4′-sulphonyl chloride (dabsyl chloride) which gives reddish brown bands suitable for quantifying by scanning at 445 nm. For this Tzeng (1983) treated a 20 mg ml^{-1} solution of protein in borate buffer at pH 9 containing 5 per cent SDS with 0·5–1·0 volumes of 10 mM dabsyl chloride in acetone, heated at 60 °C for 5 min, and then diluted the sample to an appropriate extent in buffer containing SDS, 2-mercaptoethanol, and 5 per cent glycerol and applied it to the gel directly. Excess reagent and byproducts also acted as a suitable tracking dye.

A very widely used technique is to incorporate a fluorescent label by pretreating the protein with 1-dimethylaminonaphthalene-5-sulphonyl chloride (dansyl chloride) or fluorescein isothiocyanate (FITC) (Strott-mann, Robinson and Stellwagen 1983).

The procedure described by Inouye (1971) is shown in Table 5.5. The flourescent proteins can be observed by illumination with a u.v. lamp (365 nm) during the electrophoresis so that the progress of the separation can be monitored. The band pattern can be photographed, for example, using Polaroid Type 57 film (ASA 3200) at a 1–30 s exposure at f8 with a Kodak Wratten 16 orange filter (Talbot and Yphantis 1971). The detection limit of dansyl-labelled proteins was found to be as low as 0·008 μg per band. These workers also showed that if the electrophoresis was switched off and the developed band pattern allowed to stand for 16 h and rephotographed, there was very little diffusion and band spreading. This clearly demonstrates that diffusion should not be a problem in protein separations although it is often invoked to explain poor resolution. It also explains why in preparative work (Chapter 7) reliance on simple diffusion of macromolecules out of chopped-up gel slices leads generally to disappointingly low recoveries.

More recently a number of new and very sensitive techniques for the fluorescent labelling of proteins have been introduced and these have the advantage that the reagents themselves and their hydrolysis products are non-fluorescent. Thus there are none of the interfering fluorescent bands of breakdown products which can cause difficulties in the dansylation procedure. The most widely used of these reagents is 4-phenylspiro (furan-2 [3H],1'-phthalan)-3,3'dione or fluorescamine, and the procedure of Eng and Parkes (1974) is shown in Table 5.5. If disulphide bond reduction with 2-mercaptoethanol or dithiothreitol is required this should be performed after step (3) and the sample reheated to ensure optimum binding of SDS. Fluorescamine prelabelling of peptides and proteins has become a widely used technique in SDS–PAGE and provides a simple way not only of locating separated zones after the run has ended, but also by simple u.v. illumination of observing the progress of separations (e.g. Alhanatz, Tauber-Finkelstein, and Shaltiel 1981; Strottmann, Robinson, and Stellwagen 1983) so that experiments can be continued until the desired degree of separation has been attained.

Two other reagents, 2-methoxy-2,4-diphenyl-3(2H)-furanone or MDPF (Barger et al. 1976) and o-phthaldialdehyde (Weiderkamm, Wallach, and Flückiger 1973) have also been used for prelabelling proteins for SDS–PAGE, and typical procedures are given in Table 5.5. MDPF was reported to be about 2·5 times more sensitive than fluorescamine while o-phthaldialdehyde may be as much as an order of magnitude more so (10 ng of protein could be detected on gel rods 3 mm of diameter) and it is also considerably less expensive.

A useful consequence of being able to prelabel proteins in this way is that a labelled standard protein or mixture of standards can be readily distinguished from unlabelled proteins in the sample. By running the two together any small experimental variables can be corrected for and particularly accurate measurements made of the mobilities.

Fluorescent scanning using a Gilford 250 spectrophotometer fitted with a linear transport attachment and the Gilford model 2515 fluorescence detection accessory has been applied to MDPF-labelled proteins separated by SDS–PAGE, and has been reported (Goldberg and Fuller 1978) to be superior both qualitatively and quantitively to gels stained with Coomassie Blue. Quantitative determination of collagen α-chains was linear over the range 10^{-5}–10^{-8} g applied to gel rods 5 mm in diameter.

5.6. The use of acrylamide concentration gradient gels

In Chapter 4 it was indicated that compared with PAGE in gels of uniform

TABLE 5.5
Prelabelling of proteins before SDS–PAGE separation

Method	Optimum excitation wavelength (nm)	Optimum emission wavelength (nm)
Remazol dye		
(1) Mix 0·2 ml protein (2–10 mg ml⁻¹) in 0·15 M NaCl with 0·05 ml 1 M sodium phosphate (pH 7·2–9.2)	Non-fluorescent	—
(2) Add 0·05 ml Remazol BBR (10 mg ml⁻¹) in 10 per cent SDS		
(3) Heat at 100 °C for 5 min or 56 °C for 10 min; add 1 per cent add 1 per cent 2-mercaptoethanol and reheat		
(4) Apply 4 μl samples to gels		
Dansylation		
(1) Mix protein solution (4 mg ml⁻¹) in 2 per cent NaHCO₃ with an equal volume of dansyl chloride in acetone (2 mg ml⁻¹)	340	520
(2) Incubate at 37 °C for 2 h in the dark; shake intermittently		
(3) Precipitate protein with 4 vol. acetone		
(4) Centrifuge; wash precipitate with 2 vol. acetone		
(5) Dissolve precipitate in buffer and pretreat with SDS in usual way		
Fluorescamine		
(1) Protein (50–100 μg) dissolved in 100 μl 0·015 M sodium phosphate buffer (pH 8·5) containing 5 per cent SDS and 5 per cent sucrose	390	475
(2) Heat to 100 °C for 5 min		
(3) Cool; add 5 μl fluorescamine (1 mg ml⁻¹) in acetone and shake briefly		
(4) Add 5 μl Bromophenol Blue tracking dye (5 mg ml⁻¹) if desired		
(5) Apply 10 μl sample to gel		
MDPF		
As for fluorescamine, but at (3) use MDPF (2 mg ml⁻¹) in dimethyl sulphoxide	390	480
o-Phthaldialdehyde		
(1) Mix 1 ml protein solution (0·1–50 μg) in 0.05 M sodium phosphate pH 8·5 with 2·5 μl 2-mercaptoethanol	340	460
(2) Stand for 10 min		
(3) Add 2·5 μl of 1 per cent o-phthaldialdehyde in methanol		
(4) Place in the dark for 2 h		
(5) Pretreat with SDS in the usual way and apply to gel		

concentration gels with acrylamide concentration gradients give superior resolution and sharper bands. Gradient gels also enable proteins of widely different sizes, which would otherwise require runs at two or more different values of T, to be separated on a single gel. They give particularly superior separations of relatively small components. For these reasons they have been quite widely used in SDS–PAGE separations with excellent results (e.g. Studier 1973; Mahadik, Korenovsky, and Rapport 1976; Lambin 1978; Feinstein and Moudrianakis 1984). Gradient gels are prepared as described earlier (Section 4.5) and the appropriate buffer, containing SDS if desired, is introduced electrophoretically in a pre-running step (e.g. 50 V for 4 h). Samples are prepared as usual (Section 5.3(a)) and electrophoretic conditions resemble those for non-SDS gradient gels. Lambin (1978) for example used the standard 0·01 M sodium phosphate buffer pH 7·2 containing 0·1 per cent SDS and performed the electrophoresis at 40 V for 16 h with gels about 8 cm long.

After electrophoresis and locating protein bands in the usual way, MW values are given by the relationship

$$\log \text{MW} = a \log T' + b$$

where T' is the concentration of acrylamide reached by the protein and a and b are the slope and intercept respectively of the linear regression line which is established from measurements of a number of known standard proteins run at the same time on the same gel slab. Thus in gels with any shape of concentration gradient a plot of log MW *versus* log T' should be linear (Lambin 1978; Poduslo and Rodbard 1980). When the gel concentration gradient itself is linear, to a first approximation a plot of log MW *versus* log (migration distance) is linear. These relationships have been shown by Lambin (1978) to correlate very well with observed measurements for 41 standard proteins. Glycoproteins (even those with a substantial carbohydrate content) and proteins such as papain and pepsin which bind less SDS than most other proteins, as well as proteins such as lysozyme and ribonuclease which migrate anomalously during SDS–PAGE on constant concentration gels, all migrated closely in accordance with the above relationships in gradient gel SDS–PAGE (Lambin 1978). The method was thus widely applicable for measurement of MW values over the range 13 000–10^6 using a 3–30 per cent gel gradient. Results were marginally superior when the Bis level was $C = 8·4$ per cent than when it was $C = 3·8$ per cent.

In accordance with the comments (see p. 121) made by Leach *et al.* (1980) with regard to errors in glycoprotein MW measurements by SDS–PAGE, Poduslo (1981) also concluded that the physical properties of individual glycoproteins affected the accuracy of measurments and like Lambin (1978) found that the use of gradient gels was most beneficial. Poduslo (1981) suggested that a running gel buffer composed of 0·1 M tris, 0·1 M boric acid containing 0·1 per cent SDS, and 2·5 mM Na_2 EDTA (pH 8·5) and a similar but half strength buffer for the stacking gel also helped, perhaps due to the formation of borate complexes with the carbohydrate moieties of the glycoproteins, but the difference between this and tris–glycine buffers was small.

5.7. Separation of low-molecular-weight proteins

In SDS–PAGE on conventional non-gradient gels there is an inflection in the log MW *versus* R_f plots (Dunker and Rueckert 1969) when the polypeptide MW falls below a certain critical size (about 20 000 for 5 per cent gels and 10 000 for 10 per cent gels). Fish *et al.* (1970) have suggested that the rod-shaped polypeptide–SDS particles begin to approximate to spheres as their lengths approach the magnitude of their diameters. However, the simple relationship between mobility and log MW requires all proteins to be hydrodynamically homologous as well as of constant charge to mass ratio. Also as the size decreases any difference in the intrinsic charge on the molecules, which is generally of comparatively little significance in the case of large molecules, assumes a greater importance and may influence both the binding of SDS and the net charge on the complex. The conformation of the polypeptide can also influence the binding of SDS molecules and this factor also becomes more significant at smaller MW values.

Williams and Gratzer (1971) have stated that under the standard conditions of SDS–PAGE all proteins with MW below about 6000 migrate with the same mobility regardless of the acrylamide concentration. Fortunately increasing Bis levels to $C = 10$ per cent and the addition of 8 M urea improve the practical results for small proteins, although not overcoming the theoretical objections, and to this end Swank and Munkres (1971) used a homogeneous tris-phosphate buffer system ($0 \cdot 1$ M H_3PO_4 adjusted to pH $6 \cdot 8$ with tris) containing $0 \cdot 1$ per cent SDS, and with 8 M urea also present in the buffer used to make up the gel but not in the apparatus buffer. Samples were made up in the same buffer diluted 10-fold but with 1 per cent SDS, 8 M urea, and 1 per cent 2-mercaptoethanol, and were heated to 100 °C for 1–3 min in this prior to application to the gel.

As with other PAGE and SDS–PAGE separations, so also for peptides the use of discontinuous (multiphasic) buffers gives improved results, and Anderson, Berry, and Telser (1983) introduced the system shown in Table 5.6. They selected specific high-mobility strong electroytes and buffer concentrations that permitted polypeptide stacking to occur in a temporary acetate phase and unstacking to occur in a chloride and sulphate anion separating phase without a pH discontinuity. Thus samples should be prepared in an acetate sample buffer (see Table 5.6). This has double the SDS level of that suggested by Anderson *et al.* (1983) and also contains 2-mercaptoethanol, but these are probably desirable to ensure maximum SDS binding to the sample proteins (see Section 5.2) and samples should

TABLE 5.6
The composition of buffers and gels for use with small proteins and peptides

	pH	Composition
Upper reservoir buffer (cathode)	7·8	0·074 M tris; 0·1 per cent SDS; conc. HCl to pH 7·8
Stacking gel buffer	7·8	0·2 M tris; 0·04 per cent SDS; conc. H_2SO_4 to pH 7·8
Separation gel	7·8	0·2 M tris; 0·04 per cent SDS; 8 M urea; conc. H_2SO_4 to pH 7·8
Lower reservoir buffer (anode)	7·8	0·2 M tris; 0·04 per cent SDS; conc. H_2SO_4 to pH 7·8
Stacking gel ($T = 3·125$ per cent; $C = 20$ per cent)	7·8	2.5 g acrylamide; 0.625 g Bis; Stacking gel buffer to 100 ml; 50 μl TEMED; 50 mg ammonium persulphate
Separation gel ($T = 8$ per cent, $C = 5$ per cent)	7·8	7·6 g acrylamide; 0·40 g Bis; separation gel buffer to 100 ml; 50 μl TEMED; 50 mg ammonium persulphate
Sample buffer	7·8	0·139 M tris; 1·0 per cent SDS; 20 per cent sucrose; 0·1 M 2-mercaptoethanol; glacial acetic acid to pH 7·8

also be heated to 100 °C for 1–3 min before being applied to the sample wells. Anderson *et al.* (1983) recommend that the sample wells in the gel slab be flushed out briefly with H_2O and then filled with 50 μl of sample buffer (without sucrose) before layering the 30–50 μl volumes of sample (in sample buffer with sucrose) into the bottom of the wells. This is to ensure that there is sufficient acetate to form a temporary stack and there should be 5–6 mm of acetate buffer above the sample layer in order to do this. Pyronine Y tracking dye can be added to the samples or 2 μl of a 2 per cent solution (with added sucrose) added to each well after sample application. The rest of the sample well is carefully filled with upper reservoir buffer before filling the top reservoir itself in order to avoid undue mixing of the various buffer phases. The resolving power and separations are a function of buffer composition and concentration and are not critically dependent upon gel concentration (per cent T). The system gave a liner calibration plot of log MW *versus* R_{TD}, the mobility of bands relative to the tracking dye (Pyronine Y) with proteins over the MW range 2400–92 000, in contrast to other SDS–PAGE systems which shown inflection points in the region of MW 10 000. If urea was omitted in the separation gel, then in this system also an inflection point occurs. Anderson *et al.* (1983) used gels of varying strength from $T = 5$ per cent to $T = 12$ per cent and 8–19.5 per cent or 10–18 per cent T gel concentration gradients and obtained linear plots (see Section 4.1) of MW *versus* K_r for small proteins over the MW range 1400–45 000 but two larger proteins did not fall on the linear plot.

If excessive losses of peptide during staining are suspected, fixation with formaldehyde treatment (Steck *et al.* 1980) may be beneficial (see p. 28).

Owing in part to the increased influence of the polypeptide shape and charge the accuracy is not as good as that achieved with larger proteins, but nevertheless in a study with 11 polypeptides Swank and Munkres (1971) using R_f *versus* log MW plots achieved an accuracy of $\pm$ 18 per cent. Such plots are of course sigmoidal in this as in other SDS–PAGE gel systems. A very satisfactory linear relationship of K_r *versus* MW in this system has been reported (Tracy and Chan 1979) from which the MW values of a number of standard proteins were determined with an average error of only 3·8 per cent.

5.8. Peptide mapping of proteins by SDS–PAGE

In many cases relationships between proteins cannot be proved unequivocally by electrophoretic mobility alone, especially if they are closely related, and further analysis is necessary. A rapid and convenient approach to this is peptide mapping of the peptide mixture produced by treatment of the proteins with proteolytic enzymes. One useful way of doing this has been suggested by Cleveland, Fischer, Kirschner, and Laemmli (1977). The method is particularly suitable for proteins which have been isolated by preparative electrophoresis using SDS-containing gels since removal of traces of SDS is not necessary, but it can be applied equally well to any proteins. It consists of the partial digestion of proteins by any of a number of proteolytic enzymes in an SDS-containing buffer which gives a mixture of many peptides with an MW sufficiently large for satisfactory separation by SDS–PAGE. The pattern of bands produced is characteristic of the protein and of the proteolytic enzyme used, and is highly reproducible. As little as 5–10 μg of protein can be analyzed in a few hours. The method described by Cleveland *et al.* (1977) is as follows.

Purified protein or protein from preparative electrophoretic experiments (see Sections 7.2 and 7.4) is dissolved at a concentration of 0·5 mg ml^{-1} in 0·125 M tris–HCl buffer pH 6·8 containing 0·5 per cent SDS, 10 per cent glycerol, and 0·0001 per cent Bromophenol Blue and the mixture heated at 100 °C for 2 min. Proteolytic digestions are performed typically at 37 °C for 30 min, using for example 3–30 μg ml^{-1} papain, 10–140 μg ml^{-1} chymotrypsin, or 5–130 μg ml^{-1} *Staphylococcus aureus* protease (Miles Laboratories). This last enzyme was particularly suitable because it cleaves polypeptide chains on the carboxyl side of aspartic and glutamic acid residues and generates digests which contain many appropriately sized peptide fragments. The size range of peptides can be varied of course by altering the incubation time or the proportion of proteolytic enzyme and should be optimized for any particular mixture being examined, but the most important point is that the proteins being compared should be treated identically. Subtilisin, ficin, pronase, and elastase have also been used successfully. The reproducibility of band patterns relies on the patterns being relatively insensitive to variations in incubation conditions (amount of enzyme, time, etc.) so excessively large amounts of enzyme giving extensive fragmentation in a few minutes should be avoided. After the desired incubation 2-mercaptoethanol is added to 10 per cent and SDS to 2 per cent and the samples boiled for 2 min to stop proteolysis. About 20–30 μl (10–15 μg) of each sample were loaded onto polyacrylamide gel slabs (T = 15–20 per cent, C = 2·5–3·0 per cent) for analysis by SDS–PAGE in the usual way using the buffer system (Section 5.4(b)) described by Laemmli (1970).

Bands from SDS–PAGE gels stained with Coomassie Blue R250 can also be examined by this peptide mapping procedure without prior extraction of the protein.

For this the first gel is briefly stained and destained (to avoid too long an exposure to acid conditions which could result in a small degree of acid hydrolysis), rinsed with cold water, and the individual protein bands are cut out. These are trimmed to a width of 5 mm and soaked for 30 min with occasional swirling in 10 ml 0·125 M tris–HCl buffer pH 6·8 containing 0·1 per cent SDS and 1 mM EDTA. A second SDS–PAGE slab (with a separation gel of T = 15–20 per cent, C = 2·5–3·0 per cent) is then prepared with a longer than usual stacking gel (up to 5 cm) and sample wells 5·4 mm wide (a slab thickness of 1·5 mm was best). The sample wells are filled with this same buffer and each gel slice is placed in the bottom of a sample well with the aid of a spatula. The spaces around the slice are filled by overlaying each slice with 10 μl of this same buffer containing 20 per cent glycerol. Finally 10 μl of this buffer containing 10 per cent glycerol and the required amount of proteolytic enzyme is added to each slot. Electrophoresis is performed in the usual way except that the current is turned off for 30 min when the Bromophenol Blue dyes nears the bottom of the stacking gel to allow satisfactory digestion of the protein to be completed. The initial gel slice should contain about 10 μg of protein if Coomassie Blue staining is to be used for detection, but this approach is particularly suitable for use with radioactively labelled samples and smaller amounts of material can then be analysed satisfactorily.

The use of proteolytic enzymes to digest protein to a peptide mixture which is analysed by SDS–PAGE, generally employing the Laemmli (1970) buffer system and polyacrylamide concentration gradient gels to give the best display of peptides of widely differing sizes, has become a popular method for demonstrating protein identity and is used in many studies of protein architecture (e.g. Mattick *et al.* 1983) and in protein modification reactions. It is worth remembering however that any reproducible and well-defined method of peptide generation can also be exploited in the same way for peptide mapping. Sonderegger *et al.* (1982) and Rittenhouse and Marcus (1984) used selective acid hydrolysis of Asp–Pro bonds to generate peptide mixtures, while Saris *et al.* (1983) employed hydroxylamine to specifically cleave Asn–Gly bonds. Cleavage at Trp by N-chlorosuccinimide in the presence of urea (Lischwe and Ochs 1982) or at Met by cyanogen bromide (Lonsdale-Eccles, Lynley, and Dale 1981) have also been used successfully. All these reports analysed the resulting peptide mixtures by SDS–PAGE but one potential difficulty in this is that if digestion is extensive and large amounts of relatively small peptides are produced, these may be poorly resolved and run at or close to the tracker dye marking the buffer front unless gels of very high per cent T are used. It is also possible that the presence of SDS increases the loss of small peptides from the gel during the staining and destaining stages. This is often overcome by using radiolabelled peptides and autoradiographic detection. Much useful information can be gained however by using PAGE without SDS for peptide mapping (e.g. Andrews 1984) and although no information about the molecular weight of the peptides is gained (unless gels of various per cent T are used and Ferguson plots constructed; see Chapter 4) this approach can be particularly valuable if small proteins are being fragmented and hence small peptides generated. If losses of such peptides during staining are suspected, fixation with formaldehyde (see p. 28) may be beneficial. For many purposes altogether different methods of peptide mapping, such as TLC or particularly HPLC, may be preferable to electrophoretic procedures, but although more rapid they often lack sensitivity and versatility. Two-dimensional gel electrophoresis (see Chapter 11) has also been used in peptide mapping work, Steinberg (1984) for example using the O'Farrell (1975) format of PAGIF followed by SDS–PAGE to investigate peptides in papain digests of ^{35}S-labelled regulatory subunit protein.

5.9. The use of detergents other than SDS

(a) *Cationic detergents*
Williams and Gratzer (1971) reported that highly basic proteins such as protamines precipitate in the presence of SDS and that highly acidic proteins such as the ferredoxins behave anomalously during SDS–PAGE probably owing to poor SDS binding. They therefore replaced SDS with the cationic detergent cetyltrimethylammonium bromide (CTAB). In all other respects the method was the same as the standard method described in Section 5.3 except that migration is of course towards the cathode and this should therefore be the lower electrode in the apparatus. With a number of proteins of known size on a gel of $T = 10$ per cent, $C = 5$ per cent a linear plot of R_f *versus* log MW was obtained over the range of about 10 000–40 000.

However, simple substitution of CTAB for SDS in this way results in the formation of a dense precipitate within the gel. This is probably cetyltrimethylammonium persulphate (Eley, Burns, Kannapell, and Campbell 1979). Williams and Gratzer (1971) overcame the problem of visualizing the separated bands in opaque gels by using dansyl prelabelling and fluorescent detection rather than dye staining. This of course does not circumvent the objections that the formation of the precipitate results in a non-uniform distribution of ammonium persulphate, leading to non-uniform gel polymerization, and a non-uniform distribution of CTAB, both of which might affect the protein–detergent binding equilibrium. In addition the presence of precipitate particles themselves may influence protein migration. There can also be formation of detergent–dye precipitates during any staining–destaining process. Panyim, Thitipongpanich, and Supatimusro (1977) overcame this by staining and destaining at 80–100 °C.

There are two ways in which these difficulties can be circumvented. Firstly, if ammonium persulphate is to be used as the polymerization catalyst the gels should be prepared in buffer containing no detergent. The usual sample preparation and electrophoretic procedures are then followed with detergent included in the sample and in the upper electrode compartment buffer (Eley et al. 1979). Residual persulphate in the gel moves ahead of the CTAB during the electrophoresis so they do not come into contact and no precipitate is formed. The second approach is to use different polymerization catalysts.

Marjanen and Ryrie (1974) used $T = 10$ per cent, $C = 2.67$ per cent gels containing 0·1 per cent CTAB photopolymerized with 2 μg ml^{-1} riboflavin. The buffer used throughout in both gels and electrode chambers was 0·1 M sodium phosphate pH 6·0 containing 0·1 per cent CTAB, but cacodylate (pH 6·0), citrate (pH 6·0), and succinate (pH 5·0) buffers were also satisfactory. Samples were pretreated by heating for 30 min at 70 °C in 0·01 M sodium phosphate buffer pH 6·0 containing 1 per cent CTAB and 1 per cent dithiothreitol. Methylene blue was used as the tracking dye. When 18 standard proteins were examined a linear plot of R_f versus log MW was obtained, the accuracy of MW measurement being similar to that of SDS–PAGE, but a number of membrane proteins failed to migrate, probably owing to a lack of CTAB binding.

Eley et al. (1979) in a thorough investigation of the factors affecting CTAB–PAGE used gels with T varying between 2·5 and 15 per cent with a low level of cross-linking ($C = 1.33$ per cent) made up in 0·1 M sodium phosphate pH 7·0 containing 0·1 per cent CTAB. They used photopolymerization with flavin, for which they mixed 1·5 ml of a stock solution containing 0·5 mM flavin mononucleotide and 0·05 mM EDTA with 28·5 ml acrylamide and Bis solution containing 50 μl TEMED. Samples were pretreated by heating on a boiling water bath for about 4 min in buffer containing 1 per cent CTAB and about 10 per cent 2-mercaptoethanol, and after cooling a small amount of malachite green was added as tracking dye. Staining with Coomassie Brilliant Blue R250 showed that, although plots of R_f versus log MW were correctly sigmoidal (Section 5.2), when gels with $T = 7.5$ per cent are used proteins within the MW range 36 000–96 000 fall on a straight line and MW values can be estimated with good accuracy.

Gels with $T < 7.5$ per cent should be used for proteins with MW $> 70\ 000$ and $T > 7.5$ per cent for those with MW below about 30 000.

N-Cetylpyridinium chloride (CPC) is another cationic detergent that has been used in PAGE applications. Shick (1975) reported linear plots of R_f *versus* log MW over the range 17 000–160 000 for gels with $T = 10$ per cent or $T = 12$ per cent ($C = 2.7$ per cent) made up in potassium acetate buffer pH 3.7 containing 0.01 per cent CPC in both the gel and the reservoir buffer, which was 0.3 M glycinium acetate pH 3.7. A ferric sulphate–ascorbic acid polymerization catalyst system was employed (see p. 23).

Thus it appears that both CTAB and CPC can be used to solubilize proteins, including those from membrane structures, and the resulting solutions can be applied directly to PAGE analysis. Clearly in most cases enough detergent molecules are bound to the protein molecules to produce complexes with a reasonably constant charge-to-mass ratio and with similar hydrodynamic shapes. The complexes resemble protein–SDS complexes in electrophoretic behaviour and quite accurate molecular weight values can be gained.

(b) *Non-ionic detergents*

Unfortunately both anionic and cationic detergents cause considerable disruption of the native molecular structure and consequently proteins are denatured, most enzymes are inactivated, and immunological activity is lost. There may be a partial regeneration fo activity following removal of the detergent, but this is seldom extensive and usually does not occur at all. However, many membrane proteins and enzymes can also be solubilized by non-ionic detergents such as Tween 80 or Triton X-100 which break up protein aggregates (Singh and Wasserman 1970; Clarke 1975; Erickson and Kim 1980; Nakashima, Nakagawa and Makino 1981) while still being sufficiently mild to retain enzyme activities (van den Hoek and Zail 1977).

Owing to the non-ionic nature of the detergent the intrinsic charge on the protein is important to the subsequent electrophoretic separation. This has the benefit that proteins of similar size but differing charge can be separated, which is not the case when anionic or cationic detergents are used, but it does have the disadvantage that since different proteins will not have a constant charge-to-mass ratio their MW values cannot be determined by a single electrophoretic run and a plot of R_f *versus* log MW. Rather, MW values are determined by the same method as that used for non-detergent gels, namely by the construction of a Ferguson plot and measurement of K_R (Sections 4.1 and 4.2). Hearing, Klingler, Ekel, and Montague (1976) studied 33 different soluble fibrillar and globular proteins, glycoproteins, and metalloproteins in this way using a typical discontinuous gel and basic multiphasic buffer system (Chapter 3) with separation gels of $T = 4$–15 per cent ($C = 3$ per cent). Detergent was not added to the gels routinely (0·1 per cent could be incorporated if desired) but samples were pretreated with 1 per cent Triton X-100 for several hours. A good linear plot of K_R *versus* effective molecular radius was obtained, with all the different types of proteins falling close to the same line. Plots of K_R *versus* MW were slightly less linear, at least for proteins of MW below about 50 000. Triton X-100 does not disrupt subunit structure so the native MW values are obtained. Hearing *et al.* (1976) reported that the average error in MW values was ± 10 per cent, which was a little less accurate than they found from SDS–PAGE (± 5 per cent) or PAGE in the presence of urea (± 4 per cent). The presence of Triton X-100 did not significantly affect K_R and Y_0 for a number of normally water-soluble proteins, but proteins which bound larger amounts of detergent (e.g. many membrane proteins) migrated as a function of the total size of the protein–detergent complex.

(c) Acid–urea–detergent gels

When proteins are subjected to electrophoresis as cations at low pH in the presence of non-ionic detergents they are differentially retarded depending upon their ability to bind the detergent and to form mixed micelles between the detergent and hydrophobic regions of the polypeptide chains. Thus the mobility depends upon the hydrophobic properties of the protein in addition to the size and net electrical charge (Zweidler and Cohen 1972). The extent of detergent binding is reduced by the presence of urea and the electrophoretic patterns can be altered considerably by varying the concentrations of urea and detergent. Gels of this type and with a very low degree of cross-linking (i.e. $T = 12$ per cent, $C = 0·67$ per cent) have given excellent results in the separation of histones (Alfageme, Zweidler, Mahowald, and Cohen 1974). The composition of the necessary constituents for such gels is shown in Table 5.7.

TABLE 5.7
The composition of an acid urea non-ionic detergent gel system

Gel composition	12 g acrylamide
	0.08 g Bis
	5 ml glacial acetic acid
	0·5 ml TEMED
	0·37 ml Triton X-100 or Triton DF-16
	3·8 g urea
	Dissolve components in H_2O and make up to 100 ml
	Add 0·06 g ammonium persulphate
Apparatus buffer	5 per cent acetic acid

Gels were pre-run until the current stabilized and then overlaid either with a suitable volume (0·1 ml for gel rods 5 mm in diameter) of 1 M cysteamine in 5 per cent acetic acid or with a mixture containing 8 M urea, 5 per cent acetic acid, 0·2 M cysteamine and 2 mg ml^{-1}protamine (Fernandes, Nardi, and Franklin 1978), using 1 ml for every 40 ml of gel. Pre-running is then continued at 140 V for 30–40 min to eliminate free radicals which might otherwise oxidize methionine during the subsequent separation and can significantly affect detergent binding. After this, any remaining cysteamine is flushed off the top of the gels and samples are applied in 5 per cent acetic acid containing 4–8 M urea, 4 per cent 2-mercaptoethanol and 0·02 per cent pyronine Y tracking dye.

More recently an acid–urea–detergent system has been developed which employs discontinuous gel and buffer phases (Bonner, West and Stedman 1980). The gels are prepared as indicated in Table 5.8. Not only does this method provide the superior resolution generally associated with discontinuous systems (see Chapter 3), but it actually simplifies the separation of histones for example, in that crude acid extracts can be applied directly to the gels for analysis. In addition, no pre-electrophoresis step is needed. Bonner et al. (1980) used methylene blue as tracking dye and performed the electrophoresis in slabs 150 mm wide and 0·8 mm thick run at 5–10 mA overnight. After the separation gels were stained with Coomassie Blue R250. Strips of this gel, even when stained, could be applied across the top of a second gel slab to give a two-dimensional separation (Chapter 11). If it is intended to do this, then the staining and destaining solutions should contain 0·1 per cent cysteamine hydrochloride to inhibit disulphide bond formation and the gels should not be unduly exposed to light in order to avoid photo-oxidation. As an alternative to Coomassie Blue staining Mold et al. (1983) found that Amido Black staining, destaining and exhaustive washing with water enabled silver staining to be applied successfully to these gels. They suggested that histones may have few nucleation sites for the deposition of silver particles or that these were masked by the presence of Triton X-100, but that a preliminary Amido Black staining provided such sites, so that subsequent silver staining could then be used with a sensitivity about 30-fold greater than Amido Black staining alone. For a second dimension of analysis the gel is also made up as shown in Table 5.8 except that the Triton X-100 in the stacking and separation gels is replaced by 0·15 per cent CTAB and the amount of urea is reduced to 6M (8M precipitates CTAB). The upper reservoir buffer should also contain 0·15 per cent CTAB. Two-dimensional separations are particularly helpful for complex samples such as whole cells or tissue homogenates.

An interesting variation of this method has been applied to a study of globin chain heterogeneity in a number of mammalian haemoglobins (Rovera, Magarian, and Borun 1978). The non-ionic detergent binding is proportional to the number of hydrophobic groups and the separation of a mixture of polypeptides of differing hydrophobicity thus depends on the

TABLE 5.8

A discontinuous gel and buffer system for the separation of histones

Reservoir buffer:	0·1 M glycine
(upper and lower)	1·0 M acetic acid
Stacking gel mixture:	3·3 g acrylamide
	0·16 g Bis
	6·0 ml glacial acetic acid
	0·5 ml TEMED
	0·3 ml conc. NH_4OH
	48·0 g urea
	Mix and make up to 93·5 ml with H_2O.
	Degas and immediately before use add 6·5 ml of 0·004 per cent
	riboflavin stock solution for photopolymerization
	(No Triton X-100 need be added to the stacking gel phase.)
Separation gel mixture:	15·0 g acrylamide
	0·1 g Bis
	6·0 ml glacial acetic acid
	0·5 ml TEMED
	0·3 ml conc. NH_4OH
	48·0 g urea
	Mix components and make up to 91·5 ml with H_2O. Add 2·0
	ml of 25 per cent (w/v) Triton X-100, degas and just before use
	add 6·5 ml of 0·004 per cent riboflavin solution.
	Photopolymerize.

concentrations of urea and Triton X-100. To study this Rovera *et al.* (1978) prepared slab gels with a uniform polyacrylamide composition as described above but with a 0–0·67 per cent gradient in Triton X-100 concentration. The detergent gradient was prepared in a manner analogous to the preparation of acrylamide concentration gradients (Section 4.5) but with no detergent and no sucrose in one gradient-making solution and 0·76 per cent Triton X-100 and 20 per cent sucrose in the other. After polymerization the gels were rotated through 90° so that samples were subject to electrophoresis across a Triton gradient (*cf.* transverse gradient gel electrophoresis, Section 4.7). Resolution of the globin chains was excellent, some being better separated at low Triton levels and some at higher ones. A similar approach was used by Fernandes *et al.* (1978) who used some transverse gradient gels with constant 8 M urea and a 0–16 mM gradient of Triton X-100 and others with constant 6 mM Triton X-100 and a 0–8·5 M urea gradient to investigate the optimum separation conditions of a number of membrane proteins.

(d) *Charge shift electrophoresis*

Helenius and Simons (1977) introduced a technique called charge shift electrophoresis for distinguishing between hydrophilic proteins externally bound to membranes and amphiphilic proteins with strongly hydrophobic regions that form an integral part of the membrane structure. This consists of electrophoresis in 1 per cent agarose gels, in which there is virtually no molecular sieving effect, in 0·05 M glycine–NaOH buffer pH 9·0 containing 0·1 M NaCl, and 0·5 per cent Triton X-100. The electrode buffers were the same but without detergent. Samples were examined at the same time on three gels, all with this composition except that the second also included 0·25 per cent of the anionic detergent sodium deoxycholate and the third contained 0·05 per cent of the cationic CTAB. Hydrophilic soluble proteins and peripheral membrane proteins bind comparatively little Triton X-100 but amphiphilic proteins bind much larger amounts in micelle-like clusters around the hydrophobic regions (Clarke 1975). In the detergent mixtures these clusters will be composed of both Triton X-100 and the ionic detergents and hence the net charge of the protein–detergent complexes will be modified. Thus a band due to an amphiphilic protein will have a different mobility when anionic or cationic detergent is present compared with its mobility when Triton X-100 alone is present.

5.10. Biochemical applications

In recent years detergents have gained widespread use for solubilizing complex biological systems such as membranes, virus particles, tissue preparations, etc.

When the great solubilizing power of an anionic detergent such as SDS is employed, particularly when it is used at 100 °C in the presence of 2-mercaptoethanol or dithiothreitol, large quantities of SDS are bound to the protein molecules causing complete denaturation and a dissociation into the subunits making up the native molecule. The SDS swamps the intrinsic molecular charges and produces protein–SDS complexes with an almost constant charge-to-mass ratio, so that during electrophoresis proteins separate according to size. In principle the method can be applied to any soluble protein or glycoprotein and to any protein or glycoprotein which is solubilized by the detergent treatment from a more complex environment. There is now good evidence that cationic detergents such as CTAB and CPC can act in a similar way. In the region of MW = 40 000, two proteins differing in MW by 1500–2000 should be visually distinguishable (Weber and Osborn 1975).

Apart from its obvious applicability to samples requiring detergents for solubilization (e.g. Dupuis and Doucet 1981; Kubak and Yotis 1982) it is this ability to separate the sample molecules by molecular weight differences that dictates the requirement for SDS–PAGE in most experiments. In principle it can be, and often is, applied in many of the situations where PAGE (Chapters 2 and 3) could also be used, although naturally the patterns of separated zones will differ since the basis of the separation is different and molecular charge plays an important part in PAGE separations. Thus it is often used in assessments of the purity of samples, in proving the identity of unknowns, in studies of subunit structure (e.g. Feinstein and Moudrianakis 1984), in peptide mapping (e.g. Lonsdale-Eccles; Lynley and Dale 1981; Lischwe and Ochs 1982; Sonderegger *et al.* 1982; Mattick *et al.* 1983), in studying protein hydrolysis (e.g. deterioration of food proteins during storage by indigenous proteinases or resulting from bacterial contamination) or protein synthesis occurring as the result of the translation of m-RNA (e.g. van Tol and van Vloten-Doting 1979; Willems *et al.* 1979), for identifying species of fish or animals (e.g. Corzo, Riol-Cimas, and Metendez-Havia 1984) etc. Proteins which contain disulphide bonds often exhibit different MW values when examined by SDS–PAGE in the presence and absence of reducing agents. Clearly if these are inter-polypeptide chain bonds their reduction leads to dissociation and a sharply-defined fall in MW, but even if they are intra-chain bonds their rupture leads to a greater unfolding of the polypeptide chain and a greater extent of binding of SDS, both of which alter mobility and change apparent MW. A comparison of mobilities by SDS–PAGE can therefore give useful structural information, but in complex systems of several components run separately under reducing and non-reducing conditions it is often extremely difficult to tell which band obtained under one set of conditions corresponds to the same component under the other. Two-dimensional arrangements with one reducing dimension and one non-reducing can be used but may be difficult to interpret in the case of intramolecular disulphide bonds. However, Allore and Barber (1984) have described a neat one-dimensional SDS–PAGE system using the usual (Laemmli, 1970) buffer system, electrophoretic and staining procedures. All samples were heated at 100 °C for 3 min in SDS containing sample buffer but were divided into two portions with reducing agent (0·5–5·0 per cent 2-mercaptoethanol or dithiothreitol) being added before heating to one portion and not to the other. Reduced and non-reduced portions of sample were then applied to adjacent sample slots in the gel slab and subjected to electrophoresis at the same time. As can be seen from Fig. 5.2, the proximity of the reduced and non-reduced samples, run under identical conditions, greatly facilitates interpretation. In this example the lateral diffusion of reducing agent across into lanes 3 and 7 gives partly reducing conditions in these lanes and results in a 'diagonal' banding which further aids comparison between the reduced and non-reduced forms. This is particularly helpful when intramolecular bonds are split because then the mobility differences are often comparatively small (e.g. bands 2 and 4, Fig. 5.2). When inter-chain bonds are broken and dissociation occurs mobility differences are larger (e.g. band 1* dissociated into bands 1 and 5 in Fig.

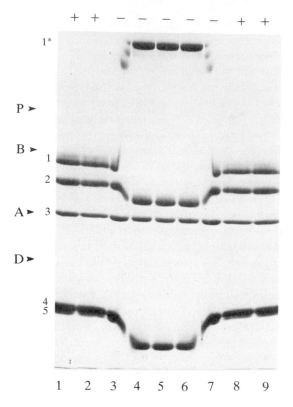

FIG 5.2. SDS–PAGE patterns of (1) IgG3 heavy chain, (2) IgG1 heavy chain, (3) pig skeletal actin, (4) kappa light chain, (5) kappa light chain derived from reduced IgG3k and (N) intact IgG3k. Protein samples were subjected to electrophoresis either without (−) prior addition of 5 per cent 2-mercaptoethanol (lanes 3–7) or with (+) its addition (lanes 1, 2, 8 and 9). Molecular weight markers were: P, phosphorylase a (94 000); B, bovine serum albumin (6800); A, actin (42 000) and D, DNase 1 (31 000). (Reproduced from Allore and Barber (1984) by permission of the authors and publishers.)

5.2) and in this case it may sometimes be possible also to see zones due to intermediate partially reduced forms in the lanes between the non-reduced and fully reduced forms, e.g. the spots below band 1* in lanes 3 and 7 (Fig. 5.2).

Because SDS–PAGE separates molecules according to size it is a highly favoured technique in the preparation of two-dimensional protein maps (Chapter 11), particularly when allied to a method separating principally according to charge differences in the other dimension. It has been used in conjunction with charge shift electrophoresis in studies of human plasma apolipoproteins from normal subjects and patients with Tangier disease or individuals with hypertriglyceridaemia (Beisiegel and Utermann 1979; Utermann and Beisiegel 1979).

Non-ionic detergents also have the merit of considerable solubilizing power, but in this case much of the native molecular structure may be retained and enzymic or immunological activity preserved (e.g. Newby, Rodbell, and Chrambach 1978; Newby and Chrambach 1979). When MW values are determined in the presence of non-ionic detergents (e.g. by the methods described in Chapter 4), it is the native MW that is measured and not the subunit MW. For proteins composed of a single polypeptide chain the two are synonymous of course. However, for more complex molecules and oligomers the combination of MW measurements in non-ionic and in ionic detergents can give valuable information on the molecular structure.

The development of PAGE under acid conditions in gels containing non-ionic detergents and urea has introduced a new parameter, namely the ability to separate two macromolecular species of similar size and charge but differing in hydrophobicity. At the present time this particular electrophoretic system appears to be unique in this regard, but we can expect that the concept will be developed and exploited in different ways to broaden further the scope of separations that can be achieved by the already very versatile techniques of electrophoresis.

6

ELECTROPHORESIS ON AGAROSE AND COMPOSITE POLYACRYLAMIDE–AGAROSE GELS. NUCLEIC ACID ANALYSIS

Gels prepared from the naturally occurring polysaccharide agar were one or the earliest media used to overcome the problems of convection associated with electrophoresis in free solution. In terms of the ability to resolve proteins in a mixture it is better than electrophoresis in free solution or on paper, similar to cellulose acetate, but not as good as starch or polyacrylamide gels. Owing to the presence of sulphate and carboxyl groups, agar gels suffer from very marked electro-endosmosis when neutral or alkaline buffers are used and in addition there may be protein loses due to absorption. Fortunately the purified polysaccharide agarose is nowadays readily available commercially and is largely free of these drawbacks. If necessary agarose can be purified further by ion-exchange chromatography (Johansson and Hjertén 1974). Agar itself is now very seldom used in electrophoretic applications. The chemistry of agarose has been discussed by Serwer (1983).

Likewise, electrophoresis on agarose gels has been eclipsed by the use of polyacrylamide in the analysis of most proteins and glycoproteins but it does remain invaluable in applications where a very large pore size and hence non-restrictive gel is required. The two major areas in which agarose gels remain widely used are in immuno-electrophoretic procedures, especially those relying on an immunodiffusion step, and in the separation of very large molecules with average hydrodynamic radius above about 5–10 nm, such as viruses, nucleic acids, lipoproteins, and some membrane proteins. Immuno-electrophoretic applications will be discussed in Chapter 8. For macromolecular separations agarose gels have advantages over polyacrylamide in that they are non-toxic and more simple to prepare. Since only a single constituent (agarose) needs to be weighed out there is much less likelihood of error and gels of highly reproducible properties are easily made. Against this, agarose gels, although quite rigid, are much less elastic than polyacrylamide and so are more easily torn and require more careful handling. Their use for electrophoretic separations has been reviewed by Serwer (1983).

With polyacrylamide gels the pore size varies inversely with the total monomer concentration T so that in order for large macromolecules to enter the gel and give satisfactory separations low values of T must be employed. As mentioned earlier (Section 2.7) gels of $T < 3 \cdot 0 – 3 \cdot 5$ per cent

are very fragile and difficult to handle. To overcome this limitation Uriel
and Berges (1966) introduced the concept of composite gels of agarose and
polyacrylamide for protein separations. This was later applied to nucleic
acid separations by Peacock and Dingman (1968). Such gels retain the
good resolving power of polyacrylamide and extend the useful MW range
considerably because the addition of agarose imparts mechanical strength
and rigidity and enables the acrylamide concentration to be reduced to
levels (e.g. $T = 1$ per cent) which would be totally unusable with
acrylamide alone. For all except the very largest molecules the agarose in
composite gels does not appear to contribute to size separation by
molecular sieving which remains proportional to the concentration of
acrylamide monomers.

Although quite widely employed for separations of appropriately sized
sample molecules the use of composite gels, which are more complicated to
prepare than gels composed only of polyacrylamide or of agarose, has been
questioned by Serwer (1983). He points out that relatively concentrated
agarose gels can have similar molecular sieving properties and resolution
ability to dilute (low per cent T) or highly cross-linked (high per cent C)
porous polyacrylamide gels, so there may not be a need for any
intermediate category of composite gels. Likewise composite gels have
been used as the supporting matrix in IEF experiments (Pino and Hart
1984), but here also they would appear to offer little advantage over
polyacrylamide alone for smaller molecules or agarose gels for very large
molecules.

6.1. The preparation of agarose gels

Typical agarose concentrations fall in the range 0·4–2·5 per cent and gels
are prepared as follows.

The weighed amount of agarose is dissolved in the buffer by heating in a boiling
water bath for a few minutes until the solution is clear and free of visible swollen
agarose particles. Possible charring caused by direct heating should be avoided. If
necessary, water is then added to restore the solution to its original volume. Almost
any buffer system (e.g. Table 2.1) can be employed, usually at an ionic strength of
between 0·03 and 0·1. The heated solution is then cooled to about 50 °C and
applied to the gel apparatus or gel mould which should have been prewarmed to a
similar temperature.

For ease of subsequent handling through staining and destaining steps it is usual
practice for gels to be mounted on a hydrophilic plastic support such as Gel
Bond™ (available from FMC Corporation, Marine Colloids Division, Rockland,
Maine, USA but also through regular suppliers of electrophoresis equipment and
chemicals such as Pharmacia Fine Chemicals AB, LKB Produkter AB, Bio Rad
Laboratories etc.). For this a sheet of Gel Bond is cut to the size required and
placed in the bottom of the gel mould, hydrophilic side (the side which can be

readily wetted) up. The agarose mixture is then poured in, covered with a glass or plastic plate and allowed to cool. Once gelled, the mould can be dismantled and the gel on the backing Gel Bond lifted out and transferred to the electrophoresis apparatus in the usual way.

The usual gelation temperature of agarose is about 45 °C although if required a low-temperature form gelling at less than 37 °C is now available commercially (e.g. from Bio Rad Laboratories).

Owing to the poor elasticity of agarose gels and the consequent difficulty of removing them from small glass tubes, the gel rod type of apparatus is seldom used. Horizontal slabs are nearly always used in immuno-electrophoretic methods, but for the macromolecular separations considered in this chapter either horizontal or vertical slab apparatus is suitable. Some designs of apparatus and methods of sample application are discussed in Section 2.6 and are equally applicable here, as are the comments relating to the conditions for the electrophoresis itself (Section 2.8). A modification of the popular Studier-type apparatus particularly suited to agarose gels has been described by Sugden, De Troy, Roberts, and Sambrook (1975).

(a) Protein and glycoprotein staining

Post-electrophoretic staining procedures described in Chapter 2 for PAGE can also be applied to agarose, but staining and destaining can often be accelerated by drying the gel slabs beforehand. This is achieved by placing the gels on a glass plate, covering them with filter paper, pressing them to a thin film (see p. 226) and drying in a current of warm air. Thin gels supported on Gel Bond can usually be air dried directly with no preliminary pressing step. Staining and destaining times of only a few minutes can then often be used. After destaining the gels can be similarly dried and kept as permanent records. Since dried agarose gels have very low absorbance in the u.v., direct scanning of unstained gels is practical also if desired (cf. Section 2.11(b)). Fluorescent detection of proteins using o-phthaldialdehyde (Section 2.9) works well with dried or undried agarose gels. Periodate oxidation methods for glycoprotein detection are usually unsatisfactory owing to the strong background staining of the agarose matrix, although a variation using thorough post-staining washing with acidic ethanol to remove background staining has been used successfully with thin agarose films (Trivedi et al. 1983).

Silver staining procedures for proteins and glycoproteins (Section 2.9a) cannot usually be applied to agarose gels, also due to high background staining, but a version specifically intended for agarose gels has been developed by Willoughby and Lambert (1983).

In their method the gel is first fixed by immersing for 30 min in a mixture consisting of 17·3 g sulphosalicylic acid, 25 g trichloroacetic acid and 25 g zinc sulphate in 500 ml solution, then soaked twice for 15 min each in H_2O and finally for 15 min in 35 per cent ethanol containing 10 per cent acetic acid. The gel should be gently agitated throughout all these steps. The gel is then covered with a sheet of filter paper wetted with ethanol (for IEF gels the presence of ampholytes gives rise to an opalescent background at this stage, so in this case Vesterberg and Gramstrup-Christensen (1984) found H_2O to be preferable to ethanol for this). The wet paper is covered with several sheets of dry filter paper and then a glass plate and a 0·5 kg weight is applied to press the gel to a thin film. The gel film is then air dried, rewashed for 5 min in H_2O, redried and stained. For the staining step two solutions, individually stable for 1–2 weeks, are combined immediately before use. Solution A is 50 g anhydrous sodium carbonate per litre H_2O. Solution B contains 2 g ammonium nitrate, 2 g silver nitrate, 10 g tungstosilicic acid, and 14 ml of 37 per cent formaldehyde per litre H_2O. Gels were stained in a specially constructed vertical plastic chamber of 100 ml capacity. For use 68 ml of B was added steadily, with stirring, to 32 ml A, poured into the chamber and the gel inserted. As B is added to A a white precipitate gradually forms which becomes grey and then, within 5–7 min, black. The gel is left in the mixture until adequately stained but should be removed before the precipitate has reached the black stage. If at this point staining is not sufficient, remove the gel, rinse briefly in H_2O, and place in a second bath of staining mixture. The staining is stopped by immersing the gel in 1 per cent acetic acid for 5 min, followed by rinsing in H_2O and drying as above. Glassware and plastic are cleaned in 50 per cent nitric acid. The stain is at least 10 times more sensitive than Coomassie Blue R250.

Since agarose gels are highly porous they are very suitable for use when immunological detection methods are employed (e.g. Castle and Board 1983).

This is achieved by overlaying the gel after electrophoresis with an appropriately diluted antiserum solution and incubating until immunofixation (immunoprecipitate formation) is complete. Depending upon gel thickness, antibody titre, temperature etc. this will vary, but 1–2 h at room temperature may be satisfactory. The gel is then rinsed briefly with tap water, pressed to a thin film, and soaked overnight at 4 °C in at least 1 l of 0·15 M NaCl to remove non-precipitated proteins. The gel is rinsed briefly, pressed again, dried, stained for 2 min with Coomassie Blue R250, and destained. In many cases the obvious alternative to agarose gel electrophoresis when immunological detection is desired will be polyacrylamide gel electrophoresis, followed by Southern blotting or electroblotting (Section 2.13) and immuno-chemical detection of the transferred antigens on the immobilizing membrane. This may be technically slightly more complex but has the advantages that the antigen is more accessible to the antibody than when it is embedded in a gel matrix, albeit a porous agarose one, and particularly that it exploits the greatly superior resolving power of polyacrylamide gels. However, arguing by analogy with blotting methods, the sensitivity of the above direct immunofixation of antigens in gels would be much improved by using enzyme-labelled or radio-labelled antibodies in place of unlabelled antibodies and Coomassie Blue staining.

(b) Nucleic acid staining

Hybridization techniques are nowadays widely employed for locating DNA and

RNA molecules and fragments immobilized on transfer blots (see p. 69), and Purrello and Balazs (1983) have shown that it can be applied directly to restriction fragments separated in agarose gels. Apart from simplicity, this avoided losses caused by incomplete transfer to immobilizing membranes of high MW material or losses by leaching or inefficient binding of low MW species. For hybridization the gels, with or without previous ethidium bromide staining, were placed on a porous polyethylene sheet or filter paper (Whatman No. 17) and dried under vacuum at 20° or 37 °C in a slab gel dryer (Bio Rad Laboratories), then soaked for 30 min in 0·5 M NaOH containing 1 M NaCl, followed by 30 min in neutralizing buffer (0·5 M tris buffer pH 7·4 containing 0·3 M NaCl) and two washes in phosphate buffer and redrying. The gel was prehybridized for at least 4 h at 37 °C in a mixture of equal volumes of formamide and 0·05 M Na phosphate buffer pH 7·0 containing 0·2 per cent SDS, 20 mM Na_2EDTA, 0·3 M NaCl, 0·03 M Na citrate, 0·1 per cent Ficoll, 0·1 per cent polyvinylpyrolidone, 0·1 per cent bovine serum albumin, and sheared denatured salmon sperm or *E. coli* DNA (400 μg ml), using 100 μl of this pre-hybridization solution for every cm^2 of gel. The ^{32}P-labelled DNA probe was then made up in this same solution without salmon sperm or *E. coli* DNA using 50 μl for every cm^2 of gel, denatured by heating at 80 °C for 10 min and added to the gel for hybridization at 37 °C for 48–72 h. The gel was then washed 6–8 times with 0·05 Na phosphate pH 7·0 buffer containing 0·3 M NaCl, 0·03 M Na citrate, and 0·1 per cent SDS at 65 °C and 2–4 times (also at 65 °C) in the same buffer diluted 20-fold but still with 0·1 per cent SDS, before being dried, covered with plastic film, and subjected to autoradiography.

Guillemette and Lewis (1983) have described the silver staining of DNA and RNA in 0·8 mm thick polyacrylamide–agarose composite gels of $T = 2.5$ per cent, $C = 5$ per cent polyacrylamide containing 0·5 per cent agarose, and with either denaturing (formaldehyde-containing) or non-denaturing buffer systems. After the electrophoresis step denaturing gels were first soaked three times, for 20 min each, in a 0·1 per cent solution of cetyltrimethylammonium bromide (CTAB) with gentle agitation in order to fix the nucleic acids and then three times (10 min each) in H_2O. This is to remove formaldehyde from the gel as otherwise it would cause a premature reduction of the silver. The washed gel is then soaked for 15 min in 0·2 per cent aqueous silver nitrate, rinsed rapidly with H_2O (< 15 s) and the pattern of separated nucleic acid zones developed with a 2 per cent sodium carbonate solution containing 0·02 per cent formaldehyde. With gels run under non-denaturing conditions, after the CTAB fixing and H_2O washing steps as above, the gels were soaked in 4 per cent formaldehyde for 30 min and then washed with H_2O three times (10 min each) before soaking in silver nitrate, washing and developing in sodium carbonate as above. With all gels development of zone patterns took about 15 min and was stopped by pouring off the developer solution and adding 0·25 per cent nitric acid. After a further 15 min this was poured off and the gel washed with H_2O. It should then be stored if required in a plastic freezer bag. The method should also be applicable to thicker gels but presumably, as with other silver staining methods for proteins, etc., the times for the various soakings and washings would need to be modified to enable the various reagents to diffuse in and out of the gel adequately.

6.2. The preparation of composite polyacrylamide–agarose gels

Composite gels can be formed in two different ways: either the warm buffer solution containing agarose and all of the ingredients for a polyacrylamide gel can be kept above 35 °C until the polyacrylamide has polymerized and then the agarose solidifies on subsequent cooling, or alternatively the solution is cooled to 20 °C so that the agarose solidifies first. Uriel and Berges (1966, 1974) used the former procedure and allowed the mixture to incubate at 45–50 °C for 30–40 min until the acrylamide had polymerized and then refrigerated for 20–40 min to solidify the agarose. They prepared gels containing 0·8 per cent agarose and 2·5–9 per cent acrylamide (T per cent) in this way. Although undoubtedly satisfactory, there seems little point in using composite gels containing 4 per cent or more of acrylamide monomer because at these concentrations gels containing polyacrylamide alone have perfectly satisfactory mechanical and separation properties and are simpler to prepare.

According to Peacock and Dingman (1968), above $T = 3$ per cent there was no noticeable difference between gels in which the agarose had gelled first and those in which the acrylamide had gelled first. However, below $T = 3$ per cent, when the polymerized acrylamide remained fluid or formed only a very weak gel, it was important that the agarose solidified first. They used this procedure for making composite gels containing 0·5–2.0 per cent agarose (routinely 0·5 per cent) and 1·0–3·5 per cent acrylamide.

Since the presence of agarose does not appear to influence the separation, the buffers, apparatus, electrophoretic conditions, staining procedures, and quantitative measurement methods described earlier can all be applied to composite gels in the same way as to polyacrylamide only gels (Chapter 2). Similarly, little useful purpose would be served here by detailing any precise formulation for composite gels since all the variations in buffer (both homogeneous and multiphasic), T, C, catalysts, types of cross-linking agent, etc. used in PAGE are in principle permissible in this case also. Suffice it to say that the general guidelines for preparing such gels (based on Peacock and Dingman 1968) are as follows:

Sufficient agarose to make gels with a final concentration of (for example) 0·5 per cent is dissolved in water by heating in a boiling water bath until a clear solution is obtained. The solution is cooled to about 50 °C and buffer of an appropriate composition and pH, acylamide, Bis (or other cross-linker), and TEMED or DMAPN is added. The mixture is adjusted to 45 °C and ammonium persulphate (e.g. 3 ml of a 1·5 per cent solution per 100 ml of mixture) is added and stirred in rapidly. The mixture is then immediately poured into the electrophoretic apparatus, which has previously been equilibrated at 20 °C or less, and the slot former is placed in position. The amount of persulphate used is comparatively low to ensure that the polymerization of the acrylamide is delayed until after the agarose has gelled, but if this is in doubt the amount of persulphate used should be reduced further.

6.3. Separation of nucleic acids

In free solution at neutral or alkaline pH nucleic acids of widely differing molecular weights possess very similar charge-to-mass ratios and do not separate. However, a gel matrix in which molecular sieving is a factor in the separation process enables the molecules to be distinguished in terms of size and shape.

The theory of the behaviour of RNA in PAGE has been discussed by Richards and Lecanidou (1971) and the influence of such factors as T, C, sample volume, sample load, buffer concentration, and joule heating has been examined. Part of their discussion in a modified form is presented below. It would apply equally well to DNA separations.

Both diffusion and electrophoresis are transport processes and involve movement of the sample molecules through the gel matrix, so it is probable that the frictional resistance offered by the matrix would be the same in both cases (Richards and Lecanidou 1971). Thus the diffusion coefficient D_0 of sample molecules in free solution, and the electrophoretic mobility u_0 in free solution (buffer) at the same temperature are diminished in the same ratio so that

$$D/D_0 = u/u_0 = \alpha \tag{6.1}$$

where u and D are the electrophoretic mobility and diffusion coefficient in the gel respectively. The retardation coefficient α depends upon the properties of the gel (particularly T and C) and upon the size and shape of the sample molecule.

The distance d moved in a gel by an RNA zone during electrophoresis is proportional to time t and the voltage gradient $g(\text{V cm}^{-1})$ so that

$$d = ugt = uit/k \tag{6.2}$$

where i is the current density (A cm^{-2}) and k the specific conductivity (Ω^{-1} cm^{-1}). This enables u to be calculated for a given species in a given gel, so that if the free-solution mobility u_0 is known α can be calculated from eqn (6.1). This in turn enables D to be calculated if D_0 is known or can be measured. In practice u_0 for RNA and denatured DNA is independent of molecular weight and is related to the viscosity of the medium, while D_0 varies inversely with the square root of the molecular weight if the RNA molecules can be considered to behave as random coils. These arguments led Richards and Lecanidou (1971) to give a value for $u_0 = 31 \times 10^{-5} \text{ cm}^2 \text{V}^{-1}$ s^{-1} in 0·05 M buffer at 25 °C, while D_0 varied from $1\cdot3 \times 10^{-6}$ to $1\cdot4 \times 10^{-7}$ $\text{cm}^2 \text{ s}^{-1}$ for RNA molecules ranging in molecular weight from 10^4 to 10^6.

Let us now consider the shape of an RNA zone during the electro-phoretic separation. This can be described by the variation of concentration c (equivalents ml^{-1}) with the displacement x, along the axis of motion, and is characterized by the maximum concentration c_m and the zone width

$2w$ at $c = c_m/2$. The load volume applied per unit cross-sectional area of the gel is v and is equal to the height of the column of sample solution containing the load (m g of RNA of charge equivalent weight M^0). The integrated area of the zone is therefore $m/\pi r^2 M^0$.

If the potential gradient in the sample solution is the same as that in the gel below it, then the starting zone width $2w_0$ at zero time is given by

$$2w_0 = \alpha v, \tag{6.3}$$

and if α and v are known or can be calculated, $2w_0$ can be determined. In practice the potential gradient in the sample layer is often greater than that in the gel so the width of the starting zone is then smaller than suggested by eqn (6.3).

If the starting zone is rectangular and diffusion the only zone-broadening effect as the zone moves through the gel, then according to Richards and Lecanidou (1971) the width of the starting zone has a negligible effect on the final zone width providing $2w_0 < (2DT)^{1/2}$ or $v < (2D_0t/\alpha)^{1/2}$. With typical runs lasting 1 h and when $\alpha = 0 \cdot 1$ load volumes of up to $0 \cdot 2$ cm cause no loss of resolution and even those of 1 cm only result in a 25 per cent increase in zone width, so that the volume in which the sample is applied is seldom a major factor determining the resolution. As long as $2w_0 < (2Dt)^{1/2}$ then the zone width is given by

$$2w = (16\,Dt\,\ln 2)^{1/2} \tag{6.4}$$

In practice, diffusion is not the only zone-broadening factor and the zone width increases with the mass of RNA within the zone. When diffusional effects are ignored, this is described by

$$2w = \left(\frac{2000\,gu\,Amt}{\mu\pi r^2 M^0}\right)^{1/2} \tag{6.5}$$

where A depends on the mobilities of the RNA and buffer ions and would equal zero if the mobility of the RNA in the gel was equal to that of the buffer anion (Richards and Lecanidou 1971). This concentration broadening is comparable with that caused by diffusion when

$$m = \frac{16D_0\mu\pi r^2 M^0\ln 2}{2000\,gu_0A} \tag{6.6}$$

For loads greater than this the concentration effect is the most important factor determining zone width, while for smaller loads diffusion is the major factor. With the small gel rods ($r = 0.238$ cm) and 0.05 M buffers used by Richards and Lecanidou this corresponds to only 0.5 μg of RNA but would be higher with buffers of higher ionic strength. A further consequence of these effects is that the zones become asymmetric, with the front becoming sharper and moving faster than the trailing part of the zone which becomes more diffuse. It is even possible for the leading edge to overtake and obscure the minor components moving just ahead of it.

During electrophoresis heat is generated in the gel by the passage of the electrical current at a constant rate of gi/J cal s^{-1}cm^{-3}. This gives rise to temperature gradients between the centre of the gel and the outer regions which are generally cooled by buffer or air. The excess temperature $\triangle H$ at a point at radius r over that at the periphery r_0 is given by

$$\triangle H = \frac{r_0^2 gi(1 - r^2/r_0^2)}{4kJ} \tag{6.7}$$

where k is the thermal conductivity of the gel. The mobility of RNA in the warm centre part of the gel is greater than in the cooler outer regions so the zones become curved, this difference in migration distance being expressed by

$$\triangle d_h = \frac{dgir_0^2 \text{ß}}{4kJ} \tag{6.8}$$

where ß is the temperature coefficient of viscosity of the solvent (ß = 0.022 for water at 25 °C). This heating effect therefore leads to distortions of the zones, but these will not be serious so long as $\triangle d_h$ does not exceed about $2w/5$ and eqn (6.8) can therefore be used to calculate the optimum current density. Heating effects can be minimized by reducing the radius of the gels but become more pronounced as the run progresses.

Many of the conclusions arrived at by Richards and Lecanidou (1971) in relation to the separation of samples of RNA apply equally to the electrophoresis of proteins in polyacrylamide gels. Likewise the influence of gel concentration (T and C) on the migration of zones of RNA or DNA is governed by the same considerations already discussed in relation to protein separations (e.g. Sections 4.1 and 4.2). Thus if it is assumed that the fraction of molecules migrating in the electric field at any given time is proportional to the fraction of the total gel volume available to them and if this fraction is equal to the retardation coefficient α, then

$$\{\ln(1/\alpha)^{1/2}\} = k(R + r) T^{1/2} \tag{6.9}$$

where r and k depend on C and R is the effective radius of the RNA molecule. For proteins R is proportional to the Stokes radius R_m and is given by

$$R_m = \frac{M(1 - \bar{v} p)}{6\pi N s \eta} \tag{6.10}$$

where M is the molecular weight, $\bar{v}$ the partial specific volume, p the density, N Avogadro's number, s the sedimentation coefficient and η the viscosity of the medium. Based on experimental data in the literature Richards and Lecanidou (1971) calculated that for RNA

$$R_m = 0.0117 M^{0.54}. \tag{6.11}$$

Examination of the function $\{\ln(1/\alpha)\}^{1/2}$ showed that as long as α lay between 0.3 and 0.95

$$\{\ln(1/\alpha)\}^{1/2} = 1.3 \, (1.14 - \alpha). \tag{6.12}$$

Thus if the Stokes radius of RNA is equal to the effective radius in eqn (6.9), which may not be the case for a flexible molecule of extended structure such as RNA, the mobility should vary approximately linearly with $M^{0.54}$. Two conclusions follow from eqns (6.9)–(6.12). Firstly there should be a linear relationship between the sedimentation coefficient and mobility, and secondly mobility should vary linearly with $T^{1/2}$ and the resolution should be better the higher the value of T that can be used. These conclusions have been verified experimentally by a number of research groups. Other workers have reported an empirical linear relationship between mobility and log M, which implies that the effective radius is a linear function of log M. While this appears to hold in most cases to give a practical method of measuring RNA molecular weights with fair accuracy, it ignores any effect of the conformation of the RNA on mobility. In fact nucleic acid conformers are often separable by electrophoresis so any measurements based on plots of mobility *versus* log M should be treated with caution.

Richards and Lecanidou (1971) found the mobilities of denatured and renatured 5S-RNA to be markedly influenced by the degree C of cross-linking of the gel. Therefore when studying the effects of variations of T on the mobility of nucleic acids it is important to maintain a constant C as in the case of work with proteins (Section 4.1).

Richards and Lecanidou (1971) also indicated that, apart from conformational effects, molecules differing by a single nucleotide should be resolvable in polynucleotides with molecular weights up to about 10 000. They thus anticipated the recent exploitation of PAGE in nucleic acid sequence analysis (see Section 6.6) which has been perhaps the most spectacular application of electrophoresis in the last few years.

The choice of gel matrix depends, as in the case of proteins, upon the molecular weight range of the nucleic acids to be separated. Nucleic acids can vary in molecular weight from 350 for a single base or one or two thousand for small restriction enzyme fragments to over 10^8. Agarose or composite polyacrylamide–agarose gels of low T are used for the larger species and PAGE with T up to 20 per cent or more for the smallest molecules and oligonucleotides. Since nucleic acids have relatively high absorbances at 260 nm their quantitative estimation by the direct densitometry of unstained gels is more practical than it is with proteins, and according to Loening (1967) as little as 0.05 μg can be detected in this way. For this it is desirable to prepare polyacrylamide gels with purified monomers and to perform a pre-electrophoresis step and/or thoroughly wash the gels in buffer before applying the sample. Likewise the purity of the nucleic acid preparations has been shown (Loening 1967) to be very important, and impure RNA for example may aggregate and stick to the top of the gel or trail during the electrophoresis.

Either multiphasic or homogeneous buffer systems can be used, the former being used particularly with dilute nucleic acid solutions, but homogeneous systems are more generally preferred. In contrast to proteins, nucleic acid is retarded and concentrated when it enters the top of the gel (Maurer 1971), so separate stacking gels and complex buffer systems are not usually required even for very high resolution work. Nucleic acid staining procedures and quantitative measurement methods have already been described in Chapter 2.

(a) RNA

Polyacrylamide gels are most usually employed for separating low molecular weight RNA (e.g. S- and tRNA and fragments of larger RNA molecules), either as an analytical technique for identifying particular species or for monitoring the progress of a preparative scheme by identifying the presence of impurities. Richards et al. pioneered such work and as continuous buffer systems they recommended a Tris–HCl buffer of pH 7·4 (1·43 g tris and 10 ml 1 N HCl per litre) or a pH 5·0 acetate buffer (4·0 g NaOH and 8·35 ml glacial acetic acid per litre) with gels with $T = 10$ per cent for the separation of S-RNA. Gels with $T = 10$ per cent can be used for 4S- and 5S-RNA and gels with $T = 15$ per cent are appropriate for smaller RNA and fragments down to small oligonucleotides. At the other end of the scale 28S-RNA migrates in gels with $T = 2·5$ per cent but not in gels with $T = 3 \cdot 0$ per cent, so for RNA larger than this composite or agarose gels must be used. Resolution in gels with $T = 2 \cdot 5$ per cent is such that two RNA differing by less than 1 S should be easily separable. Richards and Lecanidou (1971) have shown that there is a linear relationship between the logarithm of the mobility and T (as long as C is constant), as a consequence of which the resolving power between two RNA species increases as the gel concentration T increases. By exploiting the high resolving power of gels with $T = 16$ per cent ($C = 5$ per cent) for relatively small molecules Richards and Lecanidou (1974b) were able to resolve denatured and renatured forms of 5S-RNA. Phillips (1974) also clearly demonstrated that conformational changes in several small RNA species, including 5S-RNA, could be visualized by PAGE, and Peacock, Bunting, and Nishinaga (1977) have shown that even with the larger 18S-RNA different conformation states can be resolved.

Because of the large range of sizes of RNA molecules and the consequent desirability of using gels of differing T, there have been several reports of the use of polyacrylamide concentration gradient gels for RNA separation with excellent results. Gels with $T = 2$–10 per cent were used for separation of multiple forms of 23S ribosomal RNA (Schaup, Best, and Goodman 1969), gels with $T = 2 \cdot 5$–12 per cent for total RNA fractionation (Caton and Goldstein 1971), and gels with $T = 2$–20 per cent for phage T7 early RNA (Studier 1973). The preparation of gradient gels is described in Chapter 4.

In order to combat aggregation phenomena and adsorption of RNA on to the gel matrix and also to simplify the interpretation of electrophoretic patterns by removing conformational differences, the electrophoresis can be performed in the presence of denaturing agents. Various denaturing agents have been incorporated into gels, including 6–8 M urea (e.g. Dudov, Dabeva, and Hadjiolov 1976; Lehrach et al. 1977), 99 per cent formamide (e.g. Lehrach et al. 1977), 5–10 mM methyl mercuric hydroxide (Lehrach et al. 1977; Amalric, Merkel, Gelfand, and Attardi 1978), $2 \cdot 2$ M formaldehyde (Schuerch, Mitchell, and Joklik 1975; Lehrach et al. 1977; Schwinghamer and Shepherd 1980) or detergents such as $0 \cdot 1$ per cent SDS (Studier 1973; Schuerch et al. 1975). Formamide cannot be used with agarose or composite polyacrylamide–agarose gels because it dissolves the gel. If the very highly toxic methyl mercury compounds are used stringent precautions must be taken during the handling of solutions, gels, and wastes. General guidelines for safe handling procedures are given by Lehrach et al. (1977). Partly because of such difficulties, Schwinghamer and Shepherd (1980) concluded that $2 \cdot 2$ M formaldehyde was the most satisfactory denaturant for use in their work on RNA with MW = $(0 \cdot 1$–1.4$)$ $\times 10^6$ examined in composite polyacrylamide–agarose gels.

(b) *DNA*

The above comments on separation of RNA also apply in the main to the separation of DNA. Hence polyacrylamide gels, including polyacrylamide concentration gradient gels (Jeppesen 1974, 1980), are used for small DNA fragments and agarose (e.g. Serwer 1981) or composite polyacrylamide–agarose gels for larger fragments. Intact DNA molecules are usually examined on agarose gels (Fig. 6.1).

Although staining procedures for DNA have been discussed already (Section 2.9f), one interesting variation that is worth mentioning here was that described by Sharp *et al.* (1973) for DNA and DNA fragments. This is to include 0·5 μg ml^{-1} of ethidium bromide in the tris–EDTA–acetate electrophoresis buffer and in the 1·4 per cent agarose gels so that the electrophoresis is performed in the presence of ethidium bromide. This

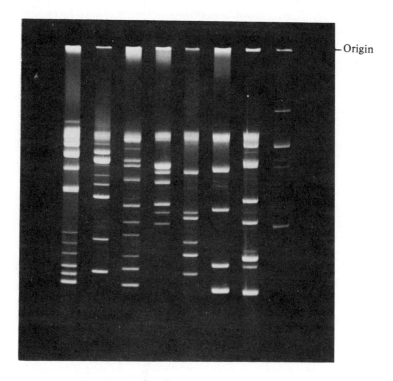

Origin

FIG 6.1. Electrophoresis of plasmid DNA from various strains of lactic *Streptococci* separated at 100 V for 2 h on a 140 mm x 140 mm x 3 mm slab of 0·7 per cent agarose gel run in a buffer of 0·04 M tris, 0·02 M sodium acetate, and 0·001 M EDTA adjusted to pH 8·2 with glacial acetic acid. After the separation the gel was ethidium bromide stained. (Photograph supplied by Dr F. L. Davies).

has the dual benefit that the progress of the separation can be observed by u.v. illumination and that no separate staining or destaining step is required after the separation. Doubtless this technique could be applied equally well to the electrophoresis of RNA and RNA fragments.

The electrophoretic behaviour of DNA, like that of RNA, is influenced not only by size and charge but also by conformation (e.g. Serwer and Allen 1984). This has been demonstrated by Johnson and Grossman (1977) who separated closed circular, nicked circular, and linear duplex forms of bacteriophage and mitochondrial DNA and showed that the logarithm of mobility is a linear function of the agarose concentration (0·6–1·4 per cent). The retardation coefficient K_R is characteristic of DNA conformation under controlled conditions. If the electrophoresis was performed in the presence of the intercalating dye ethidium bromide, the superhelix density of closed circular DNA could be estimated. By alterations in the voltage gradient and ionic strength and also by the addition of varying amounts of ethidium bromide, which causes variations in the relative mobilities of the various conformational forms, the experimental conditions can be modified to optimize their separation.

The use of agarose gel electrophoresis in studies of DNA structure and function has been applied to the replicative form of bacteriophage G4 DNA by Wheeler, Fischel, and Warner (1977). Depending upon the specific conditions of electrophoresis they were able to distinguish as single bands the following DNA species; monomeric and dimeric supercoiled circles, monomeric and dimeric relaxed circles, monomeric Form IV DNA (irreversibly denatured closed circles), monomeric Form III DNA (linear duplex), super-coiled catenated dimers, relaxed catenated dimers, half-relaxed catenated dimers, recombinant dimer figure eights, and linear *Escherichia coli* host DNA fragments. They also used agarose gels for the separation of the complementary strands of the linear duplex made by Eco RI endonuclease digestion of the bacteriophage DNA. Wheeler *et al.* and Johnson and Grossman (1977) used agarose gels varying between 0·5 and 1·5 per cent and buffers containing 40 mM tris + 5 mM sodium acetate containing either 2 mM EDTA giving a final pH of 7·8 (Johnson and Grossman) or 1 mM EDTA giving a pH of 8·2 (Wheeler *et al.*). For very large duplex DNA molecules with MW up to $1·1 \times 10^8$ and for bacteriophage particles Serwer (1981) used extremely dilute agarose gels with concentrations down to as little as 0·035 per cent in 0·05 M phosphate buffers of pH 7·4 containing 1 mM EDTA or 1 mM $MgCl_2$.

In PAGE, gels with concentration gradients (see Chapter 4) can be particularly useful when dealing with proteins of widely differing size. It is difficult to arrange for concentration gradients with agarose gels but Boncinelli *et al.* (1983) achieved the same result by preparing gels with varying thickness. The mobility of a molecule is determined by the field strength as well as the agarose concentration. The field strength is a function of the local resistance which in turn is an inverse function of the cross-section of the gel, so with a gel of constant agarose concentration if the gel mould is slanted slightly the resulting gel will be thicker at one end than at the other. This will result in a gel slab with a field strength gradient once a voltage is applied to it. Applying samples to the thin end, the smaller more rapidly moving molecules will reach areas of lower field strength and be slowed in comparison to larger molecules. Using a gel of 0·6 or 0·8 per cent agarose, 22 cm long, and varying from 5 to 9 mm in thickness, Boncinelli *et al.* (1983) were able to separate DNA fragments of very widely differing size on a single slab.

6.4. Measurement of sedimentation coefficients and molecular weights

Earlier work carried out on RNA before the influence of conformation on nucleic acid mobilities was appreciated had indicated that the electrophoretic mobility was inversely and linearly related to the sedimentation coefficient. Richards *et al.* (1965) and McPhie, Hounsell, and Gratzer (1966) reported the approximate relationships

$$S_{20,w} = 7\cdot6\text{--}6\cdot7\ R_m \qquad\qquad (T = 10\ \text{per cent},\ C = 5\ \text{per cent gels})$$
$$S_{20,w} = 13\cdot6\text{--}11\cdot3\ R_m \qquad\qquad (T = 5\ \text{per cent},\ C = 5\ \text{per cent gels})$$

where R_m is the relative mobility (relative to the anion front as usually marked by the tracking dye). The relationships are strongly dependent on both T and C, and for gels with $T = 2\cdot6$ per cent and $C = 5$ per cent Lewicki and Sinskey (1970) found that R_m was a linear function of log $S_{20,w}$. It therefore follows that R_m should be inversely related to log MW.

Plots of R_m *versus* log MW have been exploited by many groups of workers for nucleic acid measurements using polyacrylamide, composite polyacrylamide–agarose, and agarose gels (e.g. Peacock and Dingman 1968; Loening 1969). This is usually done by measuring the ratio of the migration distances (R_m values) of a series of known RNA species to that of a suitable marker substance (often the tracking dye although any other suitable standard band could be used) and plotting this ratio against log MW to produce a calibration plot for the particular gel system being used. With this the MW of unknowns can then be read off from their measured R_m values. Such measurements are most accurately made using a digital microdensitometer with computerized handling of the data (Elder *et al.* 1983; Elder and Southern 1983). As discussed in Chapter 5 for SDS–PAGE, the suitability of simple R_m *versus* log MW plots for MW estimation relies on all species, both standards and unknowns, having a constant charge-to-mass ratio and identical hydrodynamic properties. As is now known these conditions are not truly fulfilled for nucleic acids, and the above relationships for estimation of sedimentation coefficients and MW values have only been given because they are useful empirically provided that suitable qualifications are kept in mind.

For accurate MW estimations of both RNA and DNA differences arising from the state of aggregation, the extent of denaturation, and other conformation factors must be eliminated. This is done by performing the electrophoresis in the presence of denaturing agents (see Section 6.3(a)).

Dudov et al. (1976) reported that in 1·5 per cent agar gels containing 5–8 M urea a linear relationship exists between mobility (absolute or relative) and log MW over the MW range of at least 5×10^5–5×10^6 for cytoplasmic and nuclear RNA. Thus such treatment presumably cancels out conformational differences so that mobility is then related only to size. Amalric et al. (1978) used 0·8 per cent agarose gels run under more strongly denaturing conditions by including 5 mM methyl mercuric hydroxide and likewise obtained a linear relationship between mobility and log MW for a number of standard RNA covering the MW range 9×10^4–3.4×10^6.

A thorough evaluation of the electrophoretic measurement of RNA MW values has been given by Lehrach et al. (1977) who compared results under four different denaturing conditions, and Carmichael (1980) has also given a useful brief review. This was necessary because it is now known that some denaturing agents (e.g. 99 per cent formamide, 2·2 M formaldehyde) do not cause complete denaturation, particularly of GC-rich regions. When polyacrylamide gels with $T = 3·2$ per cent or less ($C = 17·5$ per cent) in 99 per cent formamide were used a linear relationship held for mobility *versus* log MW plots for MW values up to $2·6 \times 10^6$. Large ribosomal RNA with GC-rich helical regions are only correctly estimated if runs are performed at 58 °C (the GC-rich regions melt at 50 °C so this clearly demonstrates that the anomalous mobilities at lower temperatures are due to incomplete denaturation). Agarose (0·75 per cent) gels containing 10 mM methyl mercuric hydroxide gave linear plots for MW up to at least 4×10^6, but there is the disadvantage that the denaturant is highly toxic. Agarose (0·75 per cent) gels containing 2·2 M formaldehyde are also suitable for MW estimations up to at least 4×10^6, particularly if the RNA sample is pretreated under even stronger denaturing conditions (e.g. 2·2 M formaldehyde in 50 per cent formamide at 60 °C for 5 min) before analysis to ensure complete denaturation, particularly of GC-rich helical regions. These electrophoretic conditions can be extended to MW measurements of molecules which are too large to run adequately on polyacrylamide gels in 99 per cent formamide. The fourth set of conditions examined by Lehrach et al. (1977) consisted of agarose gels (0·5–3·5 per cent) containing 6 M urea run at pH 3·8 in 0·025 M citric acid, but these were less satisfactory. This was probably because under such conditions there is a specific protonation of adenine and cytosine which changes the net charges so that RNA molecules with different base compositions then have different charge-to-mass ratios. There also appeared to be other structural differences, but Lehrach *et al.* (1977) point out that, although they cannot be used for MW measurement, such gels have advantages for preparative work since they are easy to prepare and handle, are non-toxic, and biologically active RNA can easily be recovered, which is not easy with formaldehyde-containing gels.

These considerations all apply to high molecular weight RNA but should also be applicable to estimates of the MW of DNA fragments. MW values of small RNA, in which conformational differences are likely to be less important, can obviously also be estimated under denaturing conditions but simple non-denaturing conditions are more likely to be adequate than for high MW species.

For DNA molecules which are long, thin, and rigid (but not for RNA or protein molecules) Southern (1979) has shown that plots of MW *versus* the reciprocal of mobility give a more linear relationship than the usual semilogarithmic plot. This is especially true for high MW DNA and when high voltages are used, as in both these situations, the semilogarithmic plot becomes curved. Southern (1979) also gives a method for direct calculation of MW from mobilities which he claims is more accurate than graphical methods. Schaffer and Sederoff (1981) have further developed this relationship using a least squares analysis by computer and have been able to achieve values of MW (DNA fragment length) within 1 per cent when analysing samples of known base sequence.

Using agarose gels varying in concentration from 0·3 to 1·5 per cent and restriction enzyme DNA fragments, ranging in size from 47 to 6000 base pairs, Stellwagen (1983) obtained linear Ferguson plots of gel concentration versus R_f, the relative mobility, as long as this was extrapolated to zero field strength. Unknown molecular weights of all linear DNA fragments could be estimated from plots of R_f *versus* $MW^{0.8}$, provided that for large fragments over 1500 base pairs R_f was extrapolated linearly to zero field strength. Estimates for alkali-denatured single-stranded DNA fragments could be obtained from plots of R_f *versus* $MW^{1/2}$.

6.5. Restriction endonuclease maps

Restriction enzymes are endodeoxyribonucleases that recognize specific nucleotide sequences in double-stranded DNA and cleave both polynucleotide chains at these points. Nathans and Smith (1975) have pointed out that restriction enzymes fall into two broad classes. Class I enzymes, although apparently recognizing specific nucleotide sequences in DNA, cause non-specific cleavage, while Class II enzymes are cleavage-site specific and produce characteristic specific restriction fragments. The Class II enzymes are now proving as useful to the study of DNA structure and function as specific proteases have been in protein investigations.

The first step in the analysis of DNA is to produce a limit digest (a digest in which no futher fragmentation occurs on longer incubation or addition of more restriction enzyme). The resulting restriction fragments are then almost always analysed by gel electrophoresis. The fragments may vary in size from a few nucleotide pairs up to MW values of 10^6 or more. For fragments with MW $< 10^6$ polyacrylamide gels are suitable, usually with T values between 4 and 15 per cent depending upon the MW range being studied, or alternatively a gradient gel may be most useful (e.g. Jeppesen 1974; 1980). Composite polyacrylamide–agarose gels or 1–2 per cent agarose gels can be used for fragments for MW up to 3×10^6, and 0·3–0·7 per cent agarose gels are most suitable for even larger species within the MW range (3–25) x 10^6. Visualization and quantitative measurement methods for DNA fragments are the same as those described earlier (Chapter 2) for the intact nucleic acids.

Since the migration of polynucleotides in electrophoretic gels is a function of their size, a map of restriction endonuclease fragments should be calibrated so that the size of the fragments can be determined. This information is vital in studies of the structure and function of the native molecules. Calibration is performed simply by including a set of standards in at least one sample slot on the gel slabs (the use of slabs is virtually obligatory in all mapping work). For MW standards DNA or DNA fragments of known size are used, the sizes of which have mostly been determined by electron microscopy (Sharp et al. 1973) or by sequencing (Sanger et al. 1974). In practice a set of standards are usually produced by the digestion of a well-characterized DNA with one particular restriction endonuclease which gives a series of discrete fragments of known size. Mixtures of such fragments of known size are now available commercially (e.g. Pierce Chemical Co., etc.). An example of this approach is given by Prunell et al. (1977) who in their work on mitochondrial DNA used EcoRI or HpaII digests of SV40 or phage λ DNA as primary standards and generated further standard mixtures (secondary standards) which could be related to the primary standards when required. An alternative approach (Parker, Watson, and Vinograd 1977) to the preparation of standard mixtures is to produce a restriction enzyme digest limited by the presence of ethidium bromide which inhibits the digestion. When closed circular DNA is used as a substrate, DNA with one-site cleavages of one or both strands can be made by adding appropriate amounts of the dye. These permuted linear molecules are then further digested in the absence of the dye to give a series of electrophoretic bands, analysis of pairs of which can be used to determine relative MW values.

6.6. Sequencing of nucleic acids

In the last few years there have been dramatic and rapid improvements in techniques for sequencing the base composition of both RNA and DNA. Thus in the case of DNA earlier methods were improved by procedures which relied on two-dimensional maps of DNA fragments with high-voltage electrophoresis in one dimension and chromatography in the other (Ling 1972; Sanger, Donelson, Coulson, Kössel, and Fischer 1973). These were soon superceded by much more elegant and rapid procedures based on polyacrylamide gel electrophoresis of modified restriction endonuclease fragments, the so-called 'plus–minus' method (Sanger and Coulson 1975). This was replaced by the dideoxy-termination method (Sanger, Nicklen, and Coulson 1977) and the chemical cleavage method of Maxam and Gilbert (1977) both of which have been widely used and have enabled many aspects of DNA and RNA chemistry and gene structure and translation to leap forward in a most dramatic fashion. This whole area is currently evolving rapidly and any reader contemplating beginning this type of work would be advised to consult the current literature. Journals such as *Nature, Journal of Molecular Biology, Nucleic Acids Research, Proceedings of the National Academy of Sciences*, etc., publish papers dealing with nucleic acid sequences in almost every issue and many excellent review articles are also to be found (e.g. Smith 1980; Maxam and Gilbert 1980; Davies 1982; D'Alessio 1982; Deininger 1983; Hindley 1983; Meinkoth and Wahl 1984).

(a) *Sequencing of DNA*

The first step in sequencing a purified DNA molecule is to break it specifically into a number of well defined pieces with a restriction endonuclease and then to choose and prepare the piece to be sequenced. The second stage is to prepare families of related fragments from this piece by methods such as those indicated above. These fragments are then fractionated according to size differences by polyacrylamide gel electrophoresis and the resulting maps interpreted manually or by computer-linked gel scanner and data handling. The electrophoretic separation and interpretation of maps are very similar in all methods and the sequencing procedures referred to above differ principally in the way in which the DNA fragments are generated. The original references should be consulted for experimental details of the sophisticated methodology involved in this but in principal Maxam and Gilbert (1977) labelled their restriction enzyme fragments by transfer of ^{32}P from γ-labelled ATP with polynucleotide kinase after a preliminary removal of the terminal phosphate groupings of the fragments with alkaline phosphatase. The labelled mixture of fragments must then be purified and subjected to strand separation because in this method it is necessary to have available a series of purified DNA fragments labelled only on a single strand. These objectives can be achieved conveniently by electrophoretic methods. For example, James and Bradshaw (1984) first separated the radiolabelled DNA fragments from low molecular weight components of the radio-labelling reaction mixture by gel filtration on a column of Sephadex G-50, using 0·1 per cent SDS containing 2mM EDTA as the elution solvent. The DNA fragment samples were then denatured by adding solid urea to 10 M (this was based on the volume of solution collected from the column, so the final urea concentration was probably 7–8 M because no correction was made for the increase in volume on adding the urea). Samples were then heated at 65 °C for 3 min, cooled rapidly and applied to a PAGE slab gel of $T = 15$ per cent, $C = 1·67$ per cent, using a homogeneous buffer system in both gel and apparatus of 90 mM tris and 90 mM boric acid with 2·5 mM Na_2EDTA. After the electrophoretic separation gels were subjected to autoradiography and purified zones of labelled DNA fragments cut out and the DNA electroeluted for sequence analysis. In the Maxam and Gilbert (1977) procedure the fragments of DNA material were then further fragmented chemically by procedures which broke the sequence at a particular nucleotide. Partial cleavage at each base was used to give a set of fragments extending from the ^{32}P label to each of the positions of that base.

Sanger, Nicklen, and Coulson (1977) made use of chain-terminating inhibitors to form similar sets of fragments. By using a 2′,3′-dideoxyribo-nucleoside triphosphate, which unlike the usual 2′-deoxy-compound contains no 3′-hydroxyl group, the chain cannot extend further. Thus if the primer fragment and template are incubated with DNA polymerase in the presence of a mixture of the 2′,3′-dideoxy- and 2′-deoxy- derivatives as well as the other three 2′-deoxyribonucleoside triphosphates (one being labelled with ^{32}P) a set of fragments is produced, and the positions of the terminator can be deduced from the electrophoretic pattern. The complete sequence can be determined by using the 2′,3′-dideoxy derivative of each base in turn.

The original procedure required the preparation of single-stranded material from double-stranded DNA to act as a template and once this was done the most important factors were primer purity and the integrity of the 5′-end of the primer (Deininger 1983). As pointed out by Deininger the method was revolutionized by the advent of M13 single-stranded bacterio-phage cloning vectors (Sanger et al. 1980). With this the random or semi-random collection of fragments of the DNA to be sequenced is cloned into M13 bacteriophage adjacent to a DNA sequence which is separately available for use as a small primer. Suitable synthetic DNA primers are now available commercially for this 'universal primer' system.

The chemical procedure of Maxam and Gilbert (1977) has also been extensively modified and developed since its inception and forms the basis of the technique known as genomic sequencing of Church and Gilbert (1984). For this, mixtures of unlabelled DNA fragments from complete restriction enzyme and partial chemical cleavages of the entire genome are separated according to size differences on a $T = 6$ per cent denaturing polyacrylamide gel. The lanes of separated DNA fragments are electrolytically transferred to nylon membranes to which they are then cross-linked and immobilized by u.v. irradiation. Hybridization with a short ^{32}P-labelled single-stranded probe then produces the image of a DNA sequence 'ladder' extending from the 3′ or 5′ end of one restriction site on the genome. By repeated probing a single nylon membrane can be used to give several different sequences.

The sequencing gels used in all the earlier methods for the final characterization of the sets of fragments were very similar (e.g. Fig. 6.2). The gel system of Air, Sanger, and Coulson (1976) is typical and consists of slabs of polyacrylamide gel with $T = 12.5$ per cent and $C = 4$ per cent 200 mm x 400 mm in cross-section and 1·5 mm thick made up in a pH 8·3 buffer containing 10·8 tris, 5·5 g boric acid, 0·93 g EDTA, and 480 g urea per litre. Two tracking dyes, Xylene Cyanol FF and Bromophenol Blue, are generally used. In a later modification (Sanger and Coulson 1978) the superior resolution obtainable with the same gel system but using very thin gels (0·4 mm thick) has been clearly shown. The improvement stems partly from the fact that all the labelled material is physically closer to the emulsion of the autoradiographic film and partly from the fact that high currents can be used during the electrophoresis (25 mA initially at 1500 V and falling to 1300 V). Power supplies capable of high voltage outputs have been developed especially for sequencing work (e.g. the Macrodrive 5 from LKB, capable of 5000 V, 150 mA, and 200 W). This causes considerable heating and helps to keep the DNA material fully denatured. It can also lead to band distortion and cracking of the glass plates, but both of these problems are less severe when very thin gels are used. Sample slots were only 5 mm wide and this enables a sequence of at least 465 nucleotides to be read off from a single gel slab.

It is this use of thin gels that has been largely responsible for improvements in the resolution obtained since the earlier work, and gels as thin as 0·1 mm have been advocated (Garoff and Ansorge 1981) but sequencing gels are usually large (at least 40 cm long and often up to 1 m) and handling such large thin gels is not easy. Deininger (1983) considered gels of 0·35–0·4 mm most suitable for routine use. The concentration of the gel used (per cent T) has also been varied widely. The optimum of course depends upon the size range of fragments to be separated but 6–8 per cent gels are now usually used for relatively large fragments and 'ladders'

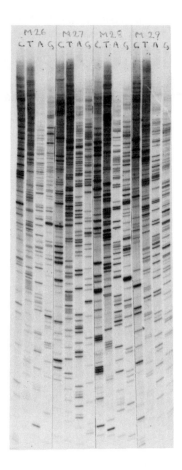

FIG 6.2. Autoradiographic pattern of a polyacrylamide sequencing gel showing four partial sequences obtained by the dideoxy method on clones from lambda DNA. (C. Howe, F. Sanger and A. R. Coulson, unpublished; photograph supplied by Dr F. Sanger.)

produced from them (e.g. Church and Gilbert 1984), while gels of up to T = 20 per cent are used with relatively small fragments (e.g. Becker and Wang 1984). Denaturing conditions however are the rule and although reagents such as formamide are quite often used, urea addition at a level of 7 or 8 M to the gel solutions is much the most usual. As referred to above gels are usually also run warm, without cooling, the optimum being 50–60 °C for most work and although temperatures as high as 80–90 °C have been used, according to Deininger (1983) this can give rise to 'fuzzy' bands. Such high temperatures of course requires that gels are run between glass, preferably Pyrex, plates and not in apparatus with plastic separation chambers. If resolution on such gels is still not adequate, gels can be dried before autoradiography which in effect reduces the average distance of radioactive molecules from the film or alternatively ^{35}S-labelled nucleotides which have shorter ß-particle pathlengths than the usual ^{32}P-label can be used. Both these modifications improve resolution during autoradiography.

The ^{35}S-label can be incorporated into DNA by using deoxyadenosine 5'-(α-[^{35}S]thio)triphosphate in place of the more usual [α-^{32}P]dATP in dideoxy sequencing reactions (Biggin, Gibson, and Hong 1983). This is an analogue of dATP in which a sulphur atom replaces a non-bridge oxygen atom at the α-phosphate grouping. The same workers also introduced the use of buffer concentration gradients into the $T = 6$ per cent, $C = 5$ per cent polyacrylamide gels used for separating 'ladders' of DNA fragments. This was achieved by making up two portions of gel solutions in 5 X, and 0·5 X concentrated tris–borate–EDTA (TBE) buffer (a 10 X concentrated stock buffer solution contained 108 g tris, 55 g boric acid, and 9·3 g Na$_2$EDTA in 1 l). A small volume (e.g.. 5–6 ml) of gel mixture in 0·5 X TBE buffer is drawn into a 25 ml glass pipette using a rubber bulb or Pro-pipette (remember, never mouth pipette toxic acrylamide solutions!) followed by an equal volume of mixture in 5 X TBE buffer. The two phases are mixed slightly by sucking up one or two small air bubbles into the pipette which then contains a very crude (but adequate) gradient in buffer concentration. The pipette is gently emptied into the gel mould which is finally filled with more gel mixture in 0·5 X TBE buffer. The apparatus buffer is 1 X TBE. The resulting gel thus has constant acrylamide concentration and mostly a constant buffer composition, but the bottom part of the gel slab has a steeply increasing buffer concentration, and hence lower electrical resistance giving a reduced field strength in this region so that the fastest moving DNA fragments (i.e. the smallest) are slowed. This means that a greater length of sequencing data can be read off a single slab than is possible when a gel with constant buffer concentration is used.

Sequence data and comparison or alignment with other nucleic acid or protein amino acid sequences are nowadays usually handled by computers (e.g. Queen and Korn 1980; Wilbur and Lipman 1983).

According to Deininger (1983) the most common artefacts observed on sequencing gels are band compressions which are caused when self-complementary sequences in the DNA fold back on themselves forming looped structures which migrate at different rates to fully denatured DNA. These structures generally involve G + C-rich sequences and can usually be detected by the presence of a number of bands with smaller spacings between them (or even superimposed) than the bands preceding them. After the compressed region band spacings are larger than average. The best way to overcome this problem, both for interpreting sequences within the compressed region and to aid detection of small compressions which could be overlooked, is to sequence both strands of the DNA. The compressions occur at slighly different positions in the two strands enabling the correct sequence to be identified. Alternatively compressions may be reduced or eliminated by increasing the denaturing power of the gels, usually by running them at higher temperature (e.g. 80–90 °C as opposed to the usual 50–60 °C) or by incorporating 25–40 per cent formamide as well as urea in the gel mixtures.

(b) *Sequencing of RNA*

As with DNA, the last few years have also seen a dramatic improvement in sequencing techniques for RNA. Earlier methods (e.g. Proudfoot, 1976; Cheng *et al.* 1976) have proved highly successful and the complete 3569 nucleotide long sequence of bacteriophage MS2 RNA was reported by Fiers *et al.* (1976). Nevertheless such methods have now been superceded by higher resolution techniques. Brownlee and Cartwright (1977) largely followed the DNA 'plus and minus' method of Sanger and Coulson (1975). In the first step the mRNA was copied in the presence of a specific primer and reverse transcriptase to give ^{32}P-labelled complementary DNA. After reannealing with excess mRNA, minus incubations (lacking one of the four deoxynucleoside triphosphates) were performed with reverse transcriptase. For plus reactions the Klenow fragment of DNA polymerase I was used in the presence of Mn^{2+} and a single deoxynucleoside triphosphate. The incubation mixtures were then separated on slabs of polyacrylamide gel with $T = 12 \cdot 5$ per cent and $C = 4$ per cent (200 mm x 400 mm and 1·5 mm thick) in tris–borate–EDTA buffer containing 8 M urea, autoradiographs were prepared, and the sequences were read off in a manner similar to that used for DNA sequencing.

Later improvements developed for sequencing DNA (e.g. thin gels, high voltages, etc.) have now been applied to RNA sequencing with similarly beneficial results. For example, they have been used in evaluating the complete nucleotide sequence of *Escherichia coli* ribosomal 16S RNA by Carbon, Ehresmann, Ehresmann, and Ebel (1979). Their paper gives full experimental details and also discusses the respective advantages and disadvantages of RNA and DNA sequencing methods.

Many aspects of RNA sequencing have been reviewed by D'Alessio (1982). Since an RNA coding for even a small protein with MW of 24 000 will contain at least 600 nucleotides and because it is difficult to sequence more than 150–200 from a labelled terminus, some preliminary specific cleavage and fractionation of the resulting fragments is required. D'Alessio (1982) recommends the use of ribonuclease H to achieve this by enzymic degradation of the RNA portion of an RNA–DNA hybrid. The enzyme is directed to specific sites by hybridization of DNA oligomers to the RNA, the sequence of the oligomer determining the frequency and distribution of cleavage sites. The enzyme attacks hybrids of four or more base pairs to give fragments having 5' terminal phosphate and 3' terminal hydroxyl groups which can be radiolabelled in the usual way and are of a size suitable for sequencing. D'Alessio (1982) then sequenced the RNA fragments by either chemical or enzymic degradation methods giving specific but limited cleavage of the nucleotide chain. The enzymic method relies on specific endoribonucleases which split the chain at a particular base, so that partial hydrolysis by treatment of samples of the RNA with endoribonucleases specific for each of the four types of base gives four digests which if applied to a sequencing gel give 'ladders' of fragments from which the RNA sequence can simply be read off. Chemical cleavage at each of the four types of base in turn achieves the same objective. Chemical cleavage is more easy to interpret and less subject to interference than the enzymic method provided reagents of suitable purity are used, but some of the reagents are noxious and the enzymic method is more rapid. With both types of digest the sequencing gels are identical and very similar to those used for DNA sequencing, so that gels are large (i.e. 40 cm long or more), contain 8 M urea and 2 mM EDTA, are typically made up in 0·1 M tris–borate pH 8·3 buffer and are run hot (55·60 °C) at 1500–20 000 V. D'Alessio (1982) recommends gels with $C = 5$ per cent and $T = 20$ per cent for sequences of up to 30 bases, $T = 10$ per cent for 10–150 bases and $T = 6$ per cent for up to 200 bases or more. Separation times are of the order of 5–7 h until the Bromophenol Blue or Xylene Cyanol tracking dyes have reached the bottom of the gel.

A rather different approach to RNA sequencing (Stanley and Vassilenko 1978) has been developed (Tanaka, Dyer, and Brownlee 1980) and appears to possess considerable potential. The general procedure (Tanaka *et al.* 1980) is summarized as follows.

Purified RNA (0·1–0·5 μg) is dissolved in 1–2 μl formamide, placed in a siliconized disposable 5 μl micropipette, sealed, and hydrolysed by heating on a boiling water bath for 5 min. The RNA sample is then blown into 50 μl of 0·3 M sodium acetate buffer pH 5·5 and recovered by ethanol precipitation. Under such conditions hydrolysis is limited to one 'hit' per molecule which generates a single internal 5'-hydroxyl group which can be radio-labelled. For this the precipitated RNA is dissolved in 5 μl 10 mM tris–HCl buffer pH 8·0 containing 10 mM $MgCl_2$, 6 mM 2-mercaptoethanol and 0·2–0·3 μl (2–3 units) of T4 polynucleotide phospho-kinase. The mixture is transferred into an Eppendorf tube containing 3 pmol dried γ-^{32}P-ATP and incubated at 37 °C for 60 min. After incubation one volume of formamide–dye mixture (prepared by mixing 100 ml formamide, deionized by treatment with a mixed-bed ion exchange resin such as Amberlite MB-1, with 20 ml 0·1 M Na_2 EDTA containing 0·02 per cent Bromophenol Blue and 0·02 per cent Xylene Cyanol FF) is added and the sample heated on a boiling water bath for 5 s. Portions of sample (2–3 μl) were separated by electrophoresis for 8 h at a constant voltage of 2·9 kV (about 35 mA) on thin slabs (800 mm x 300 mm x 0·5 mm) of 8 per cent polyacrylamide gel made up in a pH 8·3 buffer containing 10·8 g tris, 5·5 g boric acid, 0·93 g EDTA, and 420 g urea per litre. This gave a strip of gel containing a 'ladder' of separated radioactive polynucleotides, the position of the individual bands being determined by autoradiography (the gel strip can be 'mounted' on a strip of old X-ray film for ease of handling and as an aid to precise location of the bands).

A DEAE–cellulose thin-layer plate (e.g.polygram CEL 300, Macherey–Nagel GmbH) is washed by ascending chromatography with 50 mM EDTA (pH 4–5) solution and air dried, followed by a second washing in the same way with distilled water.

An origin line 0·5–0·8 cm wide and 2–3 cm in from the edge of the dried plate is then wetted by applying a strip of filter paper (Whatman 3MM) soaked in 1·0 M ammonium acetate (pH 4·5). The gel strip, still on its strip of backing film, is placed face down along this origin line and covered with a glass plate, and a weight is applied (e.g. two 1 kg weights). After standing overnight this treatment effectively transfers the polynucleotide ladder on to the thin-layer plate and the gel strip can be peeled off and discarded. The thin-layer plate is air dried and then soaked in distilled water for 2 min to remove urea and buffer salts and then dried again. The polynucleotide material is then digested enzymatically *in situ* by the following procedure. A solution of ribonuclease T_2 (0·2 units μl^{-1}), ribonuclease T_1 (0·2 units μl^{-1}), and pancreatic ribonuclease A (0·1 μg μ^{-1}) in 0·1 M ammonium acetate buffer pH 4·5 was streaked along the origin line and polynucleotide ladder with the aid of a disposable 20 μl micropipette at a proportion of about 3·5 μl cm^{-1}. The plate is covered with polythene wrapping material to prevent evaporation and incubated at 37 °C for 2–4 h. The material in each band of the ladder is thereby digested to a mixture of nucleosides but only that originating from the free 5'-terminal end will be radioactively labelled and detected by subsequent autoradiography. After the incubation step the thin-layer plate is immersed in methanol for 5–10 min and dried. The plate is then subjected to electrophoresis at right angles to the line of the polynucleotide ladder using a buffer of pH 2·3 (500 ml of 8 per cent acetic acid and 2 per cent formic acid containing 5 ml to 0·5 M Na$_2$EDTA) for 1 h at 400 V (20 V cm^{-1}). After air drying the positions of the labelled bases are determined by fluorography (see Section 2.12(c)). Since the four major bases have different electrophoretic mobilities under these conditions the final pattern should resemble the example shown in Fig. 6.3, and the base sequence can simply be read off.

6.7. Biochemical applications

Many of the areas where the use of agarose and composite polyacrylamide–agarose gels has advantages over PAGE have been indicated earlier in this chapter, but in general terms the principal merit is that these gels extend the separation of proteins and nucleic acids to a much higher molecular weight range than is possible with gels composed only of polyacrylamide. Virtually all other aspects of the technology involved are identical, and most importantly the very high resolution achievable by PAGE is retained. Electrophoresis on such gels is therefore a simple but accurate and sensitive method of analysis which can be used both for the identification of unknown substances and for monitoring the progress of a separation scheme or of suitable chemical or enzymic reactions. If required, either

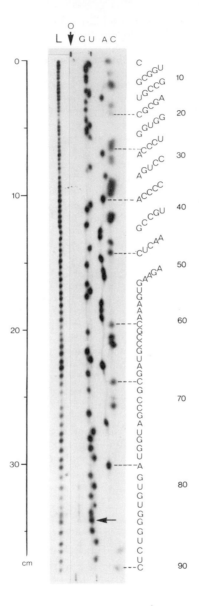

FIG 6.3 Autoradiogram showing the two-dimensional sequencing procedure applied to *Escherichia coli* 5S rRNA. First dimension (top to bottom), 4 h electrophoresis on a polyacrylamide gel with $T = 8$ per cent; second dimension (left to right, DEAE–cellulose thin-layer electrophoresis at pH 2·3. The ladder L after transfer to the thin-layer plate is aligned with the final two-dimensional pattern from which the sequence for residues 4–90 can be read off. (Reproduced from Tanaka *et al.* (1980) by permission of the authors and publishers.)

homogeneous or multiphasic buffer systems can be used. Ferguson plots can be constructed for the identification of anomalous migration behaviour or molecular weight estimation, and detergent or other denaturing agents (e.g. urea) can be included just as in PAGE.

Agarose gels have been used for the quantitative evaluation of serum lipoproteins by a gel-scanning procedure including standardization with appropriate calibration factors (Fasoli *et al.* 1978) and with 0·2 per cent of the detergent Sarkosyl (sodium N-lauryl sarcosinate) or SDS present for studies of the scrapie agent (Prusiner *et al.* 1980). Composite gels with 0·7 per cent agarose and a step gradient of 2 per cent and 3 per cent polyacrylamide have been used successfully to separate chylomicrons and the various classes of lipoproteins (VLDL, LDL, and HDL) and to identify different types of hyperlipoproteins (Moulin, Fruchart, Dewailly, and Sezille 1979). Laan and Diaz (1978) exploited electro-endosmosis and variations in it obtained by mixing various proportions of agar and agarose, to improve the electrophoretic resolution of lactate dehydrogenase isoenzymes.

In the context of nucleic acid analysis gel electrophoresis is applied not only to the identification and separation of DNA and RNA but also in conformational studies (Section 6.3). Because of very similar charge-to-mass ratios, approximate values for sedimentation coefficients and molecular weights can be obtained from simple measurements of mobility. Denaturing conditions should be used for more accurate work to overcome the effects of any conformational differences.

The ability of gel electrophoresis to separate with high resolution the breakdown fragments of nucleic acids produced by endonucleases such as the restriction enzymes has proved of immense significance in many applications. For example, it has been used by Sharp *et al.* (1973) not only to identify and assay restriction endonucleases in crude bacterial extracts and to monitor the progress of their purification, but also to identify the different restriction specificities of the enzymes present by examining the band patterns of the DNA fragments produced. The use of restriction enzymes and electrophoretic separation of the resulting fragments can also be applied to chromosome mapping by the preparation of cleavage maps or fragment maps analogous to peptide mapping of proteins. The map can be used for localizing template functions and genes and for relating nucleotide sequences of fragments to the entire genome (Nathans and Smith 1975).

As we have seen (Section 6.6), restriction enzymes and the resulting maps of fragments now provide the most powerful procedure for sequencing not merely short nucleotide segments but also complete DNA and RNA molecules (e.g. Sanger *et al.* 1977, 1978, 1982; Garoff *et al.* 1980; Houghton *et al.* 1980; Kitamura *et al.* 1981; Moss and Brownlee 1981). Once the nucleotide sequence of a DNA molecule has been established it then becomes possible simply to read off the amino acid sequence of any or all proteins for which the DNA codes. This was performed with spectacular success by Air *et al*. (1978) who calculated from the nucleotide sequence of the region of bacteriophage φX174 DNA coding for the F protein what the amino acid sequence of this protein should be and then compared it with the amino acid sequence determined experimentally. Similar nucleotide sequencing and protein amino acid sequence prediction has been reported by Porter *et al*. (1979) for an influenza virus haemagglutinin gene from cloned DNA and there have been many other examples of similar experiments since then. Work of this type on the DNA coding for adrenocorticotropin (ACTH) led to the realization that only part of the DNA molecule was transcribed in the formation of the single large precursor protein molecule which is then split up to generate the known pituitary hormones related to ACTH, and this led Nakanishi *et al*. (1979) to the discovery of at least two previously unsuspected hormones. Using recombinant DNA technology nucleotide sequences of the foot and mouth disease virus genome have been determined, from which the amino acid sequence of the virus polypeptide VPI could be deduced. From this knowledge Bittle *et al*. (1982) were able to synthesise a number of artificial peptides corresponding to parts of the VPI molecule some of which could be used to immunize experimental animals and protect them against the disease.

Cleavage maps also provide DNA fingerprints and thus can be used to demonstrate minor differences between very similar DNA species and to detect small deletions, additions, rearrangements, or substitutions for example. This can reveal such subtle modifications as evolutionary changes or recombinations between closely related strains. Restriction enzyme cleavage maps form the basis of methods of studying the localization of such chromosomal functions as the initiation and termination sites for DNA replication and transcription and the binding sites of regulatory proteins and for gene mapping (e.g. Monahan *et al*. 1977; Gannon *et al*. 1979; Hentschel, Irminger, Bucher, and Birnstiel 1980; Wilson *et al* 1980). Regions of DNA binding protein are protected from chemical or enzymatic attack by the presence of the protein, which results in a 'hole' in the sequencing results on genomic sequencing (Little 1984). This is the technique referred to as 'footprinting' (see e.g. Becker and Wang 1984) which is vital to the study of gene expression. Monogenic disorders such as ß-thalassaemia which give rise to gene deletions or modifications can now be detected by identifying restriction enzyme site polymorphisms which show up as variations in gel band patterns (Little *et al*. 1980). Thus electrophoretic maps of restriction fragments are used not only to determine the nucleotide sequences of DNA molecules but also performs a vital role in studies of their structure and function (Prunell, Strauss, and Leblanc

1977; Prunell and Bernardi 1977). Therefore it is not surprising that such maps form an integral part of experiments involving the restructuring and cloning of DNA molecules. The first stage of such experiments consists of the cleavage of DNA by restriction enzymes and the construction of a cleavage map. This provides the basis for the isolation of fragments with the desired genes or regulatory signals, perhaps by subsequent preparative electrophoresis. This is of course the field colloquially known as genetic engineering.

7

PREPARATIVE GEL ELECTROPHORESIS

There are two ways in which macromolecules separated by gel electrophoresis can be recovered for further study. The first involves performing the electrophoresis in a manner identical to that used for purely analytical evaluation and then once the component(s) of interest have been located the gel is cut up and the material extracted. The second consists of allowing the components to migrate the whole length of the gel and then collecting them at the end, usually by some suitable elution technique. Some workers (e.g. Rodbard *et al.* 1974) refer to only the latter as preparative electrophoresis, but we shall define the term in a wider context to mean those experiments in which the goal is to isolate individual components in an enhanced state of purity for further analysis and study. In this chapter we shall discuss preparative methods using polyacrylamide, agarose, or composite polyacrylamide–agarose gels. Preparative variations of the other electrophoretic and related techniques forming the subject of later chapters will be discussed in the relevant chapter.

7.1. Analysis of proteins without extraction from the gel matrix

It might appear anomalous in a chapter devoted to preparative methods, but the first question to be asked is whether the material does in fact need to be extracted from the gel for whatever further analysis is envisaged. In most cases the answer will be yes, but for proteins if only an amino acid analysis is required it may be unnecessary. For example gels can be stained with Amido Black 10B and destained in the usual way (Section 2.9(a)), and the individual stained bands can be cut out with a single-edged razor blade or scalpel (Houston 1971) and treated as follows:

The gel slices are placed directly into the tubes that will be used for hydrolysis (if gel rods 6 mm in diameter are used the corresponding slices from up to four gels can be pooled into a single tube) and 3·0 ml of 6 N HCl containing 0·1 M 2-mercaptoethanol or 0·1 per cent thioglycollate are added. The tubes are either frozen, evacuated, and sealed, or flushed with nitrogen and sealed. Hydrolysis is achieved by heating at 110 °C for 24 h. After cooling the tubes are refrigerated to precipitate the polyacrylic acid formed by hydrolysis of the gel matrix, centrifuged, and the supernatant liquid is collected and evaporated to dryness. If necessary the residual gel pieces in the centrifuge tubes can be re-extracted two or three times with 1·5 ml portions of 1 N HCl at 80 °C and these washings combined with the supernatant fractions before evaporation.

Very large amounts of ammonia are formed from the gel matrix during the hydrolysis, and according to Houston (1971) if the evaporated supernatant material is applied directly to the amino acid analyser there is a danger that there will be so much ninhydrin–ammonia complex formed that it may precipitate out and block the flow. Houston therefore advocates that as soon as the histidine peak has been recorded the flow from the ninhydrin pump should be diverted to drain for 10 min so that the ammonia can pass freely through the heating coil. In our own laboratory we have found that most of the ammonia is removed by two cycles of dissolving the dried supernatant material in a small volume (e.g. 0·5–1·0 ml) of 5 per cent Na_2CO_3 and evaporating to dryness. The resulting material can be applied directly to the analyser without difficulty.

Amido Black is completely destroyed and decolorized during the hydrolysis stage, but Coomassie Blue R250 and Xylene Brilliant Cyanine G are both stable and may interfere with the detection of the ninhydrin–amino acid complexes. It is essential to destain the gels thoroughly before slicing to remove gel impurities and particularly to remove buffer constituents such as glycine, tris, or any other components that may have been used in the electrophoresis and which react with ninhydrin. Urea and most detergents do not interfere. The presence of 1 per cent 2-mercaptoethanol or thioglycollate during the hydrolysis prevents destruction of histidine and tyrosine, and Houston (1971) reports good results with lactate dehydrogenase, bovine serum albumin, and ovalbumin. Our experience has been less satisfactory and in particular we have found that serine and glycine values are elevated probably owing to their formation during hydrolysis of the gel matrix. Nevertheless the method is useful for the preliminary identification of unknown electrophoretic components although it is probably not accurate enough for more demanding analytical work.

Some of the problems associated with the use of ninhydrin detection as above were avoided by Stein *et al.* (1974) who used fluorescence detection.

For this gels run in the presence of SDS were stained with Coomassie Blue R250 and destained by several washings in 12·5 per cent trichloroacetic acid over a period of 2 days. Slices of gel were cut out, hydrolysed, and evaporated in a very similar way to that described above, and the residual material was dissolved in pH 2·2 column buffer for direct application to the amino acid analyser. A fluorescamine detection system was used in the analyser, and because ammonia gives only a low fluorescence yield it did not interfere with the detection of lysine or histidine. The analyses reported by Stein *et al.* (1974) were very good and if equipment suitable for fluorescence detection is available this would appear to be the method of choice. The method was also applied satisfactorily to gels run in non-detergent buffer systems and in the presence of urea.

Owing to the low fluorescence yield of cysteine (about a quarter that of glutamic acid) and the fact that N-chlorosuccinimide must be added to the analyser column buffer for satisfactory determination of proline, duplicate runs are usually needed if both these amino acids are to be measured.

The method has the major advantage that it is extremely sensitive, and if a blank portion of gel is also analysed and appropriate corrections made as little as 1 μg of protein may suffice for the analysis. More routinely 10 μg of protein is used and no gel blank correction need then be made. A further advantage is that compared with the method of Houston (1971), the more sensitive stains such as Coomassie Blue R250 or Xylene Brilliant Cyanine G can be used since they are non-fluorescent and do not interfere. A very similar method has been used by Drescher and Lee (1978) who in place of fluorescamine used the o-phthaldialdehyde detection system of Benson and Hare (1975). Manabe et al. (1982) also used o-phthaldialdehyde detection but found that direct hydrolysis of polyacrylamide gel pieces led to the generation of too much ammonia from polyacrylamide decomposition for estimation of arginine and lysine. The excessive amounts of ammonia also tended to 'tail-over' into the next analytical run, and for these reasons they found it much more satisfactory to extract the proteins from the gel before analysis.

7.2. Small-scale preparations by extraction from analytical type gels

Over the years there have been many attempts to scale up for preparative purposes the types of gel system which give such excellent results on an analytical scale. The great variety of published designs of equipment for this testify eloquently to the failure in achieving the same degree of resolution that is possible analytically. Thus for accurate work requiring the separation of a particular component from others migrating reasonably close to it (say 0·5 cm or less on a typical analytical gel rod 5–7 cm long) there is at present no real alternative to running an analytical type gel and extracting the material from it. Vertical gel rods can be used if only very small amounts of material are needed (about 1–50 μg per 5 mm diameter rod) but larger slab gels, either horizontal or vertical, which can handle up to 50–70 mg are more commonly employed. The maximum amount of sample which a gel can handle depends largely upon the separation of the component of interest from neighbouring bands since a band becomes wider as the amount of material in the band increases, and eventually this loss of resolution may become unacceptable. Even if it does not, excessive sample loads will give rise to severe streaking and uneven bands.

Extraction of separated material can be achieved using either stained or unstained gels in three principal ways: (i) simple extraction with an appropriate buffer; (ii) solubilization of the gel matrix; (iii) electrophoretic elution. The first generally gives quite low yields (seldom over 60 per cent and often as little as 15–20 per cent) and the extract also contains substances leached from the gel matrix. Since these can differ widely both in terms of molecular size and charge they can be very difficult to eliminate entirely from the sample. Solubilization of the gel should theoretically give almost quantitative recoveries, but naturally the sample is accompanied by even more gel-derived material. Electrophoretic elution gives intermediate recoveries, although they can be very high, but is accompanied by a greater chance of loss of enzymic or biological activity, gives product still contaminated by gel-derived material (although less so than the other two methods), and is technically more demanding. However, the first hurdle to be overcome is the detection and localization of the component to be examined.

(a) *Detection of components—to stain or not to stain?*

The answer to this question lies in part in the objective of the experiment. Nearly all general staining procedures (Section 2.9) involve denaturation and fixing of the components in the gel. Fixing helps to prevent the separated components from diffusing and from being leached out of the gel during destaining, but the penalty lies in the fact that enzymic or biological activities are usually lost. The objection does not of course apply to many specific detection reactions, such as enzyme activity stains or the mild staining of nucleic acids with the intercalating dye ethidium bromide. A loss of activity is not prejudicial to most analyses such as N-and C-terminal amino acid analysis, total compositional analysis, molecular weight determinations, peptide mapping, sequence analysis, or studies of subunit structure, so for work of this type it may well be easiest to stain the whole gel and to slice out the relevant bands.

If the separated component is needed in its native state it is usual to cut the appropriate slices for extraction out of unstained gels. This can be done empirically by slicing up the whole gel rod or slab into segments of, for example, 1 mm thickness (larger slices may entail a significant loss of resolution) and extracting all the slices and assaying them for the particular component. As well as being very time consuming there is no guarantee that the individual cuts will be made in the best places. Thus it is preferable to localize the separated bands and excise the area of interest.

Protein zones in slab gels can be detected by passing a beam of light through the gel and observing the light and dark bands due to changes in refractive index (Fries 1976). The method is relatively straightforward but involves careful preparation of the gel slabs to avoid artefacts and requires the use of slab gels between glass plates. Unfortunately the method is not very sensitive (down to about 100 μg per band) and so cannot be used for analytical work or the detection of small amounts of impurities.

When gels are run in the presence of 8 M urea, protein zones containing a relatively large amount of material (it is some 30-fold less sensitive than

Coomassie Blue staining) can be detected by cooling the gel in air at −70 °C for 5–10 min. The urea appears to crystallize preferentially in protein containing regions of reduced free water availability which then show up as opaque areas that can easily be excised (Bachrach 1981). Proteins separated by SDS–PAGE are revealed by a blue fluorescence if gels are transilluminated through a cellophane film with light at 302 nm (Leibowitz and Wang 1984). The method works best with gels of low per cent cross-linking and samples should ideally have been treated before SDS–PAGE with dithiothreitol in place of the more usual 2-mercaptoethanol as the latter tends to give rise to a broad fluorescent band. Optimum fluoresence requires at least 5 min illumination and since it depends upon excitation of tryptophan residues some proteins containing little or no tryptophan cannot be detected. For most proteins sensitivity is about one-fifth that of Coomassie Blue staining and 10 μg usually gives a clearly visible band. presoaking the gel in buffer or fixatives eliminates the fluorescence. Alternatively, with SDS–PAGE slightly less sensitive but more general methods such as chilling (Wallace, Yu, Dieckert, and Dieckert, 1974) and precipitation of the potassium salts of the protein–SDS complexes (Nelles and Bamburg 1976) can be used.

In the former method use is made of the fact that if an individual band contains 20–40 μg or more of protein, chilling the gels for 3–5 h at 0–4 °C will precipitate the protein–SDS complex as a visible opaque white band. In the latter method gels are placed in 5 vol 1 M KCl containing 10 vol per cent acetic acid and opaque bands of the precipitated insoluble potassium salt of the protein–SDS complex develop within 10–30 min. Background opacity due to precipitated KDS is kept to a minimum by using only 0·01 per cent SDS in the upper buffer reservoir during the electrophoresis, but if even this background needs to be removed (e.g. for gel scanning) the gels can be soaked in 20 vol 25 per cent isopropanol containing 15 per cent TCA or 7·5 per cent acetic acid. Nelles and Bamburg (1976) claimed that as little as 1–2 μg of protein could be detected.

However, much the most usual method for detecting the separated bands is the use of guide strips. When this method is applied to gel slabs it consists of cutting a strip 0·5–1·0 cm wide off both sides of the slab, and for maximum accuracy and certainty of location a strip from the centre of the slab as well, and staining and destaining them in the usual way. They are then aligned with the unstained parts of the gel slab (which has meanwhile been kept in thin plastic wrapping at 4 °C) and the appropriate areas corresponding to the stained bands are cut out of the unstained portion. The only problem that may be encountered is that some shrinkage may occur in the guide strips during the staining and destaining steps making precise alignment difficult. To simplify this the guide strips can be cut off at the point reached by the tracking dye before staining, or its position can be marked by cutting a notch in the gel or inserting a short steel needle at this point. This point can then be used as a reference mark for estimating the extent of shrinkage. If shrinkage is a serious problem or is uneven (as in gradient gels for example) a series of marks can be made. One good way of doing this (Graesslin, Weise, and Rick 1976) is to punch a series of small holes down the sides of the slab 0·5–1·0 cm in from the edge at fixed intervals. The guide strips are then cut off by cutting down the gel through the middle of the holes so that both the strips and the remaining slab are left with a series of perforations down the edges which can be accurately aligned after staining.

The guide strip technique is less easy to apply to gel rods. Weiner, Platt, and Weber (1972) used a modified gel slicer of the type referred to in Section 2.12(b) to cut a thin slice off one side of the rod. Owing to some stretching of the gel, which is particularly troublesome and non-uniform in gels with $T < 5$ per cent, it was less accurate than the more usual method for gel rods which consists of running a number of rods at the same time and then staining one or two of them. Cutting out areas from the unstained gels on the basis of a comparison with the stained rods is the usual procedure but presupposes that the mobilities of all bands are the same in all the gels. It is in fact very difficult to be certain of this, since unless great care is taken there will be slight differences in the conditions between individual rods which will result in very small variations in mobility. With a batch of gels all run together and prepared from the same solutions at the same time there should not be significant compositional differences, but factors which are most likely to modify the mobility include gel length and diameter, distances of the ends of the gel tubes from the electrodes, differences in temperature arising from uneven cooling or differing heat transfer through tubes with different wall thicknesses, and variations in the pH, volume, or salt content of samples applied to each tube and sample load.

With proteins or glycoproteins run on gels in the presence of SDS the sample prelabelling procedure described in Section 5.5 is particularly helpful as the progress of the separation can be monitored and the fluorescent or prestained bands cut out as soon as the electrophoresis is stopped. Labelling in this way changes the net molecular charge, but since under these conditions the separation depends only on size and not charge the labelled molecules migrate at the same speed as unlabelled molecules. Hence only a small proportion of the sample need be labelled and will provide a suitable marker for the whole (Stephens 1975). The presence of a small proportion (as little as 1 per cent or less) of derivative molecules in the final product is unlikely to be important for most subsequent analyses (e.g. composition, amino acid sequencing, peptide mapping, etc.) but if troublesome they should usually be separable by ion-exchange techniques.

What in concept is a somewhat similar technique has also been suggested for prestaining nucleic acids for preparative work (Malhotra, Murthy, and Chaudhary 1978). This consists of reacting the nucleic acid (RNA) with 4–17 mol methylene blue per mole of RNA for 1 h at 4 °C in the electrophoresis buffer. The sample is then applied directly to the gels and the presence of this small amount of dye relative to the total molecular charge was found not to influence the mobility significantly.

(b) *Extraction of components by simple elution*

Slices of unstained gel are chopped into very small pieces, of about 1 mm^3 or less and fragmented by forcing them through a small syringe (either without a needle or fitted with a short large-bore needle) or ground into a paste in 0·15 M NaCl. The gel particles are then stirred, preferably overnight at 4 °C, with 10 vol of an appropriate buffer, although some workers have used extraction times as short as 1 h at room temperature. The mixture is then centrifuged briefly to separate the larger gel particles and the supernatant is filtered through a 0·45 μm Millipore filter (Millipore Corporation). For optimum yield the gel pieces should be re-extracted twice more in the same way and the filtered supernatants pooled and lyophilized. Yields differ considerably depending upon the substance being eluted, but the yield can sometimes be increased by freezing and thawing the gel a few times before maceration to disrupt the gel structure further. Extraction efficiency can often be improved by using strongly acidic conditions (e.g. 60 per cent acetic acid) for basic proteins such as histones or ribosomal proteins and strongly alkaline conditions for most other proteins. For example Manabe *et al.* (1982) cut small pieces (2 mm in diameter) out of a Coomassie Blue stained 2D protein mapping gel with the type of gel punch often used for making wells in gels for immunodiffusion procedures and placed them in Pyrex tubes that had been thoroughly cleaned by calcining at 500 °C for 3 h. A small portion (0.2 ml) of 0·1 M NaOH containing 2 per cent thiodiglycol was added and the mixture kept at 40 °C for 10 min. This effectively washed impurities out of the gel piece but since this had been destained in 7 per cent acetic acid the pH only rose to neutrality and little protein was extracted. The supernatant was therefore drawn off and discarded. A second 0·2 ml portion of the NaOH/ thiodiglycol solution was then added. During a further 10 min at 40 °C the pH rose to about 13 and much of the protein was extracted, so the supernatant was then drawn off, placed in another calcined tube and lyophilized. The material was then hydrolysed and amino acids analysed using fluorimetric detection with *o*-phthaldialdehyde, because with such a small piece of gel amounts of protein extracted were of course also small and a very sensitive detection method was required. This of course meant that great care had to be taken to avoid contamination at any stage, including by possible impurities from the gel or gel matrix. Using a microcomputer to calculate a similarity index comparing the amino acid composition of extracted material with that of known proteins they were able to use the method for identification of proteins on the gel maps.

The efficiency of extraction can also be improved by including urea (4–8 M) or a detergent such as 0·1 per cent SDS in the extraction buffer, although this may lead to a partial or complete loss of enzymic activity (Lacks and Springhorn 1980). If this is not a factor to be considered, the routine inclusion of SDS in the extraction buffer is to be strongly recommended. Extractions are usually carried out overnight at room temperature or 37 °C. SDS possesses strong bactericidal activity so no preservative need be added, but Bray and Brownlee (1973) suggest that it is advisable to add 1 mM phenylmethylsulphonyl fluoride to the extraction buffer to inhibit extraneous protease activity. However, since this is an inhibitor of only serine proteases the value of this particular addition relies upon such enzymes being responsible for the proteolytic activity. Protease activity can be troublesome, particularly when relatively crude tissue extracts are being studied, since proteins in the presence of SDS possess a random denatured conformation and as such are highly susceptible to proteolysis. The use of SDS in the extraction buffer is also applied to extractions of RNA from polyacrylamide and composite polyacrylamide–agarose gels (e.g. Dolja et al. 1977).

Staining procedures are usually accompanied by fixation and denaturation, so aqueous buffers are unlikely to extract more than traces of material from stained gels unless urea, or more often SDS, is present to resolubilize the substance under study. Thus by using SDS-containing buffers proteins have been satisfactorily extracted from either unstained (e.g. Weiner et al. 1972; Wallace and Dieckert 1976; Bridgen 1976) or stained (Bray and Brownlee, 1973; Drescher and Lee 1978; Gibson and Gracy 1979) gels. With care a gel slice containing as little as 5 μg protein can be sufficient for a full amino acid analysis (Drescher and Lee 1978; Sreekrishna et al. 1980), but care should be taken to make an appropriate correction for any background contributed by the gel matrix (Brown and Howard 1980).

When stained bands are extracted the extract will of course be coloured, and in order to purify the extracted protein for subsequent experiments Bray and Brownlee (1973) used the following procedure which should be generally applicable. The protein in the filtered extract solution was precipitated as the insoluble potassium salt of the protein–SDS complex by addition of 0·2 M KCl. After allowing to stand in ice for 15 min to ensure complete precipitation the mixture was centrifuged at 10 000 g for 20 min. The precipitate was then washed once with acetone containing 0·1 M HCl and once with acetone alone to remove the dye (Coomassie Blue R250) and dried in vacuo. Most gel-derived impurities in the extract are also removed by this process but small traces of detergent are likely to remain. Alternatively dye can be removed from coloured extracts by gel filtration on Sephadex G-25.

(c) Removal of SDS and other detergents

Extracted material containing SDS is suitable for further experiments such as end-group analysis, sequencing studies, peptide maps, etc., but for other work it may be desirable to remove the SDS. Most of the SDS can be removed by dissolving the lyophilized extract at an alkaline pH, precipitating the protein with 10–15 per cent TCA and removing residual SDS from the precipitate by washing with ethanol or acetone. There are many other methods of varying effectiveness for removing detergents from samples including simple dialysis, solvent extraction, gel filtration, ion-exchange chromatography, affinity chromatography, etc., and this topic has been reviewed by Furth (1980).

If extremes of pH must be avoided and if protein completely free of SDS is required Weber and Kuter (1971) have suggested the following procedure.

The protein–SDS mixture is dissolved in 0·1 M NH_4HCO_3 containing 6 M urea so that the final concentration of SDS is 0·1–1·0 per cent. The solution is then passed through a small column of Dowex 1 x 2 (or equivalent) ion-exchange resin (200–400 mesh) and protein eluted with 0·05 M NH_4HCO_3 containing 6 M urea. A settled resin bed will bind up to about 100 mg of SDS per mililitre of resin. Urea can subsequently be removed by gel filtration, dialysis, or precipitation of the protein with TCA (some dilution may be needed to reduce the urea concentration before efficient precipitation can occur).

Removal of detergent and urea in this way from enzymes denatured and inactivated by the SDS can sometimes result in a partial recovery of activity (Lacks and Springhorn 1980) the extent of which may also depend upon the quality of the SDS used (Lacks, Springhorn, and Rosenthal 1979). Prolonged dialysis alone of protein–SDS mixtures may remove much of the SDS, but it is less effective than the above methods and it is difficult to remove the last traces. In addition enzymic activity is not usually regained and only denatured proteins are obtained.

Hager and Burgess (1980), studying samples from which SDS had been largely removed by acetone precipitation of the protein, found that renaturation and regain of enzymic activity was much improved if the dry protein powder was redissolved in a small volume (e.g. 20 μl) of 6 M guanidine hydrochloride dissolved in a buffer consisting of 0·05 M tris-HCl pH 7·9 containing 20 per cent glycerol, 0·15 M NaCl, 1 mM dithiothreitol, 0·1 mM EDTA, and 0·1 mg ml^{-1} of bovine serum albumin. After standing for 15–20 min the mixture was diluted 50-fold with this same buffer without guanidine hydrochloride and renaturation allowed to progress for 1–12 h at room temperature. Clarke (1981) used a system relying on the formation of mixed detergent micelles to remove SDS by complex formation with a neutral detergent. For this solutions of SDS-protein complexes were simply mixed with an equal volume of 10 per cent Triton X-100 in 50 mM tris-H_2SO_4 buffer pH 7·7 containing 20 mM 2-mercaptoethanol followed by incubation at 37 °C. Recoveries were usually low however and seldom exceeded 10 per cent. Similar incubation of gel slices in 1 per cent Triton X-100 buffers following SDS–PAGE also led to some regain of enzymic activity. It is probably preferable in this case to remove most of the SDS from the whole gel slab by fixing and washing with isopropanol:acetic acid:H_2O (5:2:13, by volume) followed by an aqueous buffer such as phosphate buffered saline (Dupius and Doucet 1981) or with several changes of, and brief 37° incubation in, buffered 25 per cent isopropanol (Blank et al. 1982). Such treatment then permits glycoprotein localisation with ^{125}I-labelled lectins but can also lead to regain of enzymic activity. The renaturaton of SDS solubilized membrane proteins (or following SDS–PAGE) and enzymes has been discussed by Hjertén (1983b). Probably the method of choice for obtaining SDS–free proteins from SDS–PAGE gels however consists of transferrring the proteins to an immobilizing matrix such as nitrocellulose sheets by Southern blotting or electroblotting (see Section 2.13). Once the proteins have been transferred and immobilized in this way they can be freed of any contaminating residual SDS by washing the membrane or by brief continued electrophoresis in the transfer apparatus with detergent-free buffer. The proteins can then be eluted from the immobilizing membrane as indicated in Section 2.13.

Three other methods for removing SDS from protein–SDS complexes appear particularly effective.

One (Kapp and Vinogradov 1978) involves the preparation of columns of an ion-retardation resin (Bio Rad AG11A8). The columns are washed with 5 vol of 1·0 M NH_4Cl followed by 20 vol H_2O. The sample, which is dissolved in H_2O or 0·1 M sodium phosphate pH 7·2, is applied to the column and eluted with H_2O or phosphate buffer. Protein recoveries average 83 per cent, and a column of dimensions 11 mm x 96 mm containing 9·1 g of wet resin eluted at a flow rate of 40–60 ml h^{-1} could handle up to 25 mg of SDS in a 0·5 ml sample volume. Some other buffers (e.g. borate, tris–HCl) did not give such an effective separation of protein and SDS, and if the sample contains large amounts of salt a layer (e.g. about 5 ml with the above column) of Sephadex G-10 should be added above the resin bed so that the protein–SDS complex is desalted before entering the resin.

The second method (Henderson, Oroszlan, and Konigsberg 1979) is technically very simple and relies upon ion-pair extraction of the SDS into organic solvents. Two procedures were described. In the first lyophilized protein–SDS material is dissolved in a mixture of anhydrous acetone, triethylamine, acetic acid, and water (85:5:5:5 by volume). Up to 30 mg SDS will dissolve per millilitre, but the best results are obtained with 10 mg ml^{-1} or less. Either immediately or after 1 h at 0 °C a protein precipitate is formed and can be collected by centrifuging. The precipitate should be extracted twice more with this solvent and then once with anhydrous acetone alone and dried in a steam of nitrogen. Instead of using a lyophilized sample the solvent mixture, less water, can be mixed with sample solution in the ratio 20:1. The second procedure is similar but employs a mixture of heptane, tributylamine, acetic acid, and butan-2-ol (70:10:10:10 by volume) and a lyophilized sample containing no more than 20 μl H_2O per millilitre solvent mixture. The method can be applied to as little as 20 μg ml^{-1} of protein, but recoveries of enzyme activity are rather variable.

The third method has been applied to nonionic detergents such as Triton X-100 and weakly anionic detergents such as cholate and deoxycholate. It consists (Horigome and Sugano 1983) of piercing a small hole in the bottom of a conical 1·5 ml polypropylene microcentrifuge tube which is then covered with a small amount of siliconized glass wool. The tube is then filled with 0·5–1·5 ml of a thick slurry of Bio-Beads SM–2 or Amberlite XAD–2 resin in buffer, the composition of which is not critical (tris, phosphate, acetate, Hepes, bicarbonate, etc. of pH values between 6·5 and 9·0 were all satisfactory, and salt concentrations up to at least 0·5 M NaCl can also be tolerated). The tube is then inserted into a larger (e.g. 15 x 105 mm) Pyrex tube which serves as a collection tube and the whole assembly centrifuged for 1–2 min at 200 g. The buffer in the collection tube is discarded and the sample which should contain less than 2 per cent detergent added to the top of the resin. The whole sample volume should be absorbed into the bed of Bio-Beads (0·5 ml of which will absorb about 0·15 ml sample volume). After standing for 5 min the microcentrifuge tube is capped and recentrifuged at 200 g for 2 min or 400 g for 1 min. Such treatment should remove at least 97 per cent of the detergent and usually in excess of 99 per cent.

(d) *Solubilization of the gel matrix*

The preparation of gels with cleavable cross-links and the solubilization of gels containing them has been discussed already in Section 2.12(a) in relation to the quantitative measurement of radioactive labels. For preparative work the only difference is that once the gel has been solubilized the substance has to be separated from the very considerable amount of material derived from the gel. Here one must rely on such well-established techniques as gel filtration, ion-exchange chromatography, precipitation reactions, etc. The use of affinity chromatography should be particularly advantageous since many products from the gel will be charged and they will cover a substantial size range so that they may be difficult to remove entirely by other methods. Faulkner *et al.* (1982) used the solubilization of bands excised from a stained slab of BAC cross-linked gel for the successful isolation of histones. If agarose gel slabs are used for electrophoresis, materials can be readily recovered by melting and diluting the warm gel solution and collecting the protein by precipitation, ion-exchange chromatography, etc. or alternatively the gel can be digested with a mixture of hemicellulase and agarase (Cantarow *et al.* 1981).

(e) *Electrophoretic elution*

The same apparatus used for running vertical gel rods can also be used for electrophoretic elution as follows.

A glass tube identical to those used for the separation itself, and hence up to about 1 cm in diameter, is slightly constricted at one end by heating in a gas burner, and after cooling the constricted end is sealed with Parafilm. The gel slices containing the material of interest are then mixed with a little acrylamide and Bis solution (stacking gel mixture is ideal), catalysts added and the mixture poured into the tube and allowed to polymerize. The Parafilm is removed and a small sack made from dialysis tubing and filled with electrophoresis buffer (ideally 1–2 ml) is tied or fastened with rubber bands over the end of the tube (Fig. 7.1). The tube, or several of them, is then inserted into the apparatus and the sample electrophoretically eluted into the dialysis sack.

This method can be used with either stained or unstained gels and buffers with and without urea or SDS, although yields are likely to be higher and elution times shorter if SDS is included when material from

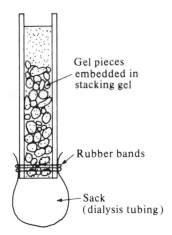

Gel pieces
embedded in
stacking gel

Rubber bands

Sack
(dialysis tubing)

FIG. 7.1. Electrophoretic elution of material from pieces of excised separation gel into free solution in a dialysis sack.

stained gels is to be collected. Stephens (1975) used this method with dansyl-labelled proteins from an SDS–PAGE experiment. Fluorescent prestaining is very helpful to electrophoretic elution experiments because it is then relatively easy to decide when elution has been completed.

A design consisting of a glass tube to one end of which a small plastic collection cup holding a disc of dialysis membrane in place could be attached has been described by An der Lan *et al.* (1983). It could be made from any size of glass tubing in order to fit whatever electrophoresis apparatus was available and was particularly effective when stacking gel or stacking buffer containing Bromophenol Blue as a marker was applied above the sample gel. In the subsequent electroelution sample proteins were eluted in the stack which is marked by the Bromophenol Blue. Electrophoresis should be continued for a further 2–3 h to ensure complete elution. A very similar approach using a rather larger column (e.g. 5–7 mm x 150–200 mm) has been described by Wu *et al.* (1982), but in this case the column was filled with 1 per cent agarose. As above, a stacking buffer system was employed and proteins collected in the stack which could be excized from the agarose gel and the proteins recovered. A still more elaborate design of apparatus for achieving the same goal by using the stacking process (which is of course the same as displacement electrophoresis or ITP) has been described by Öfverstedt *et al.* (1983) and was used for the quantitative recovery of milligram amounts of protein or DNA (Öfverstedt *et al.* 1984) from gels.

A very straightforward and elegant method that gives material almost free of possible gel contaminants has been described by Ziola and Scraba

(1976). It was applied to a number of Remazol Blue prestained proteins run in SDS-containing gels (see Section 5.5) but should be of wider applicability.

In this method the dialysis bag used in the above method is replaced by a length of tubing slightly greater in diameter than the elution tube full of gel pieces. This second tube contains a gel plug 1·5 cm thick in the bottom (made up in the SDS-containing electrophoresis buffer) overlaid with a 3 mm layer of Sephadex G-25 (Fig. 7.2). This in turn is overlaid with 1 g of hydroxyapatite (Bio-Gel HTP, Bio Rad Laboratories) suspended in electrophoresis buffer. The two tubes are brought together so that the end of the top tube is only just above the hydroxyapatite layer and the two tubes are sealed together with Parafilm, placed in the gel rod type of apparatus, and electrophoretic elution is performed at 25 mA for about 8 h (Ziola and Scraba 1976).

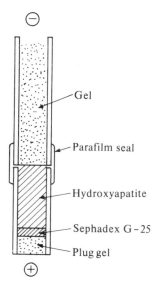

FIG. 7.2. Electrophoretic elution of material from pieces of excised separation gel into a layer of hydroxyapatite. (Reproduced with permission from Ziola and Scraba (1976)).

Coloured or fluorescent bands enable the optimum time and current combination to be determined for the particular substances being separated since some migration continues in the hydroxyapatite layer. After the electrophoresis the hydroxyapatite is transferred with a Pasteur pipette into a glass column containing a small bed of Sephadex G-25 (medium) and washed with 5 column volumes of 0·12 M sodium phosphate buffer pH 6·4 containing 1 mM dithiothreitol and 0·1 per cent SDS at a flow rate of 4–6 ml h^{-1}. The protein sample is then eluted by increasing the sodium phosphate buffer to 0·5 M, still with dithiothreitol and SDS.

A technically rather more simple approach (Otto and Šnejdárková 1981) is to polymerize some polyacrylamide gel (the composition is unimportant since it is only a supporting plug) into a gel tube and place on top of it a gel slice containing the protein zone to be eluted. Buffers used are also not critical but 0·025 M tris + 0·075 M glycine giving a pH of 8·8 would be suitable. The piece of sample gel is overlaid with buffer containing 30 per cent glycerol which in turn is overlaid with 1–2 M NaCl which is also used in the electrode compartments. Electrophoresis is then performed with the reverse of the usual polarity so that sample protein migrates upwards from the gel slice into the buffered glycerol layer which can be recovered with a pasteur pipette. In principle it can be applied to slab gel systems and SDS–PAGE as well as PAGE and if dansylated or other labelled proteins are used in preliminary runs, the times required for any particular arrangement can be established for subsequent runs with unlabelled proteins. The polyacrylamide support plug can be used at least three or four times before renewal.

Another interesting approach has been described by Méndez (1982) who used a standard type of vertical slab apparatus. The bottom part of the gel chamber was filled in the usual way with polyacrylamide gel to act as a gel plug. Once polymerized this was overlaid with 2–2·5 cm deep layer of 1 per cent agarose gel which in turn was overlaid with at least 3·5 cm of polyacrylamide separation gel, and finally with about 3 cm of stacking gel added above this. Méndez used tris–glycine buffers containing SDS in the $T = 12·6$ per cent, $C = 5$ per cent separation gel but clearly any gel concentration and buffer system with different depths of the layers could be used in the same way. Using fluorescent-labelled (fluorescamine) proteins electrophoresis was continued until the band of interest had entered the agarose layer. Electrophoresis was then stopped, the appropriate strip of agarose gel cut out of the slab and the protein recovered from the supernatant in over 90 per cent yield following high speed centrifugation to pellet the agarose in the gel. If unlabelled proteins are required labelled marker proteins can be applied to sample slots adjacent to the main sample area. If non-SDS buffer systems are used, so that fluorescent labelling is not practical, then preliminary experiments and staining of guide strips would be required to determine when the band of interest had reached the required position.

All the above methods are readily applicable to small-scale preparative experiments and have the advantage of requiring no special apparatus in addition to that required for the electrophoresis itself. Several designs of specialized equipment have also been described and may be more convenient if a large number of preparative experiments are to be performed. Other designs (Tuszynski, Damsky, Fuhrer, and Warren 1977; Green et al. 1982) can be used with stained and unstained gels and run in buffers containing urea, SDS, or non-ionic detergents in addition to detergent-free systems. The apparatus of Mardian and Isenberg (1978) should be able to handle rather larger amounts of gel and is reported to give excellent recoveries. A simple device (Allington et al. 1978) intended for electrophoretic concentration (see Section 12.10) has been used to elute proteins from gel for subsequent analysis (Bhown, Mole, Hunter, and Bennet 1980; Aebersold et al. 1984).

A rather different procedure consists of using a horizontal flat bed of Sephadex G-25 made up in gel buffer and in which a number of protein-containing strips of gel are embedded (Judd 1979). When an electric potential is applied (e.g. 150 V) at right angles to the gel strips the protein migrates into the Sephadex bed. It can be recovered by scraping the Sephadex off the backing plate, transferring it into a small column (such as a Pasteur pipette plugged with glass wool), and eluting the protein with water or a suitable buffer. A somewhat similar approach is used in an apparatus specifically intended for the electrophoretic elution of biologially active RNA and DNA (Wienand, Schwartz, and Feix 1979) which is capable of handling up to 20 gel strips or samples simultaneously.

Electrophoretic elution by isoelectric focusing in a horizontal granular gel bed is described in Section 10.9. It would appear to have great potential as a very mild, versatile, and efficient method for eluting material from polyacrylamide and agarose gels.

(f) *Extraction of RNA*

Very similar small-scale preparative methods can be applied to nucleic acid separations with only minor modifications.

For example, Dolja *et al.* (1977) separated RNA samples on polyacrylamide gels with $T = 3·15$ per cent and after the electrophoresis stained with ethidium bromide (see Section 2.9(e)). The gel slices were frozen and thawed to aid disruption and then homogenized in 20 mM glycine (pH 9·0) containing 0·05 per cent SDS (5 ml for 10 gel slices). After stirring overnight at 2–4 °C the mixture was centrifuged for 15 min at 23 000 g. The gel fragments were re-extracted twice in the same way using half the volume of glycine solution and the supernatant fractions were pooled. A little (e.g. 60 μl) 4 M KCl was added to precipitate the SDS, the mixture was centrifuged, and the supernatant was adjusted to pH 5·5–6·0. Addition of $CaCl_2$ to 0·2 M precipitated the RNA. After standing overnight at 0 °C the RNA was collected by centrifuging at 23 000 g for 15 min and dissolved in 20 mM EDTA. The solution was adjusted to pH 7·8 with tris and the RNA was precipitated with 2 vol of ethanol at –20 °C. A further cycle of dissolving in EDTA and precipitation with ethanol followed by washing in 70 per cent ethanol gives RNA free of SDS, polyacrylates, and other gel contaminants which are not usually completely removed by simple ethanol precipitation procedures.

An approach which has been used on the microgram scale for RNA separated in polyacrylamide gels (Malhotra *et al.* 1978) is to prestain the RNA with a low molar ratio of Methylene Blue (Section 7.2(a)) and then to trap the individual coloured bands as they emerge from the end of the gel rod on discs of DEAE–cellulose paper (Whatman DE-81).

For this a circle of the paper 8 mm in diameter was soaked in electrophoresis buffer and applied to the anode end of the gel tube which is also 8 mm in diameter, care being taken to avoid trapping air bubbles. The paper was held in place by pressing a small tight-fitting cap with a hole 5 mm in diameter in the bottom of it over the end of the tube (Fig. 7.3). This cap can conveniently be cut from a small disposable plastic syringe or sample vial cap. As soon as the first RNA band had

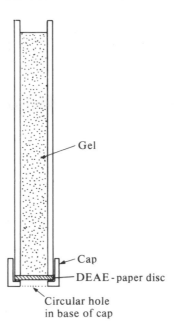

Gel

Cap

DEAE - paper disc

Circular hole
in base of cap

FIG 7.3 Electrophoretic separation of RNA on a microgram scale with entrapment of the separated material on a disc of DEAE ion-exchange paper. (Reproduced with permission from Malhotra *et al.* 1978; copyright by Academic Press.)

been absorbed onto the paper the electrophoresis was stopped, a new paper circle inserted, and the electrophoresis continued until the next RNA band had been collected, and so on. The paper circles were then washed with water and 95 per cent ethanol to remove the Methylene Blue, air dried, and RNA eluted with three successive washings of each disc with 200 μl, 200 μl, and 100 μl of 2 M triethyl ammonium bicarbonate (Smith and Khorana 1963). The washings were pooled, dialysed to remove any residual dye, and lyophilized. Two further cycles of dissolving in water and lyophilizing eliminated any remaining triethyl ammonium bicarbonate.

Very high molecular weight RNA which has been separated on low gelling temperature agarose gels can be recovered intact and in high yield (e.g. 60–90 per cent) by the following procedure (Wieslander 1979; Fourcroy 1984).

Gel slices containing the separated RNA bands are mixed with 5–10 vol of the electrophoresis buffer and warmed to 65 °C for 5 min to melt the agarose. The resulting solution is then extracted with an equal volume of buffer-saturated phenol for 15 min, centrifuged at 12 000 g for 10 min and the aqueous phase recovered. This is then re-extracted for 5 min with phenol, centrifuged, and the RNA precipitated from the aqueous layer by the addition of 0·1 vol of 1 M NaCl and 2·5 vol of ethanol.

Probably the method of choice for recovering small amounts of RNA from gel slabs involves electroblotting to DEAE paper in the same way as described by Danner (1982) for DNA (see p. 72).

(g) *Extraction of DNA*

Wheeler, Fishel, and Warner (1977) give a number of methods for extracting DNA from agarose gels

(i) Electrophoretic elution into a dialysis bag, virtually as described in Section 7.2 (d)), can be used. Recoveries are low, usually only 6–12 per cent, and the DNA is contaminated with agarose probably because agarose is slightly charged and co-migrates with the DNA into the electrophoresis bag. A specially constructed simple elution tube (Ho 1983) divided into two compartments separated by a nylon mesh for electroelution of DNA from gel pieces in the larger compartment into the buffer-filled smaller compartment (closed at the other end with dialysis membrane) may be convenient if several samples have to be extracted routinely.

(ii) A small column (8–10 cm high) of hydroxyapatite is equilibrated with 0·08 M potassium phosphate buffer. The gel slices are dissolved in saturated KI and applied to the column (gentle pressure needed). Agarose is eluted with saturated KI, the column is washed with 0·08 M buffer, and the DNA is eluted, free of agarose, with 1 M potassium phosphate. Yields are variable but generally in the range 50–75 per cent.

(iii) Gel slices are frozen to –20 °C, covered with parafilm, and squeezed either between the fingers or in a small press (Thuring, Sanders, and Borst 1975; Tautz and Renz, 1983). About 70 per cent of the gel weight is squeezed out as a clear fluid which also contains the DNA. For maximum recovery the procedure is repeated after adding a few microlitres of buffer to the residual gel pieces. The method is quick, recoveries are usually 50–70 per cent, and high-molecular-weight double- or single-stranded DNA can be handled without damage.

(iv) Gel slices are homogenized, the agarose is removed by centrifuging at 10 000 g for 10 min, and the supernatant is passed through a Millipore filter (RAWP 02500, 1·2 μm). Recoveries of relatively pure DNA are high (90–95 per cent) but this method can only be used for small DNA molecules (e.g. restriction fragments) as large molecules are damaged by shearing forces during the homogenization.

(v) Gel slices are heated at 70 °C for 10 min in 5 vol of 50 per cent formamide in 0·01 M tris and 2 mM EDTA at pH 8·0. Agarose is removed by centrifuging at 10 000 g for 10 min. The supernatant contained 80–90 per cent of the DNA, but duplex DNA is denatured so the method can only be applied to single-stranded DNA.

(vi) Gel slices are disrupted by forcing them through a syringe fitted with a 26 gauge needle, a little electrophoresis buffer is added, and the mixture is stirred overnight at 4 °C. After centrifuging at 10 000 g for 10 min, 75–80 per cent of the DNA is obtained in the supernatant together with small amounts of agarose. Passage through the syringe may cause some shearing damage.

(vii) Gel slices are dissolved in KI solution at an agarose concentration of 0·1–0.2 per cent and with sufficient KI to give a final density of 1·5 g ml^{-1}. The mixture is centrifuged at 200 000 g for 20–40 at 20 °C to form an isopycnic KI gradient (Blin, Gabain, and Bujard 1975). If the KI gradient contains 50 μg ml^{-1} of ethidium bromide the DNA can be visualized by direct u.v. illumination. Alternatively, fractions are collected and assayed for the presence of DNA or monitored for radioactivity if an isotopic label is used. Recoveries are close to 100 per cent and the DNA is entirely free of agarose. If this is used in conjunction with the syringe method (vi) to remove the bulk of the agarose first, large amounts of DNA can be rapidly processed (Wheeler et al. 1977).

Very-high-molecular-weight DNA can be recovered intact from agarose gels in good yield in a very similar way to that described above for high-molecular-weight RNA.

It consists (Wieslander 1979) of melting the agarose at 65 °C and extracting with phenol as for RNA, except that in this case the phenol extracts are not centrifuged but are adjusted to 0·2 M in NaCl and the DNA precipitated by the addition of 2·5 vol of ethanol. The DNA is redissolved in NaCl and then reprecipitated with ethanol to remove traces of phenol.

If DNA-containing gels are stained with ethidium bromide and bands located by u.v. illumination this itself can damage the DNA and cause breaks in the nucleotide strands. To avoid this Pulleyblank et al. (1975) masked gel slabs and only illuminated strips of the gel through 3 mm slots cut in the masks. DNA-containing bands are cut out and the parts of the gel that have been illuminated are discarded. The other parts are frozen and thawed three times to disrupt the gel structure and centrifuged at 40 000 g for 1 h. However, only a relatively low yield of 30–50 per cent of the DNA is recovered in the supernatant. Ethidium bromide is removed by extraction with butan-1-ol and DNA is precipitated from the aqueous phase with 2 vol of cold ethanol and collected by centrifugation at 4 °C.

Hansen (1981) has reported the use of BAC cross-linked polyacrylamide gels (see p. 53), which are solubilized by mercaptoethanol, for the separation and preparation of DNA restriction fragments for sequence analysis. Solubilized gel material was removed by ion-exchange chromatography on small columns of DEAE-cellulose.

Probably the best and most versatile method of recovering small amounts of DNA from gels however is by absorption to DEAE ion-exchange paper. This can be achieved either by the procedure desribed by Malthotra et al. (1978) for RNA (see p.192 and Fig. 7.3) or by inserting pieces of DEAE paper (Whatman DE–81) into the gel slab between zones after they have been located by ethidium bromide staining and then continuing the electrophoresis until all the DNA has been absorbed (Dretzen et al. 1981). A similar but rather better approach if several DNA bands, such as a number of restriction enzyme fragments, have to be analysed is to use electroblotting (see p. 72) in order to transfer all the bands present on the gel to a sheet of DEAE paper (Danner 1982).

For this the gel slab was first stained with ethidium bromide so that the efficiency of transfer could be readily monitored, but no doubt if a number of preliminary runs were made to establish optimum conditions this could be omitted if desired. The same buffer used in the gel was used for the electroblotting apparatus, in which two sheets of DE–81 paper were used (the second sheet to guard against overloading of the first). After transfer, DNA was recovered from slices of the paper with the aid of a small microfilter made by puncturing a 0·5 ml Eppendorf polypropylene microcentrifuge tube at both ends with a 25–gauge needle. The bottom of the tube was packed with a few millimetres of glass wool and the tube siliconized with a 5 per cent solution of dichlorodimethylsilane in chloroform to minimize DNA absorption. The wad of DE–81 paper was placed in this tube which was then nested into a larger 1·5 ml Eppendorf tube (with its cap removed) and the assembly centrifuged for 10 s. The larger tube and its contents were discarded and replaced with a clean tube. The paper pieces were washed twice by recentrifugation with 0·15 ml portions of 10 mM tris–HCl buffer pH 8·0 containing 0·15 M NaCl and 1 mM Na_2EDTA. Finally DNA was eluted by allowing the paper to stand for at least 1 h at 37 °C after adding 0·15 ml of 10 mM tris–HCl pH 8·0 buffer containing 1 M NaCl and centrifuging as above. The eluate was collected, the paper washed briefly with a further 0·15 ml of the 1 M NaCl buffer and the washings pooled with the first eluate. Ethidium bromide remains bound to the DEAE paper. DNA was collected from the 300 μl volume of eluate by precipitation with 900 μl of 95 per cent ethanol. Danner (1982) achieved overall yields from the original agarose gel of 67–96 per cent for DNA fragments varying in size from 1·2 to 12·4 kb.

7.3.   Apparatus for small-scale preparative experiments

When material is to be recovered from gels run for a fixed time by slicing the gel and using any of the extraction techniques described above, any of the types of apparatus used for analytical work can be employed (see Section 2.6). Naturally small gel rods are only suitable for the smallest-scale experiments, and even pooling slices from a number of gels can only be used for preparing 1 mg or so at most. The larger slab designs with much greater gel cross-sectional areas can be used for up to about 100 mg.

7.4.   Large-scale preparative methods with electrophoresis over a constant path length

Many preparative types of electrophoresis equipment are constant-path-length designs in which all components migrate the whole length of the gel and are collected in turn as they reach the end, usually by some form of elution or buffer flow system.

The general conduct of constant-path-length preparative electrophoresis is similar with most designs of equipment. The sample solution is layered on to the top of the gel column in a manner similar to that used for analytical-scale gel rods. The gel column may be homogeneous with the same buffer in both gel and electrode chambers, or with a different buffer (in effect a discontinuous gel system with omission of the stacking gel), or it may be a fully discontinuous gel system with stacking and separation gels and multiphasic buffer systems (*cf.* Chapter 3). The elution buffer flow is started when the tracking dye has almost reached the bottom of the gel and the column effluent collected into a fraction collector. *En route* it is monitored for u.v. absorbance with a spectrophotometer fitted with a flow cell, using absorbance at 280 nm for proteins, glycoproteins, and lipoproteins and absorbance at 260 nm for nucleic acids. If the protein is low in tyrosine and tryptophan so that absorbance at 280 nm is low, absorbance in the region of 230–210 nm can be used but is less specific for protein. A peak of u.v. absorbance due to material from the gel itself migrates quite rapidly off the gel and may interfere with monitoring, but this is relatively small if freshly made acrylamide solutions are used for gel preparation. Ultraviolet absorbance can also be misleading if two or more overlapping components are eluted since they may not be resolved in the u.v. profile. If difficulties due to any of these effects are suspected, some further monitoring of the collected fractions must be employed (e.g. analytical PAGE of selected fractions, enzyme assays, etc.).

In addition to the usual considerations of gel and buffer composition as discussed in earlier chapters, the gel length must also be optimized. For example, in fixed-time experiments a component with a mobility R_F only 0·05 that of the buffer front can be readily detected and quantified, but in fixed-distance electrophoresis it will not be eluted until $t = 20$, defining $t = 1$ as the time of elution of the front. It may therefore take a long time for slow components to be eluted and this can lead to considerable band broadening due to diffusion, which is largely time dependent. This in turn results in the slow-moving constituent being collected in a relatively large volume of elution buffer and may necessitate a concentration step before further analysis. The ease of recovery of the sample and the overall yield are therefore other important considerations in preparative exeriments which are much less significant in analytical work.

Under constant conditions the resolution between bands and the band-width both increase in proportion to the square root of time, but the separation between the bands increases in direct proportion to time (Rodbard *et al.* 1974). Thus there are diminishing returns as the gel length is increased. The choice of time depends upon the gel length and the voltage gradient, which is decided by the maximum permissible wattage that does not give rise to overheating and this in turn depends upon the ionic strength. Thus it is found that the resolution between the bands is also proportional to the square root of the field strength. Rodbard *et al.* (1974) indicate that with many designs of equipment and with elution buffer flowing at the rate of one elution chamber volume per minute electrophoresis times of up to about 5 h are appropriate. Since slow bands are eluted in larger volumes than fast bands, owing to the increased bandwidth, a variable elution buffer flow rate is beneficial. A continuous reduction in the buffer flow rate so that it is inversely proportional to R_F is the optimum correction factor.

A number of other aspects, many of them interrelated, are also important for best results in preparative PAGE. The sample load, or more precisely the load per surface area, has a marked influence on resolution, and effectively the diffusion coefficient increases as the load increases. In addition very high protein loads can cause local modifications to the pH, ionic strength, and voltage gradient within the sample stacking zone, which can result in interactions between sample components and losses due to aggregation and denaturation reactions. Rodbard *et al.* (1974) suggest that a load of 1–2 mg cm^{-2} is appropriate for the best separation of proteins, but the permissible load can be increased as the separations between the band of interest and bands due to contaminants increases. Current commercial or laboratory designs of preparative PAGE apparatus seldom have a gel cross-sectional area in excess of 15 cm^{-2} owing to the problems associated with adequately dissipating the heat produced during the electrophoresis. Therefore the total protein load that can be applied does not exceed about 100 mg and is often much less than this if good resolution is required. Thus even large-scale preparative PAGE is still quite small scale by comparison with other biochemical techniques (e.g. gel filtration, ion-exchange chromatography, centrifugal analysis, etc.).

Voltage gradient, temperature, and viscosity are other factors which must be considered in any electrophoretic experiment and can be particularly important in preparative work. The steeper the voltage gradient (V cm^{-1}) the shorter the time required for the electrophoresis and hence the less time in which diffusion can occur and the sharper the band separation. High voltages and short times are to be preferred therefore, but a compromise has to be made because high voltages increase the amount of ohmic heat produced and cause deformation of bands due to the production of thermal gradients within the gel.

The initial temperature of the apparatus (and the temperature of water circulating through cooling jackets if these are used) is another variable to be considered. Many enzymes, hormones, or other potentially heat-labile proteins are routinely run at 0–4 °C, although the actual temperature of the gel is likely to rise a few degrees during the experiment. Otherwise room temperature is usually most convenient and resolution improves at higher temperatures (Rodbard *et al.* 1974). Gels containing the ionic detergent SDS (Chapter 5) should not be cooled otherwise the detergent may precipitate out. The diffusion coefficient at 25 °C is slightly greater than at 0 °C, but owing to about a two-fold decrease in viscosity the electrophoretic mobility is much greater. Conversely, if at a given temperature viscosity is increased (e.g. by adding sucrose to the gel solutions as can be done when making gradient gels), mobility is reduced and electrophoresis is prolonged. Band spreading due to diffusion is increased with some loss in resolution compared with sucrose-free gels.

7.5.   Apparatus for preparative electrophoresis over a constant path length

Microgram quantities of proteins and nucleic acids can be prepared with two analytical-PAGE-type gel rods connected by a short sleeve of plastic tubing leaving a gap of 1–3 mm between the two gel surfaces (Fig. 7.4). Two syringe needles, preferably with sawn-off ends, are pushed into this gap through the plastic tubing so that buffer can be pumped in through one needle, across the end of the gel, and out of the other. The main disadvantage is that the geometry of the elution chamber is far from ideal for efficient flushing out of each component as it reaches the end of the gel, so that some mixing occurs and a relatively large volume of buffer is required. If one component has only slightly slower migration than another such a loss of resolution may be serious, but the method does have the

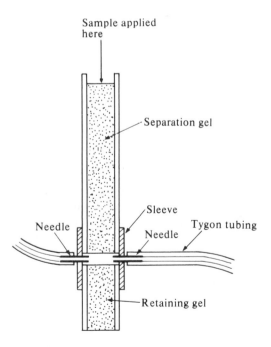

FIG. 7.4. Preparative electrophoresis of microgram quantities of proteins or nucleic acids using small gel rods and cross-flow elution after components have migrated over a fixed path length.

advantage that it requires only the standard cylindrical gel rod type of apparatus as used for analytical work. A very similar apparatus was used by Rossi (1978) for the small-scale preparation of mammalian DNA.

For small-scale preparative work fixed-time methods (Section 7.2) are generally preferred. Thus most designs of equipment for constant-path-length work are designed for relatively large amounts of material, generally up to about 100 mg, although some are capable of handling up to 1·5 g or more (e.g. Brownstone 1969; 1976; Carreira *et al.* 1980). The horizontal slab configuration may be particularly appropriate for nucleic acid separations in agarose gels or dilute polyacrylamide gels when substantial support must be given to the gel. One version has been described by Polsky, Edgell, Seidman, and Leder (1978). Most designs for use with polyacrylamide or composite polyacrylamide–agarose gels are of the vertical gel column type (e.g. Koziarz, Köhler, and Steck 1978; Hagen 1979), a configuration also generally favoured in most commercial equipment. The most critical features are as follows.

(1) Adequate cooling of the gel: temperature gradients across the direction of migration lead to curved bands and a loss in resolution, so while it is not possible to eliminate temperature gradients entirely it is clearly advantageous to minimize them.

(2) The volume and shape of the elution chamber: too large a volume leads to mixing of the separated components. To ensure good resolution it is also important that the sample component is rapidly and efficiently eluted into a fraction collector or other suitable collection device as soon as it has migrated out of the gel, and the geometry of the elution chamber is important in this respect.

(3) The porous membrane which in many designs of equipment forms one side of the elution chamber and ensures electrical contact but prevents macromolecular constituents passing out of the elution chamber into the electrode buffer: adsorption of sample material onto this membrane can result in significant losses and should be prevented.

One popular and highly versatile design is the Uniphor from LKB Produkter which, although relatively expensive, can be used with a wide range of supporting media (e.g. polyacrylamide, Sephadex, Biogel, agarose, cellulose, or a density gradient solution). It can also be used for isoelectric focusing and isotachophoresis as well as the various forms of electrophoresis. Interchangeable columns of different lengths can be used, so it is possible to be preparing a column for the next run while one run is in progress and the new column can be quickly 'plugged in' as soon as the first run has finished. Gel column heights of 5–10 cm with gels of $T = 4$–8 per cent are appropriate for many applications. The maximum protein load is of the order of 400 mg but naturally this will vary with the supporting medium and the type of experiment being performed. For most electrophoresis experiments on polyacrylamide gel it is best limited to about 80 mg.

Power supplies required for the column type of preparative electrophoresis equipment should be capable of supplying at least 500 V and up to 50 mA, but if the same apparatus is to be used for isoelectric focusing or isotachophoresis power packs giving 1000 V or more would be best.

Gel and buffer systems used for preparative work are usually identical to those already described in earlier chapters for analytical work, although formulations intended specifically for preparative experiments have been described by Duesberg and Rueckert (1965) and Jovin, Chrambach, and Naughton (1964).

It is important that the elution buffer contains the same counter ion as that used in the gel column (very often tris in basic gel buffers). The ionic strength should be somewhat higher than that in the separation gel, so that on leaving the gel phase the separated constituents enter a region of lower field strength. This slows up the rate of migration leading to the formation of sharper zones for elution and also tends to minimize adsorption on to the elution chamber membrane. If this adsorption is a problem it may be improved by incorporating urea into the elution buffer at a level of 6–8 M. Urea can of course also be used in the separation column itself, but it should be remembered that urea exists in equilibrium with cyanate which can react with the amino and sulphydryl groups of proteins in the sample. The formation of cyanate is both temperature and pH dependent, and carbamylation is negligibly slow below pH 5 or above pH 9. However, urea solutions should be deionized routinely before use by stirring 1l of 6–10 M urea for 15–30 min with 30–50 ml of a mixed-bed ion-exchange resin (e.g. Amberlite MB-1, Bio-Rex AG501-X8, or Permutit Deminrolit 225FF).

7.6. Biochemical applications

Preparative electrophoresis procedures are generally relatively small scale by comparison with other preparative methods available to the biochemist. Hence they are most often employed either as the final stage or towards the end of a preparative scheme which might well have included precipitations, centrifuging, ion-exchange chromatography, gel filtration, etc., all of which can handle much greater amounts of material.

Separation tasks faced in preparative electrophoresis usually fall into one of the following categories: (a) the separation of one or more components from a mixture containing a small number of components, or (b) the separation of a component from a complex mixture (e.g. blood serum, urine, CSF, etc.), or (c) cleaning-up an already largely purified substance by removing small amounts of contaminants. In (a) if the components are well separated from each other on examination by analytical PAGE, it will usually be relatively easy to find conditions whereby a size fractionation by gel filtration and/or a charge separation by ion-exchange chromatography will be satisfactory and will have greater handling capacity. However, if only milligram amounts of material are required any of the techniques desribed in this chapter will be appropriate and constant-path-length systems with continuous elution (Section 7.4) will be very suitable (e.g. Fig. 7.5) since the highest resolution will not be required. In (b), or in (a) when the components show only a small separation on analytical PAGE, high-resolution systems will usually prove most satisfactory. As a general rule this will militate against continuous elution equipment and the much greater separation powers of rod or slab analytical gels will need to be exploited. The separated components can then be recovered in milligram amounts from gel slices (Section 7.2), although some further fractionation may be required to remove gel-derived contaminants. For the final removal of impurities from an already largely purified substance ((c) above) continuous elution equipment will often prove suitable since high resolution may well not be needed and the substance will generally be recovered in high yield with very little contaminating material from the gel.

Virtually all the PAGE procedures described in previous chapters can be utilized in a preparative mode as well as for purely analytical procedures. Thus either homogeneous gels and buffers or discontinuous gels with

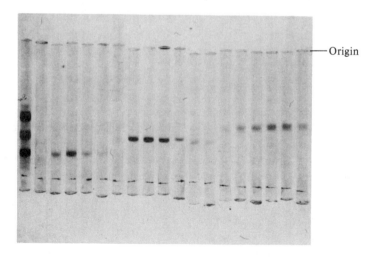

FIG. 7.5. Analytical PAGE of fractions collected from a column preparative gel apparatus. Samples of 10 mg each of ovalbumin, carbonic anhydrase, and myoglobin were applied to a column 120 mm long and 59 mm in diameter of gel with $T = 7.5$ per cent. (Reproduced from Koziarz et al. (1978) by permission of the authors and publishers.)

multiphasic buffers can be used in either the fixed electrophoresis time format or in constant-path-length methods. As a broad generalization homogeneous systems are commonly employed with constant-path-length equipment, whereas discontinuous systems are most often applied to the fixed-time approach where optimum resolution is being sought. Gels can be composed of polyacrylamide or agarose or a composite of the two, so that with very similar techniques and equipment nucleic acids and high-molecular-weight substances are as readily separable as small peptides and oligonucleotides. The basis for the preparative separation can thus be mainly related to size differences or to charge differences or both according to the composition of the gel, as in analytical PAGE.

Separations can be performed in the presence of urea or ionic and non-ionic detergents. In this context it should be remembered that proteins purified in the presence of SDS usually will retain little biological activity although they can be used as antigens (Weber and Osborn 1975). Tuszynski et al. (1977) have shown that proteins eluted from SDS gels can be further analysed by methods separating on the basis of charge (various PAGE systems) or isoelectric point (isoelectric focusing or isotacho-phoresis) if first treated with non-ionic detergents such as Triton X-100 or Nonidet NP-40. SDS–PAGE can also be combined with immuno-electrophoretic methods in two-dimensional procedures such as crossed immunoelectrophoresis if 1 per cent Triton X-100 is incorporated in the antibody-containing agarose gel used for the second dimension (Kirkpatrick and Rose 1978). A ten-fold excess or more of the non-ionic detergent can effectively compete with small residual amounts of SDS for binding sites on the protein molecules. This means that proteins that are difficult to solubilize (e.g. some membrane proteins) can be dissolved in SDS-containing buffer and then analysed by electrophoretic methods separating on the basis of differences in molecular charge in addition to size-dependent methods such as SDS–PAGE.

Unfortunately there have been some reports that preparative PAGE can result in protein modifications, including destruction of tryptophan, loss of biological activity (King 1970), and poor results in subsequent amino acid sequencing by the Edman degradation method (Weiner et al. 1972). Koziarz et al. (1978) showed that the inclusion of the antioxidant thioglycolate completely protected tryptophan. This was done by pre-running the preparative gel with sufficient thioglycolate in the electrode buffer to make the gel column have a 20 mM thioglycolate concentration after the run. Extensive destruction of tryptophan during the separation would make u.v. monitoring of column eluates inaccurate for protein measurement. Poor results in the Edman degradation reaction were shown to be due to blockage of the α-amino and perhaps ε-amino groups, probably by reactive carbonyl groupings in the gel via Schiff base formation. Since this linkage is alkali labile, running the electrophoresis at pH 9·0 or running at a lower pH of 7·4 followed by an incubation for 4 h at 58 °C in 0·5 M Quadrol at pH 9·0 prior to Edman analysis were both effective in reversing this blockage (Koziarz *et al*. 1978). Di Mari *et al*, (1982) included 1–2 mM reduced glutathione in the upper buffer reservoir and 1 mM thioglycolic acid in elution buffers in order to maintain reducing conditions in their work on the purification of tetanus toxin using a Buckler Polyprep 200 constant path-length preparative apparatus.

While these difficulties are not perhaps of wide importance outside these particular applications it is clear that the experimenter would be wise to consider the possibilty that the object of his study can be slightly altered during the preparative PAGE process.

8

IMMUNODIFFUSION AND
IMMUNO-ELECTROPHORESIS

When foreign proteins are injected into an animal under suitable conditions they behave as antigens. The animal produces an immune response and is then said to be immunized against these antigens. This response takes the form of the production of antibodies which belong to the immunoglobulin group of proteins in the blood plasma. The blood serum from immunized animals containing the antibody to the antigen under investigation is termed the antiserum. Antibodies are specific in their action and each antibody combines with a particular site on the surface of the antigen molecule. It is thus possible for two different antigens with identical or very similar binding sites to react with the same antibody, a phenomenon known as cross-reaction. This has proved useful in molecular studies for demonstrating the structural identity of certain regions in related molecules. However, in most cases the reaction of an antibody with an antigen is highly specific and results in the formation of an insoluble complex which precipitates out of the reaction mixture. At a point termed the equivalence point, all the antigen and antibody are bound in the form of precipitate. It is possible for the precipitate to dissolve in either excess antigen (e.g. with rabbit antisera) or in excess antibody (e.g. some human or horse antisera). The immunization of laboratory animals and production of antisera is outside the scope of this book and has been very adequately covered in immunological textbooks and handbooks, e.g. Herbert (1978) and Harboe and Ingild (1973). In addition a wide variety of antisera are now available commercially (e.g. Miles Seravac, Behringwerke AG, and Dakopatts A/S).

8.1. Immunodiffusion

In the classical immunoprecipitation or flocculation reactions (Ramon 1922; Dean and Webb 1926) identification and quantitative measurement was achieved by mixing a fixed quantity of antigen with serial dilutions of the antibody, or *vice versa*, since the formation of immunoprecipitates depends upon the ratio of antigen to antibody present. Precipitates are not formed when either is present in marked excess. Immunodiffusion consists in essence of replacing the series of dilutions in test tubes by diffusion through a gel. Concentration gradients of either one or both of the reactants are produced by diffusion and at a certain point in the gel the concentrations will be such that a precipitate can form. In a transparent gel this will show up as an opaque region. Samples containing two or more interacting systems will form a number of such opaque regions which will take the form of relatively sharp lines or arcs. Unless the precipitate arc is very dense it will not prevent the diffusion of other antigen or antibody components through it. Hence the number of arcs formed will correspond to the number of antigen–antibody systems that are present in amounts sufficient to form visible precipitates.

The pattern of precipitate arcs depends upon the way in which the diffusion experiment is set up. There are simple single-diffusion methods in which either the antigen or the antibody is incorporated into the gel and the other is allowed to diffuse into it, and double-diffusion methods in which the reactants diffuse towards each other. Single-dimensional, two-dimensional, and even three-dimensional (seldom used) arrangements can be designed by suitable geometrical positioning of the samples. The particular merits of many of these variations and interpretations of the patterns produced have been discussed by Ouchterlony and Nilsson (1978) amongst others, and therefore we shall confine ourselves here to some comments on the more widely used procedures.

If the pore size of the gel is much greater than the dimensions of the diffusing particles or molecules the diffusion behaviour will approximate to that in free solution. For the formation of immunoprecipitates it is usually advantageous if unhindered diffusion can occur, and it is therefore almost universal practice for agar gels of concentration 0·3–1·5 per cent to be employed as these exhibit minimal size sieving effects. (Electro-endosmosis does not occur in simple diffusion experiments with no electrophoretic step, so agar is usually used rather than the more expensive agarose.) If these are not convenient, either composite polyacrylamide–agarose gels or polyacrylamide gels of very low T may be suitable. Diffusion of antigens and antibodies proceeds readily in most cases, but once antigen–antibody complexes begin to form the gel pore size is quickly exceeded, diffusion ceases, and an immobile precipitate develops. Any apparent movement in a precipitin band is not a true movement of the band as such but rather is the result of a dissolution of precipitated material from one side of the band and a redeposition on the other. This behaviour occurs only in cases where either the antigen or the antibody is present in excess and the immunoprecipitate is soluble in the presence of the excess reactant. The band then appears to move towards the equivalence point.

The choice of buffer in the preparation of gels for immunodiffusion experiments is not critical, the only requirements being that it should not inhibit formation of the immunoprecipitate. According to Ouchterlony and Nilsson (1978) a pH range of 6·5–8·2 is best and is also least likely to induce non-specific precipitations. The concentrations of buffers and any other electrolytes can also have some effect on precipitate formation, but the solubility characteristics depend to some extent upon the particular immunoprecipitate under study. Thus many different buffer systems are doubtless perfectly suitable, but traditionally most simple immunodiffusion experiments have been conducted either in phosphate-buffered saline (0·01 M sodium phosphate buffer pH 7·1 containing 0·85 per cent NaCl) or in barbiturate buffers (e.g. 0·075 M diethylbarbiturate pH 7·2–8·5) either with or without added 0·85 per cent NaCl. Preservatives such as 0·01–0·05 per cent sodium azide or 0·1 per cent merthiolate can also be incorporated to inhibit bacterial growth during subsequent incubations.

Temperature strongly influences the rate of diffusion and hence an increase in temperature will increase the rate of precipitin formation and shorten experimental time. Although incubation temperatures within the range 0–60 °C have been successfully employed, in some cases the aggregation reaction may only take place satisfactorily in the cold so then gels should be incubated at about 4 °C.

8.2.   Analysis of antigens by immunodiffusion techniques

(a) *Single-dimension diffusion in tubes*

This is the most simple form of immunodiffusion (Oudin 1952).

The antibody solution or antiserum is mixed with melted agar solution in buffer at about 45–50 °C so that the final agar concentration is 0·3–0·5 per cent. Sufficient of this mixture is placed in a small pre-warmed glass test tube (3–5 mm internal diameter) to give a liquid column 30–50 mm high (Fig. 8.1). Tubes are cooled so that gelation occurs and then equilibrated at the desired incubation temperature. The antigen sample either in free solution or incorporated in 0·3–0·5 per cent agar is then added and the tubes sealed and incubated.

The formation of a precipitin band moving down from the interface between the gel layers demonstrates the presence in the sample of the antigen to the particular antibody incorporated in the lower gel layer. Not only can the presence of a specific antigen in an unknown sample be detected in this way but a quantitative estimate of its concentration can also be made. The displacement x of the band from the gel interface depends upon both time and antigen concentration, so that if one or more tubes containing known amounts of the antigen are incubated alongside the unknown and x is measured for all tubes after the same incubation time a standard curve can be constructed and the antigen concentration in the unknown read off. Incubation times of the order of up to 1 week are usual in such experiments. If a mixture of antibodies or an antiserum containing a number of antibodies is used in the lower gel and more than one corresponding antigen is present in the sample, then two or more precipitin bands will be observed, but the same principles apply and with suitable

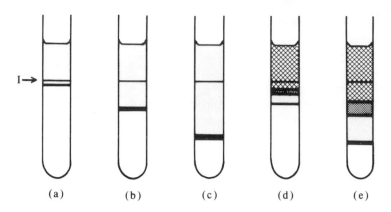

FIG. 8.1. Single-dimension immunodiffusion in glass tubes. The gel below the interface I contains antibody and that above I the antigen. Bands of immunoprecipitate form at I and move gradually downwards during the incubation period, the distance travelled depending both on the time and on the amount of antigen. Hence (a), (b), and (c) show a single antigen interacting with the antibody and could either represent increasing incubation time with similar amounts of antigen or progressively greater amounts of antigen with similar incubation times. (d) and (e) show more complex mixtures where there are three antigen–antibody interactions. The displacement of a band from the interface depends on the relative rates of diffusion of antigen and antibody, so it is even possible for a band to move upwards if the antibody should have the greater mobility but this is unusual. The shaded areas represent regions of gel where free antigen molecules are present.

standards both identification and quantitative measurement can be achieved.

(b) Single radial immunodiffusion

In this technique a layer of agar is poured into a Petri dish or on to a glass plate (e.g. of dimensions 10 cm x 8 cm, or a microscope slide) to give a gel slab of uniform thickness. It is often advisable in this and in other immunodiffusion and immuno-eletrophoretic experiments to pre-coat the glass plates to give better adhesion of the gel to the glass and to prevent leaks between gel and glass surfaces. Pre-coating is achieved by pouring a thin layer (e.g. 1 mm or less) of hot dilute agar solution (e.g. 0·1 per cent) on to the clean warm (60–80 °C) glass, spreading it over the whole surface, and allowing it to dry (on a hot plate, in a desiccator, or simply at room temperature).

For radial immunodiffusion (Ouchterlony 1958; Mancini, Carbonara, and Heremans 1965; Sakai *et al.* 1984) a melted agar solution in buffer or saline at about 48 °C is mixed with a solution of one of the reactants, generally the antibody, at the same temperature to give a final agar concentration of 0·8–1·5 per cent. A layer of this mixture 1–2 mm thick is then applied to the pre-coated plate on a flat horizontal surface and allowed to cool and gel. Gel plates can be stored in a humid atmosphere to prevent evaporation for up to 1 week at 4 °C. Holes of diameter 1–2 mm are then punched into the gel slab 10–20 mm apart to provide wells for the antigen samples. A sawn-off syringe needle attached to a water-pump vacuum line or one of the commercially available gel-cutting devices or gel punches (e.g. the System Centre from Shandon Scientific Ltd.) is used. Known volumes (up to about 5 μl, but depending upon the hole diameter and the gel thickness) of antigen solution in saline or buffered saline solution are placed in the sample wells with a micropipette or microsyringe. Standard solutions containing known amounts of antigen are also applied to other wells cut in the same slab and the whole slab is placed on filter paper moistened with buffer in a box which is then sealed and incubated (e.g. at 4 °C) for several days.

Rings of precipitate form around the antigen wells (Fig. 8.2) and as the diffusion process continues the diameter of the rings increases. The rate of diffusion and hence the ring diameter are also dependent upon the initial antigen concentration. Since antibody concentration is uniform throughout the plate and all samples on the plate are incubated together for the same length of time, the amount of antigen present can be determined by a simple comparison of the diameters of the precipitate rings produced by the unknowns and the standards. In practice an equilibrium point is reached between antigen and antibody diffusion after several days incubation and the precipitate ring does not increase further in diameter. It is at this point that measurements are usually taken and the ring diameter (or area) is linearly related to the antigen concentration and is readily calculated from the standard graph. The sensitivity range depends upon the antiserum (antibody) concentration in the gel. A low concentration gives maximum sensitivity but cannot be used for measurements where antigen is present in excess, and high concentrations permit quantitative measurement of more concentrated antigen samples but at the expense of a loss in sensitivity for more dilute samples.

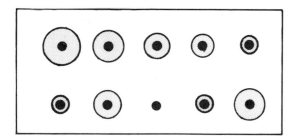

FIG. 8.2 Single radial immunodiffusion. In this technique the whole gel slab contains antiserum (antibody) and antigen is added to sample wells (filled circles). Rings of immunoprecipitate form and move outwards from the sample wells. The diameter of a ring depends on the incubation time and the antigen concentration, so after a fixed time the antigen concentration can easily be measured by comparing the size of rings from unknowns with those from known amounts of antigen.

(c) *Double-diffusion techniques*

While single radial immunodiffusion is often the method of choice for the quantitative measurement of antigens, double-diffusion methods are generally preferable for the demonstration of immunological identity. They are also useful for the analysis of more complex antigen–antibody systems.

Double diffusion can be performed in tubes (Oakley and Fulthorpe 1953) in a very similar manner to single diffusion (Section 88.2(a)), except that after addition of the antibody-containing gel a second layer about 8 mm thick of 1 per cent agar containing 1 per cent NaCl is placed on top of it. When this has gelled the antigen-containing solution or gel is added as previously. Antigen and antibody thus diffuse into the middle layer of gel from opposite directions and where the concentrations of the reactants are suitable a precipitate band is formed. Precipitate formation is usually recorded after incubation for 24 and 48 h, but if the middle layer is more than 8 mm thick longer incubation times may be needed.

It is much more usual for plate techniques to be used for double-diffusion experiments (Ouchterlony 1958). Plates are prepared essentially as described above (Section 8.2(b)) for radial immunodiffusion experiments except that no antigen or antibody is incorporated into the gel slab. Wells are then cut or punched into the gel slab for the application of antigens and antibody solutions. Several different geometrical patterns have been used for these wells (see e.g. Ouchterlony and Nilsson 1978), but the most widely used are shown in Fig. 8.3. A system of parallel slots (Fig. 8.3(a)) is really appropriate only for the demonstration of the

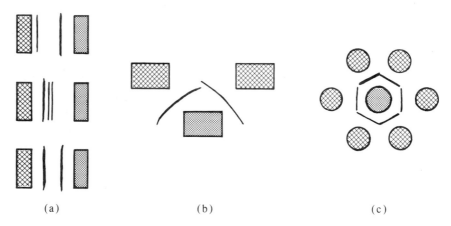

(a)   (b)   (c)

FIG. 8.3 Some common patterns of wells used for double-diffusion experiments. Shaded areas represent antibody (antiserum) wells and cross-hatched areas those for antigen samples. Bands of immunoprecipitate form between the wells as indicated by the solid lines.

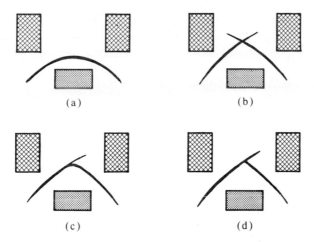

FIG. 8.4. Some typical patterns of immunoprecipitate formed with the three-well pattern. (a) Reaction type I: a smooth symmetrical arc indicates that the antigens in the two sample wells are immunologically identical. (b) Reaction type II: the lines cross showing that the antigens are different. (c) Reaction type III: related but not identical antigens give bands that partly fuse into an arc but also produce spurs. (d) Reaction Type IV: multi-specific antigens give patterns where one precipitate line inhibits the formation of the other and produces a sharp junction. (Based on Ouchterlony and Nilsson (1978) and reproduced by permission of Blackwell Scientific Publications Ltd.).

presence of a particular antigen in a sample and for a rather rough estimate of its concentration if a series of standards are analysed at the same time.

The three-well pattern (Fig. 8.3(b)) is used particularly for showing the immunological identity or otherwise of two antigen samples. These are placed in the two outer wells with antibody or antiserum in the centre well. In the original method (Ouchterlony 1958) three 10 mm x 10 mm square wells were used with gaps of 7·1 mm between the corners of the centre well and the outer wells. The development of precipitation lines is observed at intervals over a period of 1 week of incubation in a humid atmosphere at constant temperature. Some typical precipitation band patterns for balanced systems with neither antigen nor antibody present in large excess are shown in Fig. 8.4. In Fig. 8.4(a) the precipitation lines from each side fuse and form an arc, which in precisely balanced systems should be symmetrical. This indicates that the two antigens are immunologically identical. If the two antigens are different the lines cross without interacting (Fig. 8.4(b)). Antigenically related samples which are not completely identical give rise to spurred patterns (Fig. 8.4(c)), the spur being small and deviating considerably from the line of the original precipitate for closely related antigens but being more prominent and with less deviation for less closely related samples. Patterns such as Fig. 8.4(d) in which one precipitate line inhibits the formation of the other usually indicates the presence of multi-specific antigen particles in one of the samples. Further discussion of band patterns and their interpretations can be found elsewhere (e.g. Ouchterlony and Nilsson 1978).

The configuration for sample wells shown in Fig. 8.3(c) is nowadays probably the most common one for demonstrating the presence of a particular antigen, for semi-quantitative work, and for demonstrating immunological identity. The interpretation of band patterns is the same as that discussed above, namely fusion of precipitate arcs indicates identity, development of spurs shows some degree of relatedness, and a crossing of arcs shows non-identity. Conventionally, antibody solution or antiserum is placed in the central well and this configuration has the particular merit that it makes good and economical use of what may be expensive or scarce antibody solution. Of course like any other system this configuration can be inverted and antigen placed in the centre well if it is desired to screen a number of antisera for interaction with a specific antigen. This configuration is particularly suited to microanalysis for which 1–2 mm thick gels are appropriate and in which holes 1 mm in diameter are punched in a circle 20 mm in diameter around a central well 1 mm in diameter. Several such patterns can be cut in a single gel slab. Quantitative measurements are achieved by placing serial dilutions of the unknown antigen (or antibody) solution in the outer wells and noting the minimum dilution which forms a visible precipitate after suitable incubation. This is then related to a series of dilutions of a known standard solution incubated on the same plate under the same conditions.

8.3.   Immuno-electrophoretic techniques

In essence these techniques consist of a combination of an electrophoretic step with the formation of precipitates of antigen–antibody complexes. Single- and two-dimensional procedures can be used, and with different ways of applying the samples for either the first or second dimension a considerable number of variations have been described. Many of them rely upon the migration of antigens through or into an antibody-containing gel. Buffers and pH values are usually chosen so that only the antigens migrate and the antibodies either do not move at all or at most migrate only very slowly and thus remain evenly distributed throughout the gel during the whole electrophoresis.

Agar gels show very marked electro-endosmosis which effectively sweeps the antibodies through the gel. This is clearly undesirable, so although agar is commonly used in immunodiffusion work agarose is the usual medium for most immuno-electrophoretic experiments. With a 1 per cent gel and most commerical agarose preparations there is still a slight electro-endosmostic flow in the pH range 6–9 most commonly used. The extent of electro-endosmosis can be determined by applying a sample of an uncharged compound such as urea or dextran to the gel. Dextran of high molecular weight (e.g. 20 000) can be localized by precipitation on immersion of gels in 75 per cent ethanol and the displacement from the point of application is a measure of electro-endosmotic flow.

If a buffer pH of about 8·5 is chosen, electro-endosmosis is approximately cancelled out by the slow electrophoretic migration of the γ-globulins (the antibodies) which have pI values close to 8·0. As pointed out by Versey (1976) this approach can only be used for proteins showing a net electrophoretic migration towards the anode at this pH. While this holds for the great majority of proteins including most of the blood serum proteins, if necessary anodic migration can be increased or induced by blocking lysine ε-NH_2 groups to increase the net negative charge on the protein via chemical modification reactions such as acetylation, formylation, or carbamylation. Using such charge modification γ-globulins themselves can be used as antigens and studied in this way (Versey 1976). If necessary, agarose of very low electro-endosmotic flow can be prepared by ion-exchange chromatography (Johansson and Hjertén 1974).

Versey (1976) gives the following method for the formylation of proteins (sometimes known as the Sorensen titration). The protein solutions are diluted in 0·36 wt per cent formaldehyde solution to the final protein concentration that will be used in the subsequent immuno-electrophoresis. If sample volumes of 2–4 μl are used, an appropriate dilution of serum might be 1:10 for studies with IgG, IgA, or IgM for example. The protein–formaldehyde solution is incubated at room temperature for 20 min and the sample is then ready for analysis.

On a small scale carbamylation can be performed (Weeke 1973a) by mixing equal volumes (e.g. 50 μl) of protein solution at a pH of 7–8 and of 2 M potassium cyanate. After reacting at 45 °C for 30 min the mixtures are cooled to 10–15 °C and diluted with buffer to the protein concentration required for immuno-electro-phoresis. Reference proteins or sera should be treated in the same way to ensure a comparable degree of carbamylation. For larger-scale work Bjerrum, Ingild, Løwenstein, and Weeke (1973) suggest that about 70 mg of antigen or antibody protein is dissolved in 10 ml of solution containing 0·1 mmol NaCl, 2·0 mmol potassium cyanate, 0·1 mmol sodium tetraborate, and 0·04 mmol HCl. During reaction at 45 °C for about 4 h the pH is kept constant by additions of 0·2 M HCl. Carbamylation is terminated and the protein is separated from the other constituents by gel filtration on a column of Sephadex G-25.

(a) Apparatus for immuno-electrophoresis

Slab gel systems are nearly always used with the gel in the horizontal position both during sample application and during the separation stage. Apparatus of the type described in Section 2.6(c) (Fig. 2.7) are suitable, and kits of all necessary parts for the adaptation of the Shandon Model

600, LKB Multiphor, and Pharmacia FBE-3000 apparatus to immuno-electrophoretic work are available from the manufacturers. These enable agarose-covered glass plates such as microscope slides to be handled with ease. Constant-voltage power supplies with outputs of 50–100 V are adequate since gels need not usually exceed 10 cm square.

(b) *The method of Grabar and Williams (1953)*

This method consists of an electrophoretic separation followed by an immunodiffusion stage.

A 1-2 per cent agarose solution in, for example, 0·05 M veronal buffer pH 8·5 and containing 0·01 per cent merthiolate as preservative is prepared in the usual way by heating in a water bath and poured onto a pre-coated glass plate to give a gel layer about 2 mm thick. If a 75 mm x 25 mm microscope slide is used (Scheidegger 1955), the gel layer can be rather thinner (1–2 mm) and about 2 ml of mixture per slide is required. Sample wells of diameter 1·2 mm will hold approximately 1 μl. The sample well is cut about 25 mm from the cathode. After filling the well with antigen solution the slide is placed in the apparatus and the necessary electrical connections made with filter paper wicks. The gel is covered with a glass plate or other suitable cover to prevent surface evaporation and the electrophoresis is performed at 6–8 V cm^{-1} for about 40 min for a typical serum sample. A trough of dimensions about 50 mm x 2 mm is then cut for the antibody solution parallel to the line of electrophoretically separated components and at a distance of about 4 mm to one side of the sample well (or one trough on each side if desired). The trough is then filled with about 100 μl of antiserum solution and diffusion is allowed to take place in a humid atmosphere in the usual way.

The pattern of precipitate bands (Fig. 8.5) develops within about 24 h, and gels can then be pressed, washed, and stained (see Section 8.4). The application of this technique to human serum proteins (Jefferis 1975) is shown in Fig. 8.6. A very similar method was used by Hardy, Hoffman, and Ossimi (1984) in their studies of ß-glucocerebrosidase from normal and Gaucher disease samples, but in ß this case instead of electrophoresis the initial separation method consisted of IEF in agarose gel on a Gelbond support. Strips of gel between sample lanes were cut out and replaced with new agarose in phosphate-buffered saline in which troughs were then cut and filled with antibody for detection of antigens by immunodiffusion as above.

Interpretation of immunodiffusion patterns becomes rather difficult when complex mixtures are analysed with multi-specific antisera. The chief advantage of this method is that the electrophoretic step 'pulls out' the pattern and improves resolution. The electrophoretic mobility of the component giving rise to a particular precipitate arc may also be a valuable aid towards identifying it. Further identification is provided by using a monospecific antiserum or antibody solution. It may be helpful to place a

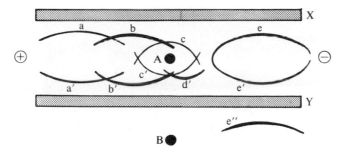

FIG. 8.5. Immuno-electrophoresis by the Grabar and Williams (1953) technique. Two antigen solutions are in the wells A and B with antisera in the troughs X and Y. After electrophoretic separation of antigen components across the plate, X and Y are filled with the antisera. Precipitin arcs such as a,a', b, and b' develop on incubation and in time may meet. If the arcs cross (c and c') this means that both antisera X and Y react with antigens in A with this electrophoretic mobility but not with the same antigen. If the arcs meet and fuse into a smooth curve (e and e') then both X and Y contain an antibody reacting with the same antigen. Comparing the two antisera X and Y, both contain antibodies against e and possibly against a and b although this is only based on electrophoretic mobility and immunological identity is not proved until the arcs meet. The antibody against c is present only in X and those against c' and d' only in Y. Comparing the antigens A and B, clearly B lacks any components corresponding to a', b', c', and d'. If a and b are subsequently found to be identical to a' and b', then obviously B lacks these too. However, it cannot be assumed that B lacks the antigen giving c because the antiserum Y does not contain the antibody to this. It would appear probable that e" may correspond to e and e' but again this only shows that there is a component in B with a similar electrophoretic mobility to one in A. Of course in addition B may contain any number of antigenic components to which Y does not contain antibodies. In other cases a single arc like this could also be produced by a pure antigen or a mixed antigen with a monospecific antibody in Y.

general multi-specific antiserum in one trough and a monospecific antiserum in the trough on the side of the separated components. The ultimate fusion of precipitate lines into a continuous curve shows serological identity. A particular antigen can also be identified by using a multi-specific antiserum pre-treated with the pure antigen. After removal by centrifugation or absorption of the resulting complex (Ouchterlony and Nilsson 1978) the antiserum will be devoid of antibody against the antigen used. If the complete antiserum is placed in one trough and the treated antiserum in the other, the absence of a band in the pattern given by the latter provides the necessary identification.

(c) 'Rocket' and 'fused-rocket' immuno-electrophesis

This can be regarded as the electrophoretic counterpart of single radial

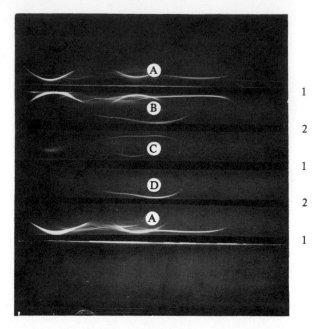

FIG. 8.6. Immuno-electrophoretic analysis: A, normal human serum; B, IgD myeloma serum; C, an IgD fraction obtained by gel filtration; D, IgD purified by DEAE cellulose ion-exchange chromatography. Antisera: 1, sheep anti whole human serum: 2, sheep anti human IgD. (Reproduced from Jefferis (1975) by permission of the author and publishers.)

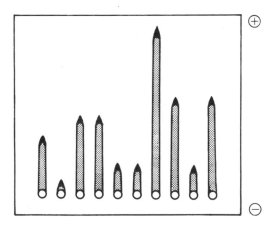

FIG. 8.7. 'Rocket' immuno-electrophoresis. The peak area and peak height are linearly related to the antigen concentration.

immunodiffusion and it is likewise particularly useful for assaying antigen solutions of unknown concentration. It can be readily applied to routine work and can be used in all cases where the antigen proteins have a molecular charge differing from that of the antibody proteins. The method (Laurell 1966) consists of the electrophoretic transport of antigen proteins through a gel slab containing the antibodies at a pH value at which the antibodies remain essentially immobile.

Thus for the analyses of serum components, for example, a 1·5-mm-thick slab of 1 per cent agarose gel containing an appropriate amount of antiserum or other antibody solution is prepared in a buffer of pH about 8·6. Gel slabs usually have dimensions between about 70 mm × 100 mm and 200 mm × 100 mm with the electrophoretic migration across the 100 mm dimension. Sample wells are then punched in a line about 20 mm from the side of the gel. Sample wells should be filled almost to the brim with antigen solution, so in a 1·5 mm thick gel wells of diameter 2 mm will take up to 3–4 μl, while wells of diameter 3 mm will hold 6–10 μl and wells of diameter 4 mm can be used with 10–15 μl or more. Since the method is quantitative it is important to measure the sample volume accurately (e.g. micropipette) and this is more easily done with 10 μl volumes than with smaller amounts. The centres of adjacent sample wells should be 7–10 mm apart, so the smaller 70 mm plates will take about eight samples whereas a plate 200 mm wide is suitable for up to about 30 samples. A number of standard samples containing known amounts of antigen are applied as well as the unknowns and electrophoresis is performed at about 10 V cm^{-1} for 2–4 h or at 2 V cm^{-1} or less overnight. The gel can then be pressed, washed, dried, and stained (Section 8.4).

A typical separation is shown in Fig. 8.7. The area enclosed by the precipitate 'rocket' at a given antibody concentration in the gel is linearly related to antigen concentraion and for a constant amount of antigen varies inversely with the gel antibody content. In practice quantitative measurement is usually achieved by measuring the peak heights and relating those of the unknowns to the standards run on the same slab. This is much more rapid than area determination by planimetry or by height times width at half height and is sufficiently accurate for most purposes. With sample volumes of only 1–4 μl and Coomassie Brilliant Blue R250 staining antigen concentrations as low as 0·1 mg l^{-1} can be measured, but the use of radiolabelled antibodies increases sensitivity by a factor of 60 or more (Kindmark and Thorell 1972). Ideally for best quantitative measurement 'rocket' heights should be about 10–50 mm. If the precipitates are too faint more antiserum should be used in the gel (1 per cent is often about right for commercial antisera), but if precipitation is very strong with precipitation 'inside the rocket' less should be used. Weeke (1973b) gives useful data on the dilution of serum samples for the quantitative measurement of major blood serum constituents and factors which may influence the formation and quantitative measurement of rockets.

'Fused-rocket' immuno-electrophoresis is a variant on the basic method and is used particularly for the demonstration of impurities and the identification of fractions obtained in separation experiments such as gel filtration or ion-exchange chromatography.

It is performed exactly as normal 'rocket' immuno-electrophoresis except that a relatively large number of sample wells (up to 40 or more on a 100 mm x 100 mm plate) are cut quite close together (4–5 mm), often in two closely spaced rows. Samples are applied in the same sequence as obtained in the separation experiment and allowed to difuse for 30 min at room temperature or 1 h at 10 °C. Overnight electrophoresis at 2 V cm^{-1} followed by the usual staining procedure gives patterns such as that shown in Fig. 8.8. When a broad non-specific antiserum is used each antigen in the sample gives rise to a continuous precipitation line, the distance of which from the sample well is proportional to the amount of antigen. For proteins of low concentration there may be losses around the sample wells due to precipitation during the diffusion step. If this is suspected a slab without antiserum should be poured and the sample wells are cut in this. The gel slab is then cut along a line 5 mm from the sample wells and the larger part of the slab discarded. The plate is then refilled with agarose gel containing antiserum. In this way diffusion occurs in the absence of antibodies and all samples have to traverse a short length of gel before encountering antibody (Svendsen 1973).

(d) *Crossed immuno-electrophoresis*

The method of single-dimension electrophoresis followed by immuno-

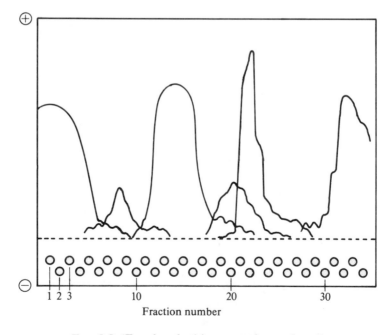

FIG. 8.8. 'Fused-rocket' immuno-electrophoresis.

diffusion proposed by Grabar and Williams (1953) suffers from the disadvantage that quantitative measurements are not easily made and 'rocket' immuno-electrophoresis is not readily applied to the analysis of complex mixtures. Crossed immuno-electrophoresis (sometimes called simply two-dimensional immuno-electrophoresis) is a combination of these techniques which can be used both qualitatively and quantitatively. First introduced by Laurell (1965) and modified by Clarke and Freeman (1968), the method consists of electrophoresis in one dimension followed by a second-dimension separation at right angles to the first consisting of electrophoresis into an antibody-containing gel slab.

A gel slab 80 or 100 mm square and 1·5 mm thick is prepared in the usual way. A sample well 2–5 mm in diameter is cut about 15 mm in from one side and filled with up to about 30 μl of test antigen mixture. For analysis of blood serum a little Bromophenol Blue can be added to stain the serum albumin as an electrophoretic marker. Electrophoresis is then performed (8–10 V cm^{-1} for about 1 h may be appropriate), after which the gel is cut 20–25 mm in from the edge and most of the gel slab removed and discarded leaving a strip containing the electrophoretically separated components on the plate. Antibody-containing gel is then poured onto the rest of the plate and after solidification electrophoresis is performed in the second dimension at 1–2 V cm^{-1} overnight. The gel is then pressed, washed, dried, and stained to give patterns such as that shown in Fig. 8.9.

The amount of antigen in a peak is proportional to the area which is obtained by planimetry or by the product of measurements of the height and the width at half height. Quantitative measurement is achieved by relating these areas to corresponding peaks obtained from a standard mixture containing known amounts of each of the antigens under study. Clearly this has to be run on a separate slab from the test sample, and although care should be taken that both slabs are run under identical conditions (preferably side by side) small differences may easily occur. The accuracy of quantitative measurements can therefore be greatly improved by adding an internal standard such as acetylated albumin, carbamylated transferrin, etc. to both samples. This internal standard should be easily identifiable on the final gel pattern and equal amounts are added to each sample so that areas can be related to it. For routine screening purposes it may suffice to express peak areas relative to the internal standard (a constant amount being added to all samples) in a whole series of test antigen mixtures (e.g. pathological sera etc.). For most proteins the lower limit of sensitivity with stained gels is about 0·1–1·0 mg l^{-1}. The size of peaks may be manipulated by changing sample volumes or the amount of antiserum; more sample, or less antibody, in the gel increases peak size and the converse decreases it.

A variant of this method is crossed hydrophobic interaction immuno-electrophoresis (Bjerrum 1978). In this the first dimensional separation is performed in a gel composed of a mixture of equal volumes of 2 per cent agarose in the chosen electrophoresis buffer and a warmed slurry of an

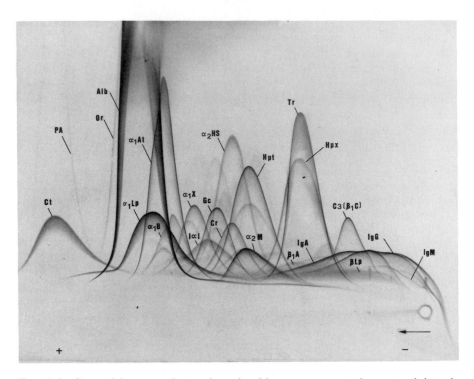

FIG. 8.9. Crossed immuno-electrophoresis of human serum antigens precipitated with rabbit polyspecific antiserum to whole human serum: Ct, carbamylated human transferrin (reference standard); PA, prealbumin; Alb, Albumin; Or, orosomucoid (α_1-acid glycoprotein); α_1LP, α_1-lipoprotein (high density lipoprotein); α_1AT, α_1-antitrypsin; α_1B, α_1-B-glycoprotein; α_1X, α_1-antichymotrypsin; IαI, inter-α-trypsin inhibitor; Gc, Gc-globulin; α_2HS, α_2HS-glycoprotein; Cr. ceruloplasmin; Hpt, haptoglobin; α_2M, α_2-macroglobulin, Hpx, haemopexin; ß1A, ß$_1$A-globulin; Tr, transferrin, IgA, immunoglobulin A; ß$_1$C, ß$_1$C-globulin (C3 complement); ßLp, ß-lipoprotein (low density lipoprotein); IgG, immunoglobulin G; IgM, immunoglobulin M. (Reproduced from Cline and Crowle (1979) by permission of the authors and publishers. © 1979, *Clinical Chemistry*)

alkyl or aryl agarose as used for hydrophobic interaction chromatography (e.g. phenyl- or octyl-Sepharose CL-4B, Pharmacia Fine Chemicals AB). The second dimensional step is performed as above except that 1 per cent Triton X-100 is incorporated into the antibody-containing gel. The method can differentiate between hydrophilic and hydrophobic proteins in crude mixtures and has been applied both to serum and to membrane proteins. It can also be used as a rapid method to predict the results of hydrophobic interaction chromatography.

(e) *Tandem-crossed immuno-electrophoresis*

This is performed in the same way as crossed immuno-electrophoresis except that two sample wells are cut 6–8 mm apart on the axis of the first-dimension electrophoresis. To avoid distortion of the electrophoretic separation, the samples are allowed to soak into the gel (a few minutes at room temperature) and the empty wells are then sealed with a drop of agarose. The method is used for the direct comparison of two antigen mixtures A and B, one of which is applied to each well. The resulting gel pattern (Fig. 8.10) will therefore consist of a series of twin peaks for all components that are common to both A and B, but clearly components that differ or are missing in one of the samples will give rise to single peaks that do not have a corresponding 'twin'. Along with rocket and crossed methods, tandem-crossed immuno-electrophoresis was useful in studies of relationships between the I/i and ABH blood group antigens of human erythrocyte membranes (Oppenheim, Nachbar, and Blank, 1983).

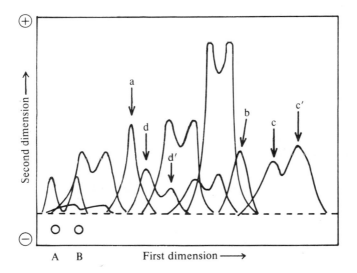

FIG. 8.10. Tandem-crossed immuno-electrophoresis in which two antigen solutions applied to the wells A and B are compared. Most components in A and B are identical and present in similar amounts, and so they give twin peaks with a lateral separation about equal to that of the two sample wells. However, peaks a and b are not twinned and so represent antigen components in A or B (one cannot distinguish which) that are not present in the other antigen solution. peaks c and c′ are given by a single antigen component that is present in lower amounts in A than in B, whereas the antigen d is present in larger amounts in A than in B.

(f) *Line immuno-electrophoresis*

In this method rectangles of sample antigen-containing gel are inserted into an antiserum-free gel slab. The slab is then cut close to the line of samples, and the larger part of the gel is discarded and replaced by anti-serum-containing gel. After electrophoresis it is found that each antigen for which the antiserum contains the corresponding antibody gives rise to a precipitin line parallel to the origin (Fig. 8.11) and at a distance from it proportional to the antigen-antibody ratio (Krøll 1973*a*). The continuity of precipitin lines between adjoining patterns from different samples provides qualitative evidence of identity, and the position gives a quantitative comparison, or if one of the samples is a standard can give absolute quantitative measurements. The method is appropriate for the comparison of complex antigen mixtures if a polyvalent antiserum is used and it is also useful when it is desirable to cut out areas of gel containing a precipitate band. These can then be used for immunization and the production of specific antisera.

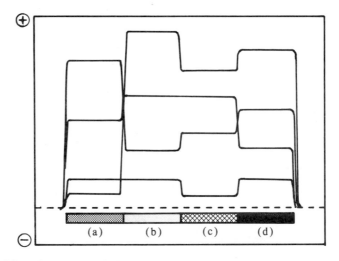

FIG. 8.11. Line electrophoresis for the comparison of four sample antigens (a), (b), (c) and (d) against a common antiserum in the gel slab.

The method can also be used for the comparison of a number of different antisera and in this case a single strip of gel containing the antigen is applied across the gel in place of the rectangles used above. Electrophoresis is then performed into areas of gel containing different antisera, which thus can be compared both qualitatively and quantitatively (Fig. 8.12).

Two further variations of line immuno-electrophoresis have been described (Krøll 1973*b*, 1973*c*), namely 'crossed-line' and 'rocket-line'.

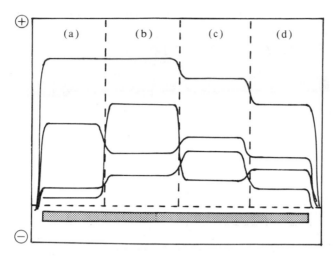

FIG. 8.12. Line immuno-electrophoresis for the comparison of four antisera in the regions (a), (b), (c) and (d) against a common antigen in the shaded strip of sample gel.

In the 'crossed-line' technique a sample of antigen(s) is first placed in a sample well and subjected to electrophoresis in the same way as for crossed immuno-electrophoresis. The strip of gel containing the separated antigen components is then placed about 8–10 mm below a further strip of gel containing the mixed sample antigens, as used in line immuno-electro-phoresis (Fig. 8.13). The antigens in both gel strips are then electrophoresed in the second dimension into agarose containing the corresponding polyvalent anti-serum and the resulting precipitate pattern consists of peaks (as in crossed immuno-electrophoresis) standing on individual baselines (Fig. 8.13). If the line immuno-electrophoresis pattern given by an appropriate standard antigen is known, this pattern permits identification of unknown peaks in a crossed immuno-electrophoresis pattern and *vice versa*. Likewise in quantitative work, if the peak area given by a known amount of antigen in the lower sample well is measured, the amount of this antigen in the strip of sample gel can be calculated from the area below the corresponding precipitate line, and again *vice versa*. A closely related technique is crossed-line immuno-electrofocusing (Guinet 1983) which is performed in the same way as the above except that isoelectric focusing (IEF) in agarose or polyacrylamide gel is used for the first step instead of electrophoresis.

'Rocket-line' immuno-electrophoresis is a one-dimensional technique in which a series of antigen sample wells are cut in antiserum-free gel, 5 mm below a strip of sample gel containing a mixture of all the antigens to be

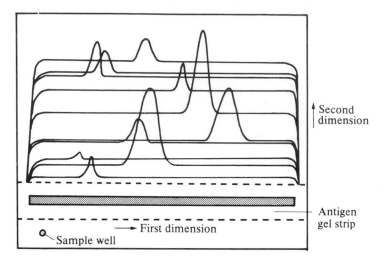

FIG. 8.13. Crossed-line immuno-electrophoresis.

added to the individual sample wells. As in the above, the resulting pattern
(Fig. 8.14) after electrophoresis into polyvalent antiserum-containing gel is
a number of lines each of which represents the sum of each particular
antigen in the sample strip and the underlying sample wells. From the line
pattern given by a known antigen it is thus possible to identify the presence
of the antigen in one or more of the individual samples applied to the wells.
As in normal 'rocket' immuno-electrophoresis the area of a peak (or peak
height) above the base line given by the appropriate line in the line pattern
can be used for quantitative determination. Krøll (1973c) states that the
ability to identify and quantify 20–30 proteins simultaneously in a number
of samples on a single plate makes the method particularly suitable in a
number of screening applications (e.g. Daussant, Renard, and Skakoun
1982).

(g) *Counter immuno-electrophoresis*

Counter immuno-electrophoresis, which is sometimes referred to as
'crossed-over' electrophoresis, should not be confused with crossed
immuno-electrophoresis (see Section 8.3(d)). Unlike all the other forms
of immuno-electrophoresis, this method (Culliford 1964) is performed in
agar gels of high electro-endosmotic flow. The conditions (i.e. pH etc.) are
chosen so that the antigen(s) moves towards the anode while the γ-globulin
antibodies are carried towards the cathode by the electro-endosmotic flow.
The gel slab is prepared in the usual way in a buffer of pH about 8·5, and

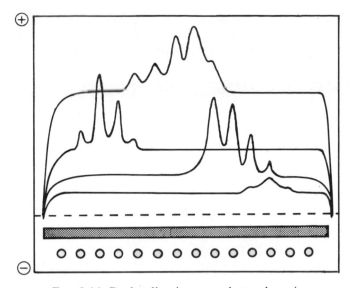

FIG. 8.14. Rocket-line immuno-electrophoresis

pairs of sample wells 1·5–2·0 mm in diameter are cut 5 mm apart. The antigen(s) is placed in the sample well on the cathode side of each pair of wells and the antibody (antiserum) in the well on the anodic side. When electrophoresis at about 5 V cm^{-1} begins, the antigen and antibody move towards each other, meet, and form a precipitin band (Fig. 8.15). The run

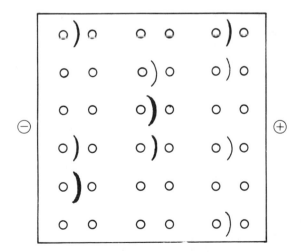

FIG 8.15. Counter immuno-electrophoresis

usually takes only about 15–20 min so the method is extremely rapid and a very large number of samples can be examined on a single gel slab. If required the gel can be stained in the usual way. The method is particularly valuable in that only very small amounts of antigen and antibody are required. The ability to deal very rapidly with many samples makes the method very attractive for screening work, although its use is confined to a purely qualitative role in the detection of unknown antigens or antibodies (e.g. in forensic work or when screening patients for the presence of particular antibodies arising from pathological conditions or allergenic reactions).

8.4.   Enhancing the sensitivity of immunodiffusion and immuno-electro- phoresis experiments by staining and labelling techniques

For many purposes the formation of an opaque precipitin arc in a transparent gel will be relatively obvious and perfectly adequate. If desired, a permanent record can be made by photography using dark-ground illumination. In addition, dark-ground viewing boxes are commercially available (e.g. Shandon Scientific Instruments Ltd.) for examining immunodiffusion and immuno-electrophoretic patterns.

(a) *Pressing, drying, and staining methods*

The sensitivity of detection can be greatly enhanced by the application of a suitable staining technique. Since antibodies are proteins all immuno-precipitates can be stained with the usual protein stains (Table 2.4). First of all it is necessary to remove non-complexed protein (antigens and antibodies) from the gel to give acceptably low background staining. This can be achieved by washing the gel for several hours or days with several changes of phosphate buffered saline (0·15 M NaCl containing 0·01 M sodium phosphate pH 7·1) or with 0·15 M NaCl alone. The gel is then covered with a filter paper and allowed to dry at room temperature overnight. Before removal the covering filter paper should be moistened to reduce adhesion of paper fibres to the gel surface.

Rather better than this method is to combine the washing with pressing and drying steps. The following procedure (Weeke 1973a) is suitable in most cases.

(1) The gel is covered with a piece of wet filter paper (avoid trapping air bubbles) which is then covered with a layer of soft blotting paper 2–3 cm thick. A slight pressure (about 10 g cm^{-2}) is applied by the addition of a thick glass plate and books, wooden blocks, etc. After 10–15 min it will be found that the gel is reduced to a thin film.

(2) The gel film is washed twice with 0·1 M NaCl for a minimum of 15 min each time. Often an overnight washing may be necessary to avoid background staining completely.

(3) Wash in distilled water for at least 15 min.
(4) Repeat pressing stage as in (1).
(5) Dry pressed gel film in a stream of hot or cold air from a hair dryer or on a hot plate at about 60 °C (unpressed gels may crack or melt if dried rapidly in this way).

Apart from the accelerated removal of uncomplexed protein during the washing step there is an additional benefit to be gained from pressing gels in that thin films can be stained and destained very rapidly. The glass plates bearing the dried gel films are placed in staining solution for 10–15 min, destained for 20–30 min using three changes of destaining solvent, and dried again as above. Although these suggestions were made (Weeke 1973a) with reference to the use of 0·5 per cent Amido Black 10B or Coomassie Brilliant Blue R250 in ethanol:acetic acid:H_2O (9:2:9 by volume), with the same solvent used for destaining, similarly accelerated staining and destaining should be expected if any other protein stain (Table 2.4) is used.

While all immunoprecipitates must of necessity react with protein stains, phosphoprotein, glycoprotein, and lipoprotein antigens can be distinguished using the appropriate staining procedures given in Tables 2.7 and 2.8 and Section 2.9(c). Periodate oxidation methods for glycoproteins cannot be used with agarose gels of course owing to interference from the gel itself. These specific stains can be followed by a non-specific staining of all the precipitate arcs on the same plate using one of the proteins stains as above. Even more specific staining procedures for various enzyme activities (as used for detection on PAGE or other gels) can often be applied to the identification of appropriate precipitate bands. Most enzymes retain their activity under the conditions of immunodiffusion or immuno-electro-phoresis and following the formation of complexes and precipitates with antibodies.

Very approximately the lower limit of detection of most protein antigens is of the order of 5–10 μg ml^{-1} of sample solution if no staining is used. Protein stains increase the sensitivity by at least an order of magnitude, with Coomassie Brilliant Blue R250 or Xylene Brilliant Cyanine G increasing sensitivity more than Amido Black 10B. Fluorescent staining with o-phthaldialdehyde (OPA) can also be used (Table 2.4) and has similar sensitivity to protein dye staining. Silver staining, as described on p.150, for protein zones in agarose gels, is also suitable.

(b) *Fluorescent and enzyme prelabelling techniques*
An alternative to staining is the prelabelling of one of the reactants, usually the antibody, with a fluorochrome.

For example, 1 mg ml^{-1} fluorescein isothiocyante (FITC) is dissolved in 0·15 M $Na_2 HPO_4$ at pH 9·0, and 1 ml of the freshly prepared solution is added to 100 mg of antibody protein contained in 4 ml or less of solution. After stirring for 60 min at room temperature, during which portions of 0·1 M Na_2PO_4 may need to be added to maintain a pH of 9·0, unreacted FITC is separated from labelled antibody by gel filtration on Sephadex G-25 (The and Feltkamp 1970).

Weigele *et al.* (1973) introduced 2-methoxy-2, 4-diphenyl-3(2H)-furanone (MDPF) for fluorescent labelling of antibodies. This has the advantage that MDPF itself and its hydrolysis products are non-fluorescent so a gel filtration step is not required.

They suggest that 150–250 mg protein in 10 ml of 0·05 M borate-buffered saline pH 9·5 is cooled in an ice bath and 1 ml of acetone containing 2–4 mg MDPF is added over a period of 60 min with stirring. After a further 60 min stirring the resulting solution can be dialysed against phosphate-buffered saline and is ready for use.

Prelabelling with OPA or fluorescamine can also be employed and suitable methods are given in Table 5.5, although in these cases the final treatment with SDS should be omitted and it may be necessary to remove excess reagent by a precipitation, dialysis, or gel filtration step to avoid subsequent reaction with antigen molecules.

A further alternative is to couple an enzyme chemically to the antibody (or antigen) molecules by cross-linking with reagents such as carbodiimide, glutaraldehyde, dimethylsuberimidate, etc. Almost any enzyme for which a suitable assay method exists could in theory be used, but a very popular one is alkaline phosphatase for which there are several simple and sensitive detection methods (see Section 2.9). In order to preserve maximum activity it is preferable to avoid drying the gel slabs after washing or unnecessarily heating them.

The fluorescent or enzyme-tagged antibody is used in place of the unlabelled antibody in any of the above techniques. After formation of the antigen–antibody precipitate the excess reagent is washed out in the usual way and the precipitates are visualized by u.v. illumination or with an enzyme activity stain. In addition to direct substitution of labelled for unlabelled antibody in the above methods, indirect (double-antibody) methods have also been used very successfully. For example (Olden and Yamada 1977) antigens are allowed to form the usual complexes with goat antisera and excess antisera are washed out. The gels are then incubated a second time with anti-goat immunoglobulin raised in rabbits. The second antiserum contains antibody molecules conjugated to horseradish peroxidase so that after washing out the unreacted antibody the precipitate bands are labelled with the enzyme activity. A fluorescent label can of course be used in the same way.

(c) Radiolabelling

Radiolabelling of immunoprecipitates can be achieved in one of two ways.

The most straightforward procedure is to incorporate a suitable label (e.g. ^3H, ^{14}C, ^{125}I, etc.) by chemical treatment of the antiserum or antibody solution (see Appendix 1). These are then substituted for all or part of the antiserum or antibody in the sample wells or within the gel matrix depending upon the type of experiment being performed. After the appropriate incubation time, gels are washed to remove non-precipitated 'background' anti-serum components, pressed and dried as above, and subjected to auto-radiography (see Section 2.12(b)).

Rather more elegantly and more economical in terms of material is double-antibody labelling. An example of this was first given by Rowe (1969) who used a sheep anti-IgE serum for detection of human IgE. After formation of the immunoprecipitates and washing, the plates were immersed for 30 min in rabbit anti-sheep immunoglobulin serum labelled with ^{131}I. After an overnight incubation unreacted labelled serum was washed out and the plates were dried and subjected to autoradiography in the usual way. By such methods antigen concentrations as low as 10–20 ng ml^{-1} can be detected and quantified.

A variation of this procedure has been described by Bigelis and Burridge (1978) and Adair, Jurivich, and Goodenough (1978) in which the second labelled antibody (antiserum) solution is replaced by a solution of *Staphylococcus aureus* protein A (Pharmacia AB). This had previously been labelled with ^{125}I by the chloramine-T procedure (see Appendix 1). Enzyme-labelled (e.g. horseradish peroxidase or alkaline phosphatase) *Staph. aureus* protein A can also be used just as successfully (e.g. Taketa 1983), followed of course by detection of the bound enzyme activity, in systems analagous to those used for detecting antigens on protein blots (Section 2.13(c)). Protein A is a cell-wall component of the bacterium which binds specifically to the Fc regions of immunoglobulins and hence with any antigen–antibody complexes within the gel. Its use has the advantage that it labels all antigen–antibody complexes, and so in a number of experiments using various antibodies or antisera it is not necessary to prepare different labelled species-specific antiglobulins.

8.5. The use of detergents

Relatively low levels of urea (e.g. 1·5 M) do not interfere with antigen–antibody interactions and may assist in solubilization of samples and preventing aggregation but are unlikely to be very effective, while higher levels will interfere with the formation of immunoprecipitates. Thus detergents are more widely used to enhance sample solubility, sometimes with the addition of urea as well (Demus and Mehl 1970; Bjerrum and Lundahl 1973; Shapshak *et al.* 1983).

(a) *Non-ionic detergents*

Non-ionic detergents such as Tween 20, Berol EMU-043, Lubrol WX, and Triton X-100 do not seriously interfere with antigen–antibody complex formation. Concentrations within the range 0·1–2·0 per cent have been used to solubilize samples and to maintain solubility during the subsequent immunodiffusion and immuno-electrophoresis experiments by simple incorporation of the detergent into the agarose gel mixture (Fig. 8.16). The detergent inhibits the non-specific precipitation of, for example, membrane proteins which may otherwise aggregate and precipitate during the separation (e.g. Bjerrum and Bøg-Hansen 1976). The binding of detergent to human erythrocyte membrane proteins has been clearly shown by Bjerrum and Bhakdi (1977). By using [14]C- or ^{125}I-labelled Triton X-100 these workers were able to make use of this fact by demonstrating that radioactive immunoprecipitates were formed which could then be located by autoradiography.

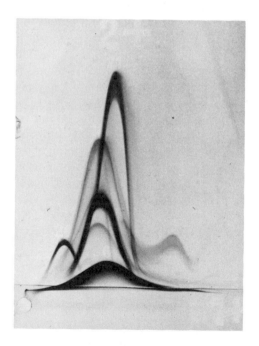

FIG. 8.16. Crossed immuno-electrophoresis pattern of human erythrocyte membrane proteins solubilized with Berol EMU-043 in agarose gels containing 1 per cent of this detergent. The sample was 20 μg membrane protein. The gel was stained for protein with Coomassie Brilliant Blue R250. (Reproduced from Bjerrum (1977) by permission of the author and publishers.)

Serum lipoproteins have been shown (Nielsen and Bjerrum 1975) not to form immunoprecipitates in the presence of Triton X-100, but rocket immuno-electrophoresis of an acidic polysaccharide from bacterial cell membranes in the presence of 1 per cent Triton X-100 has been accomplished without apparent difficulty (Owen and Salton 1976). The sensitivity limit of about 10 ng for the polysaccharide was only slightly inferior to the lower detection limit of about 3 ng for many protein antigens when the same Coomassie Brilliant Blue R250 staining procedure was used. As might be expected, glycoproteins present no problems and several immuno-electrophoretic techniques have been applied to membrane glycoproteins in the presence of non-ionic detergents (e.g. Bjerrum and Bøg-Hansen 1976; Nielsen and Bjerrum 1977; Hagen, Bjerrum and Solum 1979). Glycolipids and lipopolysaccharides can be studied by electro-immunochemical analysis in gels containing both 0·05 per cent Triton X-100 and 0·002 per cent Sudan Black (Bjerrum *et al.* 1982).

(b) *Ionic detergents*

Ionic detergents such as sodium dodecyl sulphate (SDS) bind strongly to proteins (see Chapter 5), and SDS not only interferes with precipitate formation by denaturing the antigens and antibodies but also forms charged complexes with both antigen and antibody molecules. Since both types of complex will have similar charge-to-mass ratios and since there is little molecular sieving effect in agarose gels, both antigen and antibody will migrate at similar velocities during any electrophoretic experiment and no immunoprecipitate will form. In fact the solubilizing power of SDS is such that it has been used for dissolving immunoprecipitates for their recovery (see Section 8.7) from immuno-electrophoretic gels (Norrild, Bjerrum, and Vestergaard 1977).

Proteins separated by SDS–PAGE can be localized in terms of their antigenic action if the gels are first fixed and the SDS washed out.

Bigelis and Burridge (1978) used very mild fixing conditions, consisting of soaking gels for 2 h in water:methanol:acetic acid (5:5:2 by volume). However, most authors use much more rigorous conditions such as 50 per cent trichloroacetic acid for 2 h followed by 7 per cent acetic acid (Olden and Yamada 1977) or 25 per cent isopropanol containing 10 per cent acetic acid (Olden and Yamada 1977; Adair *et al.* 1978). Fixation is followed by soaking the gel in buffer of the required pH and ionic strength for reaction with the antibody (e.g. 0·05 M tris–HCl pH 7·5 containing 0·15 M NaCl and 0·1 per cent NaN_3) and washing with buffer until the desired pH is reached. The gel is then overlaid either with antibody (antiserum) solution or with agarose gel containing the antiserum and incubated in a humid atmosphere in the usual way. Uncomplexed antiserum proteins are removed by extended washing for several days in buffered saline and the gels are stained and dried as in other polyacrylamide gel techniques.

Antigens separated by SDS–PAGE can also be detected by immunodiffusion

(Puszkin, Schook, Maimon, and Puszkin 1977). For this (Fig. 8.17) small circles of gel are cut out after the electrophoresis run or slices are cut from a cylindrical gel rod and attached to a glass slide with a drop or two of warmed agarose in buffer solution. Once gelled the rest of the slide is covered with a 2–3 mm thick layer of the agarose solution and cooled, and wells are cut with a gel-punching device. The centre line of wells are filled with antibody solution and the gels are incubated. Precipitin lines form in the same way as in the usual double-diffusion method (Section 8.2(c)) and identify which gel slice contains the antigen. To avoid distortion the polyacrylamide circles should be removed before the agarose gels are pressed, washed, and stained.

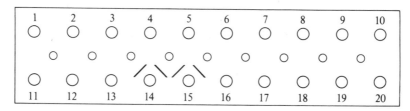

FIG. 8.17. Identification of antigens in slices cut from polyacrylamide gels by immunodiffusion as described by Puszkin *et al.* (1977). In the above example the centre line of wells are filled with antibody solution and it can be seen that gel slices 14 and 15 contain the corresponding antigen giving rise to the bands of immunoprecipitate shown.

A strip of gel cut from a rod or slab SDS–PAGE electrophoresis run can of course be embedded in agarose and an adjacent trough cut for antibody solution for analysis by an immunodiffusion stage in the same manner as the Grabar and Williams (1953) procedure (see Section 8.3(b)).

The use of SDS in the first dimension of two-dimensional immuno-electrophoretic procedures has been reported by various groups. If the second-dimension gel contains a non-ionic detergent such as Triton X-100, Nonidet NP-40, or Berol EMU-043, SDS from the first-dimension gel forms mixed micelles with the non-ionic detergent. This prevents the formation of SDS–antibody complexes, while the continued presence of (non-ionic) detergent inhibits aggregation and non-specific precipitation. This principle of competitive replacement of SDS by a non-ionic detergent (Clarke 1975) has enabled Bjerrum *et al.* (1975) to perform two-dimensional procedures such as crossed, line, and crossed-line immuno-electrophoresis using SDS in the first dimension and 1 per cent non-ionic detergent in the second. As long as SDS levels in antigen solutions did not exceed 100 nmol per sample, single-dimension rocket immuno-electro-phoresis could also be carried out satisfactorily in the presence of 1 per cent non-ionic detergent. Plumley and Schmidt (1983) suggest that when the addition of SDS to samples is necessary in order to completely solubilize whole cell extracts, membrane proteins, etc, it may be preferable to add a 12·5-fold excess (relative to the SDS level) of a non-ionic detergent to the solubilized sample itself prior to the electrophoretic separation rather than to incorporate it into the gel. Incorporation of 15 per cent polyethylene glycol (PEG) 4000 to the gels also enhanced immunoprecipitate formation. Some workers (e.g. Kessler 1981; Burnett and Chandy 1983) have suggested that the use of non-ionic detergents is unnecessary and they reported satisfactory results when antigen proteins separated by SDS–PAGE were examined by transferring gel strips directly into detergent free agarose slabs for crossed immuno-electrophoresis in the usual way. Both groups however fixed stained and destained the SDS–PAGE gels before cutting out the strips for the crossed immuno-electrophoresis stage and this will of course wash out most or all of the SDS.

Some proteins retain antigenic activity in the presence of SDS. In such cases (which can be demonstrated by using antigen with and without SDS treatment in a double-diffusion experiment) it is permissible to have a low level (e.g. 0·1 per cent) of SDS present in the second-dimension immuno-electrophoresis as well as in the first-dimension stage (Webb, Mickey, Stone, and Paulson 1977).

8.6. Immuno-electrofocusing

From what has already been discussed it will be apparent that, while electrophoresis in agarose gels is most often used for the first-dimensional separation of two-dimensional immuno-electrophoretic methods, other methods such as PAGE or PAGE in the presence of non-ionic or ionic detergents are equally applicable. Indeed most other paper or gel electrophoretic techniques are suitable. These include gel iso-electric focusing and isotachophoresis (Chapter 9).

Iso-electric focusing relies on establishing a pH gradient and focusing proteins at their iso-electric points, after which there is almost no further electrophoretic migration. This involves the application of relatively high voltages for long times and electro-endosmosis then becomes a major problem. Thus, while both stages of immuno-electrofocusing experiments have been carried out in agarose gels (see p.282) it may be preferable to use polyacrylamide gels in which there is almost no electro-endosmosis for the isoelectric focusing stage. Strips of the gel are then embedded in agarose for the second-dimension separation or immunodiffusion. Alternatively polyacrylamide gels can be used for both stages, although rather longer diffusion times should be allowed to compensate for the slower rates of diffusion in this medium. Since the diffusion rate depends on the porosity of the gels, polyacrylamide of the lowest convenient concentration should be used. For example, Weiss, Del Vecchio, and Burnett (1978) used gels with $T = 4$ per cent and $C = 10$ per cent containing 2·8 wt per cent Ampholine ampholytes for the first dimension in a horizontal gel slab apparatus.

The method has the advantage that in addition to the immunochemical information gained, one also obtains a measure of the iso-electric points of the substance of interest and any accompanying contaminants. The technique should be rather more satisfactory now that special agaroses of very low electro-endosmosis are commercially available for IEF work (see p. 282) and would therefore be appropraite for immuno-electrofocusing as well.

8.7.   Affino-electrophoresis (Affinity electrophoresis)

Although it is not strictly an immunological technique, affino-electro-phoresis is discussed here because of the many points of similarity between it and the methods described in this chapter. Lectins (phytohaemagglutinins) are proteins or glycoproteins which have the property of binding to other glycoproteins in a manner very reminiscent of the antigen–antibody reaction. While some lectins are of broad specificity, others are highly specific and only combine with glycoproteins bearing a particular oligo-saccharide grouping. Affino-electrophoresis consists essentially of replacing antibody or antiserum with a suitable lectin in any of the above immuno-electrophoretic methods. Thus Owen and Salton (1976) have successfully performed 'rocket' immuno-electrophoresis and 'rocket' affino-electro-phoresis (with 75 μg ml^{-1} of concanavalin A in place of the antiserum) on the same samples and have demonstrated very similar qualitative and quantitative results and similar levels of sensitivity. Staprans, Felts, and Butts (1983) used concanavalin A in the same way to detect plasma glycosaminoglycans by rocket affino-electrophoresis. Bjerrum and Bøg-Hansen (1976) and Nielsen and Bjerrum (1977) used concanavalin A (75 μg ml^{-1}) in the first dimension of crossed immuno-electrophoresis and antibody in the second dimension in the usual way. Glycoproteins from human erythrocyte membranes and from milk fat globule membranes were identified by comparison with the band patterns given by controls without lectin. The presence of non-ionic detergents did not affect the lectin-glycoprotein interaction.

In subsequent studies (Bøg-Hansen, Bjerrum, and Brogren 1977; Owen, Oppenheim, Nachbar, and Kessler 1977) using a number of different lectins (usually at 30–50 μg ml^{-1}) in agarose gels and a number of different glycoproteins, it was found that affino-electrophoresis peaks were usually less distinct than the corresponding immuno-electrophoresis peaks. In single-dimension separations in the presence of lectins (e.g. 'rocket affino-electrophoresis' etc.) the areas under the peaks were filled with precipitate rather than just forming an outline band of precipitate with a 'hollow' centre as an immuno-electrophoresis (Bøg-Hansen *et al.* 1977). This makes identification of components in complex mixtures difficult, but the method does have two important advantages. Firstly, unlike the usual immunological methods which rely on a retention of antigenic activity that can be lost on isolation of components or on denaturation, affino-electrophoresis is unaffected by heat, acid, or detergent denaturation of the antigens. Secondly, quantitative measurement of serum low-density lipoprotein by immunochemical methods is unsatisfactory in the presence of the detergents which may be necessary for the solubilization of lipoproteins, but affino-electrophoresis into gels containing concanavalin A was quantitative in non-ionic detergents and after denaturation with SDS.

Lectins can also be substituted for antibody in double-antibody techniques in a manner analogous to the immunological detection of specific antigens following electrophoretic separation. Separated glycoproteins can be detected by the following procedure (Olden and Yamada 1977).

Gels or gel slices are incubated for 18 h in buffered saline containing 100 μg ml^{-1} of concanavalin A. Gels are then washed thoroughly with buffer to remove unbound lectin and incubated for a further 18 h with horseradish peroxidase (100 μg ml^{-1}) which binds to available sites on the multivalent lectin. After more washing to remove free enzyme the bound enzyme is visualized by incubation with 5 mg of 3,3'-diaminobenzadine tetra-HCl plus 5 μl of H_2O_2 (30 per cent) in 10 ml of phosphate-buffered saline. To avoid excessive non-specific staining gels should be rinsed with buffered saline as soon as enzyme bands are observed (usually 5–15 s). In order to avoid using 3,3'-diaminobenzadine, a suspected carcinogen, it can be replaced with 4-chloro-naphthol (Hawkes *et al.* 1982). This is prepared as a solution (3 mg ml^{-1}) in methanol and diluted with 5 volumes of buffer just before use and then adding to the mixture the required amount of H_2O_2.

A rather similar approach to detect glycoproteins with α-D-galactopyranosyl end groups has been described by McCoy, Varani, and Goldstein (1983) and termed enzyme-linked lectin assay (ELLA) by analogy with the well-known ELISA techniques. In their work they coupled alkaline phosphatase to *Griffonia simplicifolia* I-B$_4$ isolectin with glutaraldehyde and used this enzyme-labelled lectin to measure the glycoproteins after they had been immobilized on an ELISA microtiter plate. No doubt the method could be adapted to detect glycoproteins in gels directly or in gel transfer blots (see Section 2.13), and of course by choosing other lectins with different carbohydrate specificities could also be used to locate and measure glycoproteins with other terminal saccharide groupings.

8.8. Recovery of immunoprecipitates from gels

Recovery of precipitates from gels is inevitably a very small-scale preparative procedure, but it can be useful for the production of specific antisera since a particular precipitin band can be cut out and pooled material from only a few gels provides enough for injection into experimental animals. If the band of interest, or at least part of it, cannot be cut out free of contamination by other precipitin bands, it is often possible to change the separation conditions a little. This can be achieved by changing the pH of buffers or the buffer composition, adding divalent cations (such as Ca^{2+} which affects the mobility by complex formation), adding detergents, or using a different batch of antibodies. If not clearly visible, faint bands can be detected by autoradiography if radioactively labelled precipitates are prepared. Alternatively they can be stained by overnight treatment with 0·01 per cent Coomassie Brilliant Blue R250 in water (Norrild, Bjerrum, and Vestergaard 1977) or with OPA in phosphate buffer (Table 2.4).

If required, gels can be pressed, washed, and dried before the precipitate bands are cut out. Protein can be extracted from dried gel pieces by treatment with a solution of a pH at which the particular antigen–antibody complex is soluble, by treatment with non-ionic detergent solution, or (if the retention of antigenic activity is not important) by treatment with 6 or 8 M urea or 1 per cent SDS. Norrild *et al.* (1977) subjected dried gel pieces to sonic vibration for 30 s in 0·25 ml of 0·06 M tris–H_3PO_4 buffer pH 6·9 containing 2 per cent SDS and 2·5 per cent 2-mercaptoethanol, incubated the resulting homogenates at 37 °C for 20 h, and removed agarose fragments by centrifuging. The extracted material can be used directly for further electrophoresis experiments and molecular-weight measurement by SDS–PAGE (Chapter 5).

Once the immunoprecipitate is solubilized, the antigen molecules can be separated from the antibody immunoglobulins by gel filtration, ion-exchange chromatography, etc. The purified antigens can then be subjected to further biochemical analysis in the usual way.

8.9.   Biochemical applications

Compared with other forms of gel electrophoresis, immunodiffusion and immuno-electrophoretic techniques have three major advantages. Firstly, they can be very rapid (e.g. particularly counter immuno-electrophoresis) and are readily quantifiable. Secondly, they can be extremely sensitive. For example, if methods relying on the formation of a radiolabelled immuno-precipitate are used, reasonably accurate quantitative measurement of antigens with concentrations of only one or two micrograms per litre is achievable. This high sensitivity is aided by the fact that often several antibody (Ig) molecules bind to a single antigen molecule, so that the weight of the complex formed may be a factor of 10 or more greater than that of the antigen alone. This factor is enhanced further in double-antibody techniques. Thirdly, with the aid of specific antibodies enabling one to identify a particular band as being associated with a particular antigen, it is possible both to detect and to quantify antigens very simply in highly complex samples such as blood serum, tissue extracts, etc.

Immuno-electrophoretic methods have most often been applied to proteins and glycoproteins, but some particularly good separations of serum lipoproteins have been published (Cline and Crowle 1979). With double-antibody techniques and the use of ^{125}I-labelled protein A (Section 8.4) not only is sensitivity increased but substances with only weak antigenic activity can be measured readily. The use of detergents enables membrane constituents (e.g. human platelet membrane proteins, Shulman and Karpatkin 1980; Howard et al. 1982) and viral particles to be solubilized and examined, since particularly with non-ionic detergents antigenicity is usually retained. The use of immuno-electrophoretic methods in the investigation of membrane proteins has been reviewed by Bjerrum (1977). Furthemore, antigenic activity may also survive denaturation by heat, acid, or detergent action, and in such cases immunological methods provide detection and quantitative measurement not readily available by other methods. Polysaccharides and even charged lipid micelles, provided that suitable haptens are present (Niedieck 1978), can be examined by the usual immuno-electrophoretic procedures.

Glycoproteins and some lipoproteins can also be examined by affino-electrophoresis, which has many similarities to the corresponding immuno-electrophoretic procedure with similar sensitivies. It has been applied particularly to the identification of glycoproteins in mixtures. A report in which various lectins were used in the first dimension of a system otherwise identical to crossed-immuno-electrophoresis (Kerckaert, Bayard, and Biserte 1979), applied immuno-affino-electrophoresis to the characterization of α_2-fetoproteins and compared the molecular heterogeneity in samples obtained from the rat, mouse, and human (Fig. 8.18). The same method can also be valuable in studies of oligosaccharide heterogeneity (which influences lectin binding) in glycoproteins (Hansen, Lihme, and Bøg-Hansen 1984).

When using affino-electrophoresis alone and not combined with immunological detection it should be remembered that the formation of precipitate bands showing apparent identity between two electrophoretically distinct species implies only that both species have accessible binding sites for the lectin. The sites may be similar or different. For example, concanavalin A binds to terminal non-reducing mannopyranosyl and glucopyranosyl residues as well as to internal 2–0-linked α-D-mannopyranosyl residues. Affino-electrophoresis as used in a manner analogous to immuno-electrophoresis can of course only be applied to carbohydrates which form insoluble complexes with lectins. A preliminary screening by double-

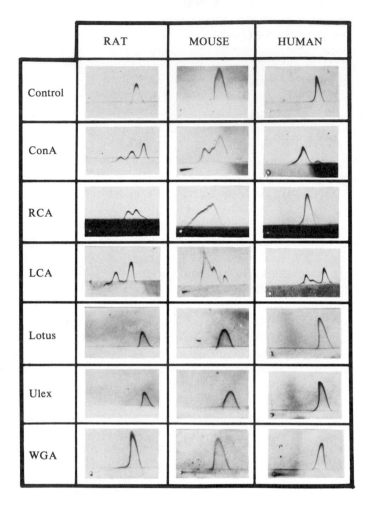

FIG. 8.18. Crossed immuno-affino-electrophoresis patterns of rat, mouse, and human α_1-fetoproteins obtained with several lectins. Control, without lectin; Con A, concanavalin A; RCA, *Ricinus communis* agglutinin; LCA, *Lens culinaris* agglutinin; Lotus, *Lotus tetragonolobus* lectin; Ulex, *Ulex europeus* lectin; WGA, wheat germ agglutinin. All lectins were included in 1 per cent agarose gel for the first dimension (left to right) at 100 μg ml^{-1}. Electrophoresis into agarose gel containing anti-α_1-fetoprotein was used for the second dimension (bottom to top). Buffers (pH 8·6) consisted of 37 mM tris, 24·5 mM sodium barbiturate, 0·36 mM calcium lactate, and 0·2 mM sodium azide. (Reproduced from Kerckaert *et al.* (1979) by permission of the authors and publishers.)

diffusion experiments with lectins in place of antibodies can be a useful guide to this (Owen *et al.* 1977).

A rather different form of affino-electrophoresis, also discussed in Section 12.8, employs sugar derivatives, in the form of 0-glycosyl polyacrylamide copolymers, immobilized in the gel matrix. In experiments analogous to conventional PAGE these gels can be used to study lectin–sugar interactions, to determine the dissociation constants of complex formation (Hořejší, Tichá, and Kocourek 1977a) and the pH dependences of these constants (Hauzer, Tichá, Hořejší, and Kocourek 1979), and to examine heterogeneity in sugar binding by lectins (Hořejší, Tichá, and Kocourek 1977b). In a similar way PAGE conducted on affinity gels containing entrapped Blue Dextran results in a decrease in the mobility compared with controls of a number of proteins and enzymes which interact with Cibacron Blue F3GA (Fig. 8.19), the blue dye of Blue Dextran (Tichá, Hořejší, and Barthová 1978). This revealed a number of isoenzyme bands with differing binding properties and in some cases the dissociation constants of the protein–dye interaction could be calculated.

These various merits of immunochemical and related techniques, in particular the ability to measure one specific component in a mixture without prior purification, have led to their widespread adoption in the clinical field for the diagnosis and confirmation of a wide range of pathological conditions. The electrophoretic methods described in this

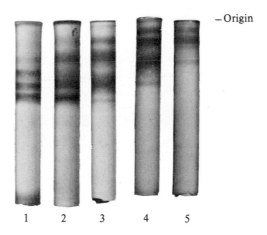

FIG. 8.19. Affinity electrophoresis of bovine heart lactate dehydrogenase on gels containing increasing amounts of Blue Dextran: 1, control; 2–5, Blue Dextran gels containing the equivalent of 4×10^{-7} M (2), 2×10^{-6} M (3), 4×10^{-6} M (4), and 2×10^{-5} M (5) of immobilized Cibacron Blue F3. (Reproduced from Tichá *et al.* (1978) by permission of the authors and publishers.)

chapter have been used widely in this context and are finding ever-increasing applications.

Immuno-electrophoretic techniques have proved to be of particular value in taxonomy and studies of evolutionary relationships between proteins (e.g. Adair *et al*. 1978). The demonstration of relatedness or otherwise by immunological identity or cross-reactivity has also proved useful in studies comparing normal and mutant membranes and proteins (e.g. Bigelis and Burridge 1978) and of normal and pathogenic cell constituents. A major application of the methods described in this chapter lies in forensic work for the identification of blood stains and other biological tissues and fluids, detection of toxins, etc. In the clinical field tissue typing and matching exploit these immunochemical techniques. They have also proved valuable for the analysis of foods, for example to demonstrate the adulteration of meat products with milk, egg, or vegetable protein (Kluge-Wilm 1978) or to show the presence of casein in heated foods (Klostermeyer and Offt 1978).

9

ISO-ELECTRIC FOCUSING

Iso-electric focusing (IEF) is a method specifically intended for the fractionation of molecular species differing only in net charge. Thus, since separation is not due to any molecular size or 'molecular sieving' effect during electrophoretic transport through the medium, optimum resolution is theoretically obtained in a gel where $T = 0$ per cent. It is therefore performed in essentially non-sieving media, such as in free solution with a density gradient, polyacrylamide gels of high porosity (low T), or in granular beds (e.g. Sephadex).

IEF is an analytical method capable of very high resolution, particularly when shallow, immobilized pH gradients are used, and can separate macromolecules differing in isoelectric point by as little as only 0·001 pH unit. It is also capable of being scaled up for small-scale preparative separations of up to about 1 g whilst still retaining good resolution (0·01–0·02 pI units). In the preparative mode it has often been regarded as useful in one of the final stages of purification regimes, but there is a strong argument for including it at a relatively early stage and quickly removing impurities that might interfere with other methods. With improvements in both materials and techniques, there is no doubt that IEF has justifiably increased greatly in popularity in recent years and looks set for still greater exploitation in future.

9.1. Iso-electric focusing: background and principles

IEF can be regarded as electrophoresis within a pH gradient. Macromolecules therefore migrate through the gradient as long as they retain a net positive or negative charge until they reach the point in the pH gradient which corresponds to their iso-electric point, where the net molecular charge will be zero, and migration ceases. The pH gradient is established with the aid of a mixture of low-molecular-weight amphoteric substances, or ampholytes. If the cathode of an electrolytic cell is placed in a solution of a strong base, the anode in strong acid, and the space between is filled with ampholyte solution, then the ampholytes will carry a net positive charge in the area close to the anode and a net negative charge near the cathode. Thus when the current is switched on they will be repelled by the electrodes and will move into the central region, each ampholyte species becoming stationary where the pH of the surrounding electrolyte equals the iso-electric point of that particular ampholyte (Haglund 1967). The most basic ampholyte thus stays closest to the basic cathodic solution and the other ampholytes arrange themselves in order of their iso-electric points, ending with the most acidic closest to the acid solution at the anode. The resulting pH gradient is then at equilibrium since any displacement of an ampholyte molecule will result in the formation of a net charge on the molecule causing it to migrate back to its iso-electric point again under the influence of the applied electric field. Provided that mixing (e.g. convection currents etc.) is prevented the pH gradient will remain stable for a considerable time.

It will be evident that the production of a good smooth pH gradient depends largely on the properties of the ampholytes used. As defined by Svensson (1961) these should have the following properties.

(1) Good buffering capacity so that they can define the pH at their iso-electric points, even in the presence of high-molecular-weight ampholytes such as proteins which may be present in the same zone.
(2) Good conductivity at their iso-electric points in order to maintain electrical conductivity in the absence of other electrolytes.
(3) Low molecular weight in order to be readily separated from the macro-molecules being studied.
(4) A composition different from the macromolecules so that they do not interfere with assay procedures.
(5) They should not react with or denature the macromolecules being fractionated.

Mixtures of many amphoteric substances such as amino acids and peptides and some amphoteric and non-amphoteric buffer components can act as suitable ampholytes (e.g. Chrambach and Nguyen 1977). The 47-component buffer mixture developed by Cuono and Chapo (1982) is available commercially (as Poly/sep 47 from Polysciences Ltd, 24 Low Farm Place, Moulton Park, Northampton, NN3 1HY, England) and such mixtures, as well as being less expensive than conventional commercial ampholytes, may have advantages if complex formation between ampho-lytes and sample constituents occurs. However, the great majority of iso-electric focusing experiments are performed nowadays with the aid of commercial ampholyte mixtures. The first of these, and still the most widely used, is marketed by LKB Produkter AB under the brand name Ampholines™. They consist of synthetic mixtures of polyaminopolycarbo-xylic acids with molecular weights mostly in the region 300–900 (Bosisio, Snyder, and Righetti 1981). Any high-molecular-weight ampholytes are undesirable because in preparative work they may not be readily separated from protein (Baumann and Chrambach 1975), and in analytical IEF gels they could result in artefactual band patterns through binding to and staining with protein dyes (Otavsky and Drysdale 1976). More recently other products have been introduced which contain sulphonic or phos-phonic acid groupings in addition to the amino and carboxylic acid groups. These products (Servalyts™, Serva-Feinbiochemica GmbH; Biolytes™,

Bio-Rad Laboratories; Pharmalytes™, Pharmacia AB) have recently been compared with the Ampholines (Fawcett 1977; Gelsema, de Ligny, and van der Veen 1979b; Låås and Olsson 1981b) and shown to have similar performance.

There were a number of experiments in the first half of this century which with the benefit of hindsight can be said to have relevance to IEF, but it was not until the work of Kolin in 1954–1955 and Svensson in 1961–1962 that the principle was shown to be both theoretically and practically sound.

Contributions relating to the theoretical basis of IEF which are scattered through the earlier literature have been collected together and presented in a concise form by Rilbe (1973), and his treatment is given below. It is derived from the simple dissociation theory of electrolytes. Considering an ampholyte HA, the cation of which is therefore H_2A^+ and the anion A^-, in aqueous solution the mass action equations apply so that

$$\frac{[H^+]\,[HA]}{[H_2A^+]} = K_1 \tag{9.1}$$

and

$$\frac{[H^+]\,[A^-]}{[HA]} = K_2. \tag{9.2}$$

We now introduce Rilbe's notation: $h = [H^+]$; $pH = -\log h$; C_+, concentration of H_2A^+; C_-, concentration of A^-, C_0, concentration of uncharged HA and zwitterionic $H^+ A^-$ (since as long as it is undissociated it does not matter whether it is uncharged or not); $C = C_+ + C_- + C_0$, total ampholyte concentration. Equations (9.1) and (9.2) then become

$$hC_0 = f_+ K_1 C_+ \tag{9.3}$$

and

$$hC_- = C_0 (K_2/f_-) \tag{9.4}$$

where f_+ and f_- are the activity coefficients of H_2A^+ and A^- respectively and the uncharged species is assumed to have an activity coefficient of unity. As a further simplification one can take

$$f_+ K_1 = K_1' \tag{9.5}$$

and

$$K_2/f_- = K_2' \tag{9.6}$$

so that eqns (9.3) and (9.4) become

$$C_+ = hC_0 / K_1'$$ (9.7)

and

$$C_- = K_2' C_0 / h$$ (9.8)

By adding them together with C_0 to obtain the total concentration C, we find

$$C_+ = \frac{h^2 C}{h^2 + hK_1' + K_1' K_2'}$$ (9.9)

$$C_0 = \frac{hK_1' C}{h^2 + hK_1' \times K_1' K_2'}$$ (9.10)

$$C_- = \frac{K_1' K_2' C}{h^2 + hK_1' \times K_1' K_2'}.$$ (9.11)

The mean valence or basicity z of the acid due to proton binding is defined as

$$\bar{z} = \frac{C_+ - C_-}{C} = \frac{h^2 - K_1' K_2'}{h^2 + hK_1' + K_1' K_2'}$$ (9.12)

This will be zero at a hydrogen ion activity h_i such that

$$h_i^2 = K_1' K_2'.$$ (9.13)

When the operator $p = -\log$ is introduced, at the iso-ionic point

$$(\text{pH})_i = (pK_1' + pK_2')/2.$$ (9.14)

In order to maintain a stable pH gradient when sample proteins are present it is important that the carrier ampholytes forming the gradient can dictate the pH at all points in spite of contributions from ionic groups on the protein molecules. The ampholytes should therefore have a substantial buffering capacity in the iso-ionic state.

The specific buffering capacity B for a weak protolyte is given by

$$B = 1/m \frac{dn}{d(\text{pH})}$$ (9.15)

where m is the amount of weak protolyte and n the amount of base or the negative of an amount of a strong acid. For a monobasic weak acid

$$B = \alpha(1-\alpha)\ln 10 \qquad (9.16)$$

where α is the degree of dissociation. It is a maximum when $\alpha = 0.5$, which corresponds to $pH = pK'$ for the weak acid, and then

$$B_{max} = \tfrac{1}{4}\ln 10. \qquad (9.17)$$

For a multivalent protolyte the specific buffering capacity is given by the derivative

$$B = -\frac{dz}{d(pH)} \qquad (9.18)$$

which for the ampholyte being considered becomes

$$B = \frac{1}{C}\,\frac{d(C_- - C_+)}{d(pH)} = \frac{d}{d(pH)}\,\frac{K_1 K_2' - h^2}{h^2 + hK_1' + K_1' K_2'} \qquad (9.19)$$

Differentiation and insertion of the relationship (9.13) at the iso-ionic point gives

$$B_i = \frac{\ln 10}{1 + (K_1'/4K_2')^{1/2}}. \qquad (9.20)$$

The relative buffering capacity b_i (the capacity in units of the maximum capacity of a monovalent weak protolyte) at the iso-ionic point is given by dividing by (9.17), so that

$$b_i = \frac{4}{1 + (K_1'/4K_2')^{1/2}}. \qquad (9.21)$$

For a bivalent protolyte the buffering capacity cannot be more than twice that of a monovalent one, so the maximum value of b_i is 2 and thus

$$K_1' \geqslant 4K_2' \qquad \text{and} \qquad \triangle pK' \geqslant \log 4. \qquad (9.22)$$

Carrier ampholytes must possess an appreciable conductivity at their iso-electric points otherwise a deep conductivity minimum would occur about the neutral point which might result in local overheating. At more extreme pH values (below 4 or above 10) the conductivity contributions of H^+ and OH^- ions are dominant and the conductivity of ampholytes becomes less important. The degree of ionization of an ampholyte is

$$\alpha = \frac{C_+ + C_-}{C} = \frac{h^2 + K_1'K_2'}{h^2 + hK_1' + K_1'K_2'} \qquad (9.23)$$

This has a value of unity at both very low and very high pH values, and has a minimum at the iso-ionic point which can be found by combining (9.13) and (9.23):

$$\alpha_i = \frac{1}{1 + (K_1'/4K_2')^{1/2}}. \qquad (9.24)$$

The conductivity contribution of an iso-ionic ampholyte is proportional to α_i, so it can be seen that ampholytes with a large pK' difference have a low conductivity in the iso-ionic state. This applies to all neutral natural amino acids which are therefore not useful as carrier ampholytes (Rilbe 1973). As pK' decreases, the degree of ionization increases to a limiting value of 0·5 which is reached for the smallest pK' difference compatible with (9.22). Comparison of (9.21) and (9.24) gives

$$b_i = 4\alpha_i \qquad (9.25)$$

and this shows that a high degree of ionization (a good conductivity) is accompanied by a good buffering capacity and *vice versa*.

In practice most carrier ampholytes in general use are polyaminopoly-carboxylic acids and hence have a number of ionizable groups at the isoionic point, so that (9.22) no longer holds and α_i does not have an upper limit of 0·5 (for large molecules, such as proteins, with many ionizable groups α_i may approach unity in spite of the fact that the net charge is zero). These facts are actually favourable for the practical application of IEF because a polyvalent ampholyte may therefore have a higher conductivity and a better buffering capacity at the iso-ionic point than a simple bivalent one.

Electrolysis of a mixture of carrier ampholytes gives rise to a smooth and stable pH gradient, the extent and shape of which is determined by a choice of ampholytes and the proportions which are iso-electric within a given pH region. Practical mixtures include as many different ampholyte species as possible because if the number of species is too small the gradient becomes a series of steps with one pH plateau for each ampholyte species. The pH gradient is always positive all the way from anode to cathode, i.e. the pH increases monotonically in the direction of the current. Although only an empirical relationship, this law of pH monotony (Rilbe 1973) appears always to hold during steady state electrolysis and no local negative pH gradients are ever found. If a complete separation of ampholyte zones occurred there would be regions of water in between them and the local pH would vary between 7 and the various iso-electric points, which would violate this law of pH monotony, so in practice it is found that ampholyte zones overlap and the pH changes smoothly from one iso-electric point to the next.

The next most significant landmark was in 1966 when Vesterberg synthesized artificial mixtures of ampholytes capable of giving good smooth pH gradients. This earlier work has been covered by a number of reviews (e.g. Haglund 1967, 1971; Vesterberg 1971; Rilbe 1973, 1977; Kolin 1977) and will not be discussed further here.

The net molecular charge on a protein, glycoprotein, or lipoprotein is determined by the content on the one hand of acidic groups (carboxyl groups of aspartic, glutamic, or uronic acids or, less commonly, phosphate and sulphate groupings) and on the other hand of basic moieties such as the amino and guanidino groupings of lysine and arginine. Thus most of these macromolecules have iso-electric points somewhere within the pH range 3–11, and in a pH gradient will migrate to their iso-electric points where they will focus in sharp bands. A number of important consequences follow from this. Firstly, as well as separating a number of proteins in a mixture, the iso-electric points of each of them can be very simply determined merely by measuring the pH in the gradient at the position of the focused band. The iso-electric point pI is an important parameter in the characterization of a macromolecule and it is most readily determined by IEF since other methods are more time consuming and require much more material. Secondly, since a protein will migrate from any point in the gradient to its pI the position of sample application is unimportant and often quite dilute sample solutions can be used. For example, in a density-gradient column, sucrose and ampholytes can be added to a dilute sample solution which is then used directly for making up the gradient. Since the focused band only occupies a small volume a concentration factor of a hundred-fold or more is easily achieved in this way. Thirdly, electrophoretic migration virtually ceases at the pI so that the band pattern stabilizes and, within limits, becomes insensitive to experimental variables such as time and applied voltage. Band patterns and measured pI values are therefore highly reproducible. Furthermore, the displacement of any ampholyte molecule, including a sample protein molecule, from its pI induces a net molecular charge of appropriate sign so that the molecule returns to its iso-electric point. Therefore unlike all other electrophoretic techniques there is no band spreading due to diffusion.

The protein pI measured by IEF is very close to the true iso-ionic point since the ionic strength within the medium of the pH gradient is very low (Vesterberg and Svensson 1966). Resolution of the method is excellent and often exceeds that obtainable in other electrophoretic methods. When unknown mixtures are being examined it is usual to use ampholytes giving a pH gradient with a wide range (e.g. pH 3–10), but once the region of particular interest has been identified narrow-range ampholyte mixtures covering 2 or 3 pH units or less can be used. Resolution in the resulting shallow gradients is often such that two macromolecular species differing in pI by only 0·02 of a unit can be clearly distinguished.

The theoretical basis of resolving power between focused zones in IEF has been summarized by Rilbe (1973) and is based on the fact that the differential equation for IEF represents a balance between electrical and diffusional mass transport so that

$$CuE = D \frac{dC}{dx}$$

where C is the protein concentration, u its mobility in $cm^2\ V^{-1}s^{-1}$, D its diffusion coefficient in $cm^2\ s^{-1}$, and E the field strength in $V\ cm^{-1}$ Focused zones are narrow and the field strength within a zone can be taken to be constant so u can be regarded as a linear function of the linear displacement x within the column. Therefore

$$u = -fx \qquad\qquad (9.27)$$

where f is a proportionality factor. Equation (9.26) can then be written as:

$$dC/C = (Ef/D)x\ dx \qquad\qquad (9.28)$$

Since E, f and D can be regarded as constants, integration gives

$$C = C(0)\ \exp(-fEx^2)/2D \qquad\qquad (9.29)$$

where $C(0)$ is the integration constant (the local concentration maximum at the level where the protein is iso-electric). This expresses a gaussian concentration distribution with inflection points x_i, at

$$x_i = \pm (D/fE)^{1/2} \qquad\qquad (9.30)$$

The proportionality factor f can be written as a derivative:

$$f = -\frac{du}{dx} = -\frac{du}{d(pH)}\frac{d(pH)}{dx} \tag{9.31}$$

Substituting this in eqn (9.30) we obtain

$$x_i = \pm \frac{D}{-E\,(du/d(pH)\,)\,(d(pH)/dx)} \tag{9.32}$$

Considering now two closely spaced zones, the pH difference $\triangle pH$ between them is determined by their separation $\triangle x$ in the pH gradient:

$$\triangle pH = \triangle x\,\frac{d(pH)}{dx} \tag{9.33}$$

The separation at which two closely spaced gaussian zones become clearly seen as partly resolved zones is proportional to the inflection points and therefore

$$\triangle x = Kx_i \tag{9.34}$$
$$\triangle(pH) = Kx_i \tag{9.35}$$

where K is a constant which depends upon the precise criteria used to define the point of resolution but is usually given the value $K = 3$. Combining (9.32) with (9.35) gives

$$\triangle(pH) = K\left\{\frac{D(d(pH)/dx)}{-E(du/d(pH)\,)}\right\}^{1/2} \tag{9.36}$$

From this it can be seen that good resolution is aided by a low diffusion coefficient and a high mobility slope ($du/d(pH)$) at the iso-electric point.

In general, proteins satisfy these criteria, but whether peptides can be focused satifactorily depends chiefly upon the mobility slope at their iso-electric points (Rilbe 1973). Good resolution is also achieved by using a high field strength and a shallow pH gradient, both of these being factors which can be varied by the experimenter within the limits imposed by his apparatus.

The overall resolution achieved does of course also depend upon the subsequent handling and analysis of the focused sample. As soon as the current is switched off diffusion spreading can occur, so it is important to measure the pH gradient and collect the separated components immediately. With density-gradient columns there is inevitably a slight loss in resolution during the draining or pumping out of the gradient solution and the very acts of collecting fractions or of slicing up gels containing focused gradients are of necessity accompanied by some loss in separating power.

It should be remembered that pI values are temperature dependent, usually decreasing with increasing temperature, so that the pH of fractions or gel slices should be measured at the same temperature as that at which the IEF experiment was performed. This temperature should be stated whenever results are presented (Vesterberg 1971). Many workers have reported pI values taken from measurements of the pH gradient recorded at 20 or 25 °C following IEF at 4 °C. Fredriksson (1977; 1978) has shown that, while pI values for both proteins and ampholytes are temperature dependent, the shifts need not necessarily coincide and errors as high as 0·2 pH may arise, especially for some weakly acidic proteins.

9.2.   Iso-electric focusing in density gradients

There are a number of descriptions of small density-gradient stabilized IEF columns of only a few ml volume (e.g. Fawcett 1977; Leaback 1977; Jackson and Russell 1984) and even IEF in capillary tubes of a few μl volume (e.g. Bispink and Neuhoff 1977). However, by far the most widely used types of apparatus are the commercially available 110 ml and 440 ml columns marketed by LKB Produkter AB. Both these columns are of the same basic design, shown diagrammatically in Fig. 9.1, and they are suitable for either analytical or small-scale preparative work. The design is almost identical to the original column design (Vesterberg and Svensson 1966), and the continued popularity of density-gradient IEF with these columns is a tribute to the soundness of this early work.

(a) Sample loading capacity

The major factors influencing the sample load-carrying ability in density gradient IEF experiments are the solubility of the protein of interest at its pI, the buffering capacity of the ampholytes at their pIs, and the carrying capacity of the density gradient (Winter and Karlsson 1976).

For analytical purposes Vesterberg (1971) recommends 5 mg or less of protein in a zone in the 110 ml column and up to 25 mg for preparative purposes. The use of a steep pH gradient (wide pH range ampholytes) gives sharp concentrated focused zones, but if zones were less sharp average protein concentrations in the zones would be lower so that more material could be focused without exceeding the solubility limit or the local

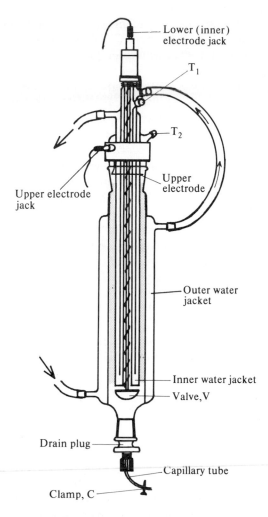

Lower (inner)
electrode jack

T$_1$

T$_2$

Upper
electrode

Upper electrode
jack

Outer water
jacket

Inner water jacket

Valve, V

Drain plug

Capillary tube

Clamp, C

FIG. 9.1. LKB Model 8100 isoelectric focusing column. Arrows show the direction of flow of the cooling water through the outer and then the inner water jackets. The shaded area is the region occupied by the sample and pH gradient; below this and surrounding the inner electrode is the dense electrode solution and above it is the light (upper) electrode solution. Reproduced by kind permission of LKB Produkter AB.

density within the density gradient. While the degree of focusing can be regulated to some extent by changes in the applied voltage, the best way of obtaining relatively broad zones without losing resolution is to use narrow-range ampholytes which give a shallow pH gradient.

If sample solubility remains a problem, solubilizing agents such as urea or non-ionic detergents can be employed. Obviously any ionic or charged additives cannot be used as they would lead to excessive electrical current and heating during the IEF run. Urea is often used at molarities of up to 8 M, but owing to the strong temperature dependence of its solubility solutions should not exceed 6 M if the IEF run is to be performed at 4 °C. Non-ionic detergents such as Tween 80, Brij 35, Triton X-100, etc. are most usually added at the level of 1–2 per cent. Ethylene glycol (33 per cent) and butanol have been used to enhance the solubility of lipoproteins and membrane proteins. The ampholytes themselves have some protecting or solubilizing effect and increasing the ampholyte concentration may also aid solubility. It should be remembered that many solubilizing reagents, particularly high levels of urea or detergent and particularly if used in combination with thiol-protective reagents (see below), are likely to cause dissociation of multimeric proteins into their constituent subunits.

For most purposes an ampholyte concentration of 1–2 wt per cent will be found suitable, and compared with the load of protein constituents this will usually give more than adequate buffering capacity.

The capacity of a density gradient to carry a protein load rises in proportion to the square of the length of the focused zone. In a relatively sharp zone 1 mm long in the 110 ml column the maximum theoretical amount is about 1 mg whilst a very wide (50 mm) zone should could carry 2·5 g. Thus it is particularly advantageous to use shallow pH gradients in preparative work since they facilitate high column loadings.

(b) Sample pretreatments

The presence of salt ions in IEF interferes with pH gradient formation, extends the running time, and can give rise to excessive heating due to high current even at moderate voltages. Local temperature differences may easily disturb the density gradient. For these reasons the recommended maximum permissible amount of salt is only 0·5 mol when using the 110 ml column (Winter and Karlsson 1976). Excessive salt in the sample must therefore be removed by gel filtration, ultrafiltration, dialysis, etc. Since many proteins have limited solubility in very low ionic strength media it may be advantageous to dialyse the samples against solutions containing 1 wt per cent of ampholytes or 1 per cent glycine, which being a zwitterion does not unduly disturb the development of a pH gradient.

(c) Density gradient components and the use of thiol-protecting reagents

Constituents used to establish the density gradient should of course be completely soluble under all the pH conditions likely to be encountered and should be chemically stable and electrically neutral at the concentrations used. They should also have little or no u.v. absorbance at wavelengths which are required for monitoring column effluents or the resulting fractions. They should not interfere with any assay procedures and should not interact in any way with sample constituents. Glycerol, sorbitol, ethylene glycol, polyvinylpyrrolidone, mannitol, Ficoll, soluble dextrans, etc. have all been used in density gradient IEF experiments, but by far the most common reagent used for this purpose is sucrose. The use of sucrose will be assumed in the following sections.

Since sucrose is of vegetable origin it may contain traces of protein impurities which naturally are focused into bands in the same manner as the proteins in the sample. They can be identified from a blank run containing no sample material, but in any case only very high purity (analytical grade) sucrose should be used. At strongly alkaline pH values (e.g. 8–11) sucrose can become partially ionized and under these conditions glycerol or sorbitol may be preferred.

Oxidation of free SH groups can occur in many preparative procedures and may lead to a loss of enzymic or other biological activity. This is not usually a problem during IEF analysis but if necessary uncharged thiol-reducing agents such as 2-mercaptoethanol, dithiothreitol, or dithioery-thritol can be added to a level of 0.1 M or more to both the gradient-forming solutions. Indeed, since these reagents will also cleave disulphide bonds they can be added together with a dissociating agent such as urea to cause the deliberate dissociation of multimeric sample proteins into their constituent subunits. This approach was followed by Andrews and Reithel (1970) for example in a study of the subunit composition of the enzyme urease.

(d) Preliminary experiments

From the above it is clear that a number of factors will influence the success or otherwise of the IEF experiment. Therefore when an unknown or poorly characterized sample is to be examined it can be helpful to carry out one or more preliminary experiments. These should give information on (a) the approximate number and pIs of the components in the sample, (b) their solubilities at their respective pIs, (c) the stability of sample components under the conditions of pH and the ionic strength that will be encountered in IEF separation, and (d) the mode of sample application.

Preliminary experiments will usually be performed with ampholytes of a wide pH range using small amounts of sample and either the density-gradient column technique or IEF on polyacrylamide gel slabs (see Section 9.3). These experiments will enable the correct choice of narrow-range ampholytes to be made for the best resolution of the component(s) of interest and will show if and at what pI solubility problems are likely to be encountered. They should also reveal whether there is a substantial loss of enzymic or biological activity which might be prevented by the incorporation of thiol-protective reagents or some form of sample pretreatment.

If very labile sample constituents are to be examined, it can be advantageous to establish the pH gradient first by pre-running the column (for 18–24 h with a 110 ml column) and then to inject the sample (with the appropriate addition of sucrose) into the pH gradient at a point close to its pI. This will both minimize the length of time required for the sample components to reach their equilibrium pIs and will protect the sample from regions of pH at which it may be unstable. The best position for sample injection can be inferred from the results of the preliminary experiments.

(e) Preparing ampholytes with 'non-standard' pH ranges

Commercially prepared ampholytes are available with wide pH ranges (e.g. 3·5–10, 2–11, etc.) or with narrow ranges covering usually 2 pH units. Ampholytes of very narrow range covering only 0·5 pH unit are available to special order (e.g. from LKB-Produkter AB). Ampholytes of different ranges can be mixed, so that for example a mixture of equal amounts of pH 4–6 and 6–8 ampholytes will give a pH gradient of 4–8.

While the available ranges therefore cover almost every eventuality, commercial ampholytes are expensive and have a finite (though long) shelf life. In particular, if only small quantities are needed, as for example for a limited number of IEF runs on polyacrylamide gel or if the correct narrow-range ampholytes are not immediately available, it may be desirable to prepare the required range from ampholytes with a broader range. This can readily be done with a density-gradient column which is merely run in the usual way as described below without any added sample. Once the pH gradient is established (after about 18 h at the running voltage in a 110 ml column) fractions are collected in the usual way and those fractions covering the desired pH range are pooled and used for the subsequent experiment(s). Since only part of the ampholytes are to be used for subsequent work, the initial ampholyte concentration for the first run should be 3–6 wt per cent. When LKB Ampholines are used it is recommended that in order to maintain good electrical conductivity around pH 7 a small amount (10 per cent of the concentration of narrow-range ampholytes) of ampholytes with the pH range 6–8 should be added to the pooled fractions. The amount of sucrose incorporated in the pooled fractions should be allowed for in calculating subsequent density gradients.

(f) Modifying the shape of the pH gradient—the use of separators

If ampholytes of overlapping pH ranges are mixed, the effect is to produce a plateau or shallowing of the gradient in the overlapping range. For example if pH 4–6 ampholytes are mixed with those of pH 5–7 the resulting gradient between pH 5 and 6 will be shallower owing to the higher ampholyte concentration in this region. This approach can be useful to increase either the resolution or the zone loading capacity within a selected part of the gradient.

Greater resolution within part of the gradient can also be achieved by the use of 'separators'. Addition of amphoretic substances such as amino acids, peptides, 5-aminovaleric acid, ß-alanine, tetraglycine, zwitterionic buffer ions, etc. flattens the pH gradient either in the vicinity of the pI of the separator or close to it and provides a controllable means of improving separations in this region. The use of separators was exploited for example in the separation of human haemoglobin subtypes α-amylase, goat antibodies, and human immunoglobulins by Brown, Caspers, and Vinogradov (1977). They found that among factors influencing the shape of the pH gradient were the nature and concentration of the separator added (usually at levels in the range of 2–20 mg ml^{-1}), the concentration of ampholytes used, and the ampholyte pH range.

(g) *The choice of polarity*

Once the pH gradient is fully established the conductivity shows a minimum between pH 6 and 7 (particularly if LKB Ampholines are used). The conductivity is also influenced by viscosity, and therefore the polarity is usually chosen so that these effects counteract each other. This is done by placing the region of minimum ampholyte conductivity in the upper less viscous parts of the column. With pH ranges lower than pH 6 the cathode is placed at the top, and above pH 6 the anode is placed at the top. In some experiments, particularly if high sample loadings are used, zones of precipitate may form in the column, and if these are above the zone formed by the component of interest particles of precipitate will fall through the density gradient and cause unwanted contamination. If this occurs the polarity should be reversed so that the zone(s) of heaviest precipitation occurs below the sample required.

The next three subsections specifically describe the use of the LKB columns for density gradient IEF separations, but many of the comments made would apply equally well to any other density gradient IEF experiments in vertical columns regardless of column design or size.

(h) *Setting-up the LKB column*

The column is set up vertically, using a plumb line if necessary, and the cooling water turned on. This should preferably be at a temperature between 4 and 10 °C.

The valve V in Fig. 9.1 is opened, and the dense electrode solution (Table 9.1) is added through the upper tube T_1 which leads into the inner electrode chamber. The gradient is then poured or pumped in through the tube T_2 keeping the tube in contact with the glass walls so that the gradient solutions flow down the wall causing the minimum disturbance. The gradient can be prepared in a number of ways. The simplest procedure is by using three channels of a multichannel peristaltic pump. One channel is used to pump light gradient solution (Table 9.2) into a stirred beaker containing the dense gradient solution (Table 9.2), and at the same time the other two channels are used to pump solution from this beaker into the column. Alternatively any gradient-making device can be used, or the gradient can be prepared manually by adding small portions (2–4 ml) of solution of gradually decreasing density until the column is filled. A table which can be used as a volumetric acid for preparing gradients manually is given by Vesterberg (1971). A precisely linear density gradient is not required since the purpose of the gradient is merely to prevent mixing by convection currents. Gradient solution is added to the column until the level is about 10 mm below the upper electrode. Finally the light electrode solution (Table 9.1) is added through T_2, care being taken not to disturb the top of the gradient.

The easiest way of applying the sample is merely to use sample solution in place of part or all of the water used for making up the gradient solutions (Table 9.2). The solutions used for making up the gradient also contain the ampholytes, and the quantities required for a column using 1 per cent ampholytes are shown in Table

TABLE 9.1

Composition of electrode solutions for 110 ml IEF column

Cathode solution	Anode solution
Anode at top of column	
15 g sucrose	1·5 ml 1 M H_3PO_4 or H_2SO_4
6 ml M NaOH or ethylenediamine	
$\quad$ H_2O to 25 ml	8·5 ml H_2O
Cathode at top of column	
$\quad$ 2·5 ml 1 M NaOH or ethylenediamine	15 g sucrose
$\quad$ 7·5 ml H_2O	4 ml 1M H_3PO_4 or H_2SO_4
	$\quad$ H_2O to 25 ml

TABLE 9.2

Composition of density-gradient solutions for 110 ml IEF column

Dense gradient solution	Light gradient solution
27 g sucrose	2·7 g sucrose
2·0 ml ampholyte solution (40 wt per cent)	0·7 ml ampholyte solution (40 wt per cent)
$\quad$ H_2O or sample solution to 54 ml	$\quad$ H_2O or sample solution to 54 ml

9.2. These should be increased proportionally when higher concentrations are required. Three-quarters of the ampholyte solution (which is usually marketed as a 40 wt per cent solution) is added to the dense gradient solution and the remainder to the light solution.

(i) *Electrical requirements and the separation in the LKB column*

Sample constituents become virtually stationary when they reach their equilibrium positions (pI) in the pH gradient so it can be appreciated that neither the applied voltage nor the experimental time are critical in IEF separations. However, it is clearly undesirable that samples should be maintained as a concentrated zone in a medium of very low ionic strength for longer than is necessary because this will promote losses in any biological activity. Furthermore, if precipitation is troublesome this usually starts as a relatively disperse particle formation which gradually aggregates, forms a more dense precipitate, and then sediments down through the gradient. For these reasons running times should be kept a short as conveniently possible.

Attainment of the equilibrium of a focused gradient and sharp sample zones depend upon the applied voltage. The higher the voltage the more rapid this is and the sharper the resolution obtained. The major limiting factor is the generation and heat which might cause convection currents and disrupt the density gradient. If this occurs it is usually visible as refractive index changes. Initially the maximum permissible power that should be passed through the columns was given as 5 W for the 110 ml LKB column and 10 W for the 440 ml column. However, it has since been concluded that the cooling capacities are greater than first thought and values of 15 W and 30 W respectively are now quoted for columns cooled with water circulating at 4 °C. The maximum permissible voltages are 1600 V and 2000 V respectively. Thus power supplies with outputs of at least 1000 V and preferably 1500 V or more and capable of up to about 30 mA are required (e.g. Bio-Rad 3000/300; LKB model 2197; Desaga Desatronic 2000/300; Pharmacia ECPS 2000/300; ECPS 3000/150 or 2500/100). Constant-voltage or, if available, constant-power settings are used.

The conductivity falls sharply during the first 3–4 h, so to avoid possible overheating it is common practice to commence with an applied voltage of 300–500 V and gradually increase this to the final running voltage as the current falls. Final currents are often as low as 1–3 mA at 1500 V in the 110 ml column. With applied voltages of 1000–1500 V experimental times of 18–24 h are usually satisfactory, but as a general rule attainment of equilibrium is slower the narrower the pH range used. Likewise, if viscosity is increased (by the addition of urea, glycerol, etc.) the necessary experimental time is also increased. The attainment of the steady state can be demonstrated by showing that the pI of the protein under study is constant under different conditions (i.e. after different focusing times) and the same pI should be obtained regardless of the point of application of the sample in the gradient.

(j) *Emptying the LKB column*

When the run is completed, the power is turned off, the valve V (Fig. 9.1) is closed to prevent dense electrode solution from flowing out of the inner electrode compartment, and the clamp C (Fig. 9.1) is opened. The column contents can then be allowed simply to drain into a fraction collector, but since different regions of the gradient have different viscosities a more constant and controllable flow rate will be achieved if the gradient is pumped out. The pumping rate should be adjusted to 60–100 ml h^{-1}. The column effluent can be passed through a flow cell for monitoring the u.v. absorbance (usually at 280 nm for proteins). Fractions of 1 ml should be collected from the 110 ml column, the collector tube size being such that the pH of each fraction can be measured with a pH meter fitted with a suitable combination electrode. Since it is temperature dependent (Fredriksson 1978), the measurement of the pH gradient in this way should be made at the same temperature as that used for the column run. Uptake of atmospheric CO_2, particularly by fractions in the alkaline region, can cause considerable errors, so pH measurements should be made as soon as possible after collection of the fractions. Delincée and Radola (1977, 1978) for example have reported that a fraction of pH 9·1 falls to 8·8 after standing in air for 16 h and at higher pH errors would be even greater.

If there is heavy precipitate formation in the lower part of the column, emptying from the top by pumping dense sucrose solution into the bottom of the column may be advantageous. Alternatively the gradient can be collected stepwise from the top by pumping out small portions each about 5 mm deep.

For many subsequent analyses the presence of ampholytes and density gradient materials in fractions is not harmful, but in some cases their removal is necessary. This is usually achieved by dialysis, ion-exchange chromatography, collection of the protein by salting out, precipitation, or gel filtration (which should be performed at high ionic strength to reduce possible electrostatic interactions between ampholytes and the protein). If the gel filtration is performed in a volatile buffer (e.g. 0·5 M acetic acid or ammonium hydrogen carbonate) a salt-free sample can be obtained by simple lyophilization.

Traces of ampholytes in protein samples isolated after IEF can be detected by thin-layer chromatography, the protein being applied to a cellulose TLC plate which is developed in a solvent of 10 per cent trichloroacetic acid (TCA), dried and sprayed with ninhydrin (Bloomster and Watson 1981). Proteins do not move from the point of application but ampholytes give a diffuse ninhydrin-positive zone ($R_f >$ 0·5) and as little as 1·25 μg can be detected.

9.3.   Iso-electric focusing in polyacrylamide

Polyacrylamide gel is a useful anti-convective medium for IEF experiments and possesses a number of advantages over density gradients in columns, particularly in analytical work. Among the major advantages are the following.

(1) It is much more rapid.
(2) A large number of samples can be run at the same time and compared, particularly if a slab gel system is used (this contrasts with at best 1 sample per 24 h with a gradient column).
(3) Only very simple equipment is needed and this can also be used for other gel electrophoretic techniques.
(4) Compared with the column method only small amounts of expensive ampholytes are needed.

(5) Very small samples consisting of only a few micrograms can be analysed.
(6) It can readily be used for one of the dimensions of two-dimensional separation systems, e.g. for protein mapping or for a subsequent immuno-electrophoretic step.

Resolution is even higher than with density gradient columns since gels can be treated (e.g. fixed, stained, etc.) immediately after focusing with no time-consuming fraction-collection step during which mixing and diffusion can occur (Fig. 9.2). Proteins differing in pI by only 0·0025 of a pH unit have been resolved (Allen, Harley, and Talamo 1974). The major disadvantage is that in preparative work it is not so easy to recover the focused protein zones, which are recovered readily and almost quantitatively from density-gradient columns. Also in polyacrylamide gel iso-electric focusing (PAGIF) pH gradient drift may be much more severe (see Section 9.3(e)) than in density gradients, where this phenomenon is usually imperceptible.

(a) *Gel composition and apparatus*

Unlike other electrophoretic applications of polyacrylamide gel in which molecular sieving plays an important role in the separation, in PAGIF any sieving effect actively slows down the attainment of the final focused band pattern and hence should be avoided. Thus highly porous gels should be used, and in the case of polyacrylamide gels with $T = 5$ per cent and $C = 3$–5 per cent are generally suitable and have quite good mechanical properties. Better mechanical properties can be obtained by using gels with $T = 7$ per cent and $C = 4$ per cent, but these should only be used for proteins with molecular weights below about 100 000. For very large proteins gels with $T = 3$·5 per cent or less are indicated and agarose (Section 9.4) or composite polyacrylamide–agarose gels (Section 6.3) can also be used. Baumann and Chrambach (1976a) advocate a stacking-gel type of composition with low T (typically 5 per cent) and high percentage cross-linking (e.g. 15 per cent) but with DATD used in place of Bis. However, it was subsequently shown (Bosisio et al. 1980) that when DATD is used as a cross-linker large amounts (up to 80–90 per cent) of the DATD remains unpolymerized in the final gel. This explains many of the curious and undesirable properties sometimes reported with DATD cross-linked gels. In view of the possibility that the gels may contain relatively large amounts of toxic monomer making them potentially unsafe to handle, it would seem that the use of DATD should be avoided for all gel applications. Bis cross-linked gels with $T = 5$–7 per cent and $C = 3$–5 per cent do in fact still retain some ability to hinder the migration of even very small molecules and the adoption of formulations similar to those used in

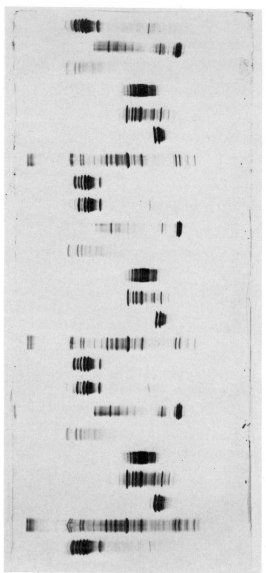

FIG. 9.2. Analysis of a number of proteins by PAGIF using a commercially available gel slab (PAG plate) containing ampholytes with a pH range 3·5–9·5 showing the excellent resolution attainable with this method. Samples: 1, haemoglobin; 2, L-amino acid oxidase; 3, β-lactoglobulin; 4, carbonic anhydrase; 5, catalase; 6, concanavalin A; 7, ovalbumin; 8, carboxyhaemoglobin. (Reproduced by permission of LKB Produkter AB.)

stacking gels in PAGE does seem to offer some advantages. If T is kept constant (and relatively low) pore size increases as C is increased above 5 per cent, and with $T = 6$ per cent gels Bisisio *et al.* (1980) found that using the protein ferritin (MW = 440 000) gels became essentially non-retarding when C reached 30 per cent and running times were halved by comparison with $C = 3$–5 per cent gels. In $T = 6$ per cent and $C = 20$ per cent gels cross-linked with Bis, about 20 per cent of the Bis remained as monomer after 1 h, but after 'ageing' overnight this fell to less than 4 per cent (no such ageing effect was seen with DATD gels). Gels with $C = 30$ per cent are brittle and difficult to handle so Bosisio *et al.* (1980) recommend casting the gels on silanized glass plates to which they adhere and this overcomes most handling problems. For most purposes when very low sieving effects are necessary gels with $T = 5$–7 per cent and $C = 15$–20 per cent probably represent a reasonably good compromize.

Gel preparation is very similar to conventional PAGE procedures (Chapter 2), except that the required amounts of acrylamide and Bis or other cross-linker are dissolved not in buffer but in H_2O or better, in 10 per cent sucrose or sorbitol (see below). TEMED is usually added, although Riley and Coleman (1968) suggest that it can be omitted as components of the ampholyte mixture can perform the same function. Sufficient ampholyte solution to give a final concentration of 1 per cent is then mixed in and the polymerization catalyst added. Specially prepared narrow-range ampholytes and separators (Sections 9.2(e) and 9.2(f)) can be incorporated if required. Either chemical polymerization with ammonium persulphate or photopolymerization with riboflavin can be used, and the quantities of catalyst and TEMED can be as described earlier (Section 2.6). However, Pharmacia workers (Låås, Olsson and Soderberg 1980) advocate omission of TEMED and careful deaeration of acrylamide solutions which then permits very low levels of ammonium persulphate to be used (200 μl of a solution containing 22·8 mg ml^{-1} for every 30 ml of gel mixture). Deionization of acrylamide solutions before catalyst addition with a mixed-bed ion-exchange resin (Amberlite MB-1 or similar) was also claimed to be beneficial. When very acidic ampholytes (e.g. pH range 2·5–4) are used with conventional amounts of TEMED and persulphate (and no deaeration step), it may be necessary to add a small amount of sodium sulphite to achieve polymerization in a reasonable length of time (see Section 2.7). For very alkaline ranges (e.g. pH 7–10) photopolymerization may be unsatisfactory. Many workers (e.g. Chrambach *et al.* 1973; Karlsson, Davies, Öhman, and Andersson 1973) advocate the addition of 10–12·5 per cent glycerol, sucrose, or sorbitol to the gel mixture to improve mechanical stability and wall adherence of the gel in glass tubes. It also greatly reduces pH gradient drift, so its use is to be recommended. Additions of urea also increase viscosity and have similar beneficial effects.

Uncharged thiol-protecting reagents and high levels of urea (e.g. up to 8 M) to cause protein dissociation can be incorporated in PAGIF gels without difficulty.

The apparatus used for PAGIF is identical to that used for PAGE. The simple and inexpensive cylindrical gel rod or vertical or horizontal slab systems can be used, and are set up as described in Chapter 2. All systems have their particular merits and their devotees, but the most widely used are the simple rod systems, which are popular for the first dimension of two-dimensional protein mapping (see Chapter 11), and the horizontal slab configuration. Since the focused bands become almost stationary in PAGIF, electrodecantation and difficulties with uneven cooling are not encountered. Consequently this is a technique in which either a simple design of horizontal slab apparatus or more sophisticated forms such as the LKB 2117 Multiphor and 2217 Ultraphor, the Shandon Model 600, the Desaphor or Mediphor from Desaga, and the Pharmacia FBE 3000 Flat Bed Apparatus possess a number of advantages. These include (1) a ready and direct comparison of a number of samples run under identical conditions, (2) simple handling of a single slab instead of a number of gel rods, (3) the ability to vary the length and thickness of the gel slab, (4) ready access to the gel slab so that the sample can be positioned anywhere, either before or after establishing the pH gradient, (5) simple measurement of the pH gradient using a microelectrode touched on to the gel surface, and (6) easy measurement of the voltage gradient if desired using voltage probes. Premanufactured gel slabs suitable for use on horizontal apparatus and containing ampholytes with various pH ranges are available commercially (e.g. LKB, Serva, etc.).

Manufacturers of horizontal slab gel apparatus generally include all the necessary accessories in terms of gel moulds, stands, clamps, instructions, etc. required for the preparation of PAGIF slabs. Naturally the details differ from one manufacturer to the next, but there is a clear trend towards thinner gels and those of 2–3 mm thickness have now been largely superceded as the 'standard' size by 0·5–1·0-mm-thick gels. Since for a given voltage and gel composition current is proportional to cross-sectional area, it follows that reducing gel thickness reduces current and hence heat generation. Cooling is also usually more efficient with thin gels, so that for both reasons they can be run at much higher voltages (field strengths) than thicker gels, which greatly improves resolution and also gives faster separation times. Staining and destaining are also more rapid and of course less reagents, including expensive ampholytes, are used so the use of thin gels has considerable advantages. Thin gels are relatively easily torn or distorted so it is usual practice to support the gel either on a plastic film (e.g. LKB PAG Moulding Sheet or cellophane) or on a silanized glass plate. For this the clean glass plate to be used in the moulding chamber is dipped for a few minutes in a 0·1 per cent solution of Silane A-174 in acetone and then allowed to air dry leaving a film of silane molecules on the plate (Bosisio *et al*. 1980). These contain reactive double bonds so the polyacrylamide gel is chemically bound to the glass plate and washing with steel wool is required if the glass is to be reused. The silane solution is stable for several months if kept dry at 4 °C in the dark.

In a typical arrangement for 1 mm thick gels (such as the Pharmacia system) the mould is assembled as shown in Fig. 9.3. The glass plate on which the gel will eventually be supported is laid flat and, if the above silanization is not used, it is wetted with a few drops of water and a piece of PAG Moulding Sheet applied. Air bubbles trapped between the sheet and the glass are removed by rolling with a rubber roller. The spacer gasket is placed in position followed by the top glass plate. If required, the inside face of this can be lightly coated with silicone fluid to prevent the gel sticking when the mould is disassembled. The mould is then clamped together, placed in a vertical position, and filled with the gel mixture using a syringe fitted with a relatively fine needle by lifting a small portion of the gasket specifically made with a small flap for the purpose. With other designs a U-shaped gasket along three sides of the mould is used and the fourth side sealed with adhesive tape after filling. A typical recipe for 30 ml of a gel mixture giving a $T = 5$ per cent, $C = 4$ per cent gel is shown in Table 9.3. The acrylamide, Bis and glycerol (or 3 g sucrose for acid pH ranges or 3 g sorbitol for alkaline pH ranges) are dissolved in H_2O and stirred for a few minutes with about 0·3 g of mixed bed ion-exchange resin (e.g. Amberlite MB–1) to remove traces of acrylic acid which can give rise to excessive gradient drift. If stock solutions of acrylamide and Bis are routinely made up and kept for a period of time, as is commonly the case, ion-exchange resin should be present in the stock bottle, but if solutions are freshly prepared it may be possible to omit ion-exchange treatment altogether. The gel mixture is then deaerated with a water pump, the ampholytes, TEMED and persulphate added and the mixture placed in the gel mould. Due to the toxicity of acrylamide and Bis monomers (see p. 9) gloves should be worn and care taken to avoid skin contact throughout these manipulations. If polymerization is uneven it

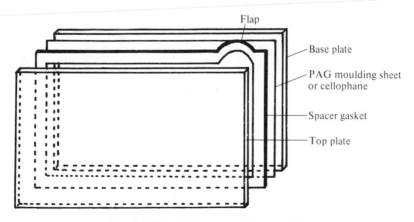

FIG 9.3. PAGIF gel mould assembly.

TABLE 9.3
Typical gel formulation required for PAGIF slab gels

Acrylamide	1·44 g
Bis	60 mg
Glycerol	3·0 ml
H_2O	to 27·8 ml
Ampholytes	2·0 ml
TEMED	10 μl
Ammonium persulphate (35 mg ml^{-1})	200 μl

may prove beneficial to flush the mould with nitrogen before adding the gel mixture.

With this type of apparatus it is very easy to make gels of almost any thickness by simply using gaskets of different thicknesses. While Neuhoff and Radola and their co-workers have been particularly active in the very successful development of ultra-thin gels of 50 or 100 μm thickness (see Section 12.1), most of the advantages of using thin gels can be gained with slightly thicker gels which are rather easier to prepare and handle. A procedure first introduced by Görg, Postel, and Westermeier (1978) has since become quite widely used. For this the gasket in an arrangement such as that described above (they used LKB Multiphor equipment) is replaced by a 'home-made' gasket cut from one, two, or three layers of Parafilm (50 cm sheet) around three sides of the mould. This gives gels of 0·12, 0·24, or 0·36 mm thickness respectively. They used gels mounted on LKB PAG Moulding sheet (applied to one glass plate as above) or cellophane which was soaked in distilled water for a few minutes, carefully spread wet on the glass plate, and dried. As it dries out it straightens into a wrinkle-free layer. In order to fill the mould two paper clips are inserted into the open side in order to make room for inserting a syringe needle and the mould about half filled with gel mixture. The paper clips are withdrawn and the clamps holding the mould together then squeeze the glass plates towards each other causing the gel mixture to spread and fill the entire mould. After polymerization they recommend that the mould is cooled for a few minutes, before opening by inserting a small spatula or knife and gently twisting it. A thin film of silicone liquid applied beforehand to the top plate to aid separation of the glass without tearing the gel may be particularly helpful with such thin gels. Commercial equipment for preparing thin gels with thicknesses ranging from 0·1 mm to 0·5 mm is now available (e.g. LKB Ultro Mould No. 2217–200). Some indication of the improved resolution obtained with thin gels, as opposed to those 1 mm thick, can be gauged from Fig. 9.4.

(b) Sample application and focusing

The sample should be relatively salt free (see Section 9.2(b)) as excessive amounts of salt lead to poor separations with 'wavy' bands. When the gel rod system is used it is better if the sample is dissolved in a solution of 1 per cent ampholytes containing 10 per cent sucrose and then layered on top of the gels as for PAGE. This is then carefully overlaid with a further 5–10 mm depth of a solution of 1 per cent ampholytes containing 5 per cent sucrose, before filling the upper electrode

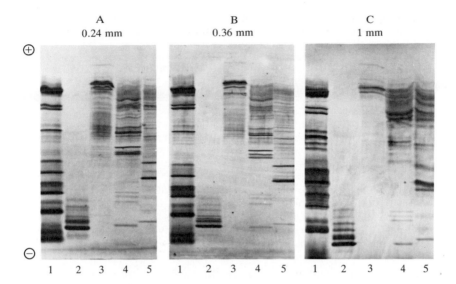

A
0.24 mm

B
0.36 mm

C
1 mm

1   2   3   4   5   1   2   3   4   5   1   2   3   4   5

FIG 9.4. The influence of gel thickness on the resolution obtainable during separation of proteins by PAGIF on $T = 7$ per cent, $C = 2.5$ per cent gels using Servalyt pH 2–11 range ampholytes followed by Coomassie Brilliant Blue G-250 staining. (A) Gel of 0.24 mm thickness, (B) 0.36-mm-thick gel and (C) 1.0-mm-thick gel. Lane (1) standard proteins, (2) trypsin, (3) soyabean lipoxidase, (4) and (5) legume seed proteins. (Reproduced from Görg, Postel and Westermeier (1978) by permission of the authors and publishers.)

chamber, to protect the sample from the extreme pH of the electrode solution. With the horizontal slab apparatus the sample can either be applied to wells preformed in the slab, or more usually is impregnated into small pieces of filter paper laid directly on the surface of the slab.

With the apparatus for gel rods the electrode chambers are filled with 0.5 M NaOH, ethanolamine, or ethylenediamine in the cathode compartment and 0.5 M H_2SO_4 or H_3PO_4 in the anode compartment, but 1 per cent ampholytes of suitably acidic or basic pH ranges can also be used for either or both electrode solutions and produce less extreme pH changes at the ends of the gel. Sponge or paper strips impregnated with these solutions are usually used in horizontal slab apparatus. Pharmacia Fine Chemicals AB recommend the use of 0.04 M glutamic acid in the anode compartment when acidic pH ranges are being investigated (0.1 M H_2SO_4 for the most acidic pH 2.5–5 range), 0.2 M HEPES when alkaline ranges are being used and 0.04 M aspartic acid for the broad pH 3–10 range. The cathode solution suggested is 1 M NaOH for alkaline pH ranges and pH 3–10, 0.2 M L-histidine for ranges below pH 6.5 and 0.2 M NaOH for the pH 5–8 range.

Various electrical conditions have been employed with gel rods, but the choice is not usually critical. Wellner and Hayes (1973) used rods 100 mm long and applied 120 V for 9 h, whereas Chrambach et al. (1973) used various times and voltages up to 800 V. Wrigley (1971) advocated that the voltage should be increased gradually up to a maximum of about 400 V with 65-mm-long rods and the current should not exceed 2 mA per rod. With salt-free samples the gradient formed in about 15 min under these conditions while focusing of proteins could take up to 2 h. When a cooled horizontal slab apparatus is used with 1–2 mm thick gels a power of up to 40–60 W can be applied, which means initial voltages in the region of 200–400 V or more and final voltages in the region of at least 1000–1200 V, so that typical running times are 2 h or less.

PAGIF gels are often run with coolant circulating at 4–10 °C, although cold water at about 15 °C is generally adequate, and indeed is to be preferred if the gel contains high concentrations of urea (to inhibit aggregation or to encourage dissociation of protein complexes) as this may crystallize out at low temperatures. Since resolution and zone sharpness are improved by a steep voltage gradient (see p. 249) it is preferable to use the highest voltage commensurate with the cooling capability of the apparatus, and voltages of up to 3000 V (averaging about 300 V cm^{-1}) have been claimed to be advantageous by Låås et al. (1980). However, resolution only varies according to the square root of the field strength (eqn 9.36) so very high voltages are not obligatory. With thin gels high voltages can be applied initially and increased more rapidly than when relatively thick gels are employed, so separation times can be reduced to 1 h or less.

It is often convenient to have a visual marker of the progress of the analysis and for this we have found it most useful to apply samples of haemoglobin (or any other coloured protein) to the end sample positions on the PAGIF slab. If further samples of haemoglobin (pI $\approx$ 6·8) are also applied in a matching position on the opposite side of the slab the two samples will migrate towards one another and when focusing is complete they should coalesce. Making allowance for the fact that haemoglobin is a relatively small protein which may reach equilibrium before some of the larger proteins that may be present in the sample, focusing should be continued for a short time after this before switching off. Cleeve and Tua (1984) have used bovine serum albumin prestained with bromocresol green (pI $\approx$ 4·6) as a visual marker in the same way.

The most common problem associated with PAGIF is the occurrence of distortions in the pH gradient. This produces wavy bands in the focused pattern and can be caused by poor electrical contact between electrodes and electrode strips, variations in gel slab thickness, uneven wetting of electrode strips, and insufficient cooling leading to local hot spots. The use of incorrect or inappropriate electrode solutions can also give rise to such distortions. The most common cause however is the presence of too much salt in the gel or in the samples. It is important therefore not to use excessive amounts of catalyst for gel polymerization and if necessary samples should be desalted by gel filtration or dialysis against water or ampholyte solution. With thin gels polymerized onto plastic backing sheets it is particularly easy to overcome problems caused by residual polymerization catalysts by simple washing procedures (Eckerzall and Connor, 1984). The gels can be prepared with acrylamide and Bis dissolved in H_2O or buffer and any of the catalyst systems described for PAGE gels (Chapter

2). Buffer and catalyst salts are washed out by immersing in several changes of H_2O and the gels are then transferred to a solution of ampholytes in H_2O and allowed to equilibrate.

Some biologically sensitive samples and enzymes are relatively unstable at the low ionic strengths prevailing in IEF separations, particularly at pH values close to their isoelectric points, and in these cases it may be beneficial to pre-run the PAGIF slab for 30–60 min, so that the pH gradient is established before the samples are applied and residence times can be kept to a minimum.

(c) Determination of the pH gradient

After focusing, the gel rods are sliced into pieces 2·5–5·0 mm long and each slice is chopped up or homogenized in a small volume (e.g. 1·0 ml) of deionized water. The pH is then measured directly with a pH meter. This same method can be used with pieces cut from a gel slab, but in this case it is quicker and more simple to measure the pH directly using a surface electrode (e.g. Ingold Type 403–30–M8 or similar). Since pH values are to some extent temperature dependent it is important that these measurements are made at the same temperature as that used during the IEF separation. Measurements at 4 °C give lower pI values than those at 25 °C and differences can be substantial, being about 0·1 pH unit at pI = 4, and of 0·5 pH unit or more above pI = 9. Fredriksson (1978) has published tables enabling appropriate corrections to be made for this temperature effect but the addition of agents such as sucrose, urea, glycerol, sorbitol, etc., can also alter the pK of ionizable groups, so it is a good rule always to measure pH (pI) values under precisely the same conditions as occur during the actual separation.

Interference in pH measurements from absorption of atmospheric CO_2 has already been referred to in Section 9.2(j) and this is potentially more troublesome in gel IEF techniques, particulary in horizontal gel slabs or granular flat beds (Section 9.5). If this is a cause of difficulty then the experiments can be performed in a nitrogen atmosphere, as for example in the PAGIF studies on histones by Valkonen and Piha (1980), and some designs of commercial equipment (e.g. LKB 2217 Ultrophor) have provision for this. If use of a CO_2-free atmosphere is not practical, or simply as a more convenient alternative, a number of marker proteins of known pI can be focused on the same gel slab. The pI of the unknown is then determined from a 'standard plot', with extrapolation if necessary. Some suitable marker proteins which are commercially available in high purity are shown in Table 9.4. IEF calibration kits of mixtures of marker proteins of known pI are now available commercially (e.g. Pharmacia Fine Chemicals AB). Very extensive tables giving pI values and molecular weights of over 1200 different proteins have been published (Righetti and Caravaggio 1976; Righetti, Tudor, and Ek 1981).

TABLE 9.4
Iso-electric points and molecular weights of some proteins suitable for use as marker proteins in IEF experiments

Protein	pI at 25 °C	pI at 4 °C	MW
Pepsin (porcine)	2·9[a], 2·8[d]		33 000
Amyloglucosidase (*Aspergillus niger*)	3·8[b], 3·5[d]		97 000
Glucose oxidase	4·2[d]		186 000
Soya bean trypsin inhibitor	4·6[d]		20 000
Ovalbumin (hen's egg)	4·7[a]		45 000
Serum albumin (bovine)	4·9[b], 5·2[a]	5·0[c]	67 000
ß-Lactoglobulin A (bovine milk)	5·1[a]	5·4[c]	36 000
ß-Lactoglobulin B (bovine milk)	5·3[a]	5·5[c]	36 000
Carbonic anhydrase (bovine erythroctye)	5·9[b]	6·2[c]	28 000
Conalbumin (hen's egg)	5·9[b]		86 000
Carbonic anhydrase (human)	6·6[d]		28 000
Haemoglobin A (bovine)	6·8		64 000
Myoglobin (equine), minor component	7·0[b], 6·9[d]	7·2[c]	17 500
major component	7·4[b]	7·6[c]	17 500
Myoglobin (sperm whale),			
minor component	7·7[b]		17 500
major component	8·3[b]		17 500
Lentil lectin–acidic band	8·2[d]		—
middle band	8·5[d]		—
basic band	8·7[d]		—
Trypsinogen (bovine pancreas)	8·7[a], 9·3[d]		24 500
Chymotrypsinogen A (bovine pancreas)	9·0[a]		23 600
Ribonuclease (bovine pancreas)	9·3[b], 8·9[a]		13 500
Cytochrome C (equine heart)	10·1[b], 9·3[a]		12 500
Lysozyme (hen's egg)	10·0[a]		14 400

[a] Righetti and Caravaggio (1976).
[b] Delincée and Radola (1978).
[c] Bours (1973).
[d] Pharmacia Fine Chemicals AB.

(d) *Staining, storage and quantitative measurement procedures*

Amido Black staining cannot be applied directly to PAGIF gels as the dye forms complexes with ampholytes and strong background staining results. If it is necessary to use this protein stain the ampholytes must first be washed out of the gels by soaking for several hours or overnight in 10–15 per cent trichloroacetic acid and using several changes of the washing solution. Amido Black staining and destaining as described in Section 2.9 can then be applied. A preliminary fixing step of soaking for 60 min in 10 per cent trichloroacetic acid may also be beneficial in Coomassie Blue R250 or Xylene Brilliant Cyanine G staining for proteins, but both of these and the periodic acid–Schiff reagents for glycoproteins (described in Section 2.9 and Tables 2.4 and 2.8) can usually be applied directly to PAGIF gels without modification and without prior removal of the ampholytes.

The Xylene Cyanine Brilliant G (Coomassie Blue G) stain made up in H_2SO_4, KOH, and TCA with the dye in the leuco form (Table 2.4 and p.30) as introduced by Blakesley and Boezi (1977) appears to be particularly suitable for use with PAGIF slabs and virtually no interference from ampholytes occurs.

Allen, Masak, and McAllister (1980) advocate staining with 0·25 per cent Fast Green FCF in 10 per cent acetic acid for 1 h (staining times from 5 min to 24 h could be used satisfactorily) followed by 24–72 h destaining in methanol: glacial acetic acid: H_2O (3:1:6 by vol). At least 72 h destaining was best for quantitative measurements. Sensitivity was about 30 per cent that of Coomassie Brilliant Blue R250. Ampholytes are not stained by Fast Green so need not be removed before staining. Bands are visible after about 3 h destaining, so with a short staining time the method can be very rapid. Most specific detection methods and enzyme activity stains developed for PAGE can also be applied to PAGIF gels.

Silver staining procedures (Section 2.9(a)) can be applied to PAGIF gels without difficulty but it is advisable to remove the ampholytes by preliminary fixing and washing steps. Confavreux et al. (1982) fixed gels for 15 min in 12 per cent TCA and then washed them 5 times for 30 min each in 500 ml of 50 per cent ethanol at 55 °C in a shaking water bath before silver staining. Allen (1980) applied silver staining to thin PAGIF gels which had already been stained with Coomassie Blue R-250 and destained in the usual way.

Most peptides down to about 15 amino acids in length are adequately fixed and stained in PAGIF gels by the Blakesley and Boezi (1977) procedure with Coomassie Blue G-250 (Righetti, Gianazza and Ek 1980) but with smaller peptides, or if poor fixation is suspected, some workers (e.g. Trah and Schleyer 1982) advocate staining with 0·15 per cent Coomassie Blue G-250 in 40 per cent methanol containing 3 per cent formaldehyde as fixative followed by destaining in 10 per cent methanol. The authors claim that polypeptides are fixed under these conditions while ampholytes remain soluble, although if the formaldehyde fixation method of Steck et al. (1980) for PAGE is used for PAGIF it appears that ampholytes are immobilized in the gel and interfere seriously with staining. Peptides, oligopeptides, proteins, etc. can of course be transferred to various immobilizing matrices just as can be done with PAGE gels (see section 2.13), and can then be located by all the usual immunochemical methods, such as the perodixase-antiperoxidase (PAP) procedure, without any interference from ampholytes. For oligopeptide detection Van der Sluis, Boer and Pool (1983) merely blotted thin gels onto glutaraldehyde-impregnated paper and detected the immobilized blots by the PAP procedure, while Johnson et al. (1983) used a more conventional electroblotting of proteins onto a nitrocellulose membrane. The latter also advocated the soaking of gels in an SDS-containing buffer before transfer because this then conferred a net negative charge to all protein species present and greatly improved transfer efficiency. It is worth noting however that Immobilines do not contain primary or secondary amino groupings so when these are used in place of conventional ampholytes for PAGIF in immobilized pH gradients (see Section 9.3(h)) gels can be sprayed directly with reagents for detecting amino groups without a requirement for a fixing step, so this should be an excellent procedure for analyusis of peptides etc, which may not be well fixed in other procedures.

After destaining very thin gels attached to glass plates can be immersed in a solution of methanol:H_2O:glycerol (70:23:3, by volume) for a few minutes, and then dried in a stream of warm air or simply allowed to air dry, giving a permanent, transparent record that remains suitable for densitometry. Similar gels on a plastic or cellophane backing are spread on a glass plate, allowed to dry for a few minutes and then attached to the glass plate with adhesive tape in order to prevent the edges from curling (Görg *et al*. 1978). Drying is then continued and when dry the gel and backing plastic can be peeled off the glass plate.

Storage of gels, quantitative measurement of focused bands, localization of radiolabels, etc. can all be performed by the identical procedures already described in Chapter 2 for PAGE analyses. This includes direct densitometry of unstained PAGIF gels run in quartz tubes by scanning at 280 nm. At this wavelength the absorbance of ampholytes is low but variable along the gel, so a blank gel with no sample should also be run (Vesterberg 1971).

(e) *Protein titration curves*

If a flat-bed PAGIF gel is focused so that all the ampholytes are at their iso-electric points their electrophoretic mobility is then very low and the pH gradient is relatively stable. When such a prefocused gel is rotated through 90° and an electric field reapplied the pH gradient is then across the gel and it remains unchanged for some time (Rosengren, Bjellqvist, and Gasparic, 1977). If a continuous band of sample protein is applied to the gel across the top of the pH gradient, subsequent electrophoresis will show the variation in mobility of the sample over a range of pH values. This procedure was originally presented as 'a simple method of choosing optimum pH conditions for electrophoresis, which could then be applied to further separations by PAGE or ITP, but the implications are very much more important than this because the resulting band pattern is in effect a representation of the protein titration curve. Righetti, together with various co-workers (Righetti *et al*. 1978a; 1978b; Righetti *et al*. 1979; Bosisio *et al*. 1980) exploited and developed this idea, showing that from the shape of the pH versus mobility curves it was possible to determine which charged amino acids had been substituted when a system of protein and its genetic variants were run together as a mixture. Within a 'family' of proteins (e.g. haemoglobin) of similar size and shape the electrophoretic mobility at a given pH could be correlated with the number of protons bound or released by the protein (Righetti *et al*. 1978a). These workers presented a number of calculated theoretical titration curves (Fig. 9.5) for a protein 'family' such as the variants of haemoglobin in which charged amino acids were replaced by other amino acids (or vice versa). When lysine (Lys) is substituted by a neutral amino acid (Fig. 9.5A) the two curves for the protein and its variant should meet before pH 10 where the Lys groups are no longer ionised so that the charge difference disappears, while below pH 9 the two curves should be parallel. Unfortunately IEF cannot be extended above about pH 10·5 to higher values where the ionization of the more strongly basic groups of arginine (Arg) is surpressed, so substitution of Arg by a neutral amino acid gives parallel curves

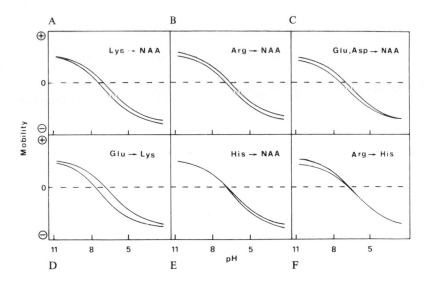

F<small>IG</small> 9.5. Theoretical titration curves for (A) substitution of a lysine residue by a neutral amino acid, (B) substitution of an arginine residue by a neutral amino acid residue, (C) substitution of a glutamic or aspartic acid residue by a neutral amino acid, (D) a double charge difference resulting from the substitution of a lysine residue by a glutamic or aspartic acid residue, or vice versa, (E) substitution of a histidine by a neutral amino acid, and (F) substitution of an arginine by a histidine residue. The dotted line represents the zero-mobility plane of the protein, which in practice is the sample application trough. (Reproduced from Righetti and Gianazza (1980) by permission of the authors and publishers.)

throughout the whole pH range (Fig. 9.5B). Substitutions of acidic aspartic (Asp) or glutamic (Glu) acids by neutral amino acids (Fig. 9.5C) lead to curves that meet at about pH 3 where the carboxyl groups become uncharged, and the curves are parallel above about pH 5. If a Lys is replaced by Asp or Glu or vice versa, so that there are two charge differences (Fig. 9.5D) the distance between the curves should be double that given by a single charge difference and at each end the curves should converge to half the separation, but not meet because only half the charge difference is lost in each case. Substitution of histidine (His) by a neutral amino acid (Fig. 9.5E) gives curves that meet at about pH 7 and only a single band is seen above this, while exchange between Arg and His also gives curves that meet at about pH 7 (Fig. 9.5F) but in this case two curves are seen above pH 7 and one below it. These theoretical conclusions were entirely confirmed by experimental results (Righetti *et al.* 1978*a*). It is important to remember, however, that these considerations apply only to charged groups on the surface of proteins exposed to the surrounding ionic environment and amino acids shielded or buried within interior hydro-phobic regions will not contribute to the titration curves.

Many proteins tend to lose stability and may become denatured under acidic conditions and this can lead to blurred curves below pH 4·0–4·5 which makes the titration of Asp and Glu difficult. Likewise there may be conformational changes affecting the number of surface groups exposed to the solvent and altering subunit interactions, binding to ligands, hydrogen bonding etc. which also affect the curves. To overcome these potential difficulties Righetti et al. (1979) introduced the idea of denaturing the proteins with 8 M urea and dithiothreitol or 2-mercaptoethanol. Because of the influence of weak acids and basis (i.e. weak salt ions) on the shape of the curves, they obtained the best results when sample proteins were first reduced and carboxymethylated and then desalted by passage through a small column of Sephadex G-25 equilibrated with deionized 8 M urea. The denatured sample proteins were then applied to the gel (also made up containing 8 M urea) as soon as possible in order to avoid possible carbamylation of Lys groups (in solution urea is in equilibrium with ammonium cyanate, and always contains traces of it, particularly at alkaline pH values, see p.23). When experiments are performed in this way sample proteins are unfolded, dissociated and have a random coil structure so the titration curve obtained depends upon the total amino acid composition and not just upon surface exposed groups.

Using this two-dimensional titration technique (in the absence of urea), by mixing a protein with a ligand it is possible to detect the formation of complexes and examine their pH dependence providing the affinity is quite high and association constants are at least in the μM range (Krishnamoorthy et al., 1978). Interactions between macromolecules can also be studied and the pH dependence of the association investigated (Righetti et al., 1978b). Much of this earlier work was essentially qualitative but in 1980, Bosisio et al. used the method for measurements of apparent free mobilities (m_o) and Ek and Righetti (1980) used it for determining protein-ligand dissociation constants (K_d). The method relied upon immobilizing the ligand by entrapment within the gel matrix and from a total of six runs it was possible not only to calculate K_d but also its variation over the pH range 3·5–10. Using affinity electrophoresis it was claimed that the same information would have required at least 42 runs. Because of the very low ionic strengths present the method was well suited to the study of protein–ligand interactions in which ionic bonds are involved and as a general rule K_d values are lower than values reported from conventional electrophoresis. Unfortunately although ionic strength is always low it is not constant across the pH gradient and it cannot be controlled as accurately as in zone electrophoresis. In spite of this and the constraints of a requirement for immobilized ligands the method does appear to have considerable potential for K_d measurements.

The method employs horizontal PAGIF gels in the original method (Righetti *et al*. 1978*a*) of $T = 7$ per cent and $C = 4$ per cent containing 2 per cent Ampholine with pH range 3·5–10·0 and also containing 5 mM aspartic acid, 5 mM glutamic acid, 5 mM Lysine and 5 mM arginine. Gels were of 2 mm thickness and used TEMED and ammonium persulphate as polymerization catalysts but were cast with a narrow sample trough 10 cm long, 1 mm wide and 1 mm deep down the middle of the gel. They subsequently used square gels (12·5 cm x 12·5 cm x 0·2 cm) and changed to much more drastic polymerization conditions (Bosisio *et al*. 1980) in which the solution containing acrylamide, Bis and Ampholines was degassed for 5 min with stirring under a vacuum of 0·1 mm Hg, the vacuum released by a stream of nitrogen (to avoid recontact with atmospheric O_2) and relatively large amounts of TEMED (50 μl 100 ml^{-1} solution) and persulphate (35 mg 100 ml^{-1} solution) added. Polymerization occurred in 5–10 min. These conditions were used with mixtures intended to give very porous gels with high percentages of Bis (i.e. $T = 6$ per cent, $C = 30$ per cent) and such rigorous conditions should not usually be necessary with lower percentage C gels. Gianazza and Arosio (1980) recommend the addition of a few grains of sodium dithionite together with the TEMED and persulphate as an aid to polymerizing gel mixtures which do not otherwise readily polymerize (although in their case the gels were only a more 'conventional' $T = 4$ per cent, $C = 4$ per cent).

For the first dimension (Fig. 9.6A), the IEF formation of the pH gradient, gels were run at a constant 10 W for 90 min, reaching 900 V at equilibrium, using paper strips soaked in 1 M NaOH and 1 M H_3PO_4 as the cathodic and anodic electrode

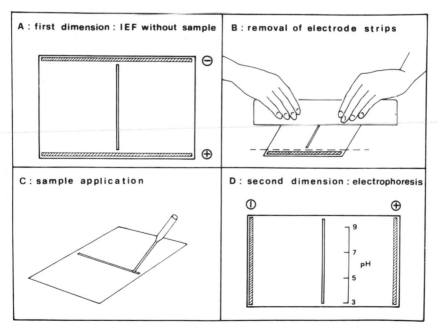

FIG 9.6. Procedure for measuring protein titration curves. (Reproduced from Righetti and Gianazza (1980) by permission of the authors and publishers.)

strips respectively. The anodic and cathodic regions are then cut off by slicing the gel along the inner side of the electrode filter paper strips (Fig. 9.6B). The gel slab is then rotated through 90° and new electrode strips applied, again using filter paper strips soaked in 1 M NaOH at the cathode and 1 M H_3PO_4 at the anode end. The sample trough which now lies across the pH gradient is filled with sample (Fig. 9.6C) and the second dimension electrophoresis stage performed by applying 600–700 V for 15–30 min (Fig. 9.6D). After this two strips of gel are cut out of the slab, parallel to the sample trough and about 5–6 cm from it in order to measure the pH gradient. These strips are then cut into segments about 5 mm long and macerated with 0·3–0·5 ml H_2O or 10 mM NaCl or KCl and the pH measured with a pH meter fitted with a combined microelectrode. These pH measurements should be made at the same temperature (by cooling if necessary) as that used for the separation (see Section 9.3(c)). Righetti et al. (1978) and Bosisio et al. (1980) recommend a temperature of 4 °C throughout. The rest of the gel slab was then stained in the usual way with Coomassie Blue G-250 (Xylene Cyanine Brilliant G) by the method of Blakesley & Boezi (1977), see Table 2.4 and p.30.

(f) *Modifying the shape of the pH gradient – 'gradient engineering'*

Just as with IEF in density gradients (see p.254), it is also possible to manipulate the shape of pH gradients in PAGIF. This can be done by mixing ampholytes of different pH ranges or employing spacers, as in density gradient experiments, by using appropriate buffer mixtures either in place of the usual commercial ampholytes or mixed with them (e.g. Nguyen and Chrambach 1980), or by altering the pH of the catholyte and particularly the anolyte (An der Lan and Chrambach 1980). Much better approaches are to alter the geometry of the gel or the ampholyte concentrations and these methods also have the merit of being generally applicable (unlike the use of spacers for example where specific spacers have to be chosen carefully for each pI range to be modified), although they can only be applied to horizontally run gel slabs. Atland and Kaempfer (1980) overlaid 0·35 mm thick base gels of $T = 5$ per cent, $C = 3$ per cent with 2 mm thick strips of $T = 8$ per cent gel of otherwise identical composition. The final concentration (weight per unit volume) of an individual ampholyte at the steady state is independent of the shape of the gel making up that unit volume. Thus modifying the profile of the gel along the pH axis alters the distance required for a fixed difference in pI, the increase ($\triangle d$) of the distance ($d_0 + \triangle d$) between two bands at any point being given by

$$\triangle d = d_0 \frac{a_1}{a_0}$$

where a_0 and a_1 are the cross-sectional areas of the base gel and overlay strip of gel respectively and d_0 is the distance when $a_1 = 0$.

Instead of increasing the thickness (and making the pH gradient steeper) in those parts of the gel of little interest, so that the rest had a flatter pH gradient, Laas and Olsson (1981a) described the converse and made those parts of the gel in which they were interested thinner. This was done by gluing strips of plastic 0·75 mm thick to the inside of the top plate of a gel mould for making 1 mm thick gels. The influence on the pH gradient of changing gel geometry in this way is indicated in Fig. 9.7. The same authors also reported a method for flattening pH gradients by altering the concentration of ampholytes, but this was technically rather more difficult. During gradient formation there is not only no net charge transport but

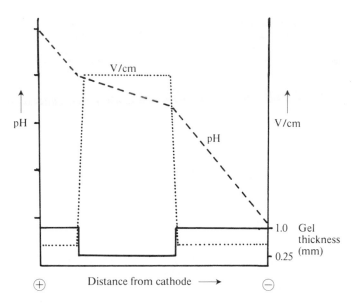

FIG 9.7. Diagram illustrating a gel thickness-modified pH gradient in PAGIF, showing the influence of gel cross-section (thickness) on field strength and the shape of the pH gradient. If the thickness of the gel is reduced to one quarter the field strength is increased about four-fold and the pH gradient flattened.

also no net mass transport, so if part of the gel initially has a lower ampholyte concentration it also will have a lower concentration once the pH gradient has formed. A lower concentration means that the molecules covering the pH range occupy a larger volume and hence the pH gradient is flattened. The slope of the gradient will depend on the relative proportions of high and low concentration regions. Unfortunately, in order to do this Laas and Olsson (1981a) had to use a stepwise gel casting procedure, using ampholytes diluted 1 part in 4 for part of the gel and once this had polymerized a further portion with ampholytes diluted 1:20 and then finally a third portion with 1:4 diluted ampholytes again. Also gels cannot be stored for long periods before use because concentration differences become reduced by diffusion, so the method is less convenient than gel thickening or thinning, in which the higher concentration of ampholytes used also gives enhanced stability to the pH gradients once formed.

(g) *Instability of the pH gradient*

The problem of a gradual drift in the pH gradient towards the cathode during PAGIF has been investigated in some detail by Chrambach *et al.* (1973), and some possible explanations for it have also been discussed by Rilbe (1977) amongst others. This drift, and the attendant formation of a plateau in the pH gradient in the region of pH 4–6, are probably due to electro-endosmosis and the transport of hydration water (Rilbe 1977). If this is caused by fixed negative charges (e.g. COO^-) on the gel matrix, there should be no flow in acid parts of the gel, a gradual increase over the range pH 3·5–5·5 as these groups become ionized, and a stronger flow in the neutral and basic parts of the gel. This would be expected to lead to shrinkage of the gel in the region around the pK of the acid groups and a flooding of alkaline parts of the gel. In agreement with this gel shrinkage in the plateau region is often observed in practice, and when using the gel rod system it may cause the acid end of the gel to be pulled into the end of the tube or a separation of the gel from the walls of the tube (Finlayson and Chrambach 1971). Separation can be minimized by precoating the tubes with high-molecular-weight linear polyacrylamide (containing no cross-linking agent), by cleaning the tubes with alcoholic KOH, and by using EDIA as the cross-linking agent instead of Bis (Finalyson and Chrambach 1971). Wall separation is a potentially serious problem since a film of liquid can develop between the gel and the wall and greatly disturb the pH gradient.

In typical PAGIF gel rod experiments the cathodic drift of the gradient is of the order of 10 mm per day (Fawcett 1977), so being time dependent the problem can be minimized by avoiding unnecessarily extended running times. The addition of sucrose, ethylene glycol, glycerol, urea, etc. to increase the viscosity also helps to minimize cathodic drift.

In polyacrylamide gel cathodic drift is seldom a seriously intractable problem, particularly if slab gel systems are used. However, if composite polyacrylamide–agarose or agarose gels, paper, or cellulose acetate strips are used (e.g. for the separation of very large molecules or for immunoisoelectric focusing), the difficulties of pH gradient instability are likely to be more severe as these matrices all contain more charged groups than polyacrylamide. The use of cellulose acetate in IEF and procedures to overcome the electro-endosmosis encountered are covered in Section 12.6(c).

(h) *PAGIF in immoblized pH gradients*

In 1982 a new PAGIF technique was introduced (Bjellqvist *et al.* 1982) in which the buffering groups responsible for the formation of the pH gradient are present as acrylamide derivatives with the general structure

$$CH_2{=}CH{-}\overset{\displaystyle O}{\overset{\displaystyle \|}{C}}{-}\overset{\displaystyle H}{\overset{\displaystyle |}{N}}{-}R$$

where R contains either a carboxylic acid or a tertiary amino group. These derivatives are copolymerized into the gel matrix together with the acrylamide and Bis and are used in place of conventional ampholytes. Under appropriate conditions the derivatives, termed Immobilines and manufactured and marked solely by LKB Produkter AB, are efficiently incorporated into the gel where they effectively determine the ionic strength and buffering capacity as well as the pH variation. The polymerization kinetics of polyacrylamide gels containing Immobilines and the efficiency of their incorporation into the gels have been discussed by Righetti, Ek, and Bjellqvist (1984).

Immobiline pH gradients are usually generated using two Immobiline solutions, one acidic and one basic, and one of them having a pK close to the mid point of the desired pH range. This is the buffering component while the other (non-buffering counter-ion) is used for pH adjustment. The first step therefore is to decide what pH gradient is to be used, and when isoelectric points of the proteins of interest are not already known it will probably be necessary to obtain at least an approximate value by a preliminary PAGIF experiment using conventional ampholytes. If this PAGIF gel contains 8 M urea and is run at 15 °C, pH readings taken with a surface electrode will be about 0·5 unit higher than for a gel run at 10 °C without urea due to the decreased H^+ ion activity, so a correction must be made for this when choosing the appropriate Immobiline. According to Application Note 321 (from LKB Produkter AB) most other common additives to PAGIF gels have only a marginal effect on pH values and no correction is needed for them.

Once the pH gradient required is known the following rule should be followed. The buffering Immobiline should have a pK value (e.g. at 10 °C in a gel) as close as possible to the pH at the midpoint of the gradient and the non-buffering counter ion should have a pK as far away from the desired pH range as possible (LKB Application Note 321). There are two exceptions to this, firstly when the desired pH gradient includes 5·2–5·5, in whole or in part, it is preferable to combine Immobiline pK 4·6 with Immobiline pK 6·2, and secondly when the gradient includes pH 7·8 a mixture of Immobiline 7·0 and Immobiline 8·5 should be used as the buffering Immobilines and Immobiline pK 3·6 as the non-buffering counter-ion. Immobilines with nominal pK values of 3·6, 4·4, 4·6, 6·2, 7·0, 8·5, and 9·3 are available from LKB and Table 9·5 shows the appropriate mixtures needed to give 30 ml of pH gradient. Amounts of Immobilines should be measured out accurately with a microsyringe as small variations can have quite a significant effect on the gradient. The Table is based on data given in LKB Application Note 321 which also includes a number of useful nomograms to enable the composition of solutions needed to give other pH gradients and it is obviously advisable to request this Application Note when ordering Immobilines from LKB Produkter AB. The pH values found when checking the starting solutions may differ from the expected values by a few hundredths of a pH unit, but not by more, and for high precision and reproducibility, particularly if very narrow pH gradients are to be used, the final pH adjustment should be made with a well-calibrated sensitive pH electrode.

Immobilines were introduced because the pH gradients are more stable than with conventional ampholytes and also because the resolution is higher, protein bands separating by as little as 0·001 pH unit can be distinguished. It is therefore usual to exploit this high resolution capability by making use of narrow (shallow) pH gradients of no more than

1 pH unit and sometimes as shallow as 0·1 pH unit, but if broader pH gradients of 1·5, 2·0, or 3·0 units are needed suitable recipes are given in Application Note 322 from LKB and even broader gradients of 2–6 units have been described (Gianazza, Artoni, and Righetti 1983; Gianazza *et al.* 1984). Gels with linear, reproducible pH gradients are prepared with the aid of a gradient gel former using equal volumes of two (light and dense) solutions containing acrylamide, Bis, and the correct proportions of Immobilines. Table 9.6 (based on LKB Application Note 321) shows typical amounts of the various constituents required to give 30 ml of gradient mixture and a gel of $T = 5$ per cent, $C = 3$ per cent. Solutions should be prepared as needed and should not be stored. When the quantities shown in Tables 9.5 and 9.6 are used the buffering capacity of the solutions is 5–6 mEq^{-1}. Suitable commercial designs of gradient gel former are available from LKB Produkter AB, Bio Rad Laboratories and doubtless other suppliers of electrophoresis equipment (see also p. 102). The LKB and Bio Rad designs are rather similar but have two and four outlet ports respectively enabling two or four gels to be prepared at the same time.

Gel moulds are prepared in the same way as for conventional PAGIF gels, and for two gels of 0·5 mm thickness 15 ml portions of light and dense solutions are added to the chambers of the gradient mixer. The magnetic stirrer, on which the gradient mixer is carefully centred, is switched on at the optimum speed

TABLE 9.5

Immobiline mixtures required to give 1 pH unit wide pH gradients

| pH range | Acidic (dense) solution | | | | Basic (light) solution | | | |
| | Buffering | | Non-buffering | | Buffering | | Non-buffering | |
	pK	μl	pK	μl	pK	μl	pK	μl
3·8–4·8	4·4	750	9·3	160	4·4	750	9·3	590
4·0–5·0	4·6	750	9·3	160	4·6	750	9·3	590
4·9–5·9	4·6	1000	6·2	720	6·2	1000	4·6	720
5·7–6·7	6·2	750	3·6	590	6·2	750	3·6	160
6·6–7·6	7·0	750	3·6	590	7·0	750	3·6	160
7·3–8·3	{ 7·0	750	3·6	980	{ 7·0	750	3·6	500
	{ 8·5	750			{ 8·5	750		
8·0–9·0	8·5	750	3·6	590	8·5	750	3·6	160
9·0–10·0	9·3	750	3·6	590	9·3	750	3·6	160

The Immobilines solutions are prepared by adding 25 ml H_2O to a whole bottle of Immobiline. This gives 0·2 M solutions. Quantities shown are those needed to make 30 ml of gradient (15 ml light + 15 ml dense solutions).

TABLE 9.6

Composition of dense and light solutions required to give 30 ml Immobiline gradients

	Acidic, (dense solution	Basic (light) solution
Acrylamide	727 mg	727 mg
Bis	23 mg	23 mg
Glycerol (87 per cent)	4·2 ml	None
Buffering Immobiline	*	*
Non-buffering Immobiline	*	*
H_2O	to 15 ml	to 15 ml

*Volumes taken directly from Table 9.4 and depending on pH gradient required.

determined previously and the polymerization catalysts added to both the light and dense gradient solutions. When the mid-point of the pH gradient is less than pH 6·5 add 140 μl TEMED solution (made up by adding 0·1 ml TEMED to 1·0 ml H_2O) and 85 μl ammonium persulphate solution (100 mg ml^{-1} in H_2O) to each chamber of the mixer, but when the mid-point of the gradient is above pH 6·5 the amount of TEMED solution should be reduced to 85 μl. Immediately open the outlet ports of the mixer and fill the gel moulds with the gradient mixture. This should take about 10 min. Disconnect the mixer and flush out all the chambers and tubing with water. The gel mould can be placed in a warm oven at 30°–50 °C to accelerate polymerisation which should take 30–60 min. When polymerized, dismantle the gel mould, remove the gel and wash for 1 h in at least 1 litre H_2O in order to remove the polymerization catalysts and any residual non-reacted Immobiline. After washing, remove excess H_2O by gently blotting the gel surface with moist tissue paper and leave the gel at room temperature until the surface is dry (LKB Application Note 321). This drying is important otherwise droplets of water will form on the surface during the electrofocusing run, and to be certain that the extent of drying is correct the gel can be weighed on removal from the mould, and after washing allowed to dry until this weight is reattained.

Methods of sample application are the same as with 'conventional' PAGIF gels, but since the pH gradient is immobilized high concentrations of salts or buffer ions do not disturb the gradient (unlike conventional PAGIF) so samples do not need to be desalted beforehand. However if a number of samples have high and differing salt contents they should be run on separate tracks of gel (LKB Application Note 321). This is done by slicing the gel and removing strips about 0·5 cm wide with a spatula, leaving individual strips (tracks) about 1 cm wide for the samples. This is necessary because protein bands from a sample with a conductivity higher than that of the surroundings will spread sideways and may migrate at an angle to the pH gradient, also disturbing neighbouring sample zones. Pre-running the gel before sample application is not usually recommended because of the low conductivity of Immobiline pH gradients once unbound ions have been removed by the electric field (about 100-fold lower conductivity than when conventional ampholytes are used in PAGIF gels). If a relatively high conductivity sample is then applied it will tend to remain at the point of application longer than desirable and precipitation may occur.

With horizontal gel apparatus such as the Multiphor (LKB Produkter AB), FBE–3000 (Pharmacia Fine Chemicals AB) or Model 1405 or 1415 (Bio Rad Laboratories) which require electrode wicks, the cathodic wick should be wetted with 0·01 M NaOH and the anodic one with 0·01 M glutamic acid. If basic pH ranges are being used the influence of CO_2 on the separation can be minimized by locating strips of paper or cloth impregnated with 1 M NaOH somewhere within the chamber, which alternatively can be flushed with nitrogen. Once samples have left the sample slot, the gels can be covered with a thin plastic sheet or wrapping. Running conditions are the same for all Immobiline pH ranges and LKB recommend an overnight run at 2500 V (about 25 mA and 5 W after initial removal of unbound ions) when electrodes are 10 cm apart. If it is known where the protein zones of interest lie, they can be made more sharp after initial focusing by moving the electrodes closer together and thus increasing the voltage gradient over the area of interest. Gel fixing and staining with Coomassie Blue R250 or Xylene Cyanine Brilliant G (Coomassie Blue G250) is performed in the usual way. Unlike conventional ampholytes Immobilines contain no primary or secondary amino groupings so after transferring the gels to a sheet of Whatman 3 MM paper, Gianazza et al. (1983) were able to detect peptides separated on Immobiline gels by spraying with ninhydrin.

The above procedure for IEF in immobilized pH gradients is essentially that described in LKB Application Note 321 (written by Bjellqvist and Ek in 1982) but very similar procedures have been reported by Bjellqvist et al. (1982), Cleve et al. (1982) and by Righetti et al. (1983). Because of the low conductivity during Immobiline PAGIF runs there are seldom difficulties with 'hot spots' in the gels, and electroendosmosis is rarely severe enough under normal conditions to give a substantial transport of water towards the cathode at low pH values or the anode at high pH values. Hence there is little flooding or drying out of parts of the gel, as is sometimes found with conventional ampholytes. If it should occur however, then addition of 20 per cent glycerol in the gel washing stage may reduce it. The Immobiline pH gradient, being covalently bound is stable, so long running times can be used, which gives proteins with low mobilities close to their pI values time to reach their equilibrium conditions. A further advantage of the use of Immobilines is that it is generally possible to load much more sample onto the gels than with ampholyte-containing gels (Bjellqvist *et al.* 1982), without causing zone distortions and this can be helpful in small scale preparative runs and for detecting minor components in analytical experiments. Rochette *et al.* (1984) report that with dilute (low per cent *T*) gels protein loads as high as 90 mg ml^{-1} of gel volume can be tolerated with little or no loss in resolution when compared to purely analytical runs. A further advantage in preparative work is that Immobilines being bound to the gel, extracted material is not contaminated with ampholytes (Gianazza *et al.* 1983). Coupled with the technical convenience of not having to desalt samples first, the major advantage of Immobiline gels must be the stability of the pH gradients which enables very narrow pH ranges to be used, resulting in much higher resolution than is practical or possible with conventional ampholytes (e.g. Rochette *et al.* 1984). These characteristics are also likely to be particularly beneficial when IEF is employed in the first dimension in two-dimensional procedures (Westermeier *et al.* 1983), but in this case gels with relatively broad pH ranges may be most useful (Gianazza *et al.* 1984). Some chromatins and histones appear to react with the charged polyacrylamide matrix of Immobiline gels (Righeti *et al.* 1983) and probably cannot be studied by this technique. All aspects of IEF in immobilized pH gradients have been reviewed by Righetti (1984).

(i) *Preparative applications*

PAGIF, particularly with slab gels, can readily be adapted to small-scale preparative work (e.g. Graesslin, Weisc, and Rick 1976). After focusing, bands are located, gels sliced, and material recovered in the same way as already described for PAGE slabs (Section 7.2). Vertical gel columns, again of the same type as used for PAGE, can also be applied in preparative PAGIF. The apparatus used can be either very similar to small-scale analytical gel rods, and tubes with internal diameters of up to 18 mm have been used (Finlayson and Chrambach 1971), or the type of column usually used with end elution (Section 7.5) can be employed. Naturally since the components will focus to an equilibrium position one would have a very long wait if end elution was employed, so the gel must be removed from the column and sliced and the material extracted as with smaller gel rods. This type of apparatus is therefore not particularly suitable for preparative PAGIF and for larger-scale preparative work IEF in beds of granulated gel (see Section 9.5) or agarose gels (Section 9.4) are preferred. After extraction ampholytes are removed from the proteins as described on p.258).

9.4. Isoelectric focusing in agarose gels

Agarose gels are much more simple to prepare than polyacrylamide gels, are non-toxic, and also, due to the very much smaller molecular sieving effect resulting from the larger pore size of the gel structure, extend the range of molecular sizes which can be studied to species with molecular weights of 10^7 or more (see Chapter 6). Although for these reasons agarose is a desirable matrix for IEF separations, because of the problems caused by high electroendosmosis it has been little used in the past. However, specially prepared agarose with very low electroendosmosis ($m_r = -0.005$) is now available commercially (Agarose-EF, LKB Produkter AB). Pharmacia Fine Chemicals AB have introduced Agarose IEF material in which stable positively charged groups have been introduced into the agarose molecules to balance the natural negatively charged groups present. Both products are emininently suitable for IEF work and descriptive literature about the products and their use is available free of charge from the respective manufacturers.

Horizontal flat-bed apparatus is best, as with this format gels are mechanically well supported. Typically gels containing 1 per cent Agarose IEF and 12 per cent sorbitol are used and commercial ampholytes added at a final dilution of 1 to 16 (Pharmacia Fine Chemicals AB). The agarose and sorbitol are dissolved by heating to close to boiling in the required volume of water, with constant stirring. The solution is then cooled to 60–80 °C, the ampholyte solution added and the mixture poured onto the supporting plate, which may be either of glass (thoroughly cleaned), plastic, or of a flexible hydrophilic plastic, such as Gel Bond^{TM} (Marine Colloids). For analytical purposes, sufficient mixture to give a gel thickness of about 1–1.5 mm is usually appropriate. Once the mixture has cooled and the gel has set, samples can be applied and IEF performed immediately, but if time permits it is better to allow the gel to harden fully for 1 h at 4 °C or overnight at room temperature. The cathode electrolyte should consist of 1 M NaOH, and 0.05 M H_2SO_4 is appropriate for the anode electrolyte in all cases except when the most basic pH ranges (e.g. 8–10.5) of ampholytes are being used, and then 0.2 M L-histidine is recommended (Pharmacia Fine Chemicals AB). A constant power setting of 15 W with a maximum voltage limit of 1500 V provides the best electrical conditions, according to these manufacturers, for all pH ranges except the extremes (e.g. 8–10.5 and 2.5–5) and typical run times are about 1.5 h, with coolant circulating through the apparatus at 10 °C. With pH 8–10.5 or 2.5–5 ampholytes the power should be limited to 6 W and voltage not permitted to rise above 600 V, so typical separation times are longer, e.g. 3 h.

Gels can be prepared containing 6–7 M urea and run the same way, although they may take longer to gel and cooling to 4 °C can be beneficial.

After focusing gels are fixed, stained, and destained in the usual way (Section 9.3d), but for rapid results, after fixing they can be pressed and dried before very brief staining and destaining times in the same way as other agarose gels (see p. 226).

Silver staining can be applied to agarose IEF gels, just as it can be for agarose electrophoresis gels (see p. 150) or PAGIF gels (see Section 9.3d), but as with the latter extensive washing to remove ampholytes before staining is required. Sano *et al.* (1984) reported poor staining when they applied typical PAGE silver staining techniques (see Section 2.9a) to agarose IEF gels and attributed this to the continued presence of ampholytes, so developed their own procedure.

This consisted of first fixing proteins in the 1-mm-thick gels by immersing for 5 min in TCA and sulphosalicyclic acid, soaking twice for 10 min each in 95 per cent ethanol, pressing, and drying onto Whatman 3 MM paper. The dry film was soaked in 2 per cent potassium ferrocyanide for 5 min and then given three 5 min washes in H_2O. The gel was then immersed in a tray of a solution warmed to 24 °C and freshly prepared by mixing 70 ml of 8 per cent sodium carbonate with 130 ml of a solution made by adding 1.9 g ammonium nitrate, 2 g silver nitrate, 10 g potassium tungstate, and 7.3 ml of 37 per cent formaldehyde in that order to 800 ml H_2O and when dissolved making up to 1 litre with H_2O. The gel was illuminated and agitated for 15 min and then immersed in 1 per cent acetic acid solution to stop development of the band pattern and dried.

After staining and destaining, dried Agarose IEF gels attached to Gel Bond can be stored permanently in a loose leaf file. They can also be quantified by scanning with a densitometer in the usual way, but the Gel Bond may fluoresce slightly so accurate scanning for fluorescent bands is not possible.

IEF in agarose gels is particularly appropriate when a second dimension immunodiffusion or immunelectrophoresis stage is to be combined with a first dimension IEF separation.

Although IEF in agarose gels has most often been used for purely analytical purposes, Chapius-Cellier and Arnaud (1981) have demonstrated its usefulness for preparative work.

They were able to apply samples of up to 1 g of blood serum protein or red cell haemolysate to 0·8 per cent agarose gel slabs about 6 mm thick in an LKB Multiphor apparatus. The sample was dissolved in a small volume (e.g. 5 ml) of warm agarose solution which was placed in a small tray to cool and gel. Once solidified, this sample gel was carefully laid on top of the main gel slab which had been prefocused for 3 h at 250 V, and then electrofocused at 15 W for a further 8 h. Ultraviolet illumination clearly revealed the position of the focused bands which were then excised and the proteins recovered by mashing the gel in H_2O with a glass rod and centrifuging. Protein content by dye-binding assay, pH, and optical density measurements were all made on the supernatant fractions.

9.5.   Iso-electric focusing in beds of granulated gel

It has been indicated above that IEF can be performed in a number of different matrices, and it is now clear that for larger-scale preparative work the matrix of choice at the present time is a flat horizontal bed of granulated gel (Winter 1977). First introduced by Radola (1975), the method (Fig. 9.8) can handle gram quantities of material and is particularly attractive in applications where IEF is employed at an early stage in a preparative scheme. It is relatively insensitive to precipitation since any precipitate that is formed as components focus at their pIs is trapped within the gel bed and has little or no further influence on the focusing of other sample constituents. After collection such precipitates usually dissolve readily on simple adjustment of pH or ionic strength.

FIG. 9.8. Iso-electric focusing of human erythrocyte haemolysate in a 70 ml bed of Sephadex G-75 Superfine: 150 mg of sample applied to a bed with a pH 6·0–8·5 gradient. (Reproduced by permission of LKB Produkter AB.)

The gel matrix should be hydrophilic and show no more than very weak electro-endosmosis. Radola (1975) showed that either beaded polyacrylamide or dextran gels could be used and that the superfine grade of Sephadex™ G-75 or G-100 was best. Winter, Perlmutter, and Davies (1975) demonstrated that ordinary Sephadex gels may contain charged water-soluble contaminants which interfere with the formation of the pH gradient. These must be washed out by suspending the swollen gel in at least 10 vol of distilled water and filtering under gentle vacuum. Specially purified and prepared products for IEF are now available commercially and require no such pre-treatment (e.g. Ultrodex™ from LKB Produkter AB; Sephadex IEF™ from Pharmacia AB).

Apparatus such as the LBK Multiphor, the Pharmacia FBE-3000, or the Desaga Desaphor and Mediphor are particularly suitable for this technique and the appropriate accessory kits are available to simplify operation.

The water content of the gel bed is very important if satisfactory results are to be obtained: too dry and the bed cracks, too wet and focused zones tend to sediment and are difficult to detect. In order to determine the optimum water level, in a preliminary experiment 100 ml of a 5 per cent gel slurry in distilled water is poured on to the gel tray and evaporated until small cracks are just visible in the gel bed. The water loss is determined by weighing the tray before and after evaporation. In all subsequent work with this same batch of gel, optimum gel beds are prepared by making up 5 per cent slurries and evaporating off only 75 per cent of this amount of water (Winter 1977). This procedure is necessary because different batches of Sephadex and other gel materials have slightly different swelling properties. While this is the approach recommended by LKB Produkter AB, other manufacturers suggest alternative ways of obtaining satisfactory gel consistencies. BioRad for example advocate drawing off excess liquid by touching dry filter papers to the edges of the gel bed, while Pharmacia suggest a careful, even sprinkling of dry gel powder over the surface of the bed (a piece of fine gauze attached over the mouth of a Sephadex bottle makes a good 'pepper-pot' for doing this). Both these approaches are far more rapid, if less precise, than the LKB method, but with due care give perfectly satisfactory results. For the IEF experiments gel beds should contain 2 wt per cent ampholytes and usually the sample is mixed into the slurry before pouring the gel bed. Alternatively a small volume of sample solution can easily be added as a zone in the middle of the gel bed, either before or after the pH gradient is established.

Running conditions of 8 W (about 400 V initially, rising to 1200 V or more) overnight or 30–40 W for 4–5 h (up to 2000 V or more) are usually adequate for a gel bed containing 100 ml of slurry, and afterwards the pH is best determined with a surface glass electrode (e.g. Ingold Type 403–30–M8).

In order to detect the focused zones a dry filter paper or a strip of cellulose acetate membrane (Frey and Radola 1982a) is placed on top of the gel bed. After a small amount of liquid has been absorbed, the paper is peeled off carefully, dried, washed three times for 15 min each with 10 per cent trichloroacetic acid, and stained with any general protein stain (Section 2.9) to give a 'print' of the separated zones. The zones are then recovered by pressing a fractionation grid into the gel bed, collecting the gel out of each section with a spatula, and transferring it to a small column, made for example from a 5 ml or 10 ml disposable syringe plugged with a small piece of glass wool or cotton wool. A series of such columns are prepared, one for each section of the fractionation grid, and the separated sample constituents are obtained by eluting each of the small columns with a small volume of water or suitable buffer. If water is used the pH gradient of the gel bed can of course be determined at this stage with a pH meter. Accompanying ampholytes can be removed from sample constituents by gel filtration, precipitation, ion-exchange chromatography, etc (see p.258).

High salt concentrations in the sample or use of Sephadex which has not been pre-washed may give rise to 'wavy' zones in the gel bed, and in the former case it may be beneficial to extract the zone of interest and to re-run it in a second experiment. If a crude material is analysed initially a similar re-running, perhaps with ampholytes of a narrower pH range, may give a greatly improved separation. Using steep voltage gradients in relatively thin (e.g. 1 mm) gel beds of Sephadex G–200 or Biogel P–60 Frey and Radola (1982b) were able to separate up to 1 g of protein with a resolution of 0·01–0·015 pI units.

9.6. Biochemical applications

Since about 1970 IEF has gained increasingly wide acceptance and has now become one of the principal techniques for analysing and characterizing proteins, peptides (Righetti and Chillemi 1978; Storring and Tiplady 1978), glycoproteins, lipoproteins, and phosphoproteins (Jonsson, Fredriksson, Jontell, and Linde 1978), cell membrane proteins (Jalkanen *et al.* 1983), isoenzyme patterns (e.g. Cassara *et al.* 1980), etc. Although less widely applied it has also been used in studies with nucleic acids (e.g. Shafritz and Drysdale 1975; Drysdale 1977) and glycosaminoglycans (e.g. Righetti, Brown, and Stone 1978). It has been most extensively employed in analytical studies, such as identifying particular components in a mixture and examining the purity of materials or monitoring the results of a purification scheme, but it is also useful for small-scale preparative work on the milligram scale in density-gradient columns or up to about 1 g with flat granular bed apparatus (e.g. Dolphin 1980; Bott, Navia, and Smith 1982; Gianazza *et al.* 1983*b*). IEF has also become the method of choice for determining iso-electric points (pI) and an extensive table showing the pIs of a large number of proteins has been published, the data all being collected from papers using IEF measurements (Malamud and Drysdale 1978). The major disadvantages appear to be the lack of solubility of some macromolecules at their pIs and the loss of enzymic or biological activity, which may occur in some cases owing to the very low ionic strength conditions. These may sometimes be alleviated by increasing the ampholyte concentration.

Measurement of pI by IEF has been used as a guide to the subsequent large-scale fractionation of unknown mixtures by ion-exchange chromatography. For example, three glycosidases from jack beans (ß-N-acetyl-glucosaminidase, α-mannosidase, and ß-galactosidase) had pI values respectively of 4·8, 6·2, and 8·0 (Li and Li 1973), so by running a DEAE-Sephadex A-50 column at pH 7·0 the ß-galactosidase was unabsorbed while the α-mannosidase was eluted by increasing the ionic strength and the ß-N-acetylglucosaminidase was eluted last of all by changing the pH to 6·0.

The majority of charged groups in globular proteins, both acidic and basic, are located on the surface of the molecules and are freely accessible to the solvent; however, some may be buried within the interior and only become accessible when the molecule is unfolded or denatured. In such cases the iso-electric point will then be altered as the protein is unfolded and these groups become exposed. Thus IEF, which is a highly sensitive technique in which changes in pI as small as 0·001 can be distinguished, can be used to study conformational changes (e.g. Ui 1971, 1973; Stinson 1977). Related to this, differences in aggregation states or alterations in net charge due to dissociation reactions can also be examined. The most common reagent employed for bringing about such conformational changes is urea, but non-ionic detergents or other uncharged substances such as glycerol, alcohols, and other water-miscible solvents have also been used. It should be remembered that high concentrations of urea, glycerol, ethanol, etc. cause apparent shifts in pH measurements owing to the reduced activity of H^+ ions, and corrections must be made accordingly in the measurement of pI values (Gelsema et al. 1979a).

If a complex band pattern is obtained with a sample that is thought on the basis of other tests to be very homogeneous, one possible explanation may be that interaction between the sample and ampholyte constituents is occurring (Hare, Stimpson, and Cann 1978; Basset et al. 1983). The way in which this problem can usually be identified is to collect the material from a single band and to resubmit it to a second IEF experiment under conditions identical to the first. If the multiple-band pattern is regenerated these types of interaction may be suspected. Alternatively protein-protein interactions, which occur in IEF experiments as elsewhere in biochemistry, can easily be misinterpreted as sample heterogeneity (Jackson and Russell 1984). Another potential source of difficulty is the possible occurrence of pH-dependent conformational changes (Cann 1979) or other pH-influenced interactions (Jackson and Russell 1984).

From the above it will be appreciated that IEF can be very useful in studies of virtually any factor that influences the net molecular charge. The method has therefore been applied for example in studies of ligand binding to a number of proteins (e.g. iron binding to transferrin (Vaneijk, Vannoort, Kroos, and Vanderheul 1978) or fatty acid binding to human serum albumin (Evenson and Deutsch 1978; Basu, Rao, and Hartsuck 1978)) and in investigating the subunit structure (including subunit exchange reactions) in multimeric proteins (e.g. Park 1973). The monitoring of the progress of chemical modification reactions of proteins, such as the carbamylation of amino groups (Bobb and Hofstee 1971) for example, is an obvious application of IEF. Both the natural microheterogeneity of proteins, glycoproteins, etc. (e.g. Lester, Miller, and Yachnin 1978) and that induced accidentally during the extraction or purification process can also often be demonstrated owing to the high-resolution capability of IEF. Similarly genetic differences between proteins have been clearly demonstrated. Some examples include the study of proteins in maize mutants (Righetti et al. 1977), phylogenetic differences in the iso-enzyme patterns of acid phosphatases and lactate dehydrogenases of closely related rodents (Kubicz and Wolanska 1977), screening of haemoglobin variants (Giuliani et al. 1978, see Fig. 11.4; Righetti et al. 1983), the resolution of the products of duplicated haemoglobin α-chain loci (Whitney, Copland, Skow, and Russell 1979), comparison of proteins from different genera of yeast (Drawert and Bednář 1979) and pea varieties (Görg et al. 1983), studies on the distribution of plasminogen allotypes in various populations (Dykes, Nelson, and Polesky 1983) the resolution of phosphoglucomutase$_1$

phenotypes for use in forensic blood typing (Pflug, de la Vigne, and Bruder 1981; Budowle 1984*b*), etc.

An interesting application of IEF is in the preparative fractionation of native and modified living mammalian cells (Manske, Bohn, and Brossmer 1977). For this a 110 ml column was set up with a density gradient composed of Ficoll and sucrose which was isotonic and hence avoided cell lysis. By using only 0·5 per cent Ampholine, injecting the sample (2 x 10^7 cells) into the middle of the column, and keeping focusing time down to only 4–5 h, which was vital for cell survival at the cost of a slight loss in resolution, cell fractions with 80–92 per cent viability were recovered. A similar approach has been used to study variations in the isoelectric points of chloroplasts during the cell cycle of *Euglena gracilis* (Brandt and Ernst 1982). A number of factors affecting the viability of cells subjected to isoelectric fractionation, in this case erythrocytes separated in a free-flow type of apparatus (see p.361), have been discussed by McGuire *et al.* (1980).

While reports of IEF in free solution, as opposed to IEF in various supporting gel matrices, are definitely in a minority they do feature prominently in attempts to scale up separations for preparative work. Several designs of equipment and different approaches have been tried but one of the most promising appears to be that referred to as recycling IEF which is claimed to be capable of fractionating gram quantities of protein in a single day and retaining reasonable resolution (i.e. proteins differing by 0·1 pI unit are separated). While these capabilities are no better than IEF in granulated gel beds or in agarose, the method appears capable of considerable development and has been used for the purification of antibodies to single band purity (Binion *et al.* 1982).

An attractive feature of IEF for clinical applications is that it is possible to examine very small amounts of tissue, such as may be provided by biopsy samples, by directly implanting the samples into the gel slab without extraction or other time-consuming pretreatment. The approach, sometimes referred to as direct tissue isoelectric focusing (DTIF), is exemplified by the work of Thompson *et al.* (1981) on human skeletal muscle in which 20 μm thick cryostat sections of biopsy material were applied directly to a horizontal agarose gel slab containing Triton X-100 to enhance protein solubility. The resulting protein patterns were revealed by Coomassie Blue staining but isoenzyme patterns were also examined by a zymogram technique. The benefit of PAGIF (or other IEF techniques) for clinical applications, including diagnosis and following the progress of treatments, is only likely to be realised if the separations can be standardized so that highly reproducible patterns are obtained and the results collected automatically with computerized data handling. A good start has been made in the case of human pancreatic secretion, using a computer controlled flying-spot gel scanner and computerized pattern matching, averaging and analysis, and data storage (Cassara *et al.* 1980; Allan *et al.* 1981).

10

ISOTACHOPHORESIS (DISPLACEMENT ELECTROPHORESIS)

ISOTACHOPHORESIS (ITP), sometimes also referred to as displacement electrophoresis, is a method which, like IEF, fractionates molecules on the basis of charge differences, and therefore, again like IEF, is most usually performed in non-sieving media where frictional resistance to passage of the molecules through the separation matrix need not be considered. ITP can be applied to both analytical problems, for which liquid filled capillary tubes or polyacrylamide gel rods or slabs of low per cent T are usually used, and for small scale preparative experiments, generally in horizontal gel beds, larger columns filled with gel or a liquid density gradient, or specialized, buffer-filled multi-compartment apparatus. For most analytical work the resolution obtainable by ITP at the present time is not as good as for many other analytical electrophoretic methods (e.g. PAGE, SDS–PAGE, IEF etc), but it does have the merit of being applicable in some situations where other methods are difficult to use or are unreliable, particularly in the field of small molecules, metabolites, etc. A further advantage is that the detection methods used in the most popular apparatus are of more general applicability than the staining or u.v. detection most commoly used in other electrophoretic procedures and this allows molecules of many different classes to be separated and detected in a single experiment. ITP has been used in a number of small scale preparative applications of up to a few grammes and it appears to have considerable potential for comparatively inexpensive scaling-up to handle much larger amounts of material, but although some effort has been given to this, progress has been slow and the potential by no means realized. Thus in summary while ITP is invaluable in some specific applications, it is a method which is holding its own, but is clearly not yet one which has shared the explosion of popularity, development, or application enjoyed by several other electrophoretic methods.

10.1. Isotachophoresis: principles and theory

Isotachophoresis (ITP) or steady state stacking is an electrophoretic technique that can be used either analytically or preparatively in capillary tubes, thin-layer equipment, or gel rods, slabs, or columns. Its designation as an independent method relies entirely upon operational considerations because the underlying theory and the separation process are in fact identical with that occurring during the stacking phase of electrophoresis in multiphasic buffer systems (Chapter 3). In the absence of a separation gel phase the stack of sample components referred to in Section 3.1 will continue to migrate in between the leading and trailing ions. Within this stack the sample constituents will be arranged in order of their mobilities in sharp zones, perhaps only a few microns thick, the concentrations of each component being determined by the concentration of the leading ion. This constitutes ITP, and indeed disc gel systems such as those described in Chapter 3 can be used for ITP separations if only a stacking gel phase is employed. For many purposes it is convenient also to make the operational distinction (Chrambach *et al.* 1981) that steady state stacking refers to the separation by inclusion of a component into the stack or by exclusion from it while the term ITP is reserved for separations occurring within the stack itself.

Theoretical discussions of the basis of ITP have been presented by a number of workers including Haglund (1970), Everaerts, Beckers, and Verheggen (1973, 1976), Routs (1973), and Mikkers, Everaerts, and Peek (1979 *a*, *b*) and extended by Everaerts and Verheggen (1983), Hjelmeland and Chrambach (1983), and Buzás, Hjelmeland, and Chrambach (1983). To give a very elementary treatment, such as that presented by Haglund (1970), during the migration in an electric field of the boundary between two salt solutions having one ion in common, the concentrations C_A and C_B of the two negative ions A^- and B^- (with the common positive counter-ion R^+) is given by

$$\frac{C_A}{C_B} = \frac{U_A}{U_A + U_R} \frac{U_B + U_R}{U_B} \frac{L_A}{L_B}$$

where U is the mobility in $cm^2\ V^{-1}\ s^{-1}$ and L is the electric charge. This applies if the salts RA and RB are fully ionized, otherwise corrections must be applied. This is the Kohlrausch equation and describes the conditions for a migrating boundary in an equilibrium state. Similar considerations also hold of course in a system of different cations with a common buffering counter-anion. The validity of the Kohlrausch regulating functions in this context has been verified experimentally (Hjertén Öfverstedt, and Johansson 1980).

Suppose now that A^- represents the leading ion and B^- the terminating ion so that the mobility U_A of A^- is greater than the mobility U_B of B^-, and suppose also that the initial concentrations C_A and C_B are equal (Fig. 10.1(a)). The starting concentration of B^- is thus greater than the equilibrium concentration given by the Kohlrausch equation, so that when the electric current is turned on the boundary between A^- and B^- moves towards the anode forming a concentration gradient in the chamber

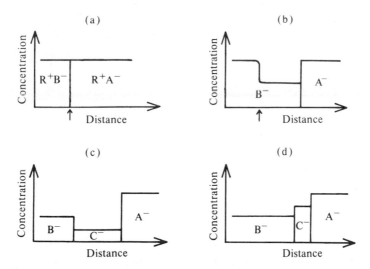

FIG. 10.1. Isotachophoresis. (a) The initial concentration of A^- and B^- at the start of the experiment (arbitrarily set equal). (b) Conditions established during the separation when A^- and B^- are in equilibrium as defined by the Kohlrausch equations. (c) Starting conditions when a zone of an intermediate ion C^- has been introduced between the leading A^- and terminating B^- ions. (d) The situation during the run when A^-, C^-, and B^- are all at equilibrium. (Based on Haglung (1970) and reproduced by permission of LKB Produkter AB).

occupied by B^-. The concentration ratio of A^- and B^- across the boundary obeys the Kohlrausch equation (Fig. 10.1(b)).

If a zone containing a third electrolyte RC is added to the system and the mobility of C^- is intermediate between that of A^- and B^- ($U_B < U_C < U_A$), this will be subject to the same equation. Thus if initially it is present at a low concentration in a relatively large volume (Fig. 10.1(c)), the steady state can only be reached by a concentration of the original zone (Fig. 10.1(d)). This arises because in the low-conductivity region of the dilute starting zone there is an increased potential gradient which tends to accelerate the C^- ions and concentrate them close to the A^-/C^- boundary. As this concentration process proceeds the conductivity of the region occupied by C^- gradually rises and the potential gradient decreases until C^- ions are migrating at the same speed as the leading A^- ions. Since the mobility of A^- is higher than that of C^-, the potential gradient required to achieve this is greater in the zone occupied by C^- ions than in the zone of A^- ions, and the difference in the potential gradient is thus inversely proportional to the difference in mobilities.

Consider now two sample ions C^- and D^- migrating at the steady state where $U_B < U_D < U_C < U_A$ (Fig. 10.2(a)). If the amounts of C^- and D^- are

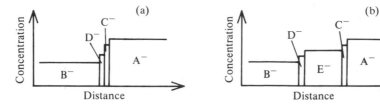

FIG. 10.2. Isotachophoresis. (a) Equilibrium conditions for a more complex sample containing small amounts of two ions C^- and D^-. These give narrow zones close to each other which are difficult to recover in a pure state. (b) The effect at equilibrium of including a further ion E^- of mobility intermediate between C^- and D^- to act as a spacer ion. C^- and D^- can now be recovered easily, but E^- should of course be chosen so that it can subsequently be readily separated from C^- and D^- since they are bound to be mixed with a little of it. (Based on Haglund (1970) and reproduced by permission of LKB Produkter AB).

small they will be concentrated according to the Kohlrausch equation into narrow zones that may only be a few microns thick. Furthermore, since these zones are immediately adjacent to each other they will be very difficult to recover separated from each other. However, if a further ion E^-, with a mobility between those of C^- and D^-, is added to the sample it will form a zone interposed between C^- and D^- (Fig. 10.2(b)), the separation depending upon the amount of the *spacer* salt RE added. This concept of spacer ions is often vital to the achievement of satisfactory preparative experiments.

From the above argument it can be concluded that the precise nature of the spacer ions is not critical as long as they possess absolute mobilities between those of the components to be separated. Svendsen and Rose (1970) introduced ampholytes, as used in iso-electric focusing, for this purpose and this has become the most widely used approach, but amphoteric and non-amphoteric buffer constituents, amino acids, etc. can also be used and may often be preferable when samples consist of mixtures of small molecules. According to Baumann and Chrambach (1976) the mobility ranges of Ampholine (LKB) ampholytes exceed those of proteins and only some proteins are satisfactorily stacked (Nguyen, Rodbard, Svendsen, and Chrambach 1977). Also it may be difficult in gel media to attain the true steady state within practical limitations of time and gel length. For these reasons Nguyen *et al.* (1977) advocate the use of a mixture of buffer constituents as spacers, which because of the multiple zones set up within the stack they termed 'cascade stacking'. Obviously a great many combinations of constituents could be used to set up cascade-stacking systems, and computer programs based on the theoretical work of Routs (1973) and Jovin (1973) are now available to describe appropriate mixtures (see Nguyen *et al.* 1977). The technique of non-equilibrium pH gradient electrofocusing (NEPHGE) as described in Section 12.2 is in essence a form of cascade stacking with ampholyte spacers.

One consequence of using Ampholines or a cascade buffer system as spacer ions is that a pH gradient is set up within the whole area of the stack between the leading and terminating ions. Owing to buffering by the counter-ion the pH gradient produced in ITP will not have the same pH range as that formed by the same ampholyte mixture in an IEF experiment. Thus, for example, proteins separated by ITP will not be subjected to their iso-electric pH, and this may have benefits in terms of solubility or stability. By replacing the buffers containing the leading and terminating ions in the two electrode compartments by strong acid and base solutions it was possible to halt the migration of the cascade stack and produce a stable pH gradient (Nguyen *et al.* 1977). Sample proteins then migrate into their iso-electric positions so that the system then appeared to be identical to IEF experiments performed with buffer mixtures in place of Ampholines (Nguyen and Chrambach, 1976; 1977). Hence this was termed 'cascade electrofocusing'. The opposite transposition of gels in which a separation had been achieved by cascade electrofocusing into a cascade-stacking system by merely changing the electrode solutions was also demonstrated. It was shown (Nguyen and Chrambach 1978) that this was particularly valuable in preparative work since the high degree of resolution and sharp zones formed during the focusing stage were maintained during elution by cascade stacking. All these experiments also demonstrate the essential identity of the physicochemical processes involved in IEF, ITP, NEPHGE, and PAGE experiments in multiphasic buffers.

Once the steady state in ITP has been reached and components are fully separated into zones the concentration of which is determined by the Kohlrausch equation, the whole region occupied by these zones (the stack) will migrate at a constant velocity. There will be no further change in concentration, order of constituent zones, or velocity so long as the separation conditions (current, pH, etc.) are unaltered. The velocity of the whole stack is determined by the absolute mobility of the trailing ion, as the slowest ion present, since any tendency of more mobile ions ahead of the trailing ion to move away from it creates a zone of lowered ionic strength. This in turn increases the local potential gradient, thus accelerating the trailing ions to 'close the gap'. A similar process tends to maintain sharp boundaries between the separated zones in the steady state. Thus if in our model the sample ions C^- should diffuse into the region occupied by the leading ions A^-, owing to the difference in mobilities, they will be entering a region of lower field strength and will be 'caught up' by the C^- zone. Likewise an A^- ion entering the C^- zone will encounter a higher field strength and will be accelerated back into the A^- zone. Thus once the steady state has been reached there is no further benefit in prolonging the experiment, or in increasing the length of the tube or gel. It therefore follows that attempts to optimize the separation process in ITP must be directed towards the transient state stage before the steady state is reached.

Resolution, which is defined as the fraction of the constituent being considered which is separated in the transient state into a pure zone, should obviously be maximized in the shortest time possible under convenient experimental conditions. For any given sample and system of buffers and spacers, four factors are particularly important here, namely the sample load, the electrical current, the properties of the counter-ion, and the pH (Mikkers *et al.* 1979*a*, *b*). As we have already seen, the concentration of any sample zone is determined by the concentration of the leading ion according to the Kohlrausch equation. Hence if there is a large amount of any constituent it can only be accommodated by a wide zone, so that during ITP zone width is directly proportional to the sample load. Analysis time is clearly inversely related to the current, but if temperature effects are ignored the separation efficiency, the length of the zone traversed before the final steady state is reached, and the load capacity are both independent of it. In practice, owing to time-dependent dispersion effects it is preferable to run at the highest convenient current for a relatively short time rather than use a longer run at low current. The efficiency of current transport is influenced by the mobility of the counter-ion, and a low mobility reduces resolution time and overall analysis time, improves separation efficiency, and increases load capability. The pH of the leading electrolyte and of the sample should not be too widely different. For anionic separations, a low pH of both leading electrolyte and sample will favour high resolution rate and separation efficiency while a high pH is preferable for cationic separations.

Two further aspects of ITP which may be mentioned here and which pertain especially to preparative applications relate to the choice of spacers and leading and trailing ions. Firstly, complex mixtures, such as many protein samples in which there are a large number of constituents each with possible microheterogeneity, can give an almost continuous spectrum of mobilities and relatively long separation times are then required (Mikkers *et al.* 1979*a*). Indeed if spacers are added, especially if in the form of a large number of Ampholine constituents, it is perfectly possible for the steady state never to be reached under practical experimental conditions. Thus while spacers may improve the ability to interpret separation patterns and may enable particular components of interest to be isolated largely free of other sample components, in true physicochemical terms their use decreases separation efficiency. Secondly, useful practical advantage can be gained if the mobilities of the leading and trailing ions are close to the mobility of the particular sample ions of interest. In this way other ions in the sample which are of no interest will migrate outside the stack in the regions of the leading or trailing ions, as in ordinary zone electrophoresis. This is in fact no different to the concept of selective stacking and unstacking already referred to in Section 3.8.

Thus, in summary, there are probably more factors that have to be optimized in order to obtain the best results in ITP than in any other electrophoretic method, but this very complexity adds to the versatility of the method. Among the factors that must be controlled are pH, ionic strength, mobilities of the leading and terminating ions, choice of counter-ion, buffering capacity, pH and composition of the sample, electrical conditions, whether to use spacer ions and choice of spacer ions, sample load, method of detection, choice of supporting matrix if any, type of apparatus to be used, etc. The actual choice made for any particular problem is usually decided by trial and error, although Baldensten (1980) has given useful theoretical and practical guidance for optimizing electrolyte systems.

10.2. Analytical isotachophoresis in capillary tubes

In this technique, which has been reviewed by Delmotte (1979), Holloway and Lüstorff (1980), and Holloway and Pingoud (1981), the separation is performed in a capillary tube of glass, PTFE (Teflon), or other plastic, with an optimum internal diameter of 0·3–0·6 mm and with length varying between about 200 and 800 mm. For rapid analyses long tubes necessitate high voltages (up to 25 kV or more) although currents may be only a few microamps. Resolution with reasonably short tubes can be achieved if a counterflow of leading electrolyte can be arranged in the opposite direction to the ITP migration, since this effectively gives the same result as increasing tube length. The apparatus required is quite complicated and highly specialized, so that by far the majority of researchers make use of commercial equipment (e.g. the Tachophor Model 2127, LKB Produkter AB). Other designs have been published by Everaerts et al. (1976) and in summarized form by Everaerts et al. (1973). This type of apparatus is nearly always used for purely analytical work as sample quantities are usually measured in micrograms, but in spite of that LKB do market an accessory for their Tachophor apparatus termed the Tachofrac which is intended for micropreparative work.

The capillary tube is filled with leading electrolyte and the sample (0·1–50 μl) injected with a microsyringe. A valve is then opened to fill the region behind the sample with terminating electrolyte. The choice of electrolytes is determined by the factors discussed above and hence depends upon the particular separation being performed. Generally ions with high mobility such as chloride, sulphate, phosphate, and cacodylate are suitable as leading ions, while the terminating ions should have low mobility (e.g. ε-aminocaproic acid, γ-aminobutyric acid, ß-alanine, glycine) and tris or Ammediol (2-amino-2-methyl-1,2-propandiol) are often used as counter-ions.

Useful systems of buffers for separating nucleotides and related substances at various pH values have been published by Holloway and Lüstorff (1980) and for the analysis of amino acids and peptides by Holloway and Pingoud (1981). It is common practice to include viscosity-increasing agents such as Triton X-100, methyl cellulose or hydroxypropyl-methylcellulose (HPMC) to reduce electroendosmosis. Holloway and Pingound (1981) recommend addition of 0·25–0·5 per cent HPMC as being the best.

Many different buffer systems have been reported and a few are shown in Table 10.1. As pointed out by Holloway and Pingoud (1981) these may not be generally optimal for all samples but they may provide a useful starting point. Buffer systems A–D have been used (Holloway and Pingoud 1981) for amino acid and peptide separations; system A (Kopwillem and Lundin 1974) being suitable for the separation of neutral and acidic amino acids as anions at high pH, system B (Holloway and Lüstorff 1980) being for anionic separations at low pH; system C (Kopwillem and Lundin 1974) is used for separation of basic amino acids as cations at high pH, while system D (Freidel and Holloway 1981) is intended for cationic separations at low pH. Systems E and F (Holloway and Lüstorff 1980) are intended for separations at pH 3·0–4·5 or 7·5–8·5 respectively of nucleic acids which can usually be separated best as anions,

TABLE 10.1
Some typical buffer systems for ITP

Buffer system	Leading ion	Terminating ion	Counter ion
A	Cl^- (10 mM)	ß-alanine (10 mM) + Ba $(OH)_2$ to pH 10	Ammediol to pH 8·6–9·7
B	Cl^- (5 mM)	Cacodylic acid (10 mM)	Histidine methyl ester to pH 5·3
C	Ba^{2+} (5 mM)	tris (20 mM) + HCl to pH 8	Valine (pH 9·9)
D	K^+ (5 mM)	ß-alanine (5 mM) + acetate to pH 5·1	Acetate to pH 5
E	Cl^- (5–10 mM)	acetate (5–10 mM)	ß-alanine (pH 3·55) or 6-aminohexanoic acid (pH 4·37)
F	Cl^- (2·5–10 mM)	ß-alanine (20 mM) + Ba $(OH)_2$ to pH 10	tris to pH 8·3
G	Cl^- (5 mM)	6-aminohexanoic acid (10 mM) + Ba $(OH)_2$ to pH 10·8	Ammediol (10·mM)

and indeed system A can also be used (Oerlemans *et al.* 1981). System G was used by Battersby and Holloway (1982) for the separation of liver cytosolic proteins and enzymes, while chemical modifications to serum albumin were assessed (Einarsson, Karlsson, and Åkerblom 1984) using a buffer system essentially the same as F but with Ba (OH)$_2$ being added to the terminating electrolyte only to pH 9·0.

The separated zones are usually detected as they move past a fixed detector by virtue of u.v. absorbance, temperature difference (thermal detectors) or by measurements of conductivity or potential gradient. Since the potential gradient varies from one zone to the next (and is proportional to the mobility) and the current is constant, the boundaries between the zones are characterized by sharp changes in temperature. These can be detected with thermocouples and converted into either an integral or a differential signal. Each detection system has its own advantages and disadvantages. Thermal detection provides a univerally useful method to indicate the complete zone pattern, but is not so sensitive as u.v. detection which, however, can only be used for measuring zones of u.v.-absorbing material. The latter is particularly suitable if non-u.v.-absorbing spacer ions such as Ampholines are used to separate u.v.-absorbing sample components, such as proteins or nucleotides (Oerlemans *et al.* 1981), because in such a case the thermal detector cannot distinguish between sample and spacer components whereas the u.v. detector can. As detectors of general utility, high resolution conductivity or potential gradient detectors probably represent the optimum at the present time. Fluorescence emission and fluorescence quenching also appear to have good potential as detection methods of wide applicability for ITP (Reijenga, Verheggen, and Everaerts 1984). The minimum zone length that can be located with a u.v. detector is about 0·1 mm, which in a 0·5 mm diameter capillary corresponds to 16 nl. If sample zones shorter than this are expected, zone length can of course be increased merely by using a lower concentration of leading ion since this determines the concentration (and hence zone length) of all other constituents. Quantification is achieved by measurements of zone length and comparison with calibration plots constructed using known amounts of the constituent being analysed measured under identical separation conditions.

The main advantages of capillary tube ITP are (a) qualitative and quantitative analysis can be carried out with only a few nanomoles of sample, (b) several separate ionic species can be measured in one experiment, (c) many different classes of sample constituent can be examined with the same basic equipment, (d) aqueous and non-aqueous (e.g. methanol, ethanol, etc.) electrolytes can be used, and (e) rapid analysis times, usually 15–60 min, can be achieved. In addition to these, in many cases no special sample pretreatment is usually needed and particularly when performed under free-flow conditions in capillary tubes there is no fixing or staining process, so ITP can be applied to small molecules such as amino acids, nucleotides, salts, metabolites, etc. which are often extremely difficult or even impossible to analyse and quantify by other electrophoretic methods.

10.3.   Other analytical and preparative isotachophoretic methods

In principle ITP can be carried out in almost all the supporting matrices and in many of the types of apparatus used for other forms of electrophoresis.

Griffith, Catsimpoolas, and Kenney (1973) described the application of ITP to the analytical separation of proteins on polyacrylamide gel rods and discussed the factors affecting separation. They point out that the method is not a simple screening technique but rather one that must be approached on an individual problem basis. Some of the limitations of the method relate to the lack of knowledge of the mobility *versus* pH relationship for the spacer ampholyte components and also the proteins under study, but there are further difficulties if unstacked components are present or if the equilibrium steady state is not attained within the length of the gel. Thus it is not only difficult to choose suitable separation conditions, but the interpretation of results is also problematical. It would appear that some constituents of Ampholines may have similar mobilities to protein constituents and migrate with them in mixed zones. Increasing Ampholine concentration to increase the separation of protein constituents caused a diffusion of the protein zones and may retard analysis owing to the large numbers of ampholytes between the leading ion and the protein with the highest mobility. Nevertheless in spite of such counterproductive aspects, spacers are nearly always used in ITP separations of protein mixtures because they greatly facilitate the interpretation of separated zones (Everaerts and Verheggan 1983). This is partly because being large molecules the molarity of proteins in samples is low so the resulting zones are very narrow and in a complex mixture extremely difficult to resolve. In addition many proteins and glycoproteins are heterogeneous with respect to molecular charge (e.g. genetic variants, differences in degree of amidation, phosphorylation, oligosaccharide attachment, sialic acid content etc.). These factors can give rise to an almost continuous spectrum of mobilities. Even with small molecules such as amino acids or nucleotides mixed zones are common in ITP experiments and with large molecules they are even more frequent. When suitable spacers are used at least part of this mixed zone formation will occur between protein and spacers and not between different proteins in the sample. Buffer constituents used for ITP in polyacrylamide gels are the same as those referred to in Section 10.2

for ITP in capillary tubes, the gels themselves being prepared in the same way as for PAGE. Since separation is based upon differences in charge and ionic mobility, gels with minimum sieving effect should be chosen so that retardation by the gel matrix does not interfere. Gels with T in the range 4–6 per cent and either a low (e.g. $C = 2\cdot5$–$3\cdot0$ per cent) or a very high (e.g. $C = 20$–40 per cent) proportion of Bis should be used to give gels of maximum porosity (see p.7). Sample application, running conditions, and other aspects of analytical gel ITP (e.g. staining, quantitative measurement, etc.) are performed as for PAGE separations (Chapter 2).

ITP can be exploited preparatively in the case of macromolecules in a flat-bed granular gel format (Brogren and Peltre 1977; Battersby and Holloway 1982) using an apparatus such as the LKB Multiphor or Pharmacia FBE–3000. The formation of the gel bed, application of the sample (preferably as a relatively narrow zone), and detection and extraction of zones are all performed in a manner similar to flat-bed granular IEF (Section 9.5), the main difference in technique being that one buffer reservoir holds electrolyte containing the leading ion and the other holds terminating ion solution.

In the procedure given by Battersby and Holloway (1982) the gel bed is prepared in exactly the same way as described in Section 9.5 for IEF experiments except that instead of the gel slurry being made up in water containing 2 per cent ampholytes, for ITP the slurry is made up with leading electrolyte buffer. The moisture content of the bed is adjusted by evaporation in the same way as for IEF gels. The anode wick is soaked in leading electrolyte buffer and the cathodic wick in terminating electrolyte of 30 gl^{-1} of 6-aminohexanoic acid adjusted to pH 8·9 with tris. The and the system prerun to establish a leading ion – terminating ion boundary. The buffer system used consisted of 20 mM HCl adjusted to pH 7·0 with tris as the leading electrolyte, in which the gel bed was prepared, with a terminating electrolyte of 30 gl^{-1} of 6-aminohexanoic acid adjusted to pH 8·9 with tris. The terminating electrolyte was used for the cathode electrode compartment and cathodic wick. A solution of 120 mM HCl adjusted to pH 7·0 with tris was used for the anode compartment and anodic wick, thus giving similar anode and terminating electrolyte ionic strengths. Depending on the gel bed size the prerun might typically be performed at a limiting current of 20 mA (initially about 400 V) and is continued until the boundary has moved about 5 cm into the gel. The boundary can often be seen visually due to differences in refractive index but can be located more reliably if a drop or two of a 0·25 per cent solution of bromophenol blue in 50 per cent methanol is added because this migrates between the leading and terminating electrolyte zones (Battersby and Holloway 1982). The flat bed apparatus kits from LKB or Pharmacia include a sample application trough and after the prerun this is pressed into the gel bed in the terminating electrolyte zone immediately behind the leading–terminating electrolyte boundary. The gel is scraped out of the trough mixed with up to about 3 ml of sample and returned to the trough. The sample applicator is removed and the separation commenced by applying (in the LKB Multiphor apparatus) a constant current of 6 mA. Initially this corresponded to about 250 V but after 18 h at the completion of the run the voltage had risen to about 900 V. Location of zones, pH measurements, fractionation of the gel bed, elution of sample material, etc. are as described in Section 9.5 for IEF experiments. Removal of buffer constituents can be by gel filtration, dialysis, precipitation of protein components, etc. and is generally more effective and efficient than the separation of ampholytes following IEF separations.

A fixed-path-length gel-column type of apparatus such as the LKB Uniphor can also be used for ITP with a polyacrylamide gel of low T in order to minimize the molecular sieving effect of the gel matrix. The polyacrylamide is prepared in the leading electrolyte buffer and a plastic column is used in the apparatus to accommodate volume changes in the gel during the course of the separation. The terminating electrolyte is placed on top of the gel and the sample layered between it and the gel. Before application the sample should be dissolved in or dialysed against the terminating electrolyte and spacer ampholytes then added. A typical experiment run at 8 mA takes about 24 h and samples are collected at the end of the gel column by a stream of elution buffer as for preparative PAGE (Section 7.5). The use of the Uniphor column for preparative ITP was first reported by Svendsen and Rose (1970) and is described in detail in *Application Note 146* (1974) available from LKB Produkter AB. Factors governing a suitable choice of spacer ampholytes for the fractionation of serum proteins by this method are given in their *Application Note 213* (1976). Up to 200–300 mg of protein can be handled in each separation.

10.4. Biochemical applications

ITP has been much less extensively used in the field of macromolecular analysis than IEF, but nevertheless it has been shown to be a technique valuable for the separation of many different classes of sample including proteins, peptides, amino acids, nucleotides, strong and weak acids and their salts, metal ions, and a wide range of other organic and inorganic ions (see for example Delmotte 1979; Everaerts *et al.* 1976; Rapášová *et al.* 1984). Zelenský *et al.* (1984) have used ITP for the direct analysis of inorganic anions such as Cl^-, NO_3^-, NO_2^-, F^-, SO_4^{2-}, and PO_4^{3-} in river water. The method was much faster than conventional analytical methods and sensitive (detection limits were 30–60 p moles), so the method appears to have great potential for water monitoring in environmental control work. According to Haglund (1970) applications of ITP have even included the separation of isotopes of potassium, rubidium, and chlorine. A useful series of *Application Notes* describing the application of ITP to a wide range of ionic substances is available from LKB-Produkter AB.

Some clinical applications of ITP include the analysis of cerebrospinal fluid (Delmotte 1977; Kjellin, Moberg, and Hallander 1977), serum proteins (Brogen 1977; Bier and Kopwillem, 1977), muscle extracts (Fig. 10.3) for AMP, ADP, ATP, NAD, and other metabolites (Gower and

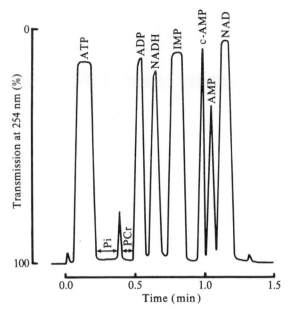

FIG. 10.3. Capillary isotachophoretic separation of a mixture of nine standard substances expected in muscle extracts: ATP, 0·8 nmol; inorganic phosphate (P$_i$), 2 nmol; phosphocreatine (PCr), 0·9 nmol; ADP, 0·4 nmol; NADH, 0·4 nmol; IMP, 1 nmol; c-AMP, 0·3 nmol; AMP, 0·2 nmol; NAD 0·9 nmol. (Reproduced from Gower and Woledge (1977) by permission of the authors and publishers.)

Woledge 1977), sweat and urine proteins (Uyttendaele *et al.* 1977), and protein–drug interactions (Sjödahl and Hjalmarsson 1978). It has been used both analytically and preparatively for human plasma proteins (Bier, Cuddeback, and Kopwillem 1977), for histoplasmin preparation (Lancaster and Sprouse 1977), in the isolation of myeloma immunoglobulins (Jefferis and Butwell 1975; Ziegler and Köhler 1976), and in antibody fractionation (Brogren and Peltre 1977). Kapadia, Vaitukaitis, and Chrambach (1981) used ITP in a small column of polyacrylamide gel to prepare milligram amounts of human chorionic gonadotrophin while Nguyen *et al.* (1981) used the same method to prepare some human growth hormones. Many of these reports merely seek to demonstrate the feasibility of ITP separations with these sample materials and it cannot be said that ITP is the method of choice, but some samples (e.g. histoplasmin, some myeloma immuno-globulins, etc.) may be difficult to prepare by other methods. In general, with proteins ITP can only be used successfully to separate sample mixtures of two or three components while more complex samples are only separated into groups with similar mobilities. The formation of mixed zones containing several components is usual rather than the exception.

Among the advantages of ITP, particularly analytical ITP in capillary tubes, are rapid analysis times, high sensitivity and reproducibility, and good resolution combined with easy quantitative measurements. Difficulties include assessing whether steady state stacking has been achieved, the formation of mixed zones, the number of factors to be taken into account for optimizing the separation, establishing the identities of particular zones or peaks, and the choice of suitable spacer ions. In the field of protein chemistry ITP has the advantages over IEF that proteins and enzymes are not necessarily exposed to environments of very low ionic strength or concentrated at their iso-electric points (which may lead to precipitation or inactivation). Unfortunately, however, few individual spacer ions exist in the mobility range of proteins, and the application of highly complex spacers such as Ampholines which contain 500 or more constituents delays attainment of the steady state and also usually results only in mixed zone separations (Nguyen and Chrambach 1979). Therefore until better spacer ions are found or produced it seems unlikely that ITP will be applied as extensively as IEF and other gel electrophoretic methods, at least within the field of macromolecular separations.

When samples contain small molecules of interest or are mixtures of large and small molecules, even at this relatively early stage of development the use of ITP may provide information which is more difficult to obtain by other methods. Good examples of this are its use in studies of the periodate oxidation of carbohydrates where quantitative measurements can be made even with limited amounts of material (Honda, Wakasa, Terao, and Kakehi 1979; Oka, Hirotsune, and Shigeta 1979), in measuring the concentrations in blood serum of metabolites such as uric acid (Verheggen et al. 1980) or of drugs such as valproate (Mikkers et al. 1980), the concentrations of citrate (Tschöpe and Ritz 1980) or of purines and pyrimidines (Oerlemans et al. 1981) in urine, and in studies on the composition of venoms (Einarsson and Moberg 1981). Using the known pK_a and absolute mobility (m_o) values of all buffer constituents and the sample ions Hirokawa et al. (1984) have been able to assess separation conditions for anaesthetic metabolites in urine by computer simulation, the results being in good agreement with observed experimental separations.

11

TWO-DIMENSIONAL ELECTROPHORESIS

Two-dimensional (2D) peptide maps (fingerprints) have been accepted and used for many years in studies involving the structure of proteins or their modification. In a similar way 2D macromolecular maps of intact proteins, nucleic acids, or polysaccharides can be most useful for the characterization of tissues, biological fluids, extracts of tissues and organs, or any other unknown mixture containing these constituents. They are also useful, or potentially useful, for identification purposes (e.g. taxonomy, forensic work, etc), for studying genetic variation and relationships, for the detection of stages in cellular differentiation and studies of growth cycles, for the examination of pathological states and the diagnosis of disease, and for many other purposes.

Nearly all the methods described in previous chapters can be regarded as single dimension macromolecular maps, and under the most favourable circumstances perhaps as many as 50–60 different components can be resolved, particularly when polyacrylamide gradient gels or IEF gels are used. However, it is more usual for resolution to be limited to 30–40 components because in most biological systems the various components are usually present in widely differing concentrations. This means that it is often not possible to apply sufficient sample for the detection of minor components without grossly overloading the major constituents. It may also be difficult to detect a minor component if it is very close to a major one. These considerations also apply to 2D maps, but here the problems associated with the reproducibility of band patterns (between samples, between experiments, and interlaboratory variation) and identification of individual components become much more demanding because by using high resolution steps for both dimensions it is possible to resolve several thousand protein zones on a single gel. Not surprisingly therefore many recent developments in 2D mapping have been related to the acquisition and handling of data.

The literature contains a great many different two-dimensional (2D) electrophoresis procedures, separating molecules on the basis of various different parameters and aimed at solving a wide variety of problems. A number of these methods have already been referred to in earlier chapters (e.g. transverse gradient gel electrophoresis in Chapter 4, peptide mapping in Chapter 5, many of the immunoelectrophoretic methods described in Chapter 8, etc), while others are discussed in Chapter 12 (e.g. methods on paper, cellulose acetate, thin-layers, etc.). Rightly or wrongly, few of these are usually thought of as 2D macromolecular maps and this term is usually confined to the separation of intact proteins or glycoproteins or large peptide fragments thereof on approximately square gel slabs, nearly always of polyacrylamide gel. 2D maps may be of high resolution or low resolution and may be prepared under denaturing or non-denaturing conditions, but before discussing these it is appropriate to consider two-stage methods.

11.1 Two-stage separations

Two-stage methods may be thought of as intermediate between one-dimensional and true 2D procedures, and consist of methods in which samples are isolated or subjected to some chemical or enzymic treatment after an initial one-dimensional separation before being reanalysed in a second one-dimensional stage. Obviously there are a great many possible protocols for two-stage separations depending not only on the procedures used for the two-separation stages but also on the treatment given to the samples in between them. At one extreme there may be no intermediate sample modification treatment at all, and in this case it will usually be best if the two stages separate sample molecules on the basis of different parameters (e.g. size, charge, or hydrophobicity), as in 2D mapping. A typical example of this minimal intermediate treatment has been given by Ruddell and Jacobs-Lorena (1982), in which a small polyacrylamide IEF gel rod (the first separation stage) was sliced into 5 mm wide pieces each of which was then homogenized in 90 μl of 2·5 mM tris–HCl buffer pH 6·8 containing 8 M urea, 2 per cent SDS, 5 per cent 2-mercaptoethanol, 10 per cent glycerol, and 0·01 per cent Bromophenol Blue and applied to the sample slots of a $T = 13$ per cent gel slab for SDS–PAGE as the second separation stage. The procedure thus approximates to a simplified version of the O'Farrell high resolution 2D method (see below), but of course the slicing of the IEF gel into relatively large pieces which are then homogenized and placed in relatively wide sample slots for the SDS–PAGE run results in a very considerable loss in resolution compared to that attainable by O'Farrell's method. More usually, however, samples for the second stage separation are obtained not by slicing a first stage gel but are fractions obtained by other methods, such as any of the preparative electrophoretic methods discussed in Chapter 7, IEF in density gradient columns or granular gel beds, electrophoresis in density gradient columns, etc. In such cases they are in concept no different to two separation stages in which the first may equally well be ion-exchange chromatography, gel filtration, or any other fractionation process.

The approach more accurately considered as two-stage electrophoresis involves some intermediate chemical or enzymic treatment of the samples in between the two separation stages, which are very often the same, although of course they need not be. Into this category fall methods such as many of the peptide mapping procedures already discussed in Section 5.8, in which sample protein zones from the first stage separation gel are treated, with or without prior extraction, with proteolytic enzymes or reagents such as cyanogen bromide (Lam and Kasper 1980), formic acid, N-chlorosuccinimide or hydroxylamine. These are all designed to fragment the proteins in the samples to peptide mixtures which are then subjected to a one-dimensional mapping in a second stage generally consisting of SDS–PAGE.

Lasne, Benzerara, and Lasne (1982) used a typical two-stage approach in their study of the role of sialic acid in causing microheterogeneity of serum thyroxin-binding globulin. In this they first performed IEF in polyacrylamide or 0·8 per cent agarose to separate the various globulin zones. Neuraminidase solution (60 μl of Sigma Type IX enzyme containing 21 units per 1 ml of activity in 50 mM acetate buffer pH 5·0) was then applied to the area of the gel slab (in the pH 4–5 region) known to contain the globulin zones. After incubating for 1 h at 37 °C in a moist box (to stop the gel from drying out) this region of the gel slab was cut out, placed on a glass plate, and a warm gel mixture containing 0·8 per cent agarose, 10 per cent sorbitol, and 3 per cent ampholytes (pH 5–7 range) poured around it, so that after cooling and solidification the globulin sample could be examined again in a second IEF stage (run at 4 °C for 3 h at 1000 V). Since this was essentially the same separation process as the first stage, the effect of neuraminidase in removing sialic acid was easy to observe and interpret.

A development (Tijssen and Kurstak 1983) of the Cleveland et al. (1977) peptide mapping procedure typifies the most sophisticated type of two-stage electrophoretic procedures and is in effect a fully 2D system with an enzyme treatment of sample components in between the two separation runs. The method is intended to facilitate the simultaneous peptide mapping of the many proteins contained in a mixture and consisted of a conventional SDS–PAGE separation of proteins in the sample mixture using a gel slab with a $T = 8$ per cent resolving gel and $T = 3$ per cent stacking gel and the Laemmli (1970) buffer system (see p. 126). A strip of gel 2–10 mm wide containing the separated proteins for the second stage separation was then soaked for 30 min in 0·125 M tris–HCl buffer pH 8·6 containing 0·1 per cent SDS. There may be some loss of sample components during this soaking but as long as it is not unduly prolonged it should not be excessive and the step is necessary for subsequent efficient stacking during the second stage. A $T = 10$–20 per cent, $C = 5$ per cent gradient polyacrylamide gel slab, again employing the Laemmli (1970) system of buffers, was made up for the second stage separation and used to fill the gel slab mould to within about 30–40 mm of the top (Tijssen and Kurstak used Bio Rad equipment giving 1·5 mm thick gel slabs). This of course is overlaid with water, or better a 0·1 per cent SDS solution in H_2O in the usual way to give a flat surface, and once the gradient gel has polymerizedd the water layer is removed and stacking gel mixture applied. Before this polymerizes the gel strip from the first stage is embedded in it across the top of the slab, taking care to trap no air bubbles and to leave at least 15 mm of stacking gel between the gel strip and the top of the separation gel. There should also be about 10 mm between the top of the gel strip and the top of the gel mould. The stacking gel and gel strip are then overlaid with 0·1 per cent SDS solution until polymerized. The gel slab is then placed in the apparatus and the electrode compartments filled. Finally an approximately 3 mm deep layer of proteinase solution in 62·5 mM tris–HCl pH 6·8 buffer containing 10 per cent glycerol, 0·1 per cent SDS, and 0·001 per cent Bromophenol Blue is applied across the top of the gel slab. With the standard 11 cm wide gel, this corresponds to about 0·6 ml of proteinaise solution and appropriate concentrations were about 0·1 μg ml^{-1} for papain (25 units mg^{-1}), 0·3 μg ml^{-1} for Staphylococcus aureus V8 protease (560 casein units mg^{-1}) or 1·25 μg ml^{-1} for α-chymotrypsin. Electrophoresis was performed at 25 V for 15 min and then 75 V until the Bromophenol Blue reached a position 2·5 mm above the start of the resolving gel. The power was then switched off and the slab allowed to stand for 30 min to permit proteolytic digestion of sample proteins within the sample stack. After this, electrophoresis was resumed in the usual way to separate the peptides generated. Slabs were stained with Coomassie Blue R250 or by silver staining. Dansylation (or presumably any other fluorescent labelling) of the initial proteins was found to be very helpful in that it enabled the progress of digestion and separation of peptides to be very easily monitored by simple u.v. illumination. A very similar approach to peptide mapping was used by Lonsdale-Eccles *et al.* (1981) in which protein-containing sample strips from the first stage SDS–PAGE, in a $T = 8·75$ per cent slab, were equilibrated with 70 per cent formic acid, incubated with 5 per cent cyanogen bromide in 70 per cent formic acid for 1 h under N_2 and washed repeatedly with 10 per cent acetic acid to remove excess cyanogen bromide. This gel was then reequilibrated with Laemmli (1970) stacking gel buffer, before being applied across the top of the stacking gel for a second stage SDS–PAGE separation of the peptides generated, using a $T = 10$–15 per cent polyacrylamide gradient slab gel.

As can be seen therefore many different types of two-stage electro-phoretic separations can be devised to suit various separation problems, but in essence the two stages are no different to the one-dimensional procedures described in other chapters. Not surprisingly, the apparatus used and methods for staining and destaining, flourescent or radio-labelling, scanning, autoradiography, data handling, etc., are the same as those used in one-dimensional procedures.

11.2 Two-dimensional (2D) separations: General considerations and apparatus

Two-dimensional maps can be prepared using virtually any combination of the one-dimensional methods already described. Clearly, if the method used for the second dimension separates components on a basis similar to that of the first-dimensional separation (e.g. PAGE at two different T or pH values), the components will tend to be clustered about the diagonal and resolution will be less than optimal. In general, the best maps are obtained when the basis for separation is different in the two dimensions (e.g. mainly on a size basis in one dimension and according to charge in the other), since this results in a more even distribution of components over the surface of the map. Fractionation on the basis of molecular charge differences can be achieved with IEF or ITP, particularly when these are conducted in relatively non-restrictive media. Electrophoresis on paper, cellulose acetate, agarose gels, composite polyacrylamide–agarose gels, and highly cross-linked ($C > 15$ per cent) polyacrylamide gels of very low T (minimum pore size is attained with 5 per cent cross-linking so that gels with high C are actually less restrictive than gels with $C = 2$–10 per cent) also give separations in which the molecular charge factor predominates. Molecular size differences are important in PAGE performed in gels of high concentration T, in gels with a polyacrylamide concentration gradient (gradient gels), and particularly in SDS–PAGE. Some degree of separation due to differences in hydrophobicity (Section 5.9(c)) can be achieved by incorporating a non-ionic detergent such as Triton X-100 and urea into the gel, since this binds to and retards hydrophobic proteins (Hoffman and Dowben 1978b; Fernandes, Nardi, and Franklin 1978).

A frequent approach to the preparation of 2D maps involves running the first-dimensional separation in a cylindrical gel rod of 3–6 mm diameter. The gel rod is then placed across the top of a slab gel for the second-dimensional separation at right angles to the first. The rod is sealed in place with either a little agarose gel or polyacrylamide gel made up in the appropriate buffer. With most designs of vertical slab gel apparatus the positioning of the gel rod across the top is relatively simple, especially since it is not necessary for the rod to be precisely aligned over the slab, but if desired special adapters to aid positioning can be made (e.g. Hoffman and Dowben 1978a). If a multiphasic buffer system is to be employed for the second dimension, the gel rod should if possible be equilibrated by washing in stacking gel buffer and should be embedded in stacking gel. This should also extend below the rod so that components migrate through 10–20 mm of stacking gel before entering the separation gel phase.

If continuous buffer systems are used in the second dimension, the diameter of the gel rod from the first dimension should not be greatly different from the thickness of the gel slab. For example with 6 mm diameter gel rods, 3 mm thick slabs give satisfactory results but when 1 mm slabs are used patterns are more blurred (Hoffman and Dowben, 1978a). With multiphasic buffer systems greater size differences can be tolerated owing to the zone-sharpening effects. Thin (1 mm) slabs are more readily dried for subsequent storage or autoradiography than thick ones and often give higher resolution due to greater permissible voltage gradients and shorter running times because of the more efficient dissipation of heat, but they are a little more difficult to handle.

Very often however a gel slab configuration is used for the first dimension separation as well as the second. In this case strips of the first dimension gel are cut out and inserted across the top of the second dimension slab in much the same way as if rod gels were used. When a continuous buffer system is used for the second dimension this sample gel slice (or indeed gel rod when one of them is used) should preferably be soaked in 10-fold diluted buffer for a few minutes to reduce its conductivity and aid zone sharpening (see p. 80), while for discontinuous systems stacking gel buffer should be used. For best resolution a relatively narrow (i.e. 1–3 mm wide) strip of gel should be cut out of the first dimension gel, although if zone sharpening or stacking is adequate this is not a very critical point. It is usual for the strip to be of the same thickness as the second dimension gel (it will often have been run in the same apparatus), so the strip is simply inserted between the glass plates and pushed down either onto the surface of the gel (for continuous buffer systems) or to within 10–20 mm of the surface (for multiphasic systems), and sealed in place with 1 per cent agarose in diluted buffer for the former case or with stacking gel in the latter. This is slightly different from the rod gel arrangement where it is normal for the rod to be placed across the top of the glass plates forming the gel mould rather than being pushed down between them. Indeed many workers advocate the bevelling of the tops of the plates inwards so that the gel rods rest in a trough.

It will be apparent from the above that the apparatus used for 2D separations is the same as that described in Chapter 2. Since slab gel apparatus is obligatory for the second dimension, and quite often used for the first dimension as well, the smaller types of power supply are not really adequate and units should be capable of 500 V and a minimum of 150 mA, but preferably of 250 mA or more.

11.3 High resolution two-dimensional (2D) procedures

Since virtually any electrophoretic method except those performed in free solution can be employed for either dimension a great variety of 2D separation methods are possible and most of them have indeed been tried at one time or another. As stated above sample components are spread over a 2D map surface most evenly when the basis of separation is different for each dimension, and clearly for the highest possible overall resolution methods of high resolution should be chosen for each step. Since no high resolution method separating purely on the basis of differences in hydrophobicity exists at the present time, the choice is narrowed to size fractionation in one dimension and charge fractionation in the other. The method of highest resolution currently available for the former is SDS–PAGE (including SDS–PGGE) and for the latter is IEF in polyacrylamide gels (PAGIF) or agarose gels (AGIF).

While this combination of separation methods was used at quite an early stage in the development of 2D macromolecular mapping it was the elegant work of O'Farrell (1975) that really demonstrated the full capabilities of this approach. He was able to resolve about 1100 different proteins from lysed *Escherichia coli* cells on a single 2D map (Fig. 11.1) and suggested

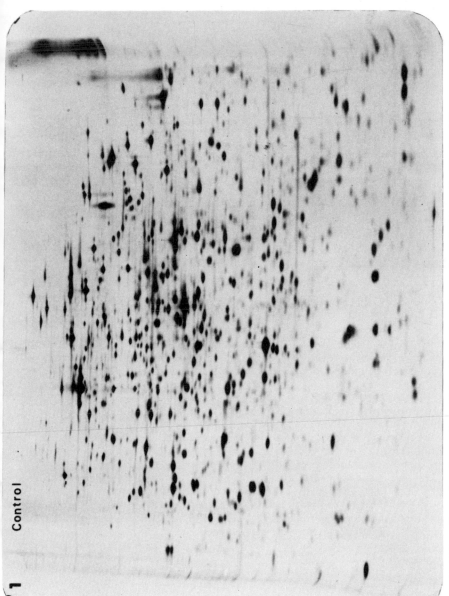

Fig. 11.1 Two-dimensional mapping of *E. coli* proteins. Approximately 10 μg (in 25 μl) of ^{14}C-labelled proteins (about 180 000 counts min^{-1}) were separated by IEF in the first dimension (left to right) and by SDS–PAGE on a 9·25–14·4 per cent exponential acrylamide concentration gradient gel in the second dimension (run top to bottom) followed by autoradiography with an 825 h exposure time. Over 1000 individual spots could be counted on the original autoradiogram (Reproduced from O'Farrell (1975) by permission of the author and publishers.)

that the maximum resolution capability may be as high as 5000 different proteins. Apart from meticulous attention to detail, major reasons for the advance in resolution obtained by O'Farrell compared to earlier workers included the use of samples labelled with ^{14}C or ^{35}S to high specific activity, and the use of thin (0·8 mm) gel slabs for the second dimension which could then be dried down easily before autoradiography. This means that labelled zones in the gel are physically very close to the emulsion of the autoradiographic film, so reducing the area of film exposed to the radiation from any one zone and keeping loss of resolution to a minimum. Autoradiography was able to detect protein zones corresponding to one part in 10^7 of the sample (usually 1–20 μg was applied initially, as higher loads caused zone spreading, although up to 100 μg could be applied). Coomassie Blue R250 staining was about three orders of magnitude less sensitive (about 0·01 μg in an individual spot) and only about 400 *E. coli* proteins could be detected. Since the procedure was intended for the total analysis of proteins, denaturing agents capable of solubilizing most proteins were present in both dimensions, urea and non-ionic detergent for the first (IEF) dimension and SDS for the second. The system does of course enable both the isoelectric point (pI) and molecular weight (MW) of any particular zone to be measured, but since the system is denaturing this will of course be the subunit pI and MW for any multimeric protein rather than the native values.

Nucleic acids can interfere with IEF separation by interacting with basic proteins and ampholytes and while this is not usually serious, it may be beneficial to remove them by selective precipitation or extraction, isopycnic centrifugation, ion-exchange chromatography, or by digestion with nucleases. O'Farrell used the last of these methods, and suspended centrifuge pellets of *E. coli* cells in 100 μl of 0·01 M tris–HCl buffer pH 7·4 containing 5 mM $MgCl_2$ and pancreatic RNase (50 μg ml^{-1}), sonicated them, added DNase to a level equivalent to 50 μg ml^{-1}, and allowed the mixture to stand on ice for 5 min. He then added solid urea to 9 M and diluted the mixture 1:1 with a lysis solution of 9·5 M urea containing 2 per cent Nonidet P-40, 5 per cent 2-mercaptoethanol, and 2 per cent Ampholines (made up of 1·6 per cent pH range 5–7 and 0·4 per cent pH 3–10). Samples should be run immediately or frozen until required. Ideally 25 μl samples should be applied to the first dimension gel and should contain at least 100 000 cpm and less than 40 μg protein.

For the first dimension O'Farrell (1975) used gel rods prepared in glass tubes 130 mm long and 2·5 mm in diameter. The constituents required to make 10 ml of gel mixture are shown in Table 11.1 and these (less the TEMED) were mixed in a flask, deaerated for 1 min, the TEMED added, and the solution loaded into the tubes to within 5 mm of the top. The gel mixture was overlaid with a little 8 M urea solution. After about 1 h (the gels should have polymerized well before this) the urea solution was replaced by a little lysis solution and the gels allowed to stand for a further hour or more. Small pieces of dialysis membrane were then fixed over the lower ends of the tubes with rubber bands and the tubes placed in the electrophoresis apparatus. Fresh portions (~20 μl) of lysis solution were added to the tops of the tubes which were then filled with 0.02 M NaOH and the apparatus

TABLE 11.1
Constituents for the first dimension PAGIF gel (O'Farrell system)

Constituent	Amount
Urea	5.5 g
30 per cent acrylamide solution:	1.33 ml
28.38 per cent (w/v) acrylamide + 1·62	
per cent Bis in H_2O	
Nonidet P-40 (10 per cent w/v in H_2O)	2·0 ml
Ampholines, pH 5–7	0.4 ml
Ampholines, pH 3–10	0.1 ml
H_2O	1.97 ml
Ammonium persulphate (10 per cent in H_2O)	10 μl
TEMED	7 μl

filled with buffers, using 0·01 M H_3PO_4 in the lower reservoir and degassed (to remove CO_2) 0·02 M NaOH in the upper reservoir. Gels were then prerun at 200 V for 15 min, 300 V for 30 min and finally 400 V for 30 min. The power was switched off and the upper buffer reservoir emptied and lysis buffer and NaOH removed from the top of the gel. The sample was then applied, overlaid with 10 μl of 9 M urea containing 0·8 per cent pH 5–7 Ampholines and 0·2 per cent of the pH 3–10 range. The tube was carefully filled with 0·02 M NaOH and the buffer reservoir also filled with this. Gels were run at 400 V for 12 h followed by 800 V for 1 h (or at least 5000 volt-hours). The gels were pushed out of the glass tubes by pressure on a syringe connected to the tube with a length of plastic tubing and then shaken for 30–120 min in 5 ml of second dimension SDS sample buffer.

For the second dimension O'Farrell (1975) used the buffer system of Laemmli (1970), so for slabs of uniform concentration running gel the recipe shown in Table 5.4 is used. The amounts of acrylamide and Bis used for the running gel could be varied easily if desired and O'Farrell used gels between $T = 5$ per cent and $T = 22·5$ per cent. He obtained better results using an exponential concave concentration gradient, generally of $T = 9$–15 or 10–14 per cent. To prepare these he used a simple two-chambered gradient mixer where the volume of the front chamber was kept constant by plugging with a stopper (see p.106). Glycerol was added to 70 per cent to the heavy acrylamide solution to increase its density. O'Farrell degassed both the light and heavy solutions and halved the level of catalysts shown in Table 5.4. The running gel mixture was pumped into the gel mould until about 25 mm from the top, then overlaid with H_2O and allowed to polymerize. The overlaying H_2O was removed and replaced with running gel buffer diluted four-fold and the gel stored overnight. The gel mould was then filled with stacking gel mixture and the top covered with a plastic strip, taking care to trap no air bubbles underneath, in order to ensure a flat top surface to the gel. When the stacking gel has polymerized the plastic strip is removed and the first dimension gel laid along the top of it and sealed in place with 1 per cent agarose gel. This is made by warming the agarose to 80 °C in SDS sample buffer (0·0625 M tris-HCl pH 6·8 containing 2·3 per cent SDS, 5 per cent 2-mercaptoethanol, and 10 per cent glycerol). After about 5 min to allow the agarose to gel the apparatus is filled with buffer, 40–50 μl of 0·1 per cent Bromophenol Blue added to the upper reservoir as tracking dye, and gels then subjected to electrophoresis until this dye reached the bottom of the gel. O'Farrell used a constant current of 20 mA with his slabs which were 134 mm wide, 146 mm long, and 0·8 mm thick and electrophoresis took about 5 h.

The two major ways in which sample proteins can be lost in the O'Farrell procedure are in the IEF stage where some proteins may not enter the gel (the additions of urea and Nonidet P-40 are designed to keep this to a minimum) or may migrate off the basic end (e.g. basic proteins such as histones, ribosomal proteins, etc.) and in the equilibration of the IEF gel prior to running on the SDS gel. The problem of basic proteins is discussed below. Depending on the identity of the protein and duration of equilibration losses resulting from this cause can be 5–25 per cent. This can be eliminated by omitting the equilibration altogether, although it may result in slight streaking in the SDS dimension. To minimize this O'Farrell (1975) recommends that the stacking gel of the SDS slab be increased from 25 mm in depth to 50 mm and that gels are run at 20 mA initially with 2 per cent SDS in the upper reservoir buffer instead of the usual 0·1 per cent. After about 20 min this buffer is replaced with the usual 0·1 per cent SDS buffer.

Since its original publication O'Farrell's technique for 2D analysis has been very widely followed, often with some minor modifications to suit the particular sample being examined. Thus the concentration and particularly the pH range of Ampholytes chosen for the first dimension should be adjusted to give a good distribution of components along the gel, and urea and non-ionic detergent concentrations can also be modified depending upon the samples or even omitted altogether if desired. On most occasions a series of preliminary experiments will be needed to establish the best conditions. The presence of urea generally unfolds sample proteins however and facilitates the binding of the maximum amount of SDS during equilibration before the second dimension, so that prolonged incubation or heating with SDS, as routinely used in one-dimensional SDS–PAGE experiments, is not necessary. Some workers (e.g. Burghes et al. 1982) have used slab gels for the first dimension PAGIF separation and applied slices to the top of the second dimension slab. Modifications to the size and acrylamide concentration of the SDS slab are also often made. Ho et al. (1979) used small commercially available gradient gel slabs (Gradipore) for the second dimension for example, whilst Johnson (1982) used slabs of almost double length (305 mm), running gel buffer adjusted to pH 8·3, not the usual 8·8, and a $T = 10$–20 per cent gradient which he claimed gave improved resolution, particularly of small MW proteins, when unequilibrated IEF gels were employed. Factors affecting the quality and resolution of 2D protein maps have been discussed by a number of workers (e.g. Tracy et al. 1982b; Burghes et al. 1982; Duncan and Hershey 1984), as a result of which it appears that it is probably beneficial to reduce the concentration of acrylamide used for the IEF gel to $T = 3·5$ per cent, to run the IEF gel at 800–1000 V for at least 10 000 volt/hours (but probably not more than about 15 000 volt/hours), and to increase the electrolyte concentrations in the buffer chambers. Concentrations of 1 M NaOH and 1 M H_3PO_4 or more have been used but Duncan and Hershey (1984) found little further improvement when the catholyte was increased above 50 mM in NaOH (which appeared to be optimal) or with more than 25 mM H_3PO_4 for the anolyte. Increasing the concentration of one electrode solution but not the other shifts the pH gradient up or down, but if it is necessary to do this it is probably preferable to extend the pH gradient by altering the composition, by adding constituents such as arginine or lysine to the catholyte and aspartic or glutamic acids to the anolyte, or to prepare appropriate cocktails of ampholytes.

Perdew *et al.* (1983) replaced the non-ionic detergent (Nonidet P-40) used in the first dimension IEF gel with a similar proportion of the zwitterionic detergent 3-[(3-cholamidopropyl)dimethylammonio]-1-propane sulphonate (CHAPS), which they claimed had superior membrane protein solubilizing properties and was particularly effective at disaggregating hydrophobic proteins. It appeared to greatly reduce the streaking sometimes observed in the IEF separation and improve the clarity of the resulting protein maps.

Among the most significant developments in high resolution 2D protein mapping reference must be made to the work of Anderson and Anderson (1977; 1978*a*, *b*) who have described apparatus (termed the ISO-DALT system) for preparing and running a large number of O'Farrell-type gels together. This greatly enhances reproducibility and comparison between the resulting protein maps as well as enabling a very large number of samples to be handled in a short time. A variation (termed BASO-DALT) and giving enhanced resolution of basic proteins by employing NEPHGE in the first dimension (see below) has also been described (Willard *et al.* 1979). The approach makes practical the Molecular Anatomy Program at the Argonne National Laboratory in the USA, the object of which is to be able to fractionate human cells and tissues with the ultimate aim of being able to describe completely the products of human genes and how these vary between individuals and in disease.

Another important development which must be mentioned is the application to 2D mapping of IEF in gels with immobilized pH gradients for the first dimension (Westermeier *et al.* 1983; Gianazza *et al.* 1984). Immobiline IEF gels are described in Section 9.3(h). These are prepared and run on Gel-Bond PAG polyester supports which after the separation are cut into strips which can then be applied across the tops of SDS–PAGE slabs for the second dimension. The use of gels with immobilized pH gradients has a number of advantages compared to conventional PAGIF. The pH range of the gradient can be very clearly defined and the range calculated to give optimum separation of sample constituents. Once formed the gradients are more stable than those formed with ampholytes so that gradients are more reproducible. Resolution is also superior, as is sample load carrying capacity, and the gradient is not disturbed by the presence of salt ions in the samples or buffers. Owing to the very low conductivity of Immobiline gradients sample zones may be sharpened by increasing the voltage gradient to very high values during the final stages of focusing without causing excessive heating. Against this their preparation is a little more technically demanding and their cost higher. Nevertheless, the advantages, particularly in terms of gradient reproducibility which will undoubtedly facilitate inter-sample and inter-laboratory comparisons will ensure that they become widely used in future for 2D protein mapping as well as for single dimension separations.

Elegant micro-versions of O'Farrell's method have been reported by Rüchel (1977) and by Poehling *et al.* (1980) in which the first-dimension IEF run is performed with gels in small capillary tubes followed by a second-dimension SDS–PAGE step on postage-stamp-size gel slabs with a $T = 1$–40 or $T = 6$–25 per cent gradient. The method was applied to the analysis of proteins from single cells (large mollusc neurons) and proteins from plant-parasitic nematodes.

Unfortunately histones and other basic proteins are generally not well separated by O'Farrell's method and the acid–urea–detergent systems described in Section 5.9c may be preferable. In O'Farrell's method they often do not enter the first-dimension IEF gel or if they do they may appear as streaks. This may be due in part to a collapse of the pH gradient at the alkaline end. For example, Giometti, Anderson, and Anderson (1979) report that even with ampholytes of very wide range intended to give a pH range of 2–11 the actual gradient after focusing in a gel was from about pH 3·0 to 8·5. It may also be due to the difficulties of completely removing DNA from histone samples, and this led Sanders, Groppi, and Browning (1980) to develop a procedure for sample preparation in which 0·4 M NaCl was added to the pH 4·8 buffer before nuclease digestion and in which a three-fold excess of protamine was then added, both steps being intended to reduce DNA–histone interactions.

To overcome the difficulties of separating such basic proteins in the first dimension O'Farrell, Goodman, and O'Farrell (1977) introduced the technique of non-equilibrium pH gradient electrofocusing (NEPHGE). For the 2D mapping of proteins using NEPHGE for the first dimension, gel rods identical to those described above for IEF separations are prepared, i.e. gels with $T = 4$ per cent and $C = 5$ per cent prepared in 9 M urea, 2 per cent Nonidet NP-40 and 2 per cent ampholytes, usually of wide pH range but narrow-range ones (e.g. pH 6–8) have been used (Sanders *et al.* 1980). NEPHGE is performed in the same way as IEF except that the electrodes are reversed and sample is applied at the acidic end rather than the basic end of the gel and the voltage applied for a shorter time than in IEF. The lower electrode is thus the cathode and the lower reservoir contains base (e.g. 0·02–0·1 M NaOH) and the upper electrode is the anode (e.g. in 0·01–0·05 M H_3PO_4 or H_2SO_4). Samples dissolved in the urea–detergent solution, with or without ampholytes, are applied to the top of the gels and overlaid with more dilute urea solution to protect the proteins in the sample from direct contact with the acid electrode solution which is then added to the upper reservoir. Different groups of workers have used different voltage conditions for the actual separation, ranging from 400 V for 2 h to 500 V for 8 h (O'Farrell et al. 1977; Giometti et al. 1979; Willard et al. 1979; Sanders et al. 1980; Horst et al. 1980), but the precise conditions will depend on the geometry of the apparatus and in particular the gel length. However, in all cases the combination of time and voltage is much less than that used for IEF (usually 15–40 per cent). Under such conditions the pH gradient may not have reached full equilibrium and certainly most proteins will not have focused at their iso-electric points. Nevertheless, they will be largely separated on the basis of charge differences as in IEF, but it must be remembered that unlike IEF no information can be gained about the true iso-electric points of proteins separated in this way.

Giometti et al. (1979) claimed that basic protein samples extracted with SDS-containing solutions could be examined directly by NEPHGE, but other workers were unsuccessful (Horst et al. 1980) and it seems possible that the non-ionic detergent Nonidet NP-40 may not always be capable of totally displacing SDS from protein–SDS complexes. Likewise, some groups have reported that acidic proteins in the same sample are not well separated during the NEPHGE analysis of basic proteins, while other groups (e.g. Giometti et al. 1979) have attained resolutions nearly as good as in conventional IEF. After separation NEPHGE gels can be stained and treated in the same way as IEF gels. For 2D mapping work of basic proteins, the first-dimension separation by NEPHGE is followed by a second-dimension SDS–PAGE separation in the usual way.

11.4 'Low resolution' two-dimensional (2D) procedures

O'Farrell's method of two-dimensional mapping, with either IEF or NEPHGE in the first dimension, employs strongly denaturing conditions so that all multi-subunit proteins are broken down to their constituent subunits. This can lead to a very large number of separated zones on the final map, particularly if subunits differ and if very complex samples of biological origin are examined.

This complexity has led to various methods of simplifying zone patterns as an aid to their interpretation and to the concept of 'low resolution' 2D protein mapping. Philosophically this is rather interesting because it might be thought of as attempting 'to put the clock back' in the sense that many of the earlier electrophoretic techniques were low resolution of necessity rather than by design! However there is no doubt that lower resolution methods do still have their place, and may offer significant advantages in terms of technical simplicity, speed (enabling a greater through-put of samples), ease, and simplicity of interpretation and hence less sophisticated and expensive equipment for scanning and data handling. Although this is of course bought at the expense of the amount of information that can be extracted from the results, there may be sufficient for the purpose in hand. A common feature of all the procedures discussed in this section is that either a relatively low resolution method (e.g. agarose gel, starch gel, cellulose acetate, paper electrophoresis, PAGE, etc.) is combined with a relatively high resolution technique (PAGIF, IEF-agarose, SDS–PAGE, PGGE, etc.) or two low resolution stages are employed. In either case the final resolution obtained is lower than would have been the case if high resolution methods had been employed for both dimensions.

At the lowest end of the resolution spectrum combinations between paper, starch gel, cellulose acetate, and agarose gel electrophoresis are nowadays seldom used. Combinations using PAGE for both dimensions were quite widely used in the early days of 2D analysis but are now limited very largely to applications where buffers of very widely different pH are used for the two dimensions or where a denaturant such as urea is present for one dimension but not the other. The 2D-PAGE system of Kaltschmidt and Wittmann (1970), modified by Lastick and McConkey (1976) to facilitate the use of small samples and reduced running times, gives a very good map of ribosomal proteins.

For this (Lastick and McConkey 1976) a first dimension gel rod (5 mm diameter, 115 mm long) of $T = 4 \cdot 1$ per cent, $C = 3 \cdot 2$ per cent polyacrylamide gel in a buffer of $0 \cdot 2$ M tris, $0 \cdot 26$ M boric acid, and $0 \cdot 01$ M Na_2EDTA pH $8 \cdot 6$ containing 6 M urea ($0 \cdot 2$ per cent TEMED and $1 \cdot 9$ mM ammonium persulphate as polymerization catalysts) was used; with the same buffer diluted three-fold in the apparatus. After installing in the apparatus the top of the gel was covered with at least $1 \cdot 5$ cm depth of an overlay of this same buffer diluted 10-fold but containing 8 M urea. The sample itself was dissolved in a portion of 10-fold diluted buffer to which 5 per cent 2-mercaptoethanol had been added, warmed to 60 °C for 10 min, cooled, a little sucrose or more urea added, and carefully layered between the overlay buffer and the top of the gel. The system thus makes use of the zone sharpening effect of lower conductivity zones produced by diluting sample buffers (see p. 80). After electrophoresis towards the cathode at 65 V for 17 h the gels are extruded, soaked for no more than 5 min in $0 \cdot 35$ M acetic acid containing 6 M urea, and 5 per cent 2-mercaptoethanol, and placed across the top of a $170 \times 130 \times 0 \cdot 8$ mm polyacrylamide gel slab of $T = 15 \cdot 5$ per cent, $C = 3 \cdot 0$ per cent made up in $0 \cdot 44$ M acetic acid, $0 \cdot 025$ M KOH buffer, pH about $4 \cdot 6$, containing 6 M urea (polymerization catalysts: $0 \cdot 5$ per cent TEMED and 10 mM ammonium persulphate). The first dimension rod gel was sealed in place with 1 per cent agarose made up in the acetic acid gel soaking solution and the second dimension then run towards the cathode at 100 V for 18 h. The system therefore separates the ribosomal proteins not only at two different pH values but also in a relatively non-restrictive gel in the first dimension, so that separation is predominantly according to charge differences, and in a much more restrictive second dimension gel so that size differences then play a greater role. Thus while not entirely according to charge or size differences in either dimension the basis of separation is sufficiently different in the two dimensions to give quite a good distribution of the proteins over the whole surface of the gel map.

In the above method 6 M urea was present in the gels for both dimensions so the conditions were mildly denaturing throughout. Other 2D arrangements with PAGE used for both dimensions may use both denaturing and non-denaturing conditions. Such an approach may have merit in studies of subunit structure and conformation. Using $T = 5 \cdot 1$ per cent, $C = 2 \cdot 4$ per cent polyacrylamide gels in $0 \cdot 09$ M tris, $0 \cdot 09$ M boric acid, $2 \cdot 5$ mM Na$_2$EDTA buffer containing 10 per cent glycerol throughout, Schumacher et al. (1983) used this approach for the detection of circular viroids and virusoids. The gel for the second dimension was identical to that for the first but included 8 M urea and was also run at 50 °C to ensure denaturing conditions.

More usually however 2D methods within this category make use of a simple, relatively low resolution step coupled with a high resolution method in the other dimension. They may be non-denaturing arrangements, such as PAGE/PGGE as used by White and Ralston (1981) in studies of water-soluble erythrocyte membrane proteins, PAGIF/PAGE as used by Karpetsky et al. (1984) in studies of nuclease activity and PAGE/PAGIF as used by Jones, Broadbent, and Gurnsey (1982) for salivary proteins. Denaturing arrangements usually employ SDS or less frequently urea or other detergents in one or both dimensions, e.g. PAGE/SDS–PAGE (White and Ralston 1981; Jones et al. 1982), PAGE/SDS–PGGE etc. Considerations of technical simplicity and a desire to use non-denaturing conditions giving relatively uncomplicated maps of proteins in the native state led Felgenhauer (1979) and Felgenhauer and Hagedorn (1980) to examine body fluid proteins using 1 per cent agarose strips run in $0 \cdot 2$ M barbital buffer pH $8 \cdot 6$ in the first dimension followed by PGGE in the second dimension. The agarose strips were placed along the top of the gradient slab and sealed in place with agarose in the usual way. Gradients of $T = 3$–25 per cent ($C = 5$ per cent) were used for larger proteins (MW of 30 000 or more) and of $T = 20$–50 per cent for small proteins down to a MW of about 3000. The gradient gels were prepared in $0 \cdot 089$ M tris–$0 \cdot 082$ M boric acid buffer (pH $8 \cdot 6$) containing $2 \cdot 5$ mM EDTA.

The method thus separates predominantly according to charge differences in the first dimension and size differences in the second. This same combination of separation factors was also employed by Lonberg-Holm et al. (1982) and Bagley et al. (1983) in studies of human plasma proteins, but in this case an agarose gel electrophoresis first dimension with a $0 \cdot 06$ M barbiturate buffer pH $8 \cdot 6$ was followed by SDS–PGGE in a $T = 5 \cdot 5$–11 per cent polyacrylamide gradient gel using the Laemmli (1970) system of discontinuous buffers. Slices of the agarose gel were applied directly to the SDS–PAGE slab without any intermediate soaking or equilibration in order to minimize losses of separated components. In spite of the fact that denaturing conditions were used in the second dimension, it was claimed that most plasma proteins were represented on the final map by single spots and that recoveries were better than with the O'Farrell high resolution procedure, although resolution was not so good and individual spots were less compact so that sensitivity was lower. The O'Farrell type of procedure dissociates many proteins into subunits and has very high resolution so that individual proteins are often split into groups of multiple or overlapping spots, the complexity of the map being increased by genetic variation and by natural microheterogeneity, often due to variations in the carbohydrate portion of glycoproteins for example. Poor migration of basic proteins and low recoveries of large proteins (MW $> 10^5$) are also problems associated with high resolution methods that Lonberg-Holm *et al.* (1982) claim are partly alleviated by their method, using which they were able to detect at least 60 separate plasma proteins and follow changes in them caused by leukapheresis, plateletpheresis or following snakebite.

11.5 High resolution non-denaturing or partially denaturing 2D procedures

At the present time IEF in either agarose of polyacrylamide gel, is the only high resolution electrophretic method capable of separating macro-molecules purely according to charge differences while PGGE and SDS/PAGE or SDS–PGGE are the methods of highest resolution separating by size differences. Thus PAGIF or AGIF combined with a PGGE separation in the other dimension represents the highest resolution 2D system capable of operating under fully non-denaturing conditions. Partially denaturing high resolution 2D systems would thus consist of AGIF or PAGIF combined with SDS–PAGE or SDS–PGGE or of PGGE combined with AGIF or PAGIF performed in the presence of denaturants such as urea, since charged detergents such as SDS interfere seriously with IEF and cannot be used. AGIF, PAGIF, and PGGE can all be performed with high concentrations of glycerol or with non-ionic detergents such as Nonidet P-

FIG 11.2. 2D gel electrophoretic maps of normal human blood serum proteins, separated (A) under non-denaturing conditions by PAGIF in the first (horizontal) dimension and PGGE in the second dimension, and (B) under fully denaturing conditions with PAGIF in the presence of 2 per cent Nonidet P-40 and 9 M urea in the first dimension and SDS–PAGE in the second. (Reproduced from (A) Manabe *et al.* (1982) and (B) Tracy *et al.* (1982) by permission of the authors and publishers. © 1982, Clinical Chemistry.)

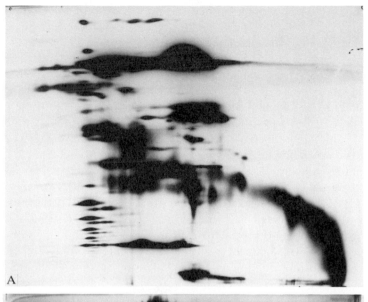

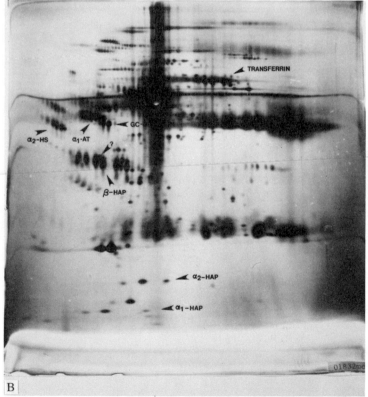

40, Triton X-100 etc. present, and these may be useful for solubilizing samples such as membrane proteins and can be considered as providing mildly denaturing or partially denaturing conditions.

Like the O'Farrell (1975) procedure the methods also provide information about the isoelectric point and MW of protein components, which in the case of the totally non-denaturing systems, will be the values for the native proteins and not of subunits. Because of this lack of dissociation, the maps contain many fewer spots than are seen in maps with fully denaturing conditions. This can be seen for example by comparing maps (Fig. 11.2) of human plasma proteins separated by PAGIF/PGGE (Manabe *et al*. 1982) with similar samples examined by the O'Farrell system (e.g. Tracy *et al*. 1982*a*) or of salivary proteins separated by the same two methods (Marshall 1984). In a clinical context at the present time it may therefore be easier to observe changes in proteins associated with disease if non-denaturing 2D mapping is used. However as automatic scanning and data-handling equipment capable of handling the greater complexity of gels run under denaturing conditions becomes more readily available, this advantage will be rapidly eroded and there is no doubt that much more significant information can be obtained from the really high resolution denaturing systems.

11.6 2D separations of nucleic acids

Nucleic acid molecules of all sizes have very similar charge densities at about neutral pH values, so that separations based on charge differences (e.g. AGIF or PAGIF) are seldom satisfactory. This seriously limits the scope for 2D separations which must therefore be based on size or conformation differences. At acid pH values, below about pH 4·5, each of the four different types of base residues making up the RNA or DNA contributes a slightly different charge so that the net charge then depends on both chain length and base composition. The base pairing phenomenon is lost at low pH values and the addition of denaturants such as urea or formamide disrupts tertiary and secondary structure. Working at elevated temperatures of very low pH also has a denaturing effect. In order to spread the nucleic acids in a mixture adequately over the surface of a 2D gel, it is obvious that as with protein mapping the basis of the separation should be as different as possible in the two dimensions.

(a) *RNA*

Many different systems have been used for RNA separations but according to de Wachter and Fiers (1982) most can be grouped into three categories:

'urea shift', 'concentration shift', and a 'pH shift' usually combined with a urea and a concentration shift.

For the urea shift approach PAGE is run in both dimensions under identical conditions of pH (usually between 4·5 and 8·5) and gel concentration but with a denaturing agent such as urea at high concentration (6–9 M) present for one of the dimensions. Discontinuous buffer systems are not usually employed with nucleic acids but sharpening of zones can be achieved by ensuring that the sample has a lower conductivity than the buffer in the gel and apparatus, so that the voltage gradient across the sample is infinitely steeper than elsewhere (see p.80). It is often convenient for the urea to be added for the second dimension so that RNA fragments containing hidden breaks migrate as base-paired complexes in the first dimension but are dissociated in the second. The method has been used to study small RNA fragments of about 13–80 nucleotides in length resulting from RNase T_1 digestion of larger RNA, for the fractionation of tRNA (on $T = 16$ per cent gels) and mRNA (on $T = 6$ per cent gels).

The concentration shift method relies on a change in T, all other conditions being the same for both dimensions, so that separation depends upon conformational differences being manifest at one or both gel concentrations. Some zone sharpening is seen as samples migrate from a low T gel strip into a high T environment, so the more dilute gel should be used for the first dimension. The method has been used to fractionate RNA in the size range 80–400 nucleotides.

The third category of a pH shift coupled with both urea and concentration shifts (de Wachter and Fiers 1972) has been the most widely used, particularly for fingerprinting RNase T_1 digests of viral RNA. The first dimension consisted of running a $T = 10·3$ per cent, $C = 0·9$ per cent gel containing 7 M urea at pH 3·3–3·5 using 0·025 M citric acid as buffer in both gel and the apparatus, with NaOH to adjust the pH to 3·5 if necessary. The catalyst mixture used to induce polymerization consisted of 0·225 mM $FeSO_4 \cdot 7H_2O$ and 1·5 mM ascorbic acid which are added to the acrylamide and Bis in the citric acid, with H_2O_2 being added to 0·012 per cent (40 μl of 30 per cent H_2O_2 per 100 ml of gel mix) immediately before the gel is poured. Under these conditions subsequent electrophoresis will separate the RNA fragments largely according to base composition and the presence of urea prevents aggregation of complementary sequences (base pairing). Gels are briefly prerun (e.g. 250 V for 1·5 h), samples applied and electrophoresis performed until the Bromophenol Blue or Xylene Cyanol FF tracking dye is close to the bottom of the slab. Strips of this gel are then cut out and either applied in the usual way across the top of a preformed slab of second dimension gel or placed in the gel mould and second dimension gel mixture poured in around it. No equilibration step is required because the pH very rapidly adjusts to that of the second dimension gel slab. This consisted of a $T = 20·6$ per cent, $C = 0·9$ per cent slab made up in 0·05 M tris adjusted to pH 8·0 with citric acid, and with 40 μl TEMED and 40 mg of ammonium persulphate per 100 ml of gel mixture as polymerization catalysts. The concentration and pH of buffers used is not critical and can be varied considerably. Steward and Crouch (1981) for example used 0·75 M tris mixed with 0·75 M boric acid and 0·022 M Na_2EDTA which gave a pH of 7·8 for the second dimension. The important point is that it should be neutral or weakly alkaline and without denaturants so that mobility is primarily a function of chain length. Lee and Fowlks (1982) used a slightly modified version of this method for fingerprinting digests of viral RNA (Fig. 11.3).

Most of these methods are rather limited in terms of the size range of RNA that

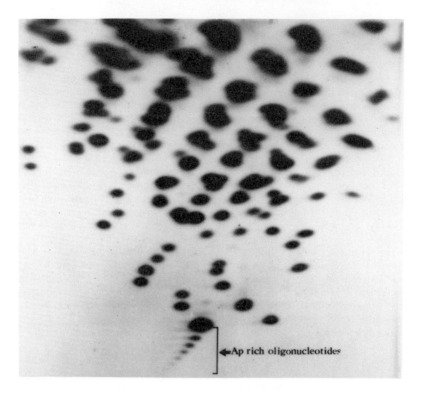

FIG 11.3 2D gel electrophoresis fingerprint of complete Rnase T_1 digest of Brome mosaic virus RNA, labelled in vivo with ^{32}P and revealed after separation by autoradiography. (Reproduced from Lee and Fowlks (1982) by permission of the authors and publishers.)

can be examined. Takeishi and Kaneda (1981) for the first dimension used gels with a $T = 3 \cdot 6$–11 per cent polyacrylamide gradient and a 0–7 M urea gradient made up in a pH $7 \cdot 7$ buffer of $0 \cdot 035$ M tris, $0 \cdot 035$ M NaH_2PO_4, and 1 mM EDTA. Strips of this gel were then laid across the top of a slab of gel with a transverse $T = 3 \cdot 6$–$22 \cdot 6$ per cent polyacrylamide gradient for the second dimension, this being made up in a pH $8 \cdot 3$ buffer consisting of $0 \cdot 089$ M tris, $0 \cdot 089$ M boric acid, $2 \cdot 5$ mM EDTA, and 5 M urea. Thus material at the $T = 3 \cdot 6$ per cent end of the first dimension gel migrated again into $T = 3 \cdot 6$ per cent gel, but this time also containing 5 M urea, while that at the other end migrated into gel of more than twice the concentration (i.e. from 11 to $22 \cdot 6$ per cent). With this system they were able to separate complex mixtures of RNA species in the size range 4 to 12 S.

(b) *DNA*

Most intact DNA molecules are too large to be fractionated on polyacrylamide gels so agarose gels are often used. The methods used for RNA are not really applicable to DNA. Even restriction enzyme fragments are usually too long for the net charge at acid pH to vary appreciably due to compositional differences. Most DNA separations are therefore based on conformational differences.

Agarose gel, typically 1 per cent in a 0·04 M tris, 0·02 M sodium acetate, 1 mM Na₂ EDTA buffer adjusted to pH 8·0 with acetic acid, is used for the first dimension (Fischer and Lerman 1980) and therefore separates the sample of DNA restriction fragments according to chain length. After electrophoresis overnight at a comparatively low wattage (e.g. 35 V for a 177 mm long slab) at 20 °C, a strip of gel is cut out and applied across the top of the second dimension gel, being sealed in place with a little warm agarose solution in the same buffer. This is a uniform $T =$ 4·2 per cent, $C = 2·7$ per cent polyacrylamide gel with a linear 0–7 M urea gradient and a linear 0–40 per cent gradient of formamide, so that components migrate from the first dimension gel into an environment of steadily increasing denaturing power. The gradient gel is prepared with a gradient maker in the usual way (see p. 102) from light and dense polyacrylamide solutions (containing 0 M and 7 M urea respectively; no sucrose or glycerol need be added to increase the density) prepared in the same tris–acetate–EDTA buffer used in the first dimension. The denaturing power of this second dimension gel is further increased by running it (usually at about 150 V for 20–30 h) at 60 °C with the aid of hot water passed through the cooling coils of the gel apparatus.

A slightly different approach has been reported by Bell and Byers (1983) who again separated restriction fragments of DNA in the first dimension by chain length differences in a 0·7 per cent agarose gel in 0·09 M tris–HCl, 0·09 M boric acid, 1 mM EDTA buffer. The gel was then ethidium bromide stained and a strip cut out and applied across the slab of second dimension gel which consisted of 1·5 per cent agarose made up in the same buffer containing 0·5 μg ml⁻¹ of ethidium bromide. The method was reported to be a general one for separating branched DNA molecules, such as replication forks and recombination intermediates, from linear forms on the basis of conformational differences.

11.7 Staining, scanning, data handling, etc. of 2D gels

The staining and destaining of 2D maps on a slab gel is no different to treating slab gels with a number of samples separated in a single dimension on them. Thus the methods discussed in Chapter 2 are entirely suitable. The amount of sample that can be applied to a gel rod or a single sample slot of a slice of a slab gel used for the first dimension is limited and all the components present should be distributed over the surface of the 2D map. Thus in a complex mixture the amounts of a large number of components will be small, so high sensitivity detection methods are called for. This effectively limits staining procedures to Coomassie Blue R250 or G250 for proteins and ethidium bromide for nucleic acids and particularly to silver staining methods for either. Silver staining was initially developed for application to 2D protein maps and most variations are optimized for use with the 1–1·5-mm-thick slab gels typical of many 2D maps. Sammons et al. (1981) compared Coomassie Blue, silver staining, and autoradiographic detection of ^{35}S-labelled human fibroblast proteins and found Coomassie Blue to be much less sensitive (about 100-fold less) than the other two methods which were of broadly comparable sensitivity. Some proteins detectable by silver staining could not be observed by autoradiography and vice versa. This was attributed to the differential uptake of ^{35}S label by different proteins during the labelling of the fibroblasts by incubating them in the presence of L[^{35}S]-methionine, the most heavily labelled being easily detectable by autoradiography but the least labelled being more easily observed by silver staining. Radiolabelling is not always practical or convenient however and it is in such cases that silver staining is most valuable.

Autoradiography remains the most sensitive detection method for proteins and nucleic acids labelled to high specific activity and was the method originally chosen when high resolution mapping was developed (O'Farrell 1975). Autoradiography and fluorography as described in Chapter 2 are most commonly used with 2D maps but double isotope labelling using ^{3}H and ^{14}C or ^{35}S can be particularly helpful for comparing maps of complex protein mixtures. This method was introduced by McConkey (1979) and subsequently updated (McConkey and Anderson 1984a, b) in the light of changes in materials used for autoradiography, such as the replacement of Kodak XR and No Screen film with XAR and Direct Exposure film respectively, the introduction of SB film and the replacement of PPO dissolved in dimethylsulphoxide with the commercially available fluor EN^{3}HANCE. The principle of double label autoradiography is that after electrophoresis the gel is impregnated with fluor to improve the efficiency of ^{3}H detection, dried and subjected to fluorography at -70 °C using XAR film. This records both the ^{3}H and the higher energy isotope (^{14}C or ^{35}S) via the light produced in the fluor when beta particles are emitted. A second autioradiogram (for example on SB film and at room temperature) is then made using film that is relatively insensitive to flashes of light. Beta particles from ^{3}H are too weak to reach the film emulsion but the higher energy beta particles from ^{14}C or ^{35}S do. Thus if this second autoradiogram is placed over a photographic negative of the first autoradiogram (fluorogram), zones on the gel corresponding to polypeptides labelled only with ^{3}H show up as white spots against the grey background of the negative while ^{14}C or ^{35}S-labelled polypeptides give black spots. Due to different relative counting efficiencies the ratio of ^{3}H-labelled material to ^{14}C or ^{35}S-labelled should be about 10:1 if approximately equal exposures on the fluorogram are desired. This ratio is affected a little by the film used and the composition of the fluor but particularly by the thickness of the gel. 'Ghost spots' on the second autoradiograph, representing faint images due to ^{3}H, can be avoided by coating the gel with a layer of ink after the fluorography step.

Protein transfer from 2D maps to immobilizing matrices by the various blotting techniques (see Section 2.13) can be readily performed by the usual methods and the components transferred to the blots detected without difficulty by staining or reaction with ^{125}I or peroxidase-labelled antibodies, as for one-dimension gel slabs. In order to handle the large numbers of gels arising in their programme Anderson *et al.* (1982) devised a device for electroblotting six gels at once in their ISO–DALT tank.

2D mapping procedures employing media containing SDS in either stage are usually accompanied by a complete loss of any enzyme activity present in the sample, so specific detection methods relying on such activity cannot be used. In the case of enzymes composed of single subunits, or of a number of identical subunits which are not separated from each other, a method has been published (Manrow and Dottin 1980) by which part of the activity can be regained enabling specific detection. For this, SDS–PAGE is used in the first dimension followed by removal of the SDS by iso–electric focusing in the presence of 7–9 M urea and 2 per cent Nonidet NP-40 in the second dimension. Urea is removed and activity regained during a subsequent incubation of the gel in a suitable buffer.

It is sometimes possible however to regain at least part of the activity with the conventional O'Farrell (1975) system by suitable incubation or soaking in aqueous buffers which removes the SDS and allows the denatured proteins to renature. Scheele *et al.* (1981) achieved this by soaking gels after electrophoresis in a large volume (900 ml) of 0·1 M tris–HCl buffer pH 8·5–9·0 containing 5 per cent glycerol. After 1 h gels could be stained by adding 100 ml of 0·3 per cent solution of Coomassie Blue G250 in the same buffer to the mixture. After a further 12–24 h at 22 °C, enough renaturation had occurred for the exocrine pancreatic enzymes being studied to be located by virtue of their enzyme activity.

As mentioned earlier, one of the major problem areas of most 2D mapping procedures lies in the extraction and handling of the vast amount of both qualitative and quantitative information that can be gained. The type of gel scanner where a strip of gel, a gel, or a slice of photographic negative is passed through a light beam, giving an absorption peak for each zone which can then be integrated to give a quantitative measure, is not really adequate for 2D maps. Perhaps because of this complexity of measuring and interpreting patterns, a still widely used approach is simply to superimpose two negatives, match them as closely as possible manually, and visually to search for spots present on one but not the other, or present in obviously different proportions. As well as being wide open to error, this is scarcely even semi-quantitative and is extremely time-consuming. Nevertheless for the laboratory handling only occasional 2D maps and looking for specific or limited changes it may be a suitable approach provided it is accepted that much other potentially useful information encompassed in the map will be missed. Moving up slightly in sophistication it is possible to use an ordinary scanning densitometer for the quantitative evaluation of 2D maps by scanning (photographic negatives or autoradiographs) in both horizontal and vertical directions and relating the response to the volume of an ellipsoid (Quitschke and Schechter 1982), but it also is time consuming and most appropriate when changes in only a few selected proteins are to be followed. It has the advantage that no expensive equipment or computers are needed.

In order to appreciate the full potential of high resolution 2D maps however it is almost essential to use some form of automatic scanning and data handling equipment. Unfortunately at the time of writing there is no adequate commercially available package of equipment for this, so various groups of researchers have devised their own arrangements and programmes. A considerable number of different approaches of greatly varying complexity have been described and are usually highly individual in that each group has its own system. Scanning is often achieved by video cameras (e.g. Mariash *et al.* 1982; Lemkin *et al.* 1982; Schneider and Klose 1983; Toda *et al.* 1984) or rotating drum scanners (e.g. Garrels 1979; Bossinger *et al.* 1979; Vo *et al.* 1981; Anderson *et al.* 1981) and the resulting signals processed by computers varying in power from simple micros (e.g. Mariash *et al.* 1982; Toda *et al.* 1984) to medium size computers such as PDP11, DEC system 10 or 20, VAX 11/780, etc. (e.g. Anderson *et al.* 1981; Miller *et al.* 1982; Vo *et al.* 1982; Lemkin *et al.* 1982; Schneider and Klose 1983; Janssen *et al.* 1983). Objectives of most workers include the gathering of the maximum quantitative information, accurately and rapidly and very often include some comparative interpretation of data from different gels. For most of this work it is helpful or essential for gel maps to be as reproducible as possible and that is an aspect of 2D mapping which has been greatly aided by the ability to prepare, run, and process several gels at once, as practised by the Argonne group.

11.8 Biochemical applications

It is not practical, and nor would it be particularly helpful, to attempt to give here an exhaustive review of all applications of 2D electrophoretic separations. Suffice it to say that the ability to prepare and display maps of the macromolecules produced by living cells and present in various tissues and fluids has attracted the collective imagination of the scientific world to such an extent in recent years that a quick glance at almost any issue of most of the major Journals in fields covering biochemical or clinical research is bound to reveal at least one article, and often many more than one, in which 2D electrophoretic separations play an important role. Most of these of course are concerned with applying established techniques to particular analytical problems and comparatively few to methodological development.

Probably the most important single facet of 2D mapping is the tremendous potential it presents to study changes occurring at the molecular level. These may be genetic differences between individuals; changes produced by disease or following therapy; variations in a single individual with time, related perhaps to reproductive cycle, diet, behaviour, etc., studies of cellular differentiation and translation of the genetic code, sex differences, age differences, and many others. When detection of a biological or enzymic activity is required it is common for non-denaturing systems to be used, but there are now many protocols for removal of SDS with some regain of activity so it is not unusual for the high-resolution O'Farrell type methods to be used even in this context. Antigen–antibody reactions for detection and measurement of specific components can readily be used in conjunction with 2D maps and are perhaps best applied to immobilizing membranes (e.g. nitrocellulose) following electroblotting transfer to them of the proteins separated in the gel map (e.g. Legocki and Verma 1981; Anderson et al. 1982).

Because of the denaturing and solubilizing power of the buffer systems employed in both stages, the O'Farrell system for high resolution 2D mapping is particularly suited to studies of membrane proteins solubilized with SDS (Ames and Nikaido 1976; Mills and Freedman 1983) or with urea and non-ionic detergents (Horst, Baumbach, and Roberts 1979). SDS interferes seriously with IEF separations, so in the former case the first-dimensional separation by IEF was made possible by the incorporation into the gels of about 6·5 M urea and 5 per cent of the non-ionic detergent Nonidet NP-40 which formed mixed detergent micelles with the SDS and essentially removed it from the pH gradient. A similar approach has been used to resolve over 200 polypeptides from human erythrocyte membranes (Rubin and Milikowski 1978; Harell and Morrison 1979) and the surface proteins of human fibroblasts (Comings and Cohen 1979).

It has been reported (Marshall and Williams 1984) that if samples are solubilized before mapping by the addition of buffers containing 2-mercaptoethanol or if this is incorporated in the equilibration buffer in which gel strips are soaked before the second dimension run, impurities present in the 2-mercaptoethanol can give rise to faint vertical streaks and horizontal bands across the final map if silver staining methods are used. The two most prominent horizontal bands are in positions corresponding to molecular weights of about 54 000 and 68 000. They can be minimized by keeping 2-mercaptoethanol levels low (e.g. 0·01 M).

An interesting application of high resolution 2D protein mapping which may become widely used after further development is for determining the amino acid composition of all the protein components in a complex mixture (Latter *et al.* 1983). This was investigated using 20 cultures of two human cell lines which were grown in 20 batches of media containing each amino acid in turn bearing a ^{14}C label (^{35}S for methionine). The radiolabelled cell extracts were mapped by O'Farrell's (1975) procedure, subjected to autoradiography on Kodak XAR-2 film, and the auto-radiograms scanned and digitized with an Optronics P-1000 film scanner. The digitized images were then processed and related to the composition of known reference proteins with a PDP 11/55 computer. Applying the technique to proteins of known composition showed that the amino acid composition could be determined with a standard error of 6 per cent over all twenty amino acids. Amino acid sequencing of polypeptides separated by 2D mapping has also been reported (Kelly *et al.* 1983), in spite of the fact that material for this had to be eluted from the gels and amounts of necessity were therefore very small.

12

MISCELLANEOUS ELECTROPHORESIS METHODS

12.1. Micro-gel electrophoresis methods

Scaling down electrophoretic methods often brings a number of advantages, such as increased sensitivity due to the reduced cross-sectional area, the ability to apply higher voltage gradients due to the more efficient heat dissipation which in turn gives shorter electrophoresis running times, more rapid fixing, staining, and destaining, a saving in reagents and reagent costs, and the ability to conserve valuable sample material or to perform analyses with amounts of sample which are too small to study by standard procedures. These advantages must be balanced against the requirement for greater technical skill and, in the case of work on a very small scale, for special equipment such as microscopes and micromanipulators.

A quarter of a century ago Edström (1956) first described an ultramicrotechnique of electrophoresis along a silk thread for the separation on a picogram (10^{-12} g) scale of nucleic acid bases from single cells. Ten years later, Matioli and Niewisch (1965) achieved good separations of haemoglobin variants from single erythrocytes on polyacrylamide fibres. Working at a similar level of sensitivity, Grossbach (1974) described a technique of more general applicability employing very fine glass capillary tubes 50 or 100 μm in diameter and 25 mm long. Gels and buffers are the same as those used in the conventional procedures (Chapters 2 and 3) but only 5–10 min is needed for the electrophoresis. Proteins are detected by fixing and staining with Coomassie Brilliant Blue R250 in the usual way. However, the tubes are so small (0·05 and 0·2 μl respectively) that micromanipulators are necessary for handling them.

Moving up slightly in scale leads to considerably greater ease of handling while retaining most of the advantages afforded by microtechniques. Probably the smallest practical scale for most work is that described by Neuhoff and co-workers. The following experiments have been performed using 5 μl capillary tubes (such as Drummond Microcaps) to which 0·1–0·5 μg of sample is applied: both analytical and preparative PAGE (Neuhoff, Schill, and Sternbach 1969, 1970), SDS–PAGE for comparisons and MW measurements (Peter, Wolfrum, and Neuhoff 1976), investigations of enzyme kinetics (Cremer, Dames, and Neuhoff 1972), studies of micro-IEF (Bispink and Neuhoff 1977), and micro-PAGE combined with crossed immuno-electrophoresis (Dames, Maurer, and Neuhoff 1972). Quantitative measurement of microgels of glycoproteins after staining with FITC-labelled concanavalin A has been achieved by fluorescence scanning

(Lane, Zimmer and Neuhoff 1979). The preparation of gel concentration gradients in capillary tubes of 1–50 μl capacity has been described by Dames and Maurer (1974). All these methods are basically only scaled-down versions of the larger-scale techniques already described, and therefore present no particular difficulties apart from a rather higher level of manual dexterity. As one progresses further up in scale the methods become no more demanding than conventional-scale experiments, and SDS–PAGE in 100 or 250 μl Drummond microcaps has recently been advocated as the method of choice for routine use (Condeelis 1977).

Micro versions of slab gels prepared with gel moulds made from two microscope slides (75 mm x 25 mm) have been applied to protein samples of 0·1–2·0 μg in a volume of 0·5–5·0 μl and the limit of detection with Coomassie Brilliant Blue staining is about 50 ng (Maurer and Dati 1972). These microslabs can be used with homogeneous or discontinuous buffer systems and for immuno-electrophoresis. Scaled-down techniques for preparing gels with a polyacrylamide concentration gradient using appa-ratus based on the original version of Margolis and Kenrick (1968) have been described, but a more recent design of apparatus has been described by Matsudaira and Burgess (1978). In this up to 21 samples can be handled per gel slab, with glass microscope slides (82 mm x 102 mm). The SDS buffer system of Laemmli (see Table 5.4) is used and as little as 20 ng of protein can be detected. As with other micro-gel methods the virtues of microslab gels include high resolution and sensitivity and rapid electro-phoresis, staining, and destaining coupled with good reproducibility and low costs of both construction and operation. The time taken for a complete analysis is as short as 2 h, and in a further hour the gels can be easily dried between two sheets of cellophane with the aid of a 5 W lamp. Micro versions of the O'Farrell (1975) system of 2D protein mapping have been referred to on p.314, but other 2D methods have also been successfully scaled down. Spiker (1980) described a modified version suitable for micro slab gels of the acetic acid–urea polyacrylamide gel system of Panyim and Chalkley (1969), which has been widely used for the analysis of histones and other very basic proteins. Davie (1982) employed Spiker's version for one dimension of a minislab 2D procedure for the rapid analysis of histones. In spite of the fact that scaling down the dimensions of a 2D map will inevitably be accompanied by a potential loss of resolution (which can both in theory and practice be improved by using the largest gels possible), Davie (1982) achieved excellent results. He used T = 15 per cent, C = 2·7 per cent resolving gels with acetic acid–urea or acetic acid–urea–Triton X-100 for the first dimension and SDS–PAGE for the second dimension or the reverse with SDS–PAGE first and one of the acetic acid–urea systems second. The gel slabs used were not very much smaller (80 x 100 x 0·5 or 0·8 mm) than those commonly used in 2D mapping however, being only about half the length and width, and of a size quite frequently used in one-dimensional methods.

With IEF in polyacrylamide gel slabs even a conservative reduction in the scale of standard procedures can give clear benefits in that heat dissipation is more efficient, permitting higher field strengths and improved resolution from shorter focusing times. Staining and destaining is also more rapid. Very thin gel slabs, only 0·02–0·1 mm thick, prepared on silanized glass plates or silanized polyester films have been advocated by Radola (1980), Kinzkofer and Radola (1981), and Frey and Radola (1982). These papers also include very thorough discussions of all aspects of IEF separations on thin gels, many of the comments being equally pertinent to conventional IEF with much larger slabs.

There is little doubt that in general micro gel methods give a substantial saving in time at all stages of the experiment and in the amount of reagents consumed. Most aspects of scaling down both single and two dimensional separations by electrophoresis methods of all types have been reviewed briefly by Poehling and Neuhoff (1980).

IEF can be performed satisfactorily on a micro-scale in density gradients as well as in polyacrylamide gels. This can be conveniently performed (Holtlund and Kristensen 1978) with the cylindrical gel rod type of apparatus used for PAGE (Chapter 2). For this the bottoms of the tubes are closed with dialysis membrane tightly fixed over them with the aid of rubber rings. Cathode solution (about 300 μl of 0·1 M NaOH in 60 per cent sucrose) is first applied and each tube is then filled almost to the top with the sucrose gradient containing sample and 1 or 1·5 per cent ampholytes. The tubes are then finally filled with the anode solution (0·25 M H_3PO_4) which is also used in the upper buffer reservoir (with cathode solution and 60 per cent sucrose in the lower one, of course). At a constant power setting of 0·1 W per tube until a voltage of 300 V is reached focusing is complete in 3–4 h. Fractions of the focused gradient can be collected from the top with the aid of a microsyringe or from the bottom by puncturing the dialysis membrane with a hypodermic needle and collecting drops.

12.2. Gel electrophoresis in organic solvents

Some hydrophobic proteins are difficult to dissolve without the aid of detergents or denaturants such as urea, and some materials, including many peptides, antibiotics, dyestuffs, steroids, etc., have very seldom been analysed by electrophoretic techniques because although charged they are not readily soluble in aqueous solvents. Ionic detergents such as SDS, CTAB, etc. and urea generally denature proteins as well as solubilize them and while some native structure and enzymic activity (if any) may be regained after their removal recovery is seldom complete. Many non-ionic detergents (e.g. Triton X-100, Nonidet P-40) absorb u.v. light strongly which makes monitoring chromatographic procedures difficult and they can be difficult to remove completely without causing precipitation of the samples. Thus it would be convenient if electrophoretic separations could be performed in organic solvents.

Formamide is similar to water in its ability to form hydrogen bonds and is quite widely used in nucleic acid analysis (Chapter 6) but most other organic solvents including alcohols, phenol, dimethylformamide, dimethyl-sulphoxide, tetramethylurea, etc. are poorly compatible with conventional polyacrylamide gels. Some of them can be used at levels of up to about 50 per cent but the gels have to be polymerized in aqueous buffers and either dried and reswollen in the solvent mixture or subjected to lengthy washing and equilibration steps, both of which procedures are relatively time consuming.

These considerations led Artoni *et al.* (1984) to investigate the suitability of N-acryloyl-morpholine (ACM) as a complete or partial substitute for acrylamide in gels. ACM is a more hydrophobic analogue of acrylamide which while still compatible with aqueous media was more suitable for use with organic solvent mixtures. The first necessity was to synthesize ACM. For this a solution of 80·72 g acryloylchloride in 300 ml of anhydrous toluene or benzene was cooled to 0–5 °C in a dry ice–acetone bath and while stirring a mixture of 90·15 g (124 ml) triethylamine and 77·6 g morpholine was added dropwise. The mixture was allowed to warm to room temperature. After standing for 2 h the mixture was filtered and the solvent removed by flash distillation in the presence of 0·5 g of hydroquinone. The residue was purified twice by distillation at a pressure of 0·1 mm Hg (boiling point 88–90 °C). Artoni *et al.* (1984) then desribed the polymerization kinetics of ACM-containing gels and properties such as pore size, viscoelastic properties, and compatability with organic solvents, showing for example that gels composed of equal amounts of ACM and acrylamide could be polymerized satisfactorily in solvents with up to at least 70 per cent dimethyl formamide. The gels were also entirely satisfactory for the electrophoretic analysis of a model peptide.

In an accompanying paper (Vecchio *et al.* 1984) the same workers showed that gels prepared with equal weights of ACM and acrylamide (T = 7·5 per cent, C = 4 per cent) were also suitable for the analysis of proteins, including water-insoluble proteins such as zein and globin chains, when used in conjunction with the solvent sulpholane. Sulpholane (thiophene) was used at concentrations up to 8 M (76 per cent by volume) and had a number of advantages when compared with urea as a solubilizing agent. For example it appeared to be entirely stable when mixed in aqueous solutions so there was no equivalent of cyanate formation and, unlike urea, proteins could be stored indefinitely in sulpholane mixtures without modification. Again unlike urea, it favoured protein structure and actually increased α-helical content. Sulpholane only very slightly reduced the efficiency of polymerization when compared to purely aqueous solvents and does not interfere with proteins or peptide detection methods. While sulpholane mixtures dissolve hydrophobic proteins, with hydrophilic proteins sulpholane acts as a mild denaturant at low concentrations and precipitates them at higher concentrations (5–7 M). Nevertheless it does appear that polyacrylamide gels containing various proportions of ACM can be used very successfully in conjunction with a number of organic solvent mixtures for the study of hydrophobic proteins, as well as hydrophilic proteins, and also for smaller hydrophobic molecules.

12.3. Starch gel electrophoresis

In the native state starch is made up of granules composed of a mixture of amylose, which is a polymer of 1,4-α linked glucose units averaging about 300 glucose units in length, and of amylopectin, which is also a glucose polymer but in this case 1, 6-α linked. When a relatively concentrated suspension of starch granules is heated in water or an aqueous buffer a viscous solution is formed which on cooling sets to a gel. The use of starch gels as an anti-convective medium for electrophoretic separations was pioneered by Smithies (1955) and its use was later reviewed by him (Smithies 1959) and by Smith (1968). The method gave a dramatic improvement in resolution compared with earlier methods such as paper or density-gradient electrophoresis and it soon became widely adopted, mainly for analytical purposes. Almost at once its use led to the discovery of genetic variations in blood serum proteins. The degree of resolution attainable by electrophoresis on starch gels is at the present time exceeded only by that attainable with polyacrylamide gels (Fig. 12.1).

Unfortunately being a natural product the composition of starch can vary and it may contain differing proportions of amylose and amylopectin which can affect is gelling ability and resolution. The optimum starch concentration therefore has to be determined in preliminary experiments for each batch of material. The concentration of starch can be altered somewhat to give gels of various consistencies and pore sizes and hence with different degrees of molecular sieving. However, it is very difficult to achieve this as reproducibly as with polyacrylamide and in addition the properties of the gels cannot be varied over anything like such a wide range. A further disadvantage is that quantitative measurement by densitometry is difficult and rather inaccurate, although starch gels can be rendered relatively transparent by soaking them first in water:glycerol: acetic acid (5:5:1 by volume) and then overnight in pure glycerol (Gratzer and Beaven 1961).

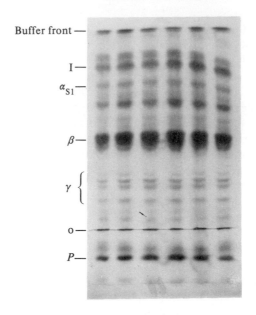

FIG. 12.1. Separation by starch gel electrophoresis of peptides and proteins in six samples of Cheddar cheese after storage at 13 °C for 6 weeks: I, α_{s1}-casein I; α_{s1}, α_{s1}-casein; ß, ß-casein; γ, γ-caseins; o, origin; P, *para-ϰ-casein*.

Primarily for these reasons, starch gel electrophoresis has been almost entirely superceded by the various forms of PAGE, although there are certain applications for which it may still be useful. For preparative work on a very small scale it has an advantage over polyacrylamide in that separated material is extracted simply and often with good yields, although it is contaminated by polysaccharide from the gel matrix. Recovery is achieved either by macerating pieces of gel with buffer or by freezing and thawing the gel pieces followed by centrifuging or merely by squeezing the sample-containing liquid out of the spongy thawed gel.

(a) *Apparatus and gel preparation*

Starch gel electrophoresis is most often performed in horizontal slabs of gel prepared in gel moulds such as those shown in Fig. 2.7 and run in simple apparatus of the type shown in Fig. 2.8. If extensive use of the technique is expected it will probably be beneficial to construct a more specialized apparatus, a good example of which has been described by Jacobson and Vaughan (1977). This design avoids the use of wicks for electrical connection, thereby circumventing the problem that wicks can take up an appreciable, variable, and generally unknown portion of the applied voltage gradient. By employing a sophisticated pulsed constant power source they achieved excellent resolution and were able to use starch gel electrophoresis to detect deletion mutations in haemoglobin and to analyse iso-enzyme patterns and restriction endonuclease DNA fragments.

Gels can be made up in most of the usual electrophoresis buffers without difficulty. A typical recipe for acid gels run at pH 2·0, either with or without the addition of 4 M urea and 2-mercaptoethanol, is shown in Table 12.1.

In order to prepare a single gel slab about 1·5 mm thick with the simple moulds shown in Fig. 2.7 the quantities shown in Tables 12.1 and 12.2 are appropriate, but they should be increased by about 50 per cent if two slabs are made up at the same

TABLE 12.1

Composition of acid (pH 2·0) starch gels with a homogeneous buffer system

Buffer (gel and electrode)	120 ml glacial acetic acid
	30 ml formic acid
	H_2O to 1 l
Gel (no urea)	8·0 g starch
	50 ml buffer
Gel (with 4 M urea)	8·8 g starch
	12·0 g urea
	40 ml buffer
	5 drops 2-mercaptoethanol

Multiphasic buffers can also be used for starch gel electrophoresis and can give resolution superior to that given by homogeneous buffers, as in PAGE. A typical alkaline (pH 8·6) system is shown in Table 12.2.

TABLE 12.2

Composition of alkaline (pH 8·6) gels with a multiphasic buffer system

Buffer (gel)	9·0 g tris
	1·8 g citric acid
	H_2O to 100 ml
Buffer (electrode)	6·0 g tris
	28·8 g glycine
	H_2O to 1 l
Gel (no urea)	4·6 g starch
	18 ml buffer
	40 ml H_2O
Gel (with 6·6 M urea)	7·2 g starch
	19·5 g urea
	6·0 ml buffer
	30 ml H_2O
	5 drops 2-mercaptoethanol

time (losses on glassware etc. are almost constant regardless of the volumes!). The buffer and water are added to the starch in a 150 ml beaker and heated on a low gas flame, stirring continuously with a glass rod. As the mixture approaches the boiling point the viscosity increases markedly and then falls sharply as heating is continued. Once the boiling point has been reached the beaker is removed from the flame, and if required the urea is added and stirred in until dissolved and the mixture is reheated to the boiling point. The hot solution is then transferred to a 250 ml buchner flask (pre-warmed with hot water) and the dissolved air is removed by evacuating the flask with a water pump and swirling vigorously. The 2-mercaptoethanol is then added and mixed in by again swirling the mixture, but this time more gently to avoid the introduction of air bubbles. The rather viscous solution is then poured into the mould which is placed on a horizontal surface and if necessary the solution is spread into the corners of the mould with the aid of a glass rod. The surface of the gel is covered with a sheet of polythene and then with the flat Perspex plate. A weight is placed on top of the plate to ensure an even thickness of gel and excess gel mixture is squeezed out of the sides of the mould and removed. After allowing to cool for about 30 min the gel is transferred to a cold room or refrigerator (4 °C). Gels are completely set and ready for use after about 4–5 h, but if urea is present storage at 4 °C overnight is usually required.

The properties of starches vary considerably but a suitable partially hydrolysed potato starch is available from Connaught Laboratories Ltd., Willowdale, Ontario, Canada. Even with this material, in which the hydrolysis conditions are carefully controlled, the properties of individual batches vary slightly and the quantities of starch needed for optimum gel properties may differ a little from those given in Tables 12.1 and 12.2.

Many workers use much thicker gel slabs (5–6 mm) and apply the samples in free solution with a microsyringe or as a slurry mixed with Sephadex into wells formed in the gel slab in much the same way as can be done with PAGE slabs. The wells are formed by short lengths of glass or Perspex rod attached to the underside of the gel mould cover plate. However, with thick gels electrodecantation can be a major problem and the samples are then not distributed evenly through the gel, which can give misleading results after staining.

(b) *Sample application and electrophoresis*

Samples are usually applied on small pieces (5–8 mm x 1·5 mm) of Whatman 3 MM filter paper. The gel slab is cut about 40 mm from one end and the paper pieces are dipped in the sample solutions. Excess liquid is removed by blotting briefly on a paper tissue or towel and the paper pieces are placed directly into the cut in the gel. Alternatively, known volumes of sample solution can be applied to the dry paper pieces with a microsyringe. When all the sample papers have been loaded (up to six in a 90-mm-wide slab if 8 mm long sample papers are used and eight to ten with 5-mm-long pieces) the gel is gently pushed together to close the cut and to ensure electrical contact and the polythene cover is replaced. Elastic bands can be placed around the gel plate to keep the cover in place and to prevent the wicks from coming loose. These wicks are cut to the width of the gel from a double thickness of medical absorbent lint, similar cloth, or five to six thicknesses of Whatman 3MM paper. The wicks are wetted with electrode buffer and the surplus liquid is squeezed out so that the wicks are very damp but not dripping wet. The wicks are then applied to the ends of the gel slab under the polythene cover so that the wicks cover the end 10–15 mm of gel. They are gently pressed into place to ensure good contact and the tank cover is placed in position. (N.B. The Perspex cover for the gel mould is not in position during the electrophoresis.)

Gels are conveniently run at 4 °C overnight at about 150–200 V for a 175 mm slab. With a single slab 1·5 mm thick the current will probably be 10–15 mA.

(c) *Staining and destaining*

After the separation the polythene cover and the side frame around the gel are removed so that the gel remains supported on the base plate only. Since at this stage the starch gel has little mechanical strength, the gel and base plate are usually immersed together in the staining mixture. With thick (5–6 mm) gels it is common practice to slice the gel horizontally before staining, which is often done with the aid of a commercially available gel slicer (e.g. Shandon Southern Ltd.), and to stain only the centre slice. With thin (1·5 mm) gels slicing is unnecessary.

Proteins

A suitable general staining mixture for proteins, glycoproteins, phospho-proteins, and lipoproteins is prepared by dissolving 0·125 g Amido Schwartz 10B (Amido Black, Naphthalene Black 12 B) and 0·25 g Nigrosin in 400 ml H_2O. Methanol (500 ml) and glacial acetic acid (100 ml) are added and the mixture is filtered. The gel, still on the base plate, is immersed in this dye mixture for 30–60 min, and whilst still in the dye the gel is separated from the base plate by carefully passing a broad knife blade between the gel and the plate. In the dye solvent the gel shrinks slightly and becomes opaque but gains considerably in mechanical strength and so is much more easily handled after staining. Destaining is achieved by washing in methanol:acetic acid:H_2O (5:1:5 by volume).

A more sensitive procedure is to use the fluorescent dye fluorescamine (Tata and Moir 1976). For this the gel is immersed in a solution of 10 mg fluorescamine per 100 ml acetone and viewed under long-wavelength (365 nm) u.v. illumination. Fluorescent bands appear within 1 min and no destaining is required. The lower limit of detection is probably below 5 μg for most proteins. Amine-containing buffers in the gel must of course be avoided when this stain is used and 0·02 M sodium borate, 0·005 M phosphate, or 0·02 M acetate buffers are satisfactory, depending upon the pH required. Electrode buffers can be similar to the gel buffer but five to ten times more concentrated.

Specific glycoprotein stains cannot be used with starch gels owing to reactions with the gel matrix.

Lipids and lipoproteins

These can be detected by immersing the gel in an Oil Red O Solution. This is prepared by mixing 20 wt per cent trichloroacetic acid solution with an equal volume of a saturated solution of Oil Red O in methanol and filtering the resulting mixture. The gel is stained overnight and destained by washing in 50 vol per cent methanol containing 10 wt per cent trichloro-acetic acid.

12.4. Paper electrophoresis

Although used in the past for the separation of proteins and other macromolecules, paper is not nowadays considered suitable for this purpose because it is not very homogeneous in terms of pore size and because some materials, particularly proteins, are absorbed. The result tends to be broad bands with poor resolution and sometimes streaking. For example, with normal blood serum only five protein bands can be seen at best owing to lack of zone sharpness.

Today the use of paper as a medium for electrophoresis is confined almost entirely to separations of small molecules such as amino acids, peptides, and nucleotides, and relatively high voltages are nearly always used. At low voltages even amino acids are only separated into three groups, namely acidic, neutral, and basic. The apparatus required for low-voltage applications is identical to that used for cellulose acetate electrophoresis (Section 12.6). High voltage paper electrophoresis has been used quited successfully for the separation of the weak acids and bases present in coal tars and crude oils (Tshabalala et al. 1981) using electrolytes containing various proportions of organic solvents. It appears also to be a very effective technique for separating the zinc dialkyl and diaryldithiophosphates commonly present in lubricating oils as multi-functional additives (Plaza 1981).

(a) Amino acids

Not all amino acids can be separated in a single run, even with high voltages, so complete resolution requires a second electrophoretic run at a different pH value from the first or a paper chromatography step. Indeed, if a spray detection reagent is used in which adjacent amino acids give different colours or shades of colour (e.g. 0·1 per cent ninhydrin in ethanol containing 5 vol per cent collidine, followed by heating at 80 °C for 1–3 min), a single paper chromatography stage using n-butanol:glacial acetic acid:H_2O (4:1:5 by volume) can enable virtually all common amino acids in a protein hydrolysate to be identified (Levy and Chung 1953) without the need for an electrophoresis stage at all.

Electrophoretic separations have been reviewed by Blackburn (1965). They are usually performed on sheets or strips of Whatman 3MM paper with volatile buffers such as 2 vol per cent formic acid in 8 vol per cent acetic acid (pH 2·0), 1 vol per cent pyridine in 10 vol per cent acetic acid (pH 3·6), or 10 vol per cent pyridine and 0·4 vol per cent acetic acid (pH 6·5). Voltages in the range 50–100 V cm^{-1} are applied for 1–6 h depending upon the length of paper, voltage, and pH.

One of two types of equipment are generally used, either a horizontal flat plate apparatus (e.g. Shandon Southern Ltd; Savant Instruments) or a liquid-filled type (e.g. Savant Instruments). In the former a water-cooled metal plate is covered with a sheet of thick polythene to provide the necessary electrical insulation and the paper electrophoresis strip is placed over this. A number of samples are applied by micropipette or microsyringe as compact spots, by repeated application and drying in a stream of warm air, along a line across the paper strip, and close to the anode with pH 2·0 buffers, or closer to the centre for more neutral buffers. The region containing the samples is slightly raised by placing a glass rod underneath the paper at this point and then spraying the paper as evenly as possible with buffer solution. Great care must be taken not to wet the line of samples excessively and this part of the paper can often be wetted most simply by allowing buffer to spread into it by capillary action from the surrounding paper. Any extra puddles of buffer on any part of the paper strip should be removed by blotting gently with a clean piece of filter paper or tissue so that the paper is uniformly damp, with a dull and not glistening surface. The glass rod is removed and the paper spread carefully over the polythene-covered metal plate so that no air bubbles are trapped between the paper and the polythene. In some designs the ends of the paper strip dip directly into the electrode buffer chambers at each end of the metal plate but more usually separate wicks made from lint or several thicknesses of Whatman 3MM paper soaked in buffer are used to make the electrical contact. Finally, the electrophoresis paper is covered with another strip of polythene and then a thick glass or metal plate.

There are also a number of designs of liquid-filled high-voltage paper electrophoresis equipment. In one of the most simple forms (e.g. Bennett 1967) two electrode compartments are filled with buffer and placed at the bottom of a chromatography tank. The samples and buffer are applied to the paper strip as above and the paper is then placed into the tank on a suitable glass or plastic frame so that the paper forms an inverted V with one end of the paper dipping into each electrode compartment. The paper is then completely covered to avoid evaporation and to provide cooling by filling the whole tank with a water-immiscible solvent. This must of course be non-conducting and unfortunately most suitable solvents are either toxic, inflammable, or both. The most commonly employed solvent is white spirit, such as Esso Varsol which is cheap, efficient, and safe provided that the usual fire precautions are taken. Finally, a glass water-filled cooling coil is placed in the top of the tank and supported from the lid directly over the paper strip since this is the region which gets warmest during the run.

For qualitative evaluation the paper strips are then dried in a stream of hot air and sprayed with 0·1 per cent ninhydrin in 95 per cent ethanol or the ninhydrin–collidine reagent referred to above. Amino acids show up as purple spots (or as various shades from grey to purple with the latter reagent), except for proline and hydroxyproline which give a yellow colour. Colour development occurs within a few hours if strips are hung in a fume cupboard at room temperature or in a few minutes if the strips are heated in an oven at 80–100 °C. Spraying with the ninhydrin reagent should be performed in a fume cupboard as ninhydrin is currently suspected of being a carcinogen.

Quantitative evaluation (Blackburn 1965) can be achieved by drawing the paper strips once through a bath of cadmium–ninhydrin reagent. This is prepared freshly before use by dissolving 1 g of ninhydrin in 112 ml of a stock solution made by dissolving 1 g of cadmium acetate in 100 ml of water and 20 ml of acetic acid and then adding 1 l of acetone (Heilmann, Barrollier, and Watzke 1957). The strip is allowed to dry in air for 30 min until most of the acetone has evaporated and then stored in the dark for 20 h in an airtight vessel over a beaker containing concentrated H_2SO_4 to maintain an ammonia-free atmosphere. After 20 h amino acids show up as red spots. These are cut out of the paper, eluted with absolute methanol, and measured quantitatively from absorption measurements at 505 nm (352 nm for proline) by comparison with readings given by known amounts of standard samples.

More recently fluorescent detection of the separated amino acids and peptides has been introduced. Tata and Moir (1976), who used veronal buffers for the separation of proteins, merely washed the paper strips in acetone and dried them before immersing the strips in a bath of fluorescamine in acetone (0.1 mg ml^{-1}) for 20 s. The strips are then air dried and viewed under u.v. illumination. In a similar procedure for amino acids and their derivatives in which pyridine buffers were used in the separation the paper strips are dried in a fume cupboard and heated at 55 °C for at least 1 h (Mendez and Lai 1975; Lai 1977). The strips are then washed first in acetone and then in 1 vol per cent triethylamine in acetone, air dried for 5 min, and sprayed with a solution of fluorescamine in acetone (0.1 mg ml^{-1}). The strips are dried again, washed in acetone, and air dried. The amino acid spots can be measured quantitatively (Cintron, Peczon, and Kublin 1978) by marking the fluoresent zones with a pencil under a u.v. lamp, cutting them out, and eluting the fluorescent material by shaking overnight at 4 °C with 1.25 ml portions of 0.2 M sodium borate buffer pH 9.0 or with 0.2 M NH_4OH (Mendez and Lai 1975). The eluates are freed of paper fibres by centrifuging and 0.75 ml portions of the supernatant are mixed with 0.5 ml of fluorescamine in acetone (0.3 mg ml^{-1}). This step is necessary because complete reaction with fluorescamine does not occur during the spraying stage. Fluorescence is measured by fluorimeter with an excitation wavelength of 390 nm and emission at 480 nm. As little as 50 pmol of most amino acids and peptides can be detected.

o-Phthaldialdehyde (OPA) has also been employed for amino acid and peptide detection and is less expensive than fluorescamine (Lai 1977). Davies and Miflin (1978) used a single spray reagent prepared by mixing just before use 25 ml of 0.05 M sodium tetraborate containing 2 drops of 2-mercaptoethanol with 20 mg of OPA dissolved in 2 ml of ethanol. Lindeberg (1976) used a two-stage technique, spraying first with 0.1 per cent OPA in acetone containing 0.1 per cent 2-mercaptoethanol and then, after waiting for 5 min, with a solution of 1 per cent triethylamine in acetone or NaOH in 60 per cent ethanol. With both methods as little as 10–100 pmol of all amino acids can be seen by u.v. illumination within a few minutes, the only exception being proline. However, this will show up after heating at 100 °C for 1 h, the lower limit of detection being about 250 pmol.

(b) *Amino acid derivatives*

Dinitrophenyl amino acids and dansyl amino acids can also be separated by paper electrophoresis, although it is less efficient than the separation of amino acids themselves, and these derivatives are best separated by thin-layer chromatography on silica gel and polyamide sheets respectively.

Whatman 3MM paper is again most suitable, and since these derivatives are slightly soluble in the solvents required for liquid-filled apparatus the use of the cooled flat-bed type of equipment is obligatory. Pyridine acetate buffers of pH 4·0–5·5 are suitable for the separation of most amino acid derivatives, although further runs at different pH values (e.g. 1 per cent ammonium carbonate (pH 9·0), 0·1 M sodium phosphate containing 0·1 M NaOH (pH 12·7), etc.) are usually required to separate all the derivatives formed from a protein digest.

(c) Peptides

The most important current application of the electrophoretic separation of peptides lies in the preparation of peptide maps. These have a variety of uses including providing a very powerful method for characterizing unknown proteins, comparing modified proteins with unmodified starting materials, and mapping the products of transcriptions of DNA and RNA (including comparisons with controls of the products of mutations and deliberate genetic modifications). They are also useful in taxonomy and the study of species relationships in which amino acid substitutions very often occur between similar proteins of one species and another. Likewise, peptide maps can be very useful in work on genetic inheritance (the many variants of haemoglobin are but one example) or for identifying individuals with a particular variant of a protein.

We have found the following procedure to be particularly convenient for the preparation of samples for peptide maps.

5 μl of trypsin (TPCK-treated, Sigma Chemical Co.) solution containing 30 mg ml^{-1} in 0·5 mM HCl is added to a solution containing 10 mg of protein in 0·1 ml of 0·1 M pyridine acetate buffer (pH 7·5) or 0·1 M ammonium bicarbonate. This gives an enzyme:substrate ratio of 1:67 and the limit of digestion is reached within 16–24 h incubation at 18 °C. The resulting peptide mixture is then sufficiently concentrated to be applied directly to paper (5 μl is a suitable amount) or to cellulose TLC plates (1–2 μl) without further treatment. Chymotrypsin can be used in the same way except that the enzyme itself is also dissolved in pH 7·5 buffer instead of the very dilute HCl which is required to dissolve trypsin adequately. The preparation of digests with pepsin is usually performed in acetic acid/formic acid buffers of pH 1·8–2·0.

Apparatus for the separation of peptides by electrophoresis on paper is the same as that used for amino acids and the methods of detection of the separated constituents are also the same. Likewise, volatile acetic acid–formic acid (pH 1·8–2·2) and pyridine acetate (pH 3·5–6·5) buffers are almost universally employed, the most generally useful pH being about 3·5.

A single-dimension run by methods virtually identical to those used for amino acids seldom gives sufficient separation of peptides for a satisfactory peptide map, so two-dimensional techniques are much more widespread. A number of different combinations of electrophoresis and/or chromatography can be employed with electrophoresis in buffers of various strengths and pH values and chromatography in various solvents.

In a typical experiment Bennett (1967) suggests that adequate separation of most peptides is given by electrophoresis at 3200 V for 1 h at 22 °C with a full-width (57 cm) sheet of Whatman 3MM paper. For the second stage chromatography is performed using the organic phase of a mixture of n-butanol:glacial acetic acid:H_2O (4:1:5 by volume) or with n-butanol:pyridine:glacial acetic acid:H_2O (90:60:18:72 by volume). As with all other forms of paper chromatography it is particularly important to ensure that the atmosphere within the chromatography tank is saturated with the vapours of all components of the solvent mixture. This is most conveniently achieved by hanging strips of filter paper down the sides of the tank, thoroughly wetting these and the bottom of the tank with solvent, and allowing the whole tank to equilibrate for some hours before inserting the sample paper.

With full size sheets of paper (46 x 57 cm) chromatograms require a running time of at least 20 h at room temperature and it may be useful to add a spot of marker dye such as phenol red to indicate the progress of the separation. For descending paper chromatography separation should be adequate when the phenol red is about 2 cm from the bottom edge of the paper.

The choice of whether electrophoresis is performed in the first dimension and chromatography in the second or *vice versa* is not critical and neither does it matter which pH is used first if electrophoresis is used in both dimensions, but in all cases the paper must be dried by heating at about 80 °C for 20 min in between the stages.

There are many variations on the basic theme in the literature, but three particularly should be mentioned, namely diagonal mapping techniques, preparative applications, and the use of radiolabelled peptides.

For *diagonal mapping* a peptide sample is applied to the centre of a piece of paper and subjected to electrophoresis so that the individual peptides will be separated in the usual way. If the paper is then rotated through 90° and electrophoresis repeated under identical conditions of pH, time, voltage, etc., the peptides will then be spread along a line running diagonally across the paper (Fig. 12.2). However, if between the two runs the paper is treated in such a way that some of the peptides are modified, then these peptides will no longer appear on the diagonal line occupied by the unaffected peptides (Fig. 12.2).

This technique was first reported by Brown and Hartley (1966) and applied to the selective identification and purification of peptides containing disulphide bridges. For this peptic digests are applied as a band to a sheet of paper and subjected to electrophoresis in the usual way. A strip is then cut off, dried, placed in a sealed jar or tank, and exposed for 2 h to an atmosphere of performic acid which is prepared by mixing 1·0 ml of 30 vol per cent hydrogen peroxide with 20 ml of 98 per cent formic acid. This breaks the disulphide bonds and oxidizes the resulting cysteine residues to cysteic acid, so that every cystine-containing peptide is then represented by a pair of cysteic-acid-containing peptides. The strip of paper is then dried in a stream of air to remove excess performic acid, stitched on to another sheet of

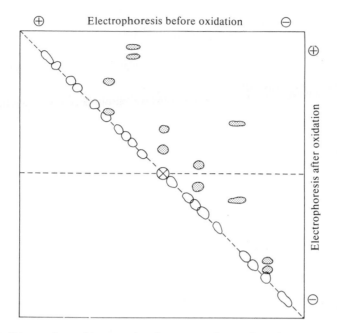

FIG. 12.2. Diagonal peptide mapping by paper electrophoresis. If the conditions are the same for the electrophoresis in both dimensions the peptides will lie on a diagonal line across the paper. However, if a chemical modification step (in this example an oxidation) is included after the first-dimension run but before the second, then those peptides which are modified will no longer appear on the diagonal after the second run. The shaded areas represent pairs of peptides produced by oxidative cleavage after the first run of cystine-containing peptides.

paper, and subjected to electrophoresis under conditions identical to those used for the original separation but at right angles to it. The cysteic-acid-containing peptides are the only peptides not lying on a 45° diagonal since they now possess an extra negative charge. The disulphide-bridged peptide can then be identified on the original sheet, cut out, and subjected to further analysis. For purely analytical work the procedure can be carried out on a square piece of paper with the performic acid oxidation being applied to the whole sheet in between the two electrophoretic separations.

Other applications of the original mapping technique have been discussed by Hartley (1970). These include spraying the paper in between separations with carboxypeptidase B. In a mixture of tryptic peptides this enzyme will remove the C-terminal lysine and arginine residues and hence change the charge of all peptides except the one derived from the C-terminus of the protein, this being the only peptide without a terminal lysine or arginine. Methionine peptides can be identified with an intermediate cyanogen bromide treatment, lysine-containing peptides and N-terminal peptides by modification with trifluoroacetic, maleic, or citraconic anhydride, histidine peptides by reaction with iodoacetamide, and arginine peptides by blocking the lysine residues and spraying the paper with trypsin in between the two separation runs. Clearly many other peptide modification reactions could be used in a similar way.

One *preparative application* of peptide maps has been referred to above, and, while it is inevitably confined to small-scale preparative work, for some purposes the method can be most valuable and will give sufficient material for subsequent analysis and characterization. Generally peptides are sufficiently soluble to be extracted simply by cutting out an area of paper containing the desired peptide, macerating it in buffer, and removing contaminating paper fibres by centrifuging or filtration. Alternatively buffer can be applied to one corner of the small piece of paper and the peptide washed into a test tube. Recoveries of peptides from paper are not very high and according to Bennett (1967) seldom exceed 20 per cent, but other workers (e.g. Tivol and Benisek 1977) refute this. The chief difficulty in preparative peptide mapping has been in the suitable detection of the separated peptides so that the appropriate areas of paper can be marked with a pencil, cut out, and the peptide eluted.

Bennett (1967) advocated a light spraying with dilute ninhydrin solution (0·025 per cent ninhydrin in a 3:1 mixture of absolute ethanol and 2 N acetic acid), but this inevitably modifies some of the peptide material and there is also the danger that peptides present in only small amounts may be overlooked. Staining with fluorescamine or one of the other fluorescent stains also inevitably modifies at least part of the protein (by reacting with NH_2 groups). This objection is overcome in the paper printing technique (Tivol and Benisek 1977) in which after the separation the still damp peptide map is covered with a dry sheet of Whatman No. 1 paper. On applying a pressure of the order of 25 g m^{-2} some liquid containing peptides is transferred to the dry sheet and in effect forms a contact print which can then be sprayed with any suitable detection reagent. By matching the print with the parent peptide map appropriate areas of the latter can then be cut out and the peptides eluted. Specific staining procedures for indoles (tryptophan), sulphur-containing amino acids, histidine, tyrosine, and arginine are given by Bennett (1967).

The use of ^{125}I *radioactively labelled* peptides in peptide mapping of proteins has been described by Bray and Brownlee (1973). Stained protein bands are cut out of polyacrylamide gels and the proteins are eluted with SDS, iodinated, and digested with trypsin. After peptide mapping the peptides are localized by autoradiography for about 16 h. Since they are very thin, paper chromatograms or peptide maps are particularly suited to autoradiography, and less than 10 μg of protein isolated from a stained PAGE band can give a satisfactory peptide map. Other radioactive labels, such as ^{14}C, ^{35}S, etc. can also be used for making autoradiographic peptide maps. Since no loss of material is incurred, this approach may be particularly useful in preparative work, especially if amounts of material are severely limited. A recent highly sophisticated application of these techniques has been in studies of the specificity of codon recognition by *E.*

coli t-RNA species (Fig. 12.3), in which the protein products formed from
[14] C-labelled amino acid precursors by messenger RNA of known base
sequence are digested with trypsin and examined by peptide mapping
(Goldman, Holmes, and Hatfield 1979; Goldman and Hatfield 1979).

(d) *Nucleotides*

In the past few years there have been considerable advances in the
separation of nucleic acids and their fragments so that most separations are
nowadays carried out on agarose or polyacrylamide gels. These have been
discussed in more detail in Chapter 6. However, the older methods of
electrophoresis on paper, cellulose acetate, or thin layers still retain some
usefulness, particularly for the separation of mononucleotides, and are
technically easier to perform.

The four major mononucleotides of RNA (AMP, CMP, GMP, and UMP) or of
DNA, in which TMP is present instead of UMP, are obtained by alkaline hydrolysis
of the nucleic acid (e.g. in 0·3 M KOH at 37 °C for 16 h), neutralized, and de-salted
with a cation exchange resin. They can then be readily separated on Whatman 3
MM paper using volatile buffers of pH 2·5–3·5, such as 10 per cent acetic acid or 5
per cent formic acid adjusted to the required pH with ammonia. The paper strips
are set up in the same way as for amino acid separations (Section 12.4(a)) and the
apparatus is the same. Typical electrophoretic conditions may be up to 50 V cm^{-1}
for 2 h. More complex mixtures of mononucleotides and small oligonucleotides
may require a two-dimensional mapping system involving chromatography (Sanger,
Brownlee, and Barrell 1965) in the second dimension with a solvent such as propan-
2-ol: conc. HCl:H$_2$O (170:41:39 by volume) or the organic phase of a mixture of
equal proportions of 2-methylpropan-2-ol and a buffer prepared by adjusting 5 per
cent formic acid to pH 3·5 with ammonia. The papers are air dried and viewed
under short-wavelength (254 nm) u.v. illumination to detect the nucleotides.

A single-dimension high-voltage electrophoretic separation of cyclic
GMP from GMP, GDP, and GTP in pH 3·5 pyridine acetate buffer
containing 10 mM EDTA on Whatman 3MM paper has been reported
(Mukerjee 1978). The method was used in the assay of guanyl cyclase
activity as this enzyme occurs in several mammalian tissues and
catalyses the formation of cyclic GMP from 'ordinary' GMP.

A particularly successful method of preparing two-dimensional oligonucleotide maps was described by Sanger and Brownlee (1967).

A narrow strip of cellulose acetate (e.g. 3 × 60 cm) is used for the first-dimension
electrophoresis. The strip is carefully moisted (see Section 12.6) with a buffer of 0·5
per cent pyridine and 5 per cent acetic acid (pH 3·5) containing urea up to 7 M if
desired as this can improve the resolution of many oligonucleotides. Surplus liquid
is removed by blotting lightly between two sheets of filter paper. The strip is placed
in the same type of apparatus as that used for paper electrophoresis (Section
12.4(a)), and the sample (10 μl containing up to about 100 μg) is then applied with a
microsyringe or micropipette at a position about 10 cm from the cathode end.

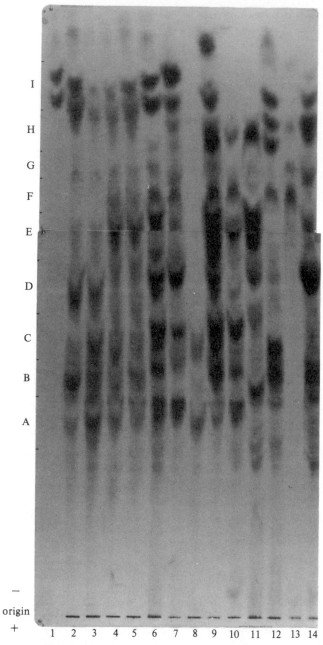

FIG. 12.3. Autoradiogram of a paper electrophoretic separation at pH 1·9 of ^{14}C-labelled peptides in tryptic digests of MS2 RNA–directed proteins synthesized *in vitro* and labelled with various different radioactive amino acids (see Goldman *et al*. 1979 for details). (Reproduced with permission from Goldman *et al*. 1979 J. molec. Biol. **129**, 567. Copyright by Academic Press Inc. (London) Ltd.)

Marker dyes such as 1 per cent xyelene cyanol FF, acid fuchsin, or methyl orange can be added to the sample and help in comparisons of different samples. Care must be taken that the cellulose acetate strip does not dry out during these operations and it can be kept moist by covering with a strip of filter paper soaked in buffer. Electrophoresis is performed at 50 V cm^{-1}. For separation in the second dimension the damp paper is then placed along one side and a few centimetres in from the edge of a sheet of Whatman DE-81 DEAE-cellulose anion-exchange paper. The strip is then covered with four thicknesses of filter paper (Whatman 3 MM) wetted with water which are covered with a glass plate. By pressing on the plate water flows from the filter paper and elutes the separated nucleotides from the cellulose acetate onto the DEAE–cellulose paper, where they are absorbed by the ion-exchange mechanism in very good yield and with little diffusion. If urea has been incorporated in the buffer used in the separation on cellulose acetate it should then be removed by rinsing the appropriate area of the DEAE–cellulose sheet briefly in 95 per cent ethanol because it interferes with the second-dimension separation. After the ethanol has been allowed to evaporate the DEAE–cellulose sheet is moistened by spraying with pH 1·9 buffer (2·5 per cent formic acid adjusted to the required pH with acetic acid) and electrophoresis is performed at 40–50 V cm^{-1} for about 4 h if a whole 46 cm x 57 cm sheet is used. In the original work and in most subsequent applications the method has been used in combination with ^{32}P-labelled nucleotides. Localization is achieved by autoradiography (e.g. Sanger and Brownlee 1967; Ford and Mathieson 1978), and quantitative measurements are made by cutting out the nucleotide-containing areas of paper and applying liquid scintillation counting.

A variation of this procedure involves a very similar first-dimensional separation by cellulose acetate electrophoresis, but using narroweɩ and shorter strips, followed by thin-layer chromatography for the second dimension (Southern and Mitchell 1971). For this the sample is transferred in a manner similar to the above on to a poly(ethyleneimine)–cellulose thin-layer plate (Machery Nagel PE1 MN 300) which is also an anion-exchange cellulose material. Thin-layer chromatography is then performed at right angles to the electrophoretic separation with a solvent of 1·5 M formic acid adjusted to pH 3·5 with pyridine.

12.5. Thin-layer electrophoresis

Virtually all the analytical separations that can be achieved by electro-phoresis on paper can also be performed on cellulose or silica gel G thin-layer plates, in a shorter time, with greater ease of handling, and with greater powers of resolution because of the more homogeneous nature of thin-layer materials. Because of the greater load capability paper may still be the medium of choice for preparative applications, but even here if material is limited thin-layer electrophoresis (TLE) is preferable.

TLE is performed in relatively simple apparatus of the type shown in Fig. 12.4 which consists essentially of a water-cooled metal base-plate with one electrode chamber on each side of it. The metal plate is covered with a sheet of thick polythene to provide electrical insulation. TLE plates can be prepared in the laboratory in an identical manner to thin-layer chromato-graphy (TLC) plates but are very often obtained from commercial sources

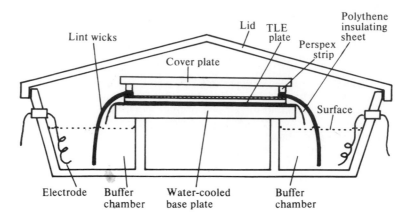

FIG. 12.4. Simple horizontal apparatus for TLE.

(e.g. Merck; EM Laboratories Inc., Elmsford, N.Y., U.S.A.; Eastman Kodak; Macherey–Nagel GmbH). They are seldom larger than 20 cm × 20 cm, so this is the most suitable base-plate size.

The samples are applied to the TLE plate with a micropipette or microsyringe as a spot not more than 5 mm in diameter, and if necessary repeated applications should be made with intermittent drying in a stream of warm air. The plate is then sprayed as uniformly as possible with buffer so that the whole plate is damp but the surface remains dull and not shining with surplus liquid. The area containing the sample should be moistened by capillary action from adjacent areas of the plate to avoid disturbing the sample zone. The plate is then placed in position and electrical contact made with the aid of moistened, but not over-wet, lint or cotton wicks. Thin glass rods or Perspex strips are then placed over the wicks and the TLE plate is covered with a glass cover plate resting on these rods or strips (Fig. 12.4) to reduce evaporation from the surface of the TLE plate. The lid of the apparatus is then placed in position and the electrophoresis is started. This lid should for safety reasons incorporate a switch that turns off the power if the lid is not properly located or is accidentally disturbed.

Because the electrophoresis is carried out for a relatively short time (typically 30–60 min) simple unstabilized power supplies are adequate. Separation voltages of up to 1000 V are used with standard 20 cm × 20 cm plates, but lower voltages (300–500 V) can also be used satifactorily if that is all that is available, although running times will of course be longer. Fahey *et al*. (1980) used 2000 V, but only for 20 min, in their separations of low molecular weight thiol compounds, such as cysteine, glutathione, thiouracil, thiosulphate, etc. on cellulose plates.

For analytical purposes (including peptide mapping), with thin layers 50–150 μm in thickness, the same buffer systems and detection methods that are used in paper electrophoresis can also be applied to TLE (e.g.

Andrews 1978). For peptide detection with fluorescamine the plate is dried, sprayed lightly with 10 vol per cent triethylamine in methylene chloride, then with 0·1 wt per cent fluorescamine in acetone, and finally again with the triethylamine spray (Gracy 1977). Fluorescence develops rapidly and is visualized by u.v. illumination (365 nm). As with other detection methods, material remaining at the origin suggests the presence of undigested protein and redigestion may be necessary. Likewise, elongated peptide spots or streaking is usually indicative of sample overloading or of excessive salt in the sample. When fluorescamine is used fluorescence fades within 1–2 days, even at 4 °C and in the dark, so as with other methods of localization the position of peptides should be marked lightly in pencil on the plate, or with felt-tip pen on the back of the plate, or traced on to paper (back illumination with light-box can be helpful).

Once the peptides have been detected, appropriate areas can be scraped off the thin-layer plate and material recovered from it so that TLE can also be used as a small-scale preparative technique. Fishbein *et al.* (1980) found that with care, contaminating amino acids were so low that it was possible to quantify amino acids accurately when less than 1 nmole of eluted peptide was present. The high recoveries possible also enabled sufficient material to be extracted from a single peptide map for sequence determination. Because of the ease of handling, TLE is particularly suitable for making two-dimensional maps such as peptide maps and the various diagonal peptide mapping procedures (Gracy 1977). When these involve both electrophoresis and chromatographic steps the chromatography exploits the same solvents as those used for paper chromatography, the only difference being that ascending elution is usually far more convenient than descending elution. Amounts of peptide required for a good map are 10–50 μg, or even less with isotope labelling, compared with 1–5 mg for maps on paper.

An example of the use of thin-layer plates for peptide mapping is given by Bryant, Nalewaik, Tibbs, and Todaro (1979). The protein in slices of stained polyacrylamide gel from PAGE analysis is iodinated with [125]I *in situ* by the chloramine T procedure. The labelled protein, still within the gel matrix, is then incubated with trypsin, the gel minced, and 70–80 per cent of the resulting labelled peptides extracted with buffer. A 2–5 μl peptide sample containing about 5×10^5 dpm of radioactivity is then spotted on to a thin-layer plate and a peptide map prepared using autoradiographic detection. If preferred, instead of autoradiography the cellulose can be scraped off the plate, mixed with scintillation fluid, and the radioactivity measured in a scintillation counter.

For complete enzymatic digestion of a protein to peptides (e.g. with trypsin) it will usually be necessary to ensure that disulphide bonds in the protein are ruptured, and a convenient method is to combine this with the incorporation of a radioactive label by using [14]C- or ^{3}H-iodoacetic acid to alkylate the cysteine residues formed (Gracy 1977). Peptides can also be labelled with ^{3}H-dansyl chloride or treated with methyl-(1-^{14}C)-acetimidate (Bates, Perham, and Coggins 1975). With such radiolabelling techniques a single PAGE gel rod can yield enough material for a good peptide map (e.g. Gibson and Gracy 1979).

TLE has been applied successfully to the quantitation of amino sugars in acid hydrolysates of glycoproteins (Farwell and Dion 1979).

The hydrolyates are reacted with dansyl chloride by dissolving the dried hydrolysates in 2 volumes of 0·2 M NaHCO$_3$ and adding 1 vol of dansyl chloride (60 mg ml^{-1}) in acetone to give a dansyl chloride:protein ratio of 9:1 by weight. After reaction in the dark at 37 °C for 30 min, 0·1 vol of 88 per cent formic acid is added to stop the reaction. Samples (2–10 μl) are applied to a 20 cm x 20 cm cellulose TLC plate and electrophoresis is applied at 500 V for 4 h with a pH 4·4 buffer composed of 0·4 per cent pyridine and 0·8 per cent acetic acid. DNS-amino sugars and basic amino acid derivatives migrate slowly towards the cathode, whereas DNS-OH and neutral and acidic amino acid derivatives move towards the anode. The amino sugar-containing area of cellulose, located by u.v. illumination, is scraped off the plate and the DNS-amino sugars eluted by mixing with 250 μl of 95 per cent ethanol. After standing overnight the mixture is centrifuged and the supernatant collected and evaporated under nitrogen. The residue is dissolved in a smaller amount of 95 per cent ethanol and applied to a silica gel TLC plate for chromatography in a solvent composed of cyclohexane:ethyl acetate:ethanol (6:4:3 by volume). Two excursions with this solvent are needed to separate the glucosamine and galactosamine derivatives completely. After separation the fluorescence is stabilized and greatly enhanced by spraying with triethanolamine: propan-2-ol (1:4 by volume).

The same workers (Farwell and Dion 1981) extended their studies to the identification of N-asparaginyl and O-seryl/threonyl glucosidic linkages to amino-sugars in glycoproteins using an extension of the above methodology. Treatment of glycoproteins with mucin-type linkages (oligosaccharide linked to serine or threonine) with alkaline sodium [^{3}H] borohydride cleaved the O-glycosidic bonds and simultaneously labelled both the sugar and amino acid components of the linkage. After acid hydrolysis and dansylation the sugar component of the linkage was identified as its dansy-hexosaminitol derivative, dansyl-galactosaminitol and dansyl-glucosaminitol being separated by TLE on silica gel plates in the presence of 0·25 per cent Na$_2$B$_4$O$_7$ (under these conditions dansyl-hexosamines remain close to the origin). The amino acid component of the glycopeptide linkage was identified by fluorography after 2D TLC of its dansyl derivative on polyamide plates. For plasma-type glycoproteins (oligosaccharide linked to asparagine), glycopeptides were produced by pronase digestion, purified by gel filtration and finally purified by dansylation and TLE for further analysis following partial hydrolysis. Thus TLE formed an essential part of their structural investigations with both classes of glycoproteins.

All the above mapping methods are performed at high voltages and are not suitable for the separation of intact proteins. However, the separation of proteins has been performed by electrophoresis on thin layers of Sephadex^{TM} (Pharmacia Fine Chemicals AB) and this can also be applied as a small-scale preparative procedure.

The superfine grade of the correct type of Sephadex for the molecular-weight range of the proteins being examined (e.g. G-150 or G-200 for most serum proteins) is used for thin-layer work. Enough of this is measured out to make 50 ml of swollen gel for each 20 cm × 20 cm plate. The Sephadex is added to any suitable buffer (e.g. 0·05 M sodium barbiturate pH 8·6 or 0·025 M sodium phosphate pH 8·6 for serum proteins), and allowed to swell overnight. The supernatant buffer is then removed and the gel stirred into a thick slurry and poured onto the plate which should be accurately level (use a spirit level!). The gel is spread evenly with a spatula, permitted to settle for a few minutes, and excess buffer is removed by touching the edges with strips of filter paper until the gel bed is firm enough not to move if the plate is inclined at 30–40 °C. The plate is then placed in the electrophoresis apparatus in exactly the same way as any other type of thin-layer plate and paper or lint wicks are used for making electrical contact. Typical separation voltages are of the order of 300 V for up to 6 h with a 20 cm square plate. Location of separated zones is usually by the paper print technique in which a piece of filter paper (Whatman No. 1) is laid on the gel bed, and when sufficient moisture to dampen the whole sheet evenly has transferred from the Sephadex bed, the paper is carefully peeled off, dried, and stained in the usual way.

Two-dimensional separations on Sephadex thin layers can be performed very easily by combining this electrophoresis separation with a gel filtration step.

This is run at right angles to the electrophoretic separation by applying filter paper wicks as shown in Fig. 12.5. The upper wick conveys buffer solution on to one end of the plate and buffer eluted off the lower end is removed by the lower wick and absorbed by a pad of filter papers or paper tissues. The progress of the separation can most easily be monitored by applying a sample of coloured or fluorescent-labelled proteins (e.g. haemoglobin, cytochrome C, DNP-proteins, DNS-proteins, etc.), Blue Dextran, or dyes to one edge (or both edges) of the plate. During the separation the whole apparatus should be covered with a lid to reduce evaporation from the surface of the plate. Gel filtration fractionates on the basis of size differences, while in a non-restrictive medium such as Sephadex the

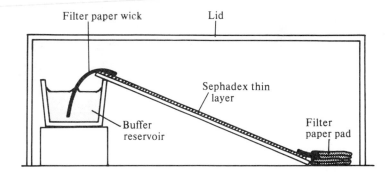

FIG. 12.5. Thin-layer gel filtration set-up which can be used as the second dimension after TLE in the first dimension or as a separation method in its own right.

electrophoretic separation is almost entirely dependent on charge differences. Thus the basis of separation is different in the two dimensions and resolution can be quite good, although largely because of diffusion effects it does not approach that possible in 2D polyacrylamide gel systems.

12.6.   Cellulose acetate

(a) *Electrophoresis in a single dimension*

Cellulose acetate is a much more homogeneous medium for electrophoresis than paper, has a much more uniform pore size, and does not absorb proteins in the way that paper does. There is therefore much less trailing of protein zones and resolution is better but still not nearly as good as with polyacrylamide gels. The resolution can be good enough (Fig. 12.6) to separate partially dephosphorylated α_{s1}-casein into individual species differing by only a single phosphate group (West and Towers 1976).

For low-voltage separations apparatus of the general type shown in Fig. 12.7 is used. The procedure used is as follows (Sargent and George 1975).

The cellulose acetate strips (e.g. Celagram, Shandon Southern Ltd.; Cellogel, Chemetron; Sartorius-Werke GmbH, Göttingen, F.R.G.) are carefully floated onto the surface of the buffer contained in a shallow tray without trapping any air bubbles below the strip. If the top surface of the strip is covered with buffer before the lower side is wetted, air is trapped in the pores of the strip. It is then very difficult to remove this air so the strip cannot be used satisfactorily. Air trapped in

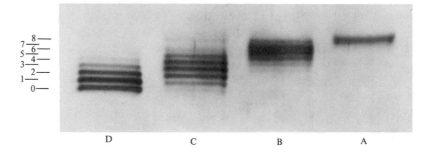

FIG. 12.6. Electrophoresis on cellulose acetate of dephosphorylated α_{s1}-casein in a pH 8·2 buffer containing 3·0 g tris, 1·2 g $Na_2EDTA \cdot 2H_2O$, 420 g urea, 2 ml 2-mercaptoethanol and 6·7 g diethylbarbituric acid per litre. A, α_{s1}-casein control; B, 30 per cent dephosphorylated; C, 63·3 per cent dephosphorylated; D, 92 per cent dephosphorylated. Resolution was achieved in 60 min at 50 V cm^{-1}. Staining was with Procion Blue. The numbers indicate the number of phosphoserine residues in each band. (Reproduced from West and Towers (1976) by permission of the authors and publishers.)

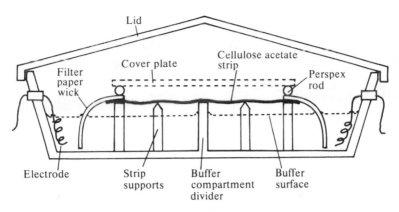

FIG. 12.7. Apparatus for cellulose acetate or paper electrophoresis.

this way is recognized by opaque white areas on the strip. After proper wetting the cellulose acetate strips are lightly blotted on filter paper to remove surplus liquid, the damp strips placed in the apparatus, and the current switched on for a 15 min pre-run to allow equilibrium conditions to be attained. If white opaque areas develop on the strip this indicates that it is drying out and the strip should be removed and rewetted with buffer. Most of the homogeneous buffers used for PAGE (Table 2.1) can be used satisfactorily with cellulose acetate. Resolution can be improved by having a discontinuity in the conductivity which is produced by using a more concentrated buffer for wetting the strip than that used in the electrode chambers (cf. West and Towers 1976).

Up to about 350 μg of sample in a volume of 10 μl can be applied to a strip 2·5 cm wide as a short streak across the strip at the origin, the position of which can be marked on the strip with a pencil before it is wetted with buffer. For serum samples examined in a buffer at about pH 8·6, the sample should be applied about one-third of the way along the strip from the cathode end. Electrophoresis is conducted at 6–8 V cm^{-1} for about 6 h if a 20 cm long strip is used, for about 3 h with one 12 cm long. If 8 per cent glycerol is included in the buffer solution to reduce evaporation from the strip, higher voltages can be used and at 40 V cm^{-1} separations on a 20 cm long strip can be achieved in as little as 20 min.

After electrophoresis some strips (e.g. Oxoid) can be dried at 100 °C for 10 min which also denatures protein samples before staining, but others (e.g. Celagram) should not be heated strongly and proteins must be denatured by a brief immersion of the moist strip in 10 wt per cent trichloroacetic acid. Dried strips should be applied to the surface of a staining bath in a similar manner to that described for wetting with buffer to avoid entrapping air, but moist strips can be immersed in staining solution directly. For staining of macromolecules such as proteins, glycoproteins and phosphoproteins, methods similar to those described in Chapter 2 are satisfactory except that dye concentrations can be rather lower. Staining and destaining times may be as short as 10 min. According to West and Towers (1976) Procion Blue H5R as a 0·1 wt per cent solution in methanol containing 2 vol per cent concentrated HCl is a very satisfactory stain for proteins on cellulose acetate. Little stain is taken up by the matrix of the strip, so that three brief washes with methanol are sufficient to give a virtually dye-free background. For quantitative measurement, stained bands are simply extracted with 0·66 wt per cent $NaHCO_3$ in 66 per cent aqueous methanol and optical densities read at 570 nm.

Cellulose acetate is readily soluble in a number of organic solvents such as acetone, dimethyl sulphoxide, etc., and this provides a simple general method of quantitative measurement. Stained zones can be cut out, the pieces dissolved in solvent, and the extinction of the resulting solution determined spectrophotometrically. For densitometry cellulose acetate strips can be cleared using a mixture of cyclohexanone and ethanol (30:70 by volume) or by treating with 30 vol per cent diacetonyl alcohol in water and drying on a glass slide in an oven at about 70 °C (Ambler 1978b). Quantitative determination of radioactively labelled bands is also very simple, and 1 mm slices of the dried strip can be counted directly in a liquid scintillation counter after mixing with a scintillant such as toluene and Liquifluor (New England Nuclear). For autoradiography the dry strip is mounted on paper, covered with cellophane, and exposed at –70 °C using pre-sensitized x-ray film (see Section 2.12) such as Kodak SB-54 for 1–4 days in a Kodak x-omatic cassette (Salmon et al.).

Low-molecular-weight substances such as amino acids, peptides, nucleotides, thiols etc. can be separated on cellulose acetate strips by exactly the same methods as those used for paper electrophoresis or TLE and using the same type of horizontal apparatus. Buffers, electrophoretic conditions, and detection methods are also identical.

The separation of glycosaminoglycans, which is of considerable clinical importance in the diagnosis of mucopolysaccharidoses (Hopwood and Harrison 1982), is a difficult problem because of the similarity of electrophoretic migration of most of the more common species. Many procedures have been published over the years, earlier single dimension systems being replaced (with little real advantage) by more complex 2D separations so that the pendulum has swung back towards improved single dimension methods. These include what might be termed two stage separations where an initial electrophoresis separation is followed by a buffer change (involving interrupting the run and washing and re-equilibration steps) before electrophoresis in the same dimension is resumed (Morris, Canoy, and Rynd 1981).

Good separations were obtained, but two technically more simple methods appear to have an advantage. The first (Schuchman and Desnick 1981) is a true single dimension method in which cellulose acetate strips which have been stored at 4 °C in 30 per cent methanol are thoroughly blotted on Whatman No. 1 filter paper and immersed for 10–15 min in 0·01 M Na_4EDTA containing 0·05 M LiCl (pH 8·4), leaving the top 6 cm not immersed in this solution. The strips are then rapidly blotted, placed in the apparatus, and samples (about 1 μl containing 1–5 μg sample) applied about 3 cm from the end of the strip in the non-EDTA/LiCl zone. This is the cathodic end of the strip and electrophoresis is then performed for 45 min at 4 °C using about 15 V cm^{-1} strip length (approx 2·5 mA cm^{-1} width). The strips were stained for 1 min in 0·1 per cent Alcian Blue 8GN and destained in two changes (5 min each) of 2 per cent acetic acid. The second method (Bertolotto and Magrassi 1984) is slightly more complex in that it is a multistage method, but only a simple blotting and a quick immersion in new solvent is required between stages. The method employed 60 × 75 mm Titan III Zip Zone cellulose acetate plates (Helena Laboratories), one end of which was dipped to a height of 15 mm in H_2O and the other in 0·1 M barium acetate buffer pH 5·0, leaving after blotting a 2–3 mm dry zone in between. Samples (1–20 μl) were applied in the middle of the water zone (the cathodic end) and electrophoresis at 200 V conducted for 3 min. The whole plate was then soaked for 1 min in 0·1 M barium acetate buffer pH 5·0, briefly blotted, and electrophoresis continued for a further 8 min. Finally the plate was immersed for 1 min in 0·1 M barium acetate pH 5·0 containing 18 per cent (by volume) ethanol and run for a further 20 min before Alcian Blue staining. Both methods give good separation of the more common glycosaminoglycans (chondro-itin-4-sulphate, chondroitin-6-sulphate, hyaluronic acid, dermatan sulphate, keratan sulphate, heparin sulphate, heparin) identification of the zones being aided if necessary by enzymic or chemical pretreatment of the samples before electro-phoresis (Bertolotto and Magrassi 1984). The sensitivty of detection may be improved (Szewczyk 1983) by taking advantage of the fact that acidic polysac-charides at acid pH precipitate bovine serum albumin. After electrophoresis the cellulose acetate strips are therefore immersed for 15 min in 100 ml of 0·05 M sodium acetate pH 4·0 containing 2mg ml^{-1} of bovine serum albumin and thoroughly washed (three times, 10 min each) in 0·15 ml sodium acetate pH 4·0. The precipitated pattern of bovine serum albumin zones was then revealed by staining for 5 min with 0·2 per cent Amido Black in 50 per cent methanol and 10 per cent acetic acid (using the same solvent for destaining), or with 0·1 per cent Coomassie Blue R250 in the same solvent. Unlike Alcian Blue and Toluidine Blue which have only low sensitivity for detecting non-sulphated polysaccharides, this procedure was claimed to be a general one for acidic polysaccharides, as little as 15–40 ng being detectable.

A silver staining procedure optimised for the detection of separated proteins on cellulose acetate strips has been reported (Fujita, Toda, and Ohashi 1984). After electrophoresis, proteins are fixed by soaking the cellulose acetate strips for 10 min or a little longer in 20 per cent sulphosalicylic acid. The strips are washed three times (5 min each) in 0·5 per cent acetic acid containing 10 per cent ethanol and then immersed in 0·4 per cent silver nitrate. After a few minutes the strips are blotted between sheets of filter paper and then floated onto the surface of a reducing solution consisting of 0·15 ml of 37 per cent formaldehyde and 8·75 mg NaBH$_4$ in 100 ml 0·6M NaOH. After 30 s the strips are pushed below the liquid surface. Protein zones are revealed in 20–30 s as brown-black spots on a yellow background and total reaction is complete in 1 min. The cellulose acetate strips are then immersed for 5 min in 0·75 per cent sodium carbonate and finally washed twice more (5 min each) with this solution. Some proteins give slightly coloured spots, the colours being more apparent when viewed with transmitted light, which can aid identification especially if relatively complex 2D patterns are being examined (e.g. Fujita et al. 1984).

Cellulose acetate is very easily used in conjunction with the detection of enzymes by the zymogram technique (e.g. Saga and Kano 1979). For this the cellulose acetate strip is simply laid on a strip of paper impregnated with a suitable buffer and substrate. After an appropriate incubation period the strips are peeled apart and the paper zymogram treated as necessary to reveal the zones of enzyme activity. The precise details naturally depend on the particular enzyme, substrate, and detection method being used. A similar approach can be used for immunofixation. In this case the paper strip is impregnated with antiserum. After incubation the paper 'print' is peeled off, washed in saline, and stained to reveal the immunoprecipitate (e.g. Johnson, 1982).

(b) *Two-dimensional procedures*

The use of cellulose acetate for the first-dimensional separation in two-dimensional oligonucleotide maps has already been described (Section 12.4), but it can also be used in other two-dimensional procedures. For example, it has been used in many immuno-electrophoretic procedures and most of those described in Chapter 8 can be performed perfectly satisfactorily with cellulose acetate electrophoresis used in the first stage. In most cases sample transfer into the agar gel for separation in the second dimension is achieved simply by placing the moist cellulose acetate strip along the origin line of the agar gel. On applying the electric current the samples migrate out of the cellulose acetate into the agar gel phase. For immunodiffusion it is adequate to place the cellulose acetate containing the separated antigens on the surface of an agar slab and to place strips of filter paper soaked in antiserum alongside it, on both sides if desired. After 1–3 days incubation in a moist chamber the immunodiffusion pattern will have developed, and the strips can be removed and the precipitin arcs photographed or stained with a protein stain in the usual way. The technique of crossed immuno-electrophoresis can be performed directly on cellulose acetate which is first wetted with buffer containing the antibody. The sample antigens are then electrophoretically migrated through it and form precipitin arcs in the usual way (Schanbacher and Smith 1974).

Cellulose acetate separation of seed storage proteins has been combined successfully (Blagrove and Frenkel 1978) with SDS–PAGE in a simple high-resolution two-dimensional procedure.

The separation is mainly according to charge differences in the relatively non-sieving cellulose acetate medium (in a 0·05 M sodium phosphate buffer pH 7·0) and is followed by placing a narrow strip of the cellulose acetate across the top of a Gradipore gradient gel which has been subjected to a pre-electrophoresis for 2 h with 0·025 M phosphate buffer pH 7·0 containing 0·15 per cent SDS. After applying the strip, SDS–phosphate buffer is added to the electrode vessels and electrophoresis for separation according to size differences in the second dimension is begun at once. With this procedure embedding the cellulose acetate in quick-gelling polyacrylamide gel on top of the slab, which has been advocated by some authors, is unnecessary.

(c) *Iso-electric focusing on cellulose acetate*

Conventional cellulose acetate membranes are not satisfactory for separations by IEF. This is probably due to the rather high electro-endosmosis that occurs. However, Harada (1975) has successfully used a special cellulose acetate strip (100 mm × 100 mm) containing an undefined surface active agent and commercially available from Fuji Photo Film Co. Ltd.

The strips are soaked in 4 per cent ampholytes containing 10 wt per cent sucrose for 5 min, briefly blotted free of surplus solution, and placed on to a glass plate on a flat-bed electrophoresis apparatus (e.g. LKB-Multiphor, Pharmacia FBE-3000, etc.), care being taken that no air bubbles are trapped between the strip and the glass. Samples absorbed on to small squares of filter paper are applied by simply placing these towards one end of the cellulose acetate strip. Before commencing the run the cathode end of the strip is moistened with a piece (5 mm × 100 mm) of filter paper soaked in 0·5 vol per cent ethylenediamine solution. A similar piece soaked in 0·5 vol per cent orthophosophoric acid is applied to the anode end and a constant voltage of 800–1000 V for 1·5–2 h is then used for the IEF separation. Needless to say care must be taken that the cellulose acetate strip does not dry out during any of these operations.

The method has also been used by Farrell, MacMartin, and Clark (1978) to examine multiple molecular forms of a number of lysosomal hydrolytic enzymes, deficiencies of which cause lethal storage diseases in infants and children.

Although clearly of practical value the inclusion of surfactants as described above is not without its hazards, and without thorough investigation one cannot be sure whether some of the heterogeneity observed is due to reaction between the sample constituents and the surfactant. A more theoretically sound approach has been taken in a series of papers from Ambler (1978 *a,b,c*) and Ambler and Walker (1979).

The cellulose acetate strips (Cellogel) in batches of 25 are washed twice in 100 ml methanol and then placed in 150 ml boron trifluoride (5·0 per cent) in methanol to methylate the free carboxyl groups. The strips are shaken at room temperature for 10 min (300 μm thick strips) or 30 min (500 μm strips) and then incubated at 45 °C for 50–60 min. After a further 10 min at room temperature the strips are washed three times in 100 ml methanol to remove excess boron trifluoride reagent and then stored in methanol until required. The methylated strips are transferred directly without drying, from the methanol into the desired pH range of ampholyte solution, containing 7–8 per cent ampholytes in 4 vol per cent glycerol solution, and allowed to equilibrate for at least 2 h. The strips are then placed in an electrophoresis tank of the type used for paper or cellulose acetate electrophoresis (e.g. Fig. 12.7), care being taken that the strips are placed non-porous side down and also that no air bubbles are trapped beneath the strip (unroll gently from one end). Wicks for electrical contact are made from strips of methylated Cellogel placed non-porous side down on top of ordinary untreated strips of Cellogel. The anode buffer used for wetting the anode wicks and in the anode electrode vessel is 0·2 M acetic acid or citric acid, and the cathodic solution is 0·2 M ethanolamine or lysine (free base). Samples of 0·8 μl are applied directly to the surface of the strip with the aid of a micro-applicator. The IEF separation is run at 500–1200 V for 1–2 h. For protein detection a brief wash in 70 vol per cent methanol removes the bulk of the ampholytes before staining with Coomassie Blue R-250. Servalyts (Serva

Feinbiochemica) were the preferred ampholytes for this work, because other commercially-available ampholytes were found to have some tendency to adhere to the membranes and give increased background colouration on staining (Ambler 1983). The use of marker proteins of known iso-electric point is particularly helpful both to determine the progress of the focusing and also to give an approximate idea of the pI of the unknown proteins. The run should be terminated as soon as samples have focused to sharp zones, because there is still some residual electroendosmosis that can cause the gradient to drift, and components with basic pI values (> 8.5) can be lost if the run is prolonged.

(d) *Applications of cellulose acetate in electrophoresis*

As a support medium for electrophoresis or IEF, cellulose acetate strips possess a number of advantages over polyacrylamide or starch gels. Most important amongst these must be the ease of handling and the minimum of pretreatment required. The relatively skilled step of preparing a gel is eliminated and all that is needed is a simple wetting of the strip with buffer, which also results in a considerable saving of time. Since cellulose acetate is prepared commercially in large batches under well-defined conditions the reproducibility in terms of separation properties from one strip to the next is excellent, far better than is generally achieved with gels. Owing to the thinness of the strip sample sizes are small, which is useful if quantities are limited, and staining, destaining, or other detection methods are very rapid. Quantitative measurement and localization of radioactively labelled material by scintillation counting and autoradiography is again particularly straightforward and reliable. In IEF work the amounts of expensive ampholytes needed are smaller than with other supporting media and the migration of high-molecular-weight proteins is not hindered by a molecular sieving effect. Cellulose acetate separations can easily be coupled with immunochemical detection methods and are applicable in most two-dimensional immuno-electrophoretic procedures. Finally only simple and inexpensive apparatus is required.

A number of these advantages are shared by electrophoresis on paper, but as has been discussed above resolution of macromolecules such as proteins and enzymes on paper is poor. Polyacrylamide gel is the medium capable of giving the highest degree of resolution, particularly when multiphasic buffer systems are used. Such buffers are not readily utilized with cellulose acetate strips, but in spite of this the resolution can be perfectly adequate for many purposes. Because of the advantages listed above the use of cellulose acetate is very popular in busy clinical laboratories and for routine screening work where the ease of handling, commercial availability of ready-to-use material, and rapidity are attributes which are particularly appreciated.

12.7. Zone electrophoresis in free solution: density gradients and Sephadex columns

Electrophoresis in the absence of any inert supporting supporting material is accompanied by considerable diffusion and is nowadays virtually confined to separations of charged particles, such as intact cells, colloidal particles, etc. The principles and instrumentation of analytical electrophoresis of cells by various methods in free solution have been briefly reviewed by Catsimpoolas (1980) and in several articles in a book edited by Righetti, van Oss, and Vanderhoff (1979). It can be performed in a closed apparatus in which the solvent phase is stationary or in a stream of moving solvent, in which case it can be referred to as a free-flow electrophoresis. However, the most widely exploited form of free-solution electrophoresis requires the use of a stationary solvent phase with a density gradient. As well as greatly reducing mixing by convection this also results in reduced diffusion owing to the increased viscosity of the medium. This approach is also particularly well suited to studies of particulate samples, but is only occasionally used to separate soluble substances. Electrophoresis on Sephadex columns is included in this section because in most cases it is arranged that the sample constituents are totally excluded from the Sephadex beads so that size fractionation does not occur and the sample is essentially subjected to free-solution electrophoresis in the solvent phase outside the beads.

(a) *Free-solution electrophoresis in a stationary solvent phase*

This is a highly specialized procedure for measurement of the mobilities and the zeta potentials of charged particles. It requires the use of relatively sophisticated apparatus dedicated to this one technique. Its use has been reviewed by Seaman (1975) and Catsimpoolas (1980). Commercially available equipment (such as the analytical particle electrophoresis apparatus from Rank Brothers, Bottisham, Cambridge, England, or the cytopherometer from Zeiss, Oberkochen, F.R.G.) provides a means of studying the movement of individual particles over a very small distance and measuring their displacements with the aid of a microscope or by closed-circuit television. Samples such as cell suspensions are either used to fill the sample chamber completely at the start of the experiment, or a few microlitres of sample suspension can be added to part of the chamber with a microsyringe. In either case a voltage is then applied across the chamber and the time measured for individual cells to travel a predetermined distance such as across a series of graticule lines within the microscope field of view. A large number of particles are observed and the results evaluated statistically. Electro-endosmotic flow of the solvent must be corrected for unless the system is a closed one in which there is a compensating backflow in some part of the apparatus (e.g. in the centre of a capillary tube). When this can be arranged it may then be possible to perform the measurements at a point where the two opposing flows match giving zero net flow (e.g. Blume, Malley, Knox, and Seaman 1978).

Apparatus of this type is often used to study processes which influence cell surface charge, such as for example the binding of lectins to murine T-cells (Blume *et al.* 1978) or treatment of cells with trypsin, neuraminidase, maleic anhydride, etc. (Ballard, Roberts, and Dickinson 1979).

A conceptually similar type of particle analysis apparatus is used for laser Doppler electrophoresis (Smith and Ware 1978). Again it is a highly specialized technique requiring dedicated apparatus, such as the Malvern-Rank Particle Charge system (Malvern Instruments Ltd., Malvern, Worcestershire, UK), in which light from a laser shining on particles moving under the influence of an electric field has its frequency Doppler-shifted. Although this effect is very small if the light beam is then combined with another part of the beam that has not been shifted in this way a beat frequency results which can then be analysed and curve-fitted by computer to give a graphical representation of the electrophoretic mobilities of the particles. The method is fast and can measure mobilities of several different types of particle of differing mobility in a single sample, but since it relies upon measuring scattered light it requires turbid solutions so cannot operate on non-turbid solutions of small molecules.

A completely different type of apparatus for free-zone electrophoresis has been described by Hjertén (1970) and is a great deal more versatile. It can be used for separating low-molecular-weight samples, e.g. simple organic and inorganic salts, amino acids, purines and pyrimidines, and nucleosides and nucleotides, and high-molecular-weight substances such as proteins and nucleic acids as well as particulate samples such as viruses, bacteria, erythrocytes, and other cells. The separation chamber consists of a horizontal quartz tube 36 cm long which is rotated about its long axis to counteract gravitational effects and to avoid the requirement for a supporting matrix. Electro-endosmosis is avoided by pre-coating the inside wall of the tube with cross-linked methyl cellulose (Methocel MC (viscosity 7000 cp), Dow Chemical Co., Michigan, U.S.A.). Separated zones are observed with a u.v. detection system consisting of a beam of u.v. light passing across the tube which is arranged so that the zones migrate through the beam once they have traversed a fixed distance. At constant voltage the identification of zones is therefore on a time basis. Typical amounts of sample are in the range 5–100 μg and the apparatus is designed so that separated zones can be recovered with a microsyringe. It can therefore be used as a very-small-scale preparative apparatus as well as an analytical apparatus. Substances which do not absorb u.v. light can be detected by performing the electrophoresis in a buffer which does absorb it. Migrating zones obey the Kohlrausch conductivity regulating function, and hence the presence of a non-absorbing substance in a zone is compensated for by a reduction in the concentration of the carrier buffer so that a peak of diminished absorbance is seen. Generally similar concepts, equipment and experiments have been described by Mikkers, Everaerts, and Verheggen (1979) and by Jorgensen and Lukacs (1981).

(b) *Free-flow electrophoresis*

This is a preparative procedure for the separation of charged particles on a relatively large scale. Although it can be used analytically and separation times can be as short as 30 s (Hannig 1978), the resolution of soluble substances (e.g. proteins) is very poor by comparison with other electrophoretic methods. When applied preparatively to soluble samples, better results can be obtained just as easily and more cheaply by ion-exchange chromatography which can also be scaled up for much larger amounts of sample if needed. It appears that now we are able to perform electrophoresis in space, where owing to lack of gravity the problems of convection and sample sedimentation are no longer with us, this technique will be capable of giving much better results (Giannovario, Griffin, and Gray 1978; Vanderhoff and van Oss 1979)! Meanwhile, and for those of us who will remain resolutely earthbound, its use is confined to the separation of cells and particles for which better methods do not exist or are not convenient. A major advantage in cell-separation work is that retention of cell viability is very good (e.g. Plank *et al.* 1983). Smolka, Margel, Nerren, and Rembaum (1980) have described the separation of human erythrocytes by this method. This work was extended by the introduction of polyacrolein microspheres (Kempner, Smolka, and Rembaum 1982) which can be coated with antibody and bound to cells, resulting in alterations in their electrophoretic mobility. This enables mixed populations of cells, such as human B and T lymphocytes, with similar native mobilities to be separated (providing suitable antibodies are available to coat the micro-spheres) both analytically and preparatively (Smolka, Kempner, and Rembaum 1982).

The apparatus consists of a separation chamber, typically about 500 mm long, 100 mm wide, and 0·5–1·0 mm thick, through which a film of buffer is pumped at a constant rate. At the same time an electric potential of for example 100–150 Vcm^{-1} is applied across the chamber perpendicular to the direction of buffer flow. The sample is applied through a port close to the start of the chamber, and as it is carried through individual components are deflected across the chamber, the most highly charged being displaced further than those with lower charges. Thus if sample is continuously pumped into one end of the chamber, components migrate along a series of diagonal lines and are collected at the far end into different fractions. Although the method is simple in basic concept, the apparatus is quite complex and expensive and the technique is highly specialized. Many factors can influence the separation achieved and they are well covered in brief reviews by Hannig (1978; 1982) and a more comprehensive review by Zeiller, Löser, Pascher, and Hannig (1975). These reviews also cover a number of applications, particularly to the separation of cells. Theoretical aspects have been discussed by Hannig, Wirth, Meyer, and Zeiller (1975).

A particularly novel design of apparatus where the solution in a rotating separation chamber is stabilized by a velocity gradient has been described by Thomson (1983). The apparatus is designed for continuous operation with soluble samples and residence times are short (typically ~ 1 min) so throughput is high and as much as 100 g protein can be separated per hour. Resolution is not very good however, again certainly no better than that obtainable by ion-exchange chromatography with far cheaper apparatus. The equipment could also be used for separation of mixtures of intact cells and subcellular organelles (Thomson 1983).

Several continuous-flow designs of apparatus for the preparative separation of proteins by isoelectric focusing have been described (e.g. Jonsson & Rilbe 1980; Basset, Froissart, Vincendon, and Massarelli 1980; Fawcett, 1983) and have similar advantages to free-flow electrophoresis designs in terms of high throughput of sample but also severe disadvantages in terms of resolution when compared to the smaller scale separations described in Chapter 8.

(c) *Zone electrophoresis in density gradients*

This is also primarily a preparative method for cells, viruses, and other particulate samples, but it has been used for soluble materials of much lower molecular weight and also for purely analytical work, although for both these types of application other electrophoretic methods are usually preferable.

If zone electrophoresis is performed in a simple buffer solution there is a considerable loss in resolution owing to diffusion and sedimentation of the separated zones. The influence of both these factors is greatly reduced by carrying out the electrophoresis in a density gradient. Diffusion is dependent on viscosity, so the most advantageous situation is achieved by preparing the gradient with the aid of inert non-conducting molecules which also increase the viscosity as much as conveniently possible. Sucrose is by far the most frequently used substance for making up density gradients, but glycerol, mannitol, etc. are also suitable. The density gradient need not be accurately linear so long as it is adequate to prevent convection currents, and any of the methods described in Section 9.2(h) for preparing density gradients for IEF columns is perfectly suitable for zone electrophoresis work. Major advantages of the technique in preparative work are that collection of the separated zones is very simple (e.g. with a syringe, a pump, or by draining into a fraction collector), recovery is nearly always quantitative, and it is generally very easy to separate the sample components from the sucrose-containing supporting medium by dialysis, dilution and centrifuging, precipitation, etc.

Zone electrophoresis in density gradients is almost always performed in vertical columns. These can range in size from small tubes similar to those used for gel rods in PAGE (e.g. Korant and Lonberg-Holm 1974; Eisenbach and Eisenbach 1979) to relatively large columns capable of handling about 100 mg, such as the Uniphor system (LKB Produkter), and to even larger columns capable of separating 1 g or more of sample (e.g. Walters and Bont 1979). A typical sucrose gradient may be from about 10 per cent to about 50 per cent, but much shallower gradients may be beneficial for certain applications such as separations of intact cells. Blad-Holmberg (1979) for example used an approximately 8–14 per cent gradient in studies of the surface charges of membranes and subcellular particles and the way in which these charges were modified by treatment with various enzymes. The steeper the density gradient the greater its stabilizing power against convection and gravitational sedimentation of sample zones and hence the greater the sample load-carrying capacity. When homogeneous buffer systems are used throughout, buffers such as those described in Chapter 2 (Table 2.1) are suitable. Once the density gradients have been prepared the samples, which are often made 5–10 per cent in sucrose, are simply added directly to the top of the gradient with the aid of a syringe or a Pasteur pipette and are then in turn overlaid with electrode buffer. When multiphasic buffers (Chapter 3) are used the sample should be made up in the buffer used in the electrode chambers (e.g. tris–glycinate) and not in the buffer used for the gradient-making solutions (e.g. tris–HCl). Electrophoretic conditions vary with the sample, buffers, and columns used, but with a column such as the Uniphor may typically be about 200 V for 16–18 h.

When small columns are used both the setting-up times and electrophoresis times are shorter. In the most simple procedure glass tubes 5–6 mm in internal diameter and 100–200 mm in length are sealed at one end with a piece of dialysis tubing secured with a tight-fitting ring cut from a suitable length of polythene or polyvinyl chloride (PVC) tubing. The tubes are then placed in the same sort of apparatus as used for PAGE gel rods (Fig. 2.3), sealed-end downwards, and filled to within about 10 mm of the top with the sucrose gradient. The sample is applied with a microsyringe and overlaid with upper electrode buffer. The same buffer is used in both upper and lower chambers but sucrose should be added to the lower chamber buffer in the same proportion as that in the most dense part of the gradient. In their work on the separation of proteins and virus particles in tubes 200 mm long Korant and Lonberg-Holm (1974) applied a constant current of 4 mA per gradient and electrophoresis was complete in 3·5 h, by which time a marker of phenol red had migrated the whole length of the gradient.

Separation times are further reduced when shorter tubes are employed and Eisenbach and Eisenbach (1979) separated membrane fragments at 100 V constant voltage for only 2–3 h in 100 mm tubes. Short separation times are clearly one benefit of small-scale experiments, and this can be an important factor in the preservation of the viability of cells or the retention of enzymatic or biological activity. A further advantage of this scale of operation is that it is a simple matter to run 10, 12, or more identical or different samples at the same time in the same apparatus.

A recent development has been the introduction of electrophoresis in iso-osmolar density gradients, first reported by Boltz, Todd, Streibel, and Louie (1973) but subsequently extensively used and further developed by Catsimpoolas and co-workers (e.g. Catsimpoolas and Griffith 1977). Such gradients should be particularly beneficial in retaining cell viability (e.g. Plank *et al.* 1983; Omenyi and Snyder 1983).

They are achieved by preparing a 2·5–6·25 per cent gradient of Ficoll™ (400 000 MW, Pharmacia Fine Chemicals AB) which is also an inverse 6·35–5·72 per cent gradient of sucrose in a buffer of close to physiological pH. This buffer, of pH 7·4 and ionic strength 0·023 M, is composed of 0·20 g KCl, 1·5 g Na_2HPO_4, 0·02 g KH_2PO_4, 0·12 g sodium acetate, and 10·0 g glucose in 1 l distilled water. Before use, 25 g of Ficoll should be dissolved in 200 ml of this buffer and dialysed overnight at 4 °C against at least 5 l of the same buffer. The pre-treatment reduces subsequent clumping of the sample cells. The dialysed Ficoll solution is then made up into a concentrated (10 per cent) stock solution which is used to make the gradient solutions as required.

(d) *Zone electrophoresis in Sephadex™ columns*

Like the other methods already described in this section, zone electrophoresis in columns of Sephadex can be applied to samples in true solution, but unlike them it is not really suitable for particulate materials owing to the filtering effect of the Sephadex bed. While this can be controlled to some extent by varying the grade of Sephadex (superfine grades will naturally exert a much greater filtering effect than coarse grades), the mobility of all but the smallest colloidal particles is likely to be seriously hindered.

The method is used exclusively in preparative work as the amount of band spreading due to diffusion, adsorption effects, electro-endosmosis and to the filtering process all conspire to make for relatively low resolution. The resolution is in fact no better and often not as good as can be achieved by gel filtration, but since the basis by which separation occurs is charge rather than size, in contrast to gel filtration, the method does have some value.

Apparatus of the same type as that used for zone electrophoresis in free solution can usually be used for this method also because the only real difference between the two procedures is that in one case a sucrose gradient is used to prevent convection and reduce diffusion, whereas in the other Sephadex beads serve to prevent convection (but not to reduce diffusion since the buffer viscosity is not increased). The same homogeneous or multiphasic buffers can be used in both methods. A very simple design of apparatus is shown in Fig. 12.8.

It consists of a glass or plastic column 300 mm long and 18 mm in internal diameter with a porous polyethylene disc cemented onto the lower end. A 50–70

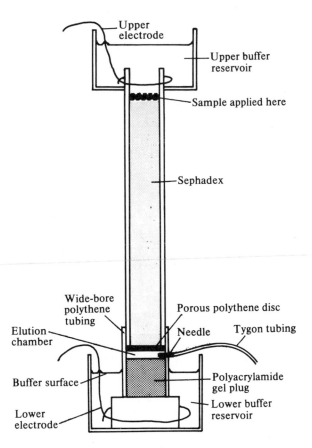

FIG. 12.8. Apparatus for electrophoresis in columns of Sephadex. The same apparatus can also be used with a (sucrose) density gradient in place of the Sephadex column.

mm long piece of thick polythene tubing of larger diameter is placed over this end and a short length of sawn-off syringe needle is pushed through it just below the porous disc (Fig. 12.8). The upper electrode vessel is made from the bottom half of a plastic reagent bottle in which a hole just large enough to push the column through is cut. The column is cemented in place making sure that there is no leakage around the column itself.

For setting up, the bottom of the column is immersed in a 50 ml beaker containing a 7–10 per cent acrylamide solution (e.g. $T = 7·5$ per cent, $C = 5$ per cent) in buffer to which TEMED and ammonium persulphate have just been added. The top of the column is quickly closed with a rubber bung and air is withdrawn through the sawn-off needle with a syringe attached to a length of Tygon tubing. Sufficient air is withdrawn to draw the acrylamide up to within about 2 mm of the porous disc but taking care not to cover the end of the needle. The acrylamide polymerizes forming a gel plug. The column is carefully removed from the beaker, taking care not to disturb the plug unduly, excess polyacrylamide is cut off, and the column is then placed in a larger beaker (600 ml) of buffer containing one electrode and a sponge pad as a mechanical support for the polyacrylamide plug (Fig. 12.8). (N.B. It is important that the buffer in the beaker does *not* cover the needle otherwise a short circuit is formed, the needle becomes hot, and burning may occur around the needle causing a leak.) The column is then packed with Sephadex in the usual way and excess solvent flows to waste out of the tubing attached to the needle below the porous disc. Finally the top of the column above the Sephadex bed and the upper electrode chamber are filled with buffer. When multi-phasic buffers are used the sample and electrode buffers are of course of a different composition to that in the column. The sample, containing 2–5 per cent sucrose, is layered on to the column with the syringe and the upper electrode is placed in position. The electrodes are made from loops of platinum wire encircling the ends of the column. Electrophoresis is typically performed at 50–70 V overnight but higher voltages and shorter times, e.g. 300 V for 2–3 h, can also be used. After the electrophoresis the sample is allowed to drain out through the needle below the porous disc and flow via a u.v. monitor into a fraction collector.

Columns of this general type can be set up with Sephadex G-25 or G-15 so that all sample constituents are totally excluded from the beads and separation is entirely according to charge, as for example described by Whitehead, Kay, Lew, and Shannon (1971). Alternatively some gel filtration effect, and hence size separation, can also be introduced during the elution stage if for example Sephadex G-200 is used (Tedesco, Bonow, and Mellman 1972). A very similar arrangement has been used for the preparative separation of proteins by IEF in a Sephadex column (Wahrmann, Gros, Piau, and Schapira 1980), the G-15 column being set up in 10 per cent glycerol, 2 per cent ampholytes, and 1 per cent Lubrol PX, and run at about 1000 V for 7 h. However, the use of horizontal granular gel beds (see Section 9.5) is probably preferable for IEF work in most cases.

12.8.   Affinity electrophoresis

Affinity electrophoresis (or affino-electrophoresis) is a technique whereby sample constituents, usually proteins, are separated not only by virtue of mobility differences under the influence of an applied electric field, but also at the same time by interactions between the sample and specific affinity ligands. The specific case of glycoproteins separated with matrices containing lectins has already been discussed in Section 8.7, but the general concept is more widely applicable and can theoretically cover almost any case where a specific binding between two different types of molecule occurs. It could take the form of electrophoresis of a protein in the presence of a small ligand (e.g. lectins in the presence of various sugars, enzymes in the presence of inhibitors or substrates, serum albumin in the presence of fatty acids, etc.), so long as the size, shape, or charge of the complex differs from that of the uncomplexed protein and results in a change in electrophoretic mobility. However, more usually the ligand is immobilized within the gel matrix. In either case the strength of the affinity binding can be estimated from the proportions of complexed and free protein when the electrophoresis is performed with different concentrations of ligand. The theory of measurements of the dissociation constants of affinity interactions in this way has been presented by Takeo and Nakamura (1972) and by Hořejší (1979). It has been applied by various workers to such studies as the interaction between phosphorylase and glycogen, α-amylase and starch, concanavalin A and sugars, and dextran-specific myeloma protein and dextrans. The theory, calculation of dissociation constants, factors influencing affinity patterns (e.g. buffers, temperature, ligand concentrations, etc.) and analytical and preparative applications have been reviewed by Hořejší (1981), Hořejší and Tichá (1981b), Matoušek and Hořejší (1982), and by Takeo (1984).

In their method Hořejší et al. (1977a,b) and Hořejší (1979) prepare a macromolecular derivative of the ligand to be studied (e.g. coupled to a high molecular weight dextran or soluble O-glycosyl polyacrylamide copolymers etc) which can act as the immobilized ligand, and this is added to the acrylamide mixture before polymerization. After polymerization this macromolecular ligand is physically trapped in the gel matrix and immobilized. However, in addition to this immobilized ligand they also include in the gel a free ligand which may be different from the immobilized one. Interaction of the migrating sample protein molecules with the free ligand reduces the retardation caused by interactions between protein molecules and the immobilized ligand. Thus from measurements of the mobility of the protein zone at different concentrations of both the immobilized and free ligands, the dissociation constants for both the protein-free ligand and the protein-immobilized ligand complexes can be calculated.

Consider a protein zone which moves a distance x in the control gel in time t and which in the same time moves a distance x' in the affinity gel containing both free and immobilized ligand. If it can be assumed that (Hořejší 1979) (1) the protein concentration in terms of molarity is much lower than that of the immobilized or free ligands, (2) the complex formed between the protein and the free ligand has the same mobility as that of the free protein, (3) the immobilized ligand is completely immobile so that the complex formed between it and the protein has zero mobility, and (4) complex formation is rapid compared with the movement of the protein zone through the gel, then the time t_1 that the protein is in the free state or complexed with free ligand is given by

$$t_1 = tx'/x \tag{12.1}$$

The time t_2 spent as a complex with immobilized ligand is

$$t_2 = y(x-x')/x \tag{12.2}$$

Thus at any particular moment

$$\frac{[P] + [PL_f]}{[PL_i]} = \frac{t_1}{t_2} = \frac{x'}{x-x'} \tag{12.3}$$

and

$$K_i = \frac{[P][L_i]}{[PL_i]} \qquad\qquad K = \frac{[P][L_f]}{[PL_f]} \tag{12.4}$$

where $[P]$, $[L_i]$ and $[L_f]$ are the concentrations of free protein, uncomplexed immobilized ligand, and free ligand respectively in the moving zone, and $[PL_i]$ and $[PL_f]$ are the concentrations of the complexes btween protein and immobilized ligand and free ligand respectively. K and K_i are the dissociation constants for the protein-free ligand complex and the protein-immobilized ligand complex. If assumption (1) is true, then $[L_i]$ is approximately equal to the total concentration $[L_I]$ of immobilized ligand in the affinity gel and $[L_f]$ is approximately equal to the total concentration $[L_F]$ of free (low-molecular-weight) ligand in the affinity gel. Thus eqn (12.4) can be written as

$$K_i = \frac{[P][L_I]}{[PL_i]} \qquad\qquad K = \frac{[P][L_F]}{[PL_f]} \tag{12.5}$$

Substituting $[PL_i]$ and $[PL_f]$ from eqn (12.5) into eqn (12.3) gives

$$\frac{x'}{x-x'} = \frac{K_i}{[L_I]}\frac{(1 + [L_F])}{K} \tag{12.6}$$

This equation is the simple basic equation used by Hořejší et al. (1977*a*) for the calculation of dissociation constants from affinity electrophoresis when the four assumptions indicated above are justified. While the errors implicit in these assumptions are usually small and the results obtained are reasonably accurate, there are other cases where one or more of these assumptions do not hold. Hořejší (1979) also derives equations for calculating dissociation constants under more general conditions and discusses the experimental consequences that follow from them.

A development has been the extension of affinity electrophoresis to a study of hydrophobic bonding with hydrophobic ligands incorporated into the gel matrix (Nakamura, Kuwarhara, and Takeo 1979). Just as the electrophoretic mobility of phosphorylase was retarded by ligand binding when glycogen was entrapped within the gel (Takeo and Nakamura 1972), so too a number of different phosphorylases were retarded by hydrophobic interactions when synthetic alkyl dextrans were immobilized within the gel. In a similar way this was used to determine the dissociation constants of the hydrophobic bonding involved. Incorporation of hydrophobic acrylamide derivatives in polyacrylamide gels to give media of varying hydrophobicity has also been used (Chen and Morawetz 1981).

The procedure for carrying out affinity electrophoresis with an immobilized ligand is no different from the conventional forms of PAGE described in Chapters 2 and 3 except that a small proportion of the affinity ligand is added to the solutions used for making up the polyacrylamide gel. When gelation occurs the ligand is physically trapped within the gel. When small ligands are used which would not be trapped in this way, they should either be uncharged so that they do not migrate out of the gel during the electrophoresis (e.g. sugars) or else they should also be included in the electrode buffers to that concentrations within the gel remain at a constant level. Polyacrylamide gels have been employed most frequently but agarose has also been used. For example, Caron, Faure, Keros, and Cornillot (1977) bound γ-globulin to Sepharose 4B by treatment with cyanogen bromide and then incorporated the resulting immuno-adsorbent into a 1·2 per cent agarose gel to study the equilibrium between human serum albumin and its immobilized antibody by affinity electrophoresis.

Hořejší *et al.* (1982) also discussed a number of methods suitable for immobilising ligands to prepare the affinity matrix, this being the most crucial part of the affinity electrophoresis process. The coupling of ligands to beaded agarose (Sepharose) by various techniques originally developed for affinity chromatography followed by their incorporation into the gel matrix is a good general procedure. It overcomes the difficulties associated with synthesizing polymerizable derivatives (e.g. allyl derivatives) of ligands, or with derivatizing a large carrier macromolecule (protein or polysaccharide) that can be easily immobilized by entrapping in the gel;

both of which procedures are often accompanied by undesirable side reactions. They preferred treatment with periodate for activation of Sepharose for coupling as this gave rise to fewer cross-links within the Sepharose particles themselves than other activation methods and did not therefore raise the melting point of the resulting ligand complexes. For preparation of these complexes Hořejší et al. (1980) therefore used the following procedure.

Washed Sepharose 4B (5 ml) was added to 5 ml of 0·5 M NaIO$_4$ solution and the suspension agitated for 2 h at room temperature. The activated Sepharose beads were washed thoroughly with water and could then be stored at 4 °C for a 'long time' if required. For coupling, 2 ml of this oxidized gel was washed thoroughly with 0·5 M sodium phosphate buffer pH 6·0 and then added to 2 ml of the same buffer containing 20 mM ligand and 20 mM NaCNBH$_3$. The mixture was agitated for 24 h at room temperature, washed thoroughly with 0·1 M tris-HCl buffer pH 8·2 and the ligand-Sepharose complex can then be stored at 4 °C in this same buffer containing 0·1 per cent NaN$_3$ as preservative until required. Magnetic stirring bars may fragment Sepharose particles excessively so other methods of gentle stirring, or agitation with shaking devices, are preferable.

Using affinity electrophoresis methods Hořejší et al. (1980) separated not only glycoproteins, lectins, and enzymes but also double-stranded phage DNA fragments, using acriflavine as the immobilized ligand. When coupling this to Sepharose by the above procedure, Hořejší et al. (1982) found that it was best if the initial coupling was done without NaCNBH$_3$. After the stage of washing with tris-HCl buffer NaCNBH$_3$ was then added to 10 mM and the mixture agitated for a further 24 h before rewashing with tris-HCl and storing as above.

Johnson, Metcalf, and Dean (1980) used a very similar approach but in this case immobilized the ligand (Cibacron Blue F3GA dye) to Sepharose 6B which was then melted in a boiling water bath and added to the solutions used for making polyacrylamide gels. The resulting composite gels of 4 per cent agarose and $T = 5$ per cent polyacrylamide were used for analysing proteins with affinity for this dye. This same approach was also used by Hořejší et al. (1982) who stated that melting of the ligand–agarose complex and mixing with gel solutions give rise to a very homogeneous affinity matrix. If warming should be avoided, it is also entirely satisfactory to mix the ligand coupled to beaded gel materials (Sepharose, Sephadex, Biogel, etc.) directly into the polyacrylamide gel-forming solutions without prior warming, although this naturally gives a much less homogeneous distribution of immobilized ligand.

Just as affinity electrophoresis may be regarded as the affinity analogue of PAGE, so IEF has its counterpart in affinity isoelectric focusing (Hořejší and Tichá 1981a), for which a major requirement is that the gels used should have very low electroendosmosis. Provided this condition is met IEF in gels containing immobilized ligands results in a shift in the position of the bands of the proteins which interact with the ligand. The technique is a qualitative one that can be used both for detecting ligand-binding proteins and, after purification, for confirming their homogeneity in terms of ligand-binding capability.

For this technique Hořcjší *et al.* (1982) used gels consisting of either 2 per cent IEF agarose or $T = 5\cdot27$ per cent, $C = 5$ per cent polyacrylamide containing 2 per cent Ampholytes (pH range 3·5–10), and less than 0·5 per cent ligand-Sepharose 4B complex (more than this giving rise to unacceptable electroendosmosis).

The apparatus used for affinity electrophoresis and affinity isoelectric focusing, details of technique, staining, quantitative measurement, recovery of separated zones, etc. are all otherwise identical to those described earlier for non-affinity methods, and most of the homogeneous or multiphasic buffer systems should also prove suitable.

12.9. Electrophoretic desorption in affinity chromatography and in immuno-adsorption

Recovery of samples bound to affinity matrices is sometimes a difficult feature of experiments involving affinity chromatography, affinity adsorption, or immuno-adsorption, particularly if competing ligands are not readily available or are very costly. Very high salt concentrations and the use of chaotropic agents such as detergents, sodium thiocyanate, urea, or guanidine hydrochloride have been used to overcome high-affinity interactions, but they may result in unacceptable products with total loss of biological or enzymic activity. Just as electrophoresis has been combined with biospecific interactions in procedures which we now term affinity electrophoresis, so electrophoresis can also be applied to the preparative elution of material from affinity matrices. It will overcome even the interactions between haptens and immuno-absorbents under conditions far more mild than the use of chaotropic reagents (Morgan, Brown, Leyland, and Dean 1978). The influence of a number of factors, such as desorption current, matrix thickness, buffer molarity, temperature, etc., have been discussed by Morgan, George, and Dean (1980).

As an analytical procedure (Morgan *et al.* 1978) the sample bound on to the affinity matrix is simply loaded directly on to the top of the stacking gel of a typical Ornstein and Davis type of PAGE system with multiphasic buffers (see Chapter 3) and subjected to electrophoresis in the usual way. The separated sample zones are detected by the usual methods of staining with quantitative measurement by gel scanning or by slicing into sections for scintillation counting if a radiolabelled protein is present.

For use as a preparative method, Morgan, Slater, and Dean (1979) designed a simple sample cell (Fig. 12.9) made out of Perspex. The volume of the central chamber can be adjusted by pushing the end sections together during assembly. The sample bound on to the affinity matrix is

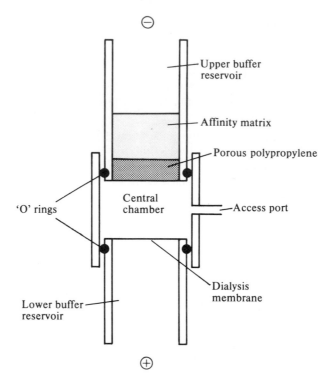

FIG. 12.9 Cell for the preparative electrophoretic desorption of material from affinity matrices and immuno-absorbents. (Reproduced from Morgan *et al.* (1979) by permission of the authors and publisher.)

loaded into the upper tube, the diameter of which should be such as to give the largest possible area of contact between the affinity material and its support. This upper supporting disc should be made of porous polypropylene as the diffusional characteristics appear to be important, and nylon mesh, for example was not so satisfactory. Desorption is performed at room temperature and under constant-current conditions with a polarity chosen so that the sample migrated into the central chamber. The kinetics of the desorption process can be examined by withdrawing samples at intervals through the access port with a microsyringe. Recoveries of protein of about 85 per cent are achieved with this device.

A new technique for electrophoretic desorption by flat-bed IEF has been devised (Vesterberg and Hansen 1978; Haff, Lasky, and Manrique 1979; Lasky and Manrique 1980). This method employs the horizontal granular-gel flat-bed arrangement described in Section 9.5 with apparatus such as the Pharmacia FBE-3000 or LKB Multiphor. The bed (in this case washed

Sephadex G-200 Superfine grade is used) is set up with ampholytes of wide pH range (e.g. pH 3–10 etc.) and focused into a pH gradient in the usual way. A narrow rectangular zone across the pH gradient is scooped out and replaced with affinity gel–sample complex. This zone should be at a part of the pH gradient several pH units away from the iso-electric point of the material to be desorbed. The power is then switched on again (the precise conditions needed will depend upon the affinity of the bound material for the immobilized ligand on the affinity gel matrix) and the bound sample is desorbed. It migrates to its iso-electric point and focuses to a sharp band. It is recovered in the same way as normal preparative flat-bed IEF samples (Section 9.5). Desorption is claimed to be rapid and very mild so that biological and enzymic activities are well preserved. The method is claimed to be flexible and with a high capacity. It has been applied to separating antibodies bound to immobilized antigens and *vice versa*, to proteins bound to immobilized lectins, to proteins bound to hydrophobic gels, and to the elution of proteins from agarose and polyacrylamide gels. It is powerful enough to separate even such strongly bound complexes as serum albumin and Blue Sepharose, haptoglobin and haemoglobin-Sepharose and trypsinogen and immobilized soyabean trypsin inhibitor.

12.10. Electrophoretic concentration

The very fact that charged macromolecules in solution move under the influence of an electric field enables them to be concentrated from regions of unhindered mobility into small areas where further movement is restricted. In its most simple form apparatus for carrying out such electrophoretic concentration may consist merely of a filter funnel with a dialysis bag tied over the stem. The arrangement shown in Fig. 12.10. was suggested by K. Graham of Universal Scientific Ltd.

The sample, in buffer and containing 2–10 per cent sucrose, is placed in the funnel, the lower end of which is immersed in buffer, also containing sucrose, in the lower electrode chamber. The upper electrode chamber is then filled with sucrose-free buffer, which is also layered over the sample solution, taking care to avoid excessive mixing. Homogeneous buffer systems are usually, but not necessarily, used and the pH and electrical polarity are chosen so that the component of interest in the sample migrated downwards towards the lower buffer chamber and is concentrated in the dialysis bag. If the current is temporarily switched off first, the progress of the concentration can be followed by withdrawing a portion of solution from the dialysis bag with a syringe for analysis, optical density measurement, etc.

A 50-fold concentration can easily be achieved in a few hours, and when the desired degree of concentration has been reached the dialysis bag can be tied off and the concentrated sample dialysed to remove sucrose and buffer salts. The method is particularly suited to relatively large volumes of

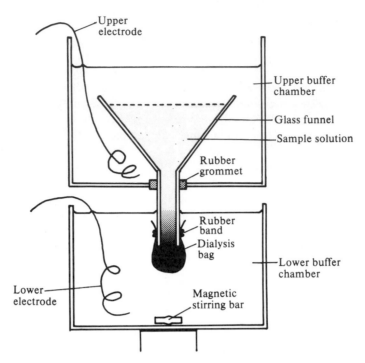

FIG. 12.10. Simple apparatus for electrophoretic concentration of dilute sample solutions. The shaded area represents the sample solution.

dilute samples, such as may be obtained during tissue extractions, but it can also be applied on a much smaller scale to fractions from chromatographic columns, ion-exchange separations, density gradients, preparative electrophoresis columns, etc.

A more sophisticated variation on this theme has recently been described by Wachslicht and Chrambach (1978). In their design the dialysis bag is replaced by a threaded Plexiglas sleeve glued to the bottom of the funnel with epoxy adhesive. This is screwed into a threaded cup at the base of which is a porous glass membrane (Corning No. 7930), thus making a permanent and re-usable apparatus. Another difference is that the funnel is partly filled with a polyacrylamide stacking gel ($T = 5$ per cent, $C = 15$ per cent) containing 3 per cent bacitracin to prevent non-specific binding of sample protein to the gel matrix and improve recoveries.

The sample is made up in stacking gel buffer (see Chapter 3) containing 10 per cent sucrose and a little Bromophenol Blue tracking dye. Electrophoresis is performed at 10 mA until the tracking dye, which marks the position of proteins concentrated within the stack, has migrated into the cup (4–7 h). After a further 0·5–1·0 h the current is turned off and the sample collected by unscrewing the cup.

Other devices have been described for electrophoretic concentration. One suggested by Posner (1976) utilized the electrodecantation phenomenon, and in his design a central vertical sample cell is flanked by an electrode chamber on either side. The three compartments are separated by walls that include a section of membrane, so that upon applying a voltage the macromolecules in the sample move towards the appropriate electrode and concentrate at the surface of the membrane. Owing to the resulting increase in density they sediment down into a small trough cut in the base of the sample cell. To avoid excessive heating and convection currents the applied voltage is kept low (e.g. 50–100 V). Samples of proteins, lipoproteins, and RNA (20–100 ml) can be concentrated up to about 50-fold in 6–12 h, with recoveries generally over 90 per cent.

A particularly convenient device for concentrating smaller volumes of dilute sample has been described by Allington et al. (1978).

In this the ends of an inverted U-shaped sample cup are sealed with membranes which are held in place by acrylic rings cut to fit snugly over the arms of the sample cup. The cup is then filled with sample (the top of the cup is open) and simply placed across two buffer-filled electrode chambers with one arm dipping into each chamber. On applying a voltage macromolecules migrate into one arm of the sample cup and can be collected with a syringe after disconnecting the power. Concentrations of up to 50–fold with recoveries approaching 100 per cent can be obtained in 1–3 h. Therefore the method is fast and samples are not subjected to harsh conditions, so biological and enzymic properties are well preserved. In addition no assembling or dismantling of apparatus is needed, no gels need to be prepared, and the device is ready for instant use. A number of units can be run side by side using the same electrode chambers.

The device is now commercially available from ISCO. It has also recently been used (Bhown et al. 1980) for the prepartive electrophoretic elution of proteins from SDS–PAGE gels. For this slices of gel containing the protein band of interest are chopped up and the pieces placed in the larger (sample) cup of the device, and the protein is recovered from the smaller cup with a syringe after performing electrophoresis as above.

13

CLINICAL APPLICATIONS

IN previous chapters we have seen how a wide variety of electrophoretic techniques can be used for the qualitative and quantitative analysis of macromolecules such as proteins, nucleic acids, etc. It therefore follows that any factor which causes a change in the macromolecular composition of a solution should be detectable by such techniques. In the field of clinical investigation there are two major areas in which electrophoretic separations are applied. Firstly, we have situations in which it is necessary to examine the change in one or at most a few known constituents. This most often occurs when the objective is primarily that of a test to confirm a clinical diagnosis, to distinguish between a small number of possible diagnoses, or to screen a number of individuals for a particular pathological condition or propensity to develop such a condition, i.e. to detect high-risk individuals. Nearly all clinical applications of electrophoresis fall within this general category at the present time. The second major area of application is at the moment only in its infancy and consists of the use of a macromolecular profile as the primary and perhaps sole diagnostic procedure for any unknown pathological condition. The ultimate goal must be the ability to prepare a macromolecular map from a drop of blood or a milligram or two of tissue and then to give a clinical diagnosis of whatever may be wrong.

Virtually all of the techniques described in the other chapters can be and have been applied to clinical problems, and a number of such uses have already been referred to where appropriate. However, one must be careful to distinguish between those methods which can be applied in the rarefied atmosphere of the research laboratory, where highly skilled staff are well versed in the use of perhaps very specialized equipment, and those methods which are well suited to routine use in the environment of a busy pathology laboratory.

An important aspect of the effective application of electrophoretic methods is the development of expertise in all aspects of the technique(s) being employed. This may sound rather obvious, but the interpretation of results can be quite demanding, particularly with the very-high resolution procedures, as many different factors may influence the final outcome. For example, with PAGE separations it is important to know how the relative mobilities of protein bands depend upon gel concentration T and cross-linking C and upon the pH and ionic strengths of the sample itself and of the buffers used; what the differences are when multiphasic buffers or homogeneous buffer systems are employed, how the composition of the stacking gel influences the separation of larger constituents, whether dissociating agents should be added to samples or gels, what is the influence of excessive salt in the sample, etc. The electrical conditions, running time, staining, and destaining will clearly affect the results, but perhaps less obviously there may be differences between running at a relatively high voltage for a short time and a lower voltage for a longer time owing to viscosity changes caused by temperature differences from the amount of ohmic heating. Considerable experience of both the technique and of how the sample behaves under differing conditions may be required to interpret the final band pattern correctly.

To add further complications, a considerable number of proteins occur in several different genetic forms which may differ in size or molecular charge or both; the composition of blood serum proteins for example differs even in a single individual at various stages throughout life. It is markedly affected by pregnancy but is also influenced to a lesser degree by other factors such as excessive exposure to heat or cold, stress, diet, etc. In those disease states that can be shown to produce changes in protein band patterns (e.g. of serum), both the appearance and the extent of any modification in the pattern will also depend upon the stage of the disease. Given all these possible qualifying factors one might be forgiven for concluding that little useful information could be gained, but fortunately this turns out to be grossly over-pessimistic and electrophoretic methods, including relatively demanding techniques such as two-dimensional mapping procedures, are of rapidly increasing importance to the clinical world in which they have already proved to be of immense value.

13.1. Sample handling

From the above comments it is clear that the single most important factor is the strict attention to reproducibility in technique. This of course includes the preparation and handling of samples, which should be examined as soon as possible after collection because it is known that protein band patterns of blood and tissue extracts change during storage, even at low temperatures. Part of this change can be due to physical processes such as aggregation, precipitation, etc., but of greater importance are changes caused by the action of enzymes such as proteinases, lipases, and other hydrolases which are always present in samples of this type. Storage should therefore be minimized, but if unavoidable samples should be refrigerated. For more prolonged storage deep-freezing or lyophilization may be necessary, but it must be remembered that both these processes promote physical changes such as protein aggregation (e.g. the aggregation of caseins in milk samples). In addition many enzymes are not stable to freezing and thawing, so that the samples are not then suitable for enzyme activity measurements or iso-enzyme analysis. Different forms of an enzyme activity (iso-enzymes) can have different stabilities to freezing and thawing, to heat treatment, to preservatives, to attack by endogenous proteinases, or just to length of storage. Some lactate dehydrogenase isoenzymes for example (McKenzie and Miller 1983) are not even stable to refrigeration temperatures, so in this case room temperature storage (for no more than 3 days) must be used. To avoid misleading results it is therefore always very important to ensure that the sample treatment is reproducible before analysis.

Allen (1978) recommends that blood serum should be collected as soon as the blood has clotted. Samples for assay of enzyme activities, clotting factors, and lipoproteins should be prepared from chilled blood and then stored at $-70\,°C$ if immediate examination is not possible. Samples intended only for lipoprotein analysis can be stored at $4\,°C$ for up to 48 h. Haemoglobin samples can be kept at $-20\,°C$ for 2–3 weeks after treatment with potassium cyanide and glycerol. Urine can be concentrated by freeze dialysis and stored at $-70\,°C$ (Allen 1978). Any samples of blood, urine, other body fluids, or tissue that are to be repeatedly examined over a prolonged period should be divided before storage at $-70\,°C$ into several portions which can be withdrawn as required in order to avoid repeated freezing and thawing of the whole sample.

13.2. Interpretation of the patterns of separated zone

Problems of variability in patterns arising from differences in the properties of nominally identical gels can be minimized by preparing gels in large batches, and some of the factors most relevant to the reproducible preparation of gels have been discussed by Maurer and Allen (1972). Slab gel systems are nearly always used, so that individual samples can easily be compared with each other and with standards run at the same time under identicial conditions. Providing they are suitably overlaid with buffer and are never permitted to dry out, after the initial 'maturing' period of about 24 h during which time traces of residual acrylamide gradually polymerize the properties of the gel remain almost constant for periods as long as a year or more. For accurate clinical work gel to gel variation can be still further minimized by the use of commercially available products with excellent standards of reproducibility, such as the polyacrylamide gradient gels from Universal Scientific (Gradipore) or Pharmacia, PAG-Plates for IEF from LKB and agarose plates (Universal Electrophoresis Film Agarose) from Corning Medical, Medfield MA 02052, U.S.A. or (Panagel) from Worthington Diagnostics, Freehold, N.J., U.S.A. The cost of these products is naturally greater than laboratory prepared counterparts, but this should be offset against the valuable saving in time and equipment for preparation.

With gels of good quality and reproducibility, optimum sample handling, and careful control of electrophoretic conditions, difficulties in interpretation are likely to be inherent in the samples themselves. For example, following PAGE analysis of blood serum the γ-globulins form a complex series of bands which tend to be rather diffuse and to form a general background in the first part of the gel. Upon this background are superimposed the bands due to a number of other components (e.g. α_2-macroglobulin, ß-lipoprotein, the haptoglobulin bands, etc.). Quantitative evaluation then becomes almost impossible and even qualitative assessment of the γ-globulins is difficult. A number of other overlaps between different proteins may occur (e.g. albumin and trypsin inhibitor, ceruloplasmin and haemopexin, etc.), and to obtain a complete and accurate picture of the composition of a blood sample more than one run is required using different separation conditions or even different methods, 2D methods often being particularly appropriate.

In order to be able to make meaningful comparisons on a day-to-day basis it is also necessary to have some reference system so that the bands obtained can be identified readily and accurately. A number of groups of workers have achieved this by taking the distance from the origin (or the start of the separation gel in discontinuous systems) to the centre of the albumin band as 100 and expressing the positions of all other bands as a percentage of this. The approach also has the benefit that the band patterns are then described in a digital form which may be helpful for subsequent computer analysis. Other teams have taken the leading edge of the albumin band or have used the transferrin band for the same purpose. An alternative approach is deliberately to add a known amount of a purified standard protein to the sample, preferably one with a clearly different mobility to components in the sample. An example of such an internal standard, which can be used not only as a qualitative reference point but also as a reference for quantitative measurements, is the addition of carbamylated human transferrin to samples for a very wide range of immuno-electrophoretic procedures (Axelsen, Krøll, and Weeke 1973).

13.3. Blood proteins

The classical separation of blood serum proteins by electrophoresis on paper gives at best five diffuse bands due to albumin, α_1-globulins, α_2-globulins, ß-globulins, and the γ-globulins respectively, but it is sometimes difficult to distinguish the α-globulin zone from that of albumin. Cellulose acetate (Ambler and Rodgers 1980) gives similar results, although the zones are rather sharper and are more easily distinguished, while agar gel electrophoresis followed by immunodiffusion (Grabar and Williams 1953) resolves 20–30 components. Examination of serum by PAGE with multiphasic buffers generally gives 30 or more bands, similar to the number found by crossed immuno-electrophoresis (see Fig. 8.9), but by PAGE on gradient gels or PAGIF nearer to 60 components can be seen.

For the general analysis of blood serum proteins by PAGE, gels within the concentration range of about 5–7 per cent (i.e. $T = 5$–7 per cent; $C = 1{\cdot}5$–5$\cdot$0 per cent) are most suitable. A multiphasic buffer system comprising tris-HCl pH 8$\cdot$9 in the separation gel, tris–HCl or tris–phosphate pH 6$\cdot$9 in the stacking gel ($T = 3$ per cent, $C = 20$ per cent), and tris–glycine pH 8$\cdot$3 in the electrode chambers is particularly suitable. The densitometric scan of the proteins from normal serum separated on such gels after staining and destaining in the usual way (Section 2.9) might resemble that shown in Fig. 13.1. Normal patterns like this and the identities of individual bands on them have been established by many groups of workers, often in studies with several hundred individual

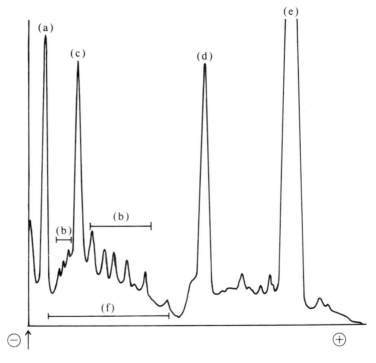

FIG. 13.1. Diagram of a typical densitometer scan of the serum proteins of normal blood separated by PAGE. The arrow marks the start of the separation gel. The identities of the major peaks are as follows: (a) ß-lipoprotein; (b) a number of peaks due to haptoglobin polymers found on both sides of peak (c) which is α_2-macroglobulin; (d) transferrin; (e) serum albumin, the γ-globulins occupy the region (f) and are not seen as discrete peaks in this system.

samples. A similar approach has been used to establish genetic variants of proteins such as the haptoglobins and the haemoglobins. Much of this work has been reviewed by Maurer (1971), Hoffmeister (1974), and Allen (1978).

The diagnosis of disease likewise involves the screening of a very large number of samples in order to establish differences between the patterns of normal and pathological sera (for example Hoffmeister (1974) examined over 10 000 samples). Unfortunately it is not practical to give a detailed review here of the application of all the various methods described in earlier chapters to the diagnosis of pathological conditions. Some idea of the magnitute of such a task can be gained from the fact that a comprehensive review on immuno-electrophoretic methods by Verbruggen (1975) lists no fewer than 505 references, most of which refer to papers with some clinical significance. Since then doubtless hundreds more will have been published, and this only includes the area covered by the techniques described here in one chapter!

Considerable information on the effects of disease on the plasma proteins can be gained from books such as Putnam (1965) and Ritzman and Daniels (1975) as well as the review articles mentioned above. Allen (1978) includes a useful table which is reproduced here in Table 13.1 and indicates the importance of a number of blood serum proteins in clinical diagnosis, following their separation and quantitative measurement by a single-dimension PAGE run. He also gives a table of enzymes which are of importance in diagnosis, either from their qualitative absence or presence, from quantitative measurements of levels of activity in a particular band or from the development of iso-enzyme patterns. Measurements of enzymic activity in separated zones obtained by PAGE or PAGIF are particularly useful as diagnostic aids because of the specificity of detection procedures and their extreme sensitivity. The release into the serum of an iso-enzyme which can be attributed to a particular organ or tissue is one of the most sensitive and unequivocal pieces of evidence currently available to us of damage to that organ or tissue resulting from a pathological condition. The clinical relevance of iso-enzyme assays has been reviewed by Burlina (1979).

Much of the earlier work on PAGIF as a diagnostic aid is covered by Latner (1975). Since that time there have been rapid developments that appear likely to continue. Amongst these are the development of IEF on cellulose acetate (see Section 12.6(c)) and the commercial introduction by LKB Produkter AB of thin PAGIF gels on flexible Mylar backing film. Both these advances give materials that can be handled far more easily than simple gel slabs in the clinical laboratory situation. They also make it far easier to perform iso-enzyme printing, immunoprinting (Arnaud et al. 1977), or other replicating techniques (Narayanan and Raj 1977), so that a

TABLE 13.1
Human serum proteins separated by PAGE and of importance to clinical diagnosis

Protein	Biological function	Importance in diagnosis
Acid α_1-Glycoprotein	?	Increase: neoplasms, inflammations, rheumatoid arthritis Decrease: liver diseases
Prealbumin	Thyroxine-binding globulin, retinol-binding globulin	Decrease: liver diseases
Albumin	Osmotic function, protein pool, binding of ions, dyes, etc.	Decrease: kidney diseases, neoplasms, liver diseases, chronic inflammations
α_1-Antitrypsin	Proteinase inhibitor	Increase: inflammations, liver diseases Decrease: liver diseases
Gc-Globulin	Three genetic types	
Ceruloplasmin	Copper-binding globulin, oxidase	Increase: pregnancy, neoplasms Decrease: Morbus Wilson
C-Reactive protein	Phagocytosis promoting activity	Increase: inflammatory conditions
Hemopexin	Haeme binding	Decrease: haemolytic anaemias
Transferrin	Iron binding	Decrease: neoplasms, inflammations, paraproteinaemias, nephrosis
β_1-A-Globulin	Complement factor	Increase: infections, toxoplasmosis Decrease: auto-immune diseases

Haptoglobins
  Type 1–1
  Type 2–1
  Type 2–2
	Binding of haemoglobin three genetic types	Increase: inflammations, neoplasms, infections, PCP, nephrosis
		Decrease: liver diseases

α_2-Macroglobulin	Proteinase-inhibitor	Increase: liver disease, diabetes, nephrosis
α-Lipoprotein HDL	Transportation of cholestrol	Increase: lipid metabolism disturbances Decrease: liver diseases, Tangier disease
Pre-β-lipoprotein LDL	Transportation of lipids	Increase: type-3, 4 hyperlipoproteinaemia
β-Lipoprotein LDL	Transportation of lipids	Increase: nephrosis, type-2 hyperlipoproteinaemia Decrease: liver diseases
Chylomicron VLDL		Increase: type-1 hyperlipoproteinaemia
γA-Globulin	Antibody	Increase: liver diseases, chronic infections, PCP Decrease: paraproteinaemias, antibody deficiency syndrome
γG-Globulin	Antibody	Increase: liver diseases, chronic infections, myeloma, PCP Decrease: paraproteinaemias, antibody deficiency syndrome
γM-Globulin	Antibody	Increase: chronic infections, macroglobinaemia Waldenström, liver diseases

Reproduced from Allen (1978) by permission of the authors and publishers number of different tests can be carried out on the same strip of gel (e.g. protein staining, glycoprotein staining, enzyme activity location, immuno-chemical identification, etc.). Transfer to immobilizing membranes (protein blotting; see Section 2.13) is a very useful alternative for this however.

After a rather slow beginning we have now reached the stage where electrophoretic methods of all types are becoming widely accepted in the routine clinical laboratory, and this increased interest is providing a great impetus to research into improving techniques and into the interpretation of results. The resulting information explosion is currently based largely on single-dimension electrophoretic separations, either on their own or coupled with a subsequent immunochemical identification as in the quantitative measurement of α-fetoprotein by electro-radio-immunoassay (Sykes and Dennis 1977), the detection of immunoglobulin abnormalities by crossed immuno-electrodiffusion (Shulman 1979), or the crossed immuno-electrophoresis of α_1-lipoproteins (Cline and Crowle 1979). Fluorescent fluorescein-conjugated antisera have been applied successfully to the identification of proteins on stained and dried electrophoresis strips (Fenton and Shine 1979). A combination of electrophoretic and immuno-logical properties can also be useful in studies of iso-enzyme patterns, e.g. alkaline phosphatase (Guilleux, Hayer, Thomas, and de Bornier 1978; Saga and Kano 1979), creatine kinase (van Lente and Galen 1978), etc., and may also be combined with enzyme activity measurements or specific staining for enzyme activity.

Although modifications to protein and iso-enzyme patterns caused by a very large number of different disease states have been reported in the last few years, mention should be made of two particular areas in which electrophoretic methods have made considerable contributions and in which there is considerable interest at the present time. These are the study of serum lipoproteins in the differential diagnosis of the various forms of hyperlipoproteinaemia (Fig. 13.2) and the phenotyping of serum α_1-antitrypsins. The former have a suggested relationship to atherosclerosis, coronary artery disease, and possibly cancer.

Gels which are very non-restrictive are used for the separation of molecules as large as the serum lipoproteins, and typical procedures employ electrophoresis on agarose (e.g. Demacker et al. 1978; Hoeg et al. 1983), composite agarose–polyacrylamide (e.g. Moulin et al. 1979), or polyacrylamide gels of low ($T = 3 \cdot 0$–$3 \cdot 5$ per cent) concentration (e.g. Muñiz 1977; Terebus-Kekish, Barclay, and Stock 1978). Amongst other approaches which have been applied to the study of human serum lipoproteins are crossed immunoelectrophoresis (Rerabek 1977; Cline and Crowle 1979), flat-bed IEF (e.g. Marcel, Bergseth, and Nestruck 1979), PAGIF (e.g. Warnick, Mayfield, Albers, and Hazzard 1979), charge-shift electrophoresis (Utermann and Beisiegel, 1979), and 2D mapping

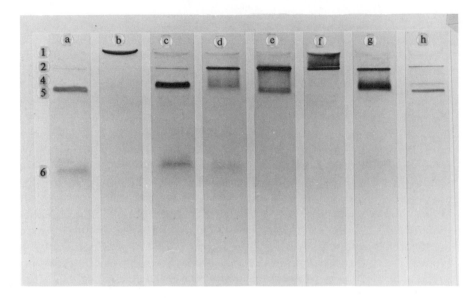

FIG. 13.2. Identification of different types of hyperlipoproteinaemia by electrophoresis on composite gels containing 0·7 per cent agarose throughout and $T = 2$ per cent, $C = 15$ per cent polyacrylamide in the upper (sample) part and $T = 3$ per cent, $C = 15$ per cent polyacrylamide in the lower part. The gel cross-linking agent was diallytartardiamide and a homogeneous buffer system of 0·074 M tris and 0·12 M glycine containing 9 per cent sucrose was used throughout. Samples were prestained with succinylated Sudan Black B and were (a) normal, (b) type I, (c) type II, (d), type III, (e), type IV, (f) type V, (g) type VI or IIb, and (h) normal with lipoprotein Lp(a). 1, chylomicrons; 2, VLDL; 4, lipoprotein Lp(a); 5, LDL; 6, HDL. (Reproduced from Moulin et al. (1979) by permission of the authors and publishers.)

(Chapter 11) by O'Farrell's procedure (Zannis and Breslow 1980). Several different techniques, including charge-shift electrophoresis, SDS–PAGE, IEF, and immunodiffusion were all employed by Beisiegel and Utermann (1979) in their studies on the isolation and characterization of apolipoproteins of hypertriglceridaemic individuals. This provides a good illustration of the fact that various groups of workers often apply different techniques or different variations of a technique to the same problem. It may often be difficult to judge the particular merits or disadvantages of each approach unless deliberate attempts are made to compare them, as for example has been done in this case by Kuba, Lippel, and Frantz (1979) who compared four methods for the diagnosis of Type III hyperlipoproteinaemia.

The conclusion that can be drawn from the above is that there may not be just one way of solving a separation problem, but a number of different methods that are capable of giving good and useful information if they are properly executed. Thus it may not be a sensible allocation of time and resources to seek the very best way of performing the most thorough separation attainable, but rather it may be preferable to choose the technically most simple or most rapid and convenient method that will give the required amount of information. The time saved can then be spent on applying the method to the problems. In the clinical laboratory cellulose acetate is often chosen as the supporting matrix for electrophoretic separations on the grounds of rapidity and ease of handling (e.g. Schmidt 1980) in spite of the fact that superior resolution and of course more information can be gained from other procedures.

The second area of considerable current interest involving electrophoretic methods, namely α_1-antitrypsin ($\alpha_1 P_i$) phenotyping, contrasts with the field of serum lipoproteins in that earlier PAGE methods have now been very largely superceded by IEF and especially PAGIF. At least 56 phenotypes and 30 different alleles are now known (Allen 1978). The α_1-P_i system is a valuable genetic marker used in population genetics as well as being important by virtue of the finding that certain variants and deficiencies are associated with a predisposition to a number of disease states. These include juvenile arthritis, non-allergic asthma, infantile cirrhosis, and pulmonary emphysema, as well as other inflammatory diseases (Jeppsson 1977a; Arnaud et al. 1977; Alten, Oulla, Arnaud, and Baumstark 1977; Allen 1978; Charlionet et al. 1979). These groups of workers all used PAGIF, but IEF on cellulose acetate (Harada 1975) has also been used successfully for separating most of the variants. According to Allen (1978) other methods such as radial immunodiffusion, trypsin inhibitor capacity measurements, and acid starch gel electrophoresis are not as accurate, are more liable to misinterpretation, and are not as fast or convenient as PAGIF. It is also the only technique capable of fully resolving the microheterogeneity of the α_1-P_i system, particularly if it is performed (Görg et al. 1983a; Weidinger and Cleve 1984) in immobilized pH gradients (see Section 9.3(h)). The best pH range to use for this work is 3·5–5·0 (Fig. 13.3) (Charlionet et al. 1979; Sesboüé, Martin, and Lebreton 1981; Klasen and Rigutti 1982).

PAGIF, and other IEF methods, are very suitable for haemoglobin typing (Fig. 13.4). A pH gradient from 6·0–8·0 is most appropriate for the separation of both normal and abnormal types, and the use of commercially produced PAGIF gel slabs for haemoglobin typing has been discussed by Jeppsson (1977b). Haemoglobins possess the advantage of being coloured so that usually no staining or destaining is necessary, and in addition it is particularly easy to tell when the migrating zones have reached their equilibrium positions because they then focus into a very sharp zone at

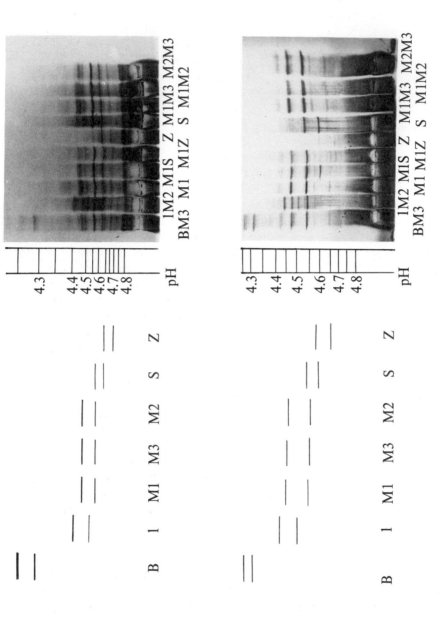

FIG. 13.3. Polyacrylamide gel iso-electric focusing (PAGIF) patterns of several α_1 antitrypsin phenotypes. The upper and lower gels show the results obtained with two different preparations of ampholytes which were synthesized and fractionated to give mixtures forming gradients in the pH ranges 4.2–5.3. The pH scale in the centre was measured on the gel plate with a contact electrode after focusing. Major bands are identified diagramatically on the left. (Reproduced from Charlionet *et al*. (1979) by permission of the authors and publishers.)

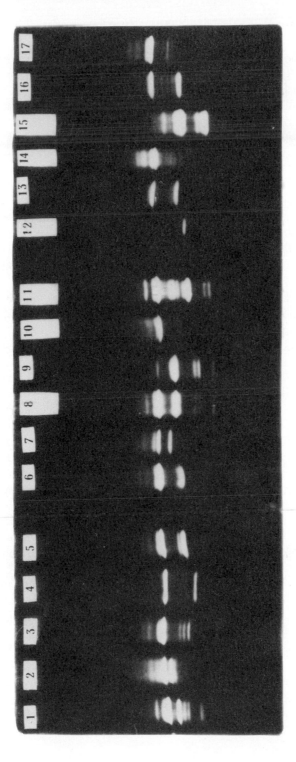

Fig. 13.4. Haemoglobin typing of electrophoretically slow variants by PAGIF on $T = 5$ per cent, $C = 3$ per cent gels containing 2 per cent ampholytes (pH 5–9). Samples were (1) HbF + HbA, (2) Hb Waco, (3) HbS Travis, (4) Hb Hasharon, (5) Hb Philadelphia. (6) Hb Russ, (7) Hb Kempsey, (8) HbG San José (heterozygous), (9) HbG San José (double heterozygous), (10; Hb Lepore Boston, (11) Hb Gavello, (12) Hb Ferrara, (13) Hb Punjab, (14) HbQ India, (15) HbC + HbS, (16) HbA + HbS, and (17) Hb Normal. (Reproduced from Giuliani *et al.* (1978) with permission of the authors and publishers.)

their iso-electric points. Because of this, haemoglobin is often applied as a standard to a PAGIF gel slab as an indicator protein to show when the pH gradient is established and focusing is completed. The method is useful in the study of various haemoglobinopathies and erythrocytosis, and as an index of the control of carbohydrate metabolism in diabetes mellitus since one variant (HbA_{1C}) is greatly increased when the disease is poorly controlled (Jeppsson, Franzén, and Nilsson 1978; Simon and Cuan 1982). By including 0·2 M ß-alanine and 0·2 M 6-aminocaproic acid as spacers with Ampholines of pH range 6–8 Cossu *et al.* (1984) transformed the usual 2 pH unit gradient into a shallow 0·6 pH unit gradient spanning the range pH 6·7–7·3. This improved the resolution between HbA and HbA_{1C} by a factor of three, greatly improving densitometric measurement of the latter with conventional gel scanners. The method also permitted complete resolution between HbA_{1C} and the F and F_{ac} bands of foetal haemoglobins. Alten *et al.* (1977) calculate that in PAGIF screening of babies for abnormal haemoglobins (e.g. sickle cell anaemia etc.), 16 man hours of working time and $50 for materials including ampholytes is sufficient for the haemoglobin analysis of 1000 samples. However, other electrophoretic methods can also give useful information on haemoglobin variants, and the use of cellulose acetate, for example, has been advocated (Rich, Ziegler, and Grimley 1979) for the separation of HbA and HbA_2, the latter being an important diagnostic criterion for individuals who are heterozygous for ß-thalassemia. Separation of globin chains on cellulose acetate has also been used for ß-thalassemia diagnosis (Boccaci, Massa, and Tentori 1981). Many aspects of electrophoretic and chromatographic methods which are applicable to the diagnosis of haemoglobinopathies have been discussed by Basset, Braconnier, and Rosa (1982). More recently, Rochette *et al.* (1984) have demonstrated that using the superior resolving power of PAGIF in gels with immobilized pH gradients it may be possible to separate haemoglobin variants (e.g. Hb San Diego and HbA) that can scarcely be distinguished by PAGIF with conventional ampholytes. It is reasonable to hope that this technique will further enhance our diagnostic capabilities in this area.

Isotachophoresis (ITP) is currently the only electrophoretic method capable of analysing low-molecular-weight ionic substances as well as medium- and high-molecular-weight species. Consequently, it has been applied (Mikkers, Ringoir, and de Smet 1979) to the analysis of uremic blood samples, and according to these authors was rapid, reliable, and inexpensive, gave qualitative and quantitative information on several different components in the same sample, and was well suited to use as a screening technique.

13.4. Other body fluids and tissues

(a) *Cerebrospinal fluid*

The cerebrospinal fluid (CSF) of normal subjects has a broadly similar protein composition to the blood serum but there are significant differences. Most importantly, the protein content of CSF is in the range 15–40 mg per 100 ml, whereas in blood serum it is closer to 7–8 g per 100 ml. This means that if PAGE is used with homogeneous buffer systems the CSF must first be concentrated, usually by ultrafiltration, lyophilization, acetone precipitation, etc., but this can induce the formation of artefacts. With multiphasic buffer systems the zone-stacking effect makes prior concentration unnecessary, but sample sizes of up to 2·0 ml may be necessary although 0·5 ml is more usual. Such large samples may make it beneficial to use relatively long stacking gels above the separation gel.

When compared (Felgenhauer 1974) with normal blood serum, normal CSF contains relatively lower amounts of haptoglobin 1-1, and haptoglobin polymers, especially higher-order polymers, are not detectable. The third postalbumin and the second post-transferrin bands are typical of CSF while ß-lipoprotein is virtually absent. A comparison of CSF and blood serum proteins by ITP (Delmotte 1979) is shown in Fig. 13.5. The immunoglobulins of CSF are mostly composed of IgG and are usually found in the stacking

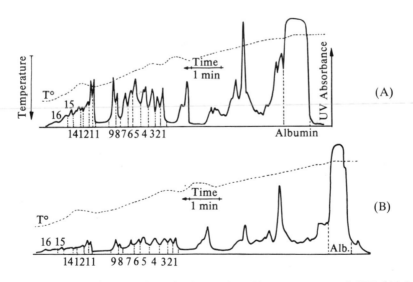

FIG. 13.5. Capillary isotachophoresis of proteins of human serum and CSF: (A) 0·6 µl human serum; (B) 2 µl of 15-fold concentrated CSF from the same subject. Separation time, 25 min. (Reproduced from Delmotte (1979) by permission of the author and publishers.)

gel and are poorly stained. α_2-Macroglobulin is always present but only in small amounts. Felgenhauer (1974) concluded that it was therefore possible to distinguish those diseases (such as acute inflammations of bacterial or viral origin and tumours or vascular diseases of the central nervous system) in which the blood–CSF barrier was disturbed from diseases in which new proteins are synthesized within the central nervous system itself (such as syphilitic diseases, multiple sclerosis, and subacute sclerosing leucencephalitis). In the former case, which includes most neurological diseases, the CSF protein level increases and begins to approximate blood serum in composition. The high-molecular-weight proteins (α_2-macroglobulin, ß-lipoprotein, haptoglobin polymers, etc.) increase in relative concentration while the contribution of small proteins such as prealbumin become relatively less important, and bands such as the third postalbumin band typical of CSF become undetectable. In the latter case the new proteins synthesized within the central nervous system are mostly locally produced immunoglobulins (e.g. Casey *et al.* 1981), so that although the total CSF protein content is again elevated the composition remains different from that of the blood serum. It is of course possible to obtain a combination of a strong local reaction and a disturbance of the blood–CSF barrier, and this can occur for example in the late stages of meningitis, myeloma infiltration, and mycotic brain infections.

Most of these protein changes can be detected by PAGIF just as easily as by PAGE, with the added advantage that the low protein concentrations in CSF are no longer a problem owing to the concentrating effect of IEF. Methodological aspects and the protein composition of normal CSF have been discussed by Nilsson and Olsson (1978). The lipid composition of native lumbar liquor has recently been studied by cellulose acetate electrophoresis (Kleine and Enders 1979). Unconcentrated CSF has been examined by crossed immuno-electrophoresis on cellulose acetate (Schmidt 1979; 1980) and at least 22 proteins have been quantified starting with only 40 μl of sample.

Silver staining procedures for detecting proteins and peptides (Tables 2.5, 2.6, and 2.7) are claimed to be 100 times more sensitive than Coomassie Blue R250 staining, and as such are particularly useful in the study of dilute biological fluids such as CSF. After two-dimensional electrophoretic separations of unconcentrated CSF and amniotic fluid the method revealed complex patterns of proteins that were not visible with Coomassie Blue staining (Switzer *et al.* 1979). Allen (1980) and Confavreux *et al.* (1982) applied silver staining to PAGIF patterns of unconcentrated CSF from patients with slipped disc, Friedreich's ataxia and multiple sclerosis.

Gels requiring minimum preparation are favoured in many clinical laboratories, and the combination of agarose gels (especially now that they are commercially available) for electrophoresis or IEF with silver staining is often the preferred technique for protein analysis in unconcentrated CSF (e.g. Link and Kostulas 1983; Mehta, Mehta, and Patrick 1984; Sano *et al.* 1984), now that satisfactory silver staining procedures for application to agarose gels have been developed (see p.150).

An interesting study in which these techniques have been applied to CSF from both normal subjects and patients suffering not only from multiple sclerosis but also from various other inflammatory and non-inflammatory neurological diseases has been reported by Harrington *et al.* (1984). In general, quantitative changes in certain zones in the 2D protein maps enabled each disease state to be distinguished from the others and in some cases there were also qualitative changes (e.g. the appearance of additional immunoglobulin light chain components).

(b) *Urine*

Analysis of the urinary proteins shows that normal urine contains some 30 zones detectable by PAGIF while that of patients with severe renal damage may contain 40 or more. As with CSF, urine is also a rather dilute protein source, so for most electrophoretic investigations the proteins have to be concentrated beforehand by ammonium sulphate precipitation, ultrafiltration, lyophilization, etc. Alternatively sensitive silver staining procedures can be applied to gels on which samples of unconcentrated urine have been run, and this is probably the best approach currently available.

The majority of myeloma patients show greatly elevated levels of β_2-microglobulin in their urine, as do patients with tubular deficiencies or following renal transplantation (Allen 1978). The progress of therapy with cytostatic drugs can be judged from the disappearance of the Bence–Jones proteins. The analysis by either PAGE or PAGIF of urinary glycoproteins, proteins of blood serum origin, and glycosaminoglycans can be most useful in diagnosing and following the effects of therapy in many kidney diseases, such as nephrosis, glomerulonephritis, uremic–haemolytic syndrome, and tubular insufficiency. Many such diseases result in the excretion in the urine of antigens of glomerular basement membrane origin which can be detected immunologically (e.g. Bygren, Wieslander, and Heinegård 1979) by, for example, crossed immuno-electrophoresis, tandem crossed immuno-electrophoresis, etc. The origins of proteins and enzymes in urine and the effects of various diseases on their presence, absence, or concentration levels have been briefly reviewed by Anderson, Anderson, and Tollaksen (1979).

Although not yet generally accepted outside the research environment, ITP has been proposed (Kodama and Uasa 1979; Tschöpe and Ritz 1980; Oerlemans *et al.* 1981) as a very useful method for urine analysis. The

method was claimed to be simple to use, faster than thin-layer chromatography, and unlike gas–liquid chromatography required no sample
pretreatment. These authors analysed directly 0·1 μl samples of urine and
considered ITP would be a good screening technique for the detection of
inborn errors of metabolism such as isovaleric acidemia. Since ITP gives
information on both high- and low-molecular-weight solutes, it seems
reasonable to conclude that it could be very helpful in any investigation of
kidney function which might result in changes in urine composition or
concentration.

Two-dimensional maps of the proteins present in human urine have now
been reported (Anderson *et al.* 1979; Edwards *et al.* 1982), and in the
medium-term future this approach could form the basis for diagnosis of any
pathological conditions which cause changes in the pattern of urinary
proteins excreted. Indeed, Clark, Kricka, and Whitehead (1980) suggest
that maps of this kind may be useful for the early diagnosis of rheumatoid
arthritis, while Edwards *et al.* (1982) reported the presence in urine and
prostatic tissue of a protein characteristic for patients with prostatic
adenocarcinoma.

Cellulose acetate electrophoresis (see Section 12.6(a)) has been used for
the analysis of glycosaminoglycans in urine (Hopwood and Harrison 1982).
These were first concentrated by precipitation with cetylpyridinium
chloride, applied to Titan III Zip Zone cellulose acetate plates (Helena
Laboratories) for electrophoresis, and the separated zones revealed by
Alcian Blue staining. The method was considered by the authors to be very
suitable for screening large numbers of urine samples, for the detection
and diagnosis of various different mucopolysaccharidoses.

(c) *Other fluids and tissues*

Clinical applications of electrophoretic methods to other body fluids and
tissue samples are not widespread, although they are of increasing
importance. It is known that parotis, cystic fibrosis, rheumatoid arthritis,
and some other diseases, as well as some drugs, affect the secretion and
composition of saliva and these changes can be detected by PAGE and by
PAGIF. Saliva is another rather dilute fluid so proteins are often
concentrated by precipitation (e.g. Jones, Broadhurst, and Gurnsey 1982).
If unconcentrated samples are employed and gels stained with Coomassie
Blue, relatively large volumes (0·1–0·2 ml) must be added to each sample
slot for PAGE analysis. This objection does not hold for immunochemical
techniques, and crossed immunoelectrophoresis and crossed immunoisoelectric focusing with Coomassie Blue staining of the resultant immunoprecipitate patterns have both been used to study the variations in salivary
protein composition between individuals and in a single person with time

(Eckersall and Beeley 1980). IEF of course has a considerable ability to concentrate a dilute protein solution into a narrow zone, so in this technique the dilute nature of saliva is less of a problem, and Sano *et al.* (1984) used IEF in agarose gels to study proteins in saliva as well as urine, CSF, tears, sweat, and gingival fluid. They further enhanced sensitivity by employing silver staining and considered that a protein concentration of 80 mg l^{-1} was sufficient for analysis, although concentrations in the range 200–600 mg l^{-1} were preferable. Marshall (1984) examined salivary proteins by 2D mapping under both denaturing and non-denaturing conditions with both silver and Coomassie Blue staining. There are various reports of the electrophoretic analysis of proteins and enzymes in a very wide range of body fluids (Felgenhauer and Hagedorn 1980), cells and tissues including sweat (Marshall 1984), tears (Janssen and van Bysterveld 1981), seminal fluid (Edwards, Tollaksen, and Anderson 1981), pancreatic secretions (Cassara *et al.* 1980; Allen *et al.* 1981), milk (Anderson, Powers, and Tollaksen 1982*a*), amniotic fluid (Jones and Spragg 1983), erythrocytes, lymphocytes (Willard and Anderson 1981), leukaemic cells (Lester *et al.* 1982; Anderson *et al.* 1983), platelet membranes (McGregor *et al.* 1981), muscle proteins (Giometti *et al.* 1980; Young and Davey 1981; Thompson *et al.* 1981), etc. from both normal and pathological sources. If desired, serum proteins can be removed before analysis by immunoabsorption to accentuate differences in the tissue or fluid-specific proteins (Dermer and Edwards 1983).

A particularly interesting technique is the development of direct tissue IEF (Saravis, O'Brien, and Zamcheck 1979; Thompson *et al.* 1981; 1982) as this avoids any preliminary extraction or fractionation of sample material before analysis, and enables analysis to be carried out on tiny amounts of material such as thin cryostat sections of needle tissue biopsy samples. It also appears to circumvent the loss of protein components by denaturation or proteolysis by idigenous proteinases which may occur during extraction steps. The method consists of allowing 20-μm-thick tissue sections to adhere to the hydrophilic sides of small pieces of Gel Bond film which are then applied directly to the surface of thin agarose gel slabs for IEF. The gels contained 0·5 per cent Triton X-100 to aid solubilization of proteins from the samples, but since the gel medium was otherwise non-denaturing various enzyme and isoenzyme patterns could be revealed by activity staining, as well as dye or silver staining for the total protein pattern.

At the present time most of the work described above has not reached the stage of true clinical exploitation. In nearly all cases the sample numbers are too restricted or the analyses too limited for proper evaluation of the signifance of the findings for diagnosis of disease or the monitoring of therapeutic regimes, but many of these findings have considerable potential for the future.

13.5. Two-dimensional macromolecular maps

The discussion here in a separate section of two-dimensional mapping methods as opposed to the one-dimensional procedures considered above is a somewhat arbitrary distinction because all samples that are suitable for electrophoretic analysis can be examined by either one- or two-dimensional techniques. The former are more rapid, easier to perform, and more economical in terms of materials, generally require only simple equipment, and give results which are easier to interpret, while the latter are more complex and more demanding in all respects but can give much more information. Perhaps the distinction should be made between the methods that are currently practised in the clinical laboratory, which are nearly always one-dimensional separations, and those 2D methods which are now on the threshold of clinical application and can be expected to be predominant in a few years time. The potential of 2D electrophoretic analysis for clinical purposes has been discussed in more detail by Young and Tracy (1983) and by Kossmann (1984).

Either the one or two-dimensional approach will require extensive automation for the full potential to be realized. In the case of one-dimensional electrophoretic analysis this is now with us. For example, the Olympus HITE system for cellulose acetate electrophoresis of serum proteins (Olympus Corporation Clinical Instruments) automates the whole procedure from sample application to preparation of report froms, can handle 350 samples a day, and requires only 50 μl of serum per sample.

The development of two-dimensional macromolecular maps for clinical purposes has concentrated on protein analysis, often in blood serum samples. Apart from the immuno-electrophoretic procedures (Chapter 8) most studies employed the method of O'Farrell (1975), or only slightly modified versions of it. This involves (see Section 11.3) IEF separation according to molecular charge in the first dimension followed by SDS–PAGE in the second dimension, separating according to size differences. Either method alone in a one-dimensional separation is theoretically capable of resolving up to about 100 proteins, so together they should be capable of resolving about 10 000 proteins in a two-dimensional system and perhaps as many as 40 000 after further development. The human genome is thought to code for about 30 000–50 000 proteins or protein subunits (protein products of genetic translations), probably less than 10 per cent of which appears in any one type of cell. As pointed out by Anderson *et al.* (1979), these facts point to the almost incredible conclusion that we are now in a position to consider describing and measuring the complete composition of every tissue and fluid of man at the molecular level, and how this composition varies in health and disease—nothing less than the anatomy and pathology of man in molecular terms!

There are of course years of work ahead before this goal is achieved, but a start has already been made and the results will in time revolutionize much of medicine and biochemistry. A major contribution in this area has come from the work of the Molecular Anatomy Program of the Argonne National Laboratory (Illinois, U.S.A.), the objective of which is to catalogue human gene products in order to explore differentiation, carcinogenesis, and related phenomena. To this end Anderson and Anderson (1978a, b) describe apparatus and techniques for preparing large numbers of gels and for running them simultaneously under identical conditions. Over 100 IEF gels can be prepared and loaded by one operator in one morning and as many as 100 slab gels subsequently run together in the second dimension (Anderson et al. 1979). This approach has been used to map human plasma proteins (Anderson and Anderson 1977; Anderson et al. 1982; Rosenblum, Neel, and Hanash 1983) and other human tissues and fluids (see Sections 13.4(b) and 13.4(c)).

As pointed out by these workers, a number of problems have to be overcome to enable the full potential of macromolecular mapping to be exploited for clinical diagnosis and the control of therapy, and these fall into four main areas. The first is that in order to handle a large number of samples on a routine basic the system must be automated. This has been achieved for one-dimensional separations as we have seen already, and doubtless in the relatively near future two-dimensional gel maps will also be prepared automatically. Indeed, some of the steps involved can be largely automated at the present time.

The second area concerns the identification of individual spots on the maps produced. Because the conditions used are strongly dissociating, multi-subunit proteins will be broken down into their constituent subunits, which if they are not identical will be separated from each other. We are then faced with the difficulty of deciding which subunits were originally associated and what is the identity of the parent protein. If pure proteins are available for use as standards the problem is easily resolved, but often they are not and then one must rely upon identification based on physical data (e.g. pI, molecular weight, etc.), immunochemical reactions, physical properties (precipitation, heat denaturation, inactivation, etc.), or in some cases genetic variants (Anderson and Anderson 1979). When a protein can be obtained in a highly purified form it should be possible to radiolabel it and add it to the unlabelled sample under investigation, so that after the separation a comparison of the autoradiograph with the protein map obtained by silver staining should reveal the protein of interest unequivocally, and possibly also its variants and subunits. However, while desirable it may not be essential to identify a spot unequivocally, for if the presence or absence of a spot can be clearly associated with a certain disease then the actual identity of the component may be of lesser importance.

The third problem area is that of the detection and quantitative measurement of the separated components on the map. Most research applications of two-dimensional mapping have employed radioactive labelling of proteins and subsequent measurement of them by auto-radiography, but this may not always be feasible for clinical work so that staining procedures are then required. Coomassie Blue staining has often been used for this purpose to date, but silver staining techniques with sensitivities comparable to autoradiography seem likely to take over almost entirely in future. Of course maps can be selectively stained for particular components if required. For example, Horst *et al.* (1979) used ^{125}I-labelled concanavalin A to 'stain' glycoproteins in maps of fibroblast plasma membranes. Selective stains can often be combined with a subsequent general protein staining step and can assist in the identification of unknown spots. With a small number of maps, patterns can often be matched visually and differences detected, but this is not adequate for large numbers or when maps are being used routinely. Generally, stained gels are photographed carefully and as reproducibly as possible for subsequent matching of the photographic images or scanned with video cameras.

One approach to the comparison of complex mixtures is double radioactive labelling (see Section 11.7) followed by two-stage auto-radiography (McConkey 1979; Walton, Styer, and Gruenstein 1979; Choo, Cotton and Danks 1980; McConkey and Anderson, 1984*a*, *b*). The resulting pairs of maps are useful in phylogenetic studies, in the comparison of the proteins synthesized by different types of cells in the same organism or one cell type at different stages of the cell cycle, in the comparison of tumour cells or virus-infected or virus-transformed cells with normal cells, in the diagnosis and study of metabolic abnormalities, in genetic diseases such as cystic fibrosis, in genetics and genetic polymorphism, etc.

The last major area of activity concerns data gathering and handling. Firstly there must be some points of reference on the gel map. This is best achieved by the use of internal standards for pI and molecular weight, and this aspect has also received some attention from the Argonne group. The gel photographs are suitably enlarged and scanned so that the co-ordinates and intensities of every spot can be measured and transferred to a computer for the normalization of spot patterns and comparisons of one map with another. Just as advances have recently been made in the other three areas, so also in this area development is proceeding rapidly. Capel, Redman, and Bourque (1979) and Anderson *et al.* (1979) reported semi-automated methods of gel scanning to collect the mobility data, followed by computer normalization and averaging. More highly automated and very sophisticated procedures, based on the use of a drum scanner to measure the autoradiographic films and with fully computerized data handling followed (Garrels 1979; Bossinger *et al.* 1979). Garrels was able to resolve over 800 proteins on a single map and used the method to study protein synthesis in two clonal cell lines at various stages in their growth cycles. He found that the two lines could be distinguished clearly at all stages of growth from their patterns of cell-specific proteins and by their characteristic rates of synthesis of many of the shared proteins. These methods have in turn been superceded by some of those referred to in Section 11.7 and doubtless further developments will continue rapidly.

While these examples are as yet of purely research interest and are not immediately related to the solving of clinical problems, they provide a striking pointer to the way in which protein mapping for clinical diagnosis will progress in the next few years. Given the continued existence of the human race, the end is not now seriously in doubt for the technology is already within our grasp and only the time scale remains to be decided. When the era of 'science fiction medicine' dawns depends only upon the amount of resources we are prepared to devote to it.

APPENDIX 1

METHODS FOR THE RADIOLABELLING
OF PROTEINS AND NUCLEIC ACIDS

A1.1. Proteins

A1.1.1. ^{125}I or ^{131}I

(a) *The chloramine-T method (e.g. Das et al. 1977)*
- (i) Protein (100 μg) and carrier-free Na^{125}I (40 mCi) are mixed in a total volume of 300 μl of 0·2 M potassium phosphate buffer pH 7·5.
- (ii) 40 μl of chloramine-T (50 mg ml^{-1}) is added and the mixture is stirred vigorously for 1 min at 4 °C.
- (iii) The reaction is stopped by the addition of 40 μl of sodium metabisulphite (100 mg ml^{-1}) and 40 μl of KI solution 0·7 g ml^{-1}).
- (iv) The labelled protein is separated from unreacted Na^{125}I by gel filtration on a small column (e.g. 100 mm x 10 mm) of Sephadex G-10, G-15, or G-25 with a buffer of 0·02 M potassium phosphate pH 7·5 containing 0·2 M NaCl.

Comments

Because of its more favourable radiochemical properties and longer half-life ^{125}I is generally to be preferred to ^{131}I but the method is of course the same for both. In (i) the strength and composition of the buffer is not critical and phosphate buffers of 0·1 to 1·0 M have been used, as have tris–EDTA buffers, but the pH should be between 7·0 and 8·0. High concentrations of urea (up to 8 M), NaCl (up to 2 M) or 0·1 per cent sodium dodecyl sulphate (SDS) can be included in the buffer to cause protein unfolding or dissociation and increase the efficiency of labelling. Smaller proportions (up to 10-fold less) of chloramine-T have been used by some groups in step (ii), and reaction times as long as 20 min at room temperature or as short as a few seconds before the addition of excess bisulphite and KI in step (iii) have been reported. However, the conditions of the reaction are highly oxidizing, and short reaction times, preferably at 4 °C, reduce undesirable side reactions and loss of biological or enzymic activity in the protein being labelled. In step (iv) it is advisable to prewash the Sephadex column with a solution of the unlabelled protein being studied before gel filtration of the labelled material. This avoids adsorption of the labelled material on to the Sephadex and the glassware by blocking the available binding sites and greatly improves yield, especially if very small amounts of labelled protein are being separated. Almost any convenient buffer is suitable for use in the gel filtration step. As an alternative to gel filtration a carrier protein such as 1 ml of 1 mg ml^{-1} serum albumin can be added followed by 0·2 ml of 50 per cent trichloroacetic acid. The precipitated protein is separated from other components in the reaction mixture by centrifuging for 20 min at 10 000 g and the protein pellet is washed with acetone containing 0·1 M HCl, then with acetone alone, and finally air dried.

(b) *The lactoperoxidase method (Phillips and Morrison 1970; Lamas 1979)*

- (i) Each millilitre of the reaction mixture contains the following constituents:

protein sample (1 nmol), 2·5–5·0 μg lactoperoxidase ($A_{412}/A_{280} = 0·8$, Sigma Chemical Co.); 50–100 nmol of Na^{131}I or Na^{125}I; 50–60 nmol of phosphate buffer pH 7·0; 0·5–1·0 mg glucose monohydrate; 0·5–1·0 μg of glucose oxidase (Type IX, 25 000 units g^{-1}, Sigma Chemical Co.).

(ii) This mixture is incubated at 37 °C for 20–30 min.

(iii) The reaction is stopped by adding phosphate buffer containing 1·5 mg ml^{-1} of unlabelled NaI or an ice-cold solution of 1-methyl 2-mercaptoimidazole to 15 mM.

(iv) The labelled protein is separated from the reaction mixture by ion-exchange chromatography, gel filtration, or precipitation.

Comments

The purpose of the glucose and glucose oxidase is to generate hydrogen peroxide, but if preferred they can be omitted and small amounts (e.g. 25 μl) of 30 per cent H_2O_2 can be added directly. Further portions should be added every 5 min to ensure maximum labelling. Differing extents of iodination can be obtained by varying the incubation time or temperature (step (ii)) or by varying the concentration of lactoperoxidase and/or the H_2O_2 concentration (Lamas 1979). The chloramine-T method gives very rapid, almost instantaneous, labelling, but the lactoperoxidase method is more efficient and the reaction conditions can be chosen to label cell surface proteins selectively for example (Phillips and Morrison 1970; Comings and Cohen 1979). In addition the reaction mixture is less strongly oxidizing than with chemical methods, such as those involving chloramine-T or iodine monochloride (McFarlane 1958), so that labelling is less likely to be accompanied by loss of enzymic or biological activity.

(c) *The immobilized lactoperoxidase method* (David 1972; David and Reisfeld 1974)

While the lactoperoxidase procedure possesses several advantages over chemical methods of radioiodine labelling, it has the disadvantage that protein contaminants are added to the sample protein and may be difficult to separate from it. Self-iodination of the lactoperoxidase may also occur. These problems can be overcome by the use of lactoperoxidase immobilized onto Sepharose beads prepared by the method of March, Parikh, and Cuatrecasas (1974).

(i) One volume of a slurry of washed Sepharose 4B (Pharmacia AB) consisting of equal volumes of gel and water is mixed with 1 vol of 2 M sodium carbonate.

(ii) The mixture is stirred rapidly and 0·05 vol of a solution of cyanogen bromide (2 g ml^{-1}) in acetonitrile is added and vigorous stirring continued for 1–2 min.

(iii) The mixture is poured on to a sintered glass filter funnel and washed successively with 5–10 vol of 0·1 M sodium bicarbonate (pH 9·5), water, and 0·2 M sodium bicarbonate pH 9·5.

(iv) The slurry is then filtered under vacuum to a moist compact cake and transferred to a plastic bottle containing 1 vol of 0·2 M sodium bicarbonate pH 9·5 and lactoperoxidase (2·3 mg ml^{-1} of settled Sepharose beads).

(v) Coupling is allowed to proceed at 4 °C for 20 h.

(vi) Residual reactive sites on the beads are then blocked by adding glycine or ethanolamine to 1 M and allowing the mixture to sand for at least 4 h at room temperature.

(vii)   The beads are then washed successively with 20 vol each of 0·1 M sodium
        acetate pH 4·5, 2 M urea, and 0·1 M sodium bicarbonate pH 10. All of
        these solutions also contain 0·5 M NaCl.
(viii) The Sepharose-bound lactoperoxidase preparations can then be stored
        for considerable periods at 4 °C in phosphate-buffered saline pH 7·4
        containing 10^{-5} M merthiolate preservative.

The cyanogen bromide solution used in step (ii) is most easily prepared by adding
50 ml of dry redistilled acetonitrile to a 100 g bottle of cyanogen bromide and
storing at –20 °C when not in use. The amount of cyanogen bromide used can be
varied to give differing degrees of activation. Activation and coupling can be
performed at room temperature, but this is less efficient than that performed at 4
°C. The insolubilized lactoperoxidase beads are then used for radio-iodination as
follows (Griffiths and Huang 1979).

(i)   The beads are washed three times with phosphate buffer pH 7·4 or 0·01 M
        tris–HCl buffer pH 7·0 to remove the merthiolate.
(ii)   An appropriate volume of beads (generally 10–15 μl) is added to 250 μl of
        the protein sample (e.g. 1 mg ml^{-1}) in the same buffer, followed by 2·5 μl
        of carrier KI (2mM) to give a final iodide concentration of about 10^{-5} M.
(iii)   2 μl of Na^{125}I (100 μCi μl^{-1}) is then added and iodination initiated by the
        addition of 15 μl of H$_2$O$_2$ (2 mM). Further portions of 5 μl of H$_2$O$_2$ are
        added every 5 min. Generally a 20–30 min reaction time is appropriate.
(iv)   The reaction is terminated by the addition of 10 μl of 2·5 M sodium azide.
(v)   The insolubilized lactoperoxidase is removed by low-speed centrifuging,
        and the labelled proteins can be recovered from the supernatant by
        precipitation with trichloracetic acid to 10 per cent if the amounts of
        protein are great enough. Alternatively, the protein can be separated
        from free ^{125}I by dialysis or, probably better, the reaction mixture
        obtained from step (iv) can be applied directly to the top of a small column
        of Sephadex G-25 or Bio-Gel P30 for separation by gel filtration, using for
        example 30 per cent acetic acid for equilibration and elution

Comments

This procedure has also been used for the specific labelling of cell surface proteins,
for which it is particularly suitable since the bulk of the lactoperoxidase–sepharose
particle prevents its access to non-surface proteins and to buried tyrosine residues.
The reaction can be carried out in the presence of 4–8 M urea or 0·1 per cent SDS
to denature and unfold the protein molecules if it is considered desirable to expose
the internal tyrosines of globular proteins to iodination. If the labelled protein is to
be recovered in step (v) by precipitation, yields are generally improved by the
addition of small amounts of an unlabelled carrier protein, such as serum albumin
or immunoglobulin, before the addition of the precipitant. Improvements in yield
are most marked if only very small quantities of labelled protein are present.
Alternative procedures using Iodogen (Tuszynski et al. 1983) or Iodo-beads
(Markwell 1982) as oxidizing agents instead of lactoperoxidase may be more simple
and have advantages if denaturing conditions or detergents are needed to solubilize
sample proteins.

(d) The method of Bolton and Hunter (1973)

Some proteins contain no tyrosine residues and hence cannot be radiolabelled with
^{125}I directly, while others may be damaged by the conditions of the iodination
reaction and lose antigenic or enzymatic activity. In these cases it may be

advantageous to radiolabel an acylating agent such as a hydroxysuccinimide ester which can be separated from the iodination reaction mixture and subsequently reacted with the amino groups of the protein under study.

For this 0.2–0.25 μg of N-succinimidyl 3-(4-hydroxyphenyl) propionate was iodinated by the chloramine-T method and the iodinated product separated by adding 200 μg of carrier KI in 10 μl buffer followed by 5 μl of dimethylformamide and solvent extracting into benzene (two 0.25 ml portions). Owing to the instability of the hydroxysuccinimide ester under the reaction conditions, the procedure must be as rapid as possible and 20 s or less should suffice from the time of Na ^{125}I addition to the solvent extraction stage. The iodinated ester is recovered by evaporation of the benzene under vacuum and if kept dry is stable for at least 5 days.

The protein to be labelled (5 μg) is dissolved in 10 μl of 0.1 M borate buffer pH 8.5, added to the dry iodinated ester (about 3 mol per mole of protein) and the mixture shaken at 0 °C for 15 min. Excess ester is then destroyed by adding 0.5 ml of 0.2 M glycine in 0.1 M borate buffer pH 8.5. After 5 min at 0 °C the ^{125}I-labelled protein is separated from the glycine conjugate, 3-(4-hydroxyphenyl) propionic acid, and other products of the reaction mixture by gel filtration in 0.5 M phosphate buffer pH 7.5 containing 0.25 per cent gelatin to prevent absorption losses on the column. Serum albumin and other plasma proteins should not be used to prevent absorption losses as some low-molecular-weight labelled components of the reaction mixture bind to them.

A1.1.2. Tritium (^3H)

(a) *Reductive methylation (e.g. Kumarasamy and Symons 1979)*

  (i)   100 mCi of ^{3}H-potassium borohydride (13 Ci mmol^{-1}, Radiochemical Centre, Amersham) is dissolved in 100 μl of 0.1 M KOH and 5 μl aliquots dispensed into separate vials, lyophilized, and stored *in vacuo* over NaOH pellets at room temperature.

  (ii)   Just before use the contents of each vial are dissolved in 30 μl of 0.01 M KOH.

  (iii)   The protein to be labelled (5 μg) is dissolved in 50 μl of 0.2 M sodium borate buffer pH 9.0 and cooled in ice.

  (iv)   3 μl of 0.02 M formaldehyde is added to the cold protein solution, and after waiting for 30 s 1 μl (166 μCi) of the tritiated borohydride solution in 0.01 M KOH is added.

  (v)   Allow to stand at 0 °C for 20 min and then add 2 μl of unlabelled 0.01 M NaBH$_4$.

  (vi)   After a further 20 min at 0 °C the reaction is stopped by the addition of 200 μl of 0.4 M sodium phosphate buffer pH 5.8 containing 0.1 M glycine.

  (vii)   A carrier protein such as bovine serum albumin (50 μg) is added, and this and the labelled protein are precipitated by the addition of 3.0 ml of 15 per cent trichloroacetic acid.

  (viii) After standing at 0 °C for 30 min, the precipitate is collected by centrifuging and washed four times with 3 ml of ice-cold diethyl ether:acetone (3:1 by volume).

Comments

The method reductively methylates protein amino groups so the specific activity of the products depends largely upon the lysine content, but most proteins are labelled to a level of about $(1-5) \times 10^5$ cpm μg^{-1}. The presence of 10 per cent glycerol, 0·01 M 2-mercaptoethanol, or 0·5 per cent SDS caused some inhibition of the reaction, and tris–HCl buffers should be avoided as they also lead to slightly lower levels of ^3H incorporation. If more convenient, unlabelled borohydride can be used in step (iv) and unlabelled formaldehyde replaced by ^3H-formaldehyde.

(b) Low pressure tritiation (Lively, Moran, and Powers 1979)

This is a low-pressure gaseous-phase method in which the protein is exposed to $^3H_2$ gas and ^3H exchanges for ^1H mainly in the N–H and O–H groupings in the protein. The method is mild and probably applicable to almost all biological macromolecules, but it does require specialized apparatus.

Protein (3–100 mg) is lyophilized from water or dilute (e.g. 1 mM) buffer solution on to a semi-cylindrical stainless steel plate (9 cm x 15 cm) which is then placed in a stainless steel reaction chamber. This is evacuated to 1×10^{-6} Torr. Tritium gas stored on 10 per cent palladium on charcoal is then admitted by heating the palladium until the pressure reaches 1×10^{-3} Torr (approximately 5 mCi of ^3H). The protein is allowed to remain in this atmosphere for 6–100 h, the excess ^3H is readsorbed on to the palladium–charcoal mixture, and the reaction chamber is re-evacuated to 1×10^{-6} Torr. After at least 12 h the vacuum is released and the labelled protein dissolved in 0·1 M tris–HCl buffer pH 7·2. Readily exchangeable ^3H (usually about 95 per cent of the total absorbed) is removed by dialysis at 4 °C and the protein is purified by affinity chromatography or gel filtration.

(c) Hydrogen–tritium exchange (Voordouw and Roche, 1975; Barksdale and Rosenberg 1978)

    (i)   A reaction mixture might contain the following: protein (2–5 mg), 0·1 M NaCl, 0·01 M tetraethylenepentamine–HCl buffer pH 9·0 or 0·05 M borate buffer pH 9·5, and 1–5 mCi of tritiated water per millilitre.

    (ii) The mixture is incubated at 25 °C for 24 h.

    (iii) Labelled protein is separated from unexchanged tritiated water by gel filtration on a column (e.g. 1·5 cm x 12 cm) of Sephadex G-25 in the buffer used in step (i).

Comments

^3H in the solvent exchanges for ^1H present in the peptide bond and as primary amide hydrogens. It is a reversible reaction, so the label is quite rapidly lost again in solution (e.g. perhaps reduced to 25 per cent in 2 h). Therefore this method is only applicable if the sample is to be examined immediately, but it does have the advantage of being extremely mild and of causing virtually no damage to the protein, loss of activity, etc.

(d) Alkylation with ^3H-iodoacetate or iodacetamide

See Section A1.1.3(a)

A1.1.3. ^{14}C

(a) Alkylation with ^{14}C-iodoacetate or iodacetamide (cf. Gurd 1967)

    (i)   Protein (30–100 μg) is dissolved in 0·1 M tris–HCl buffer pH 8·0 containing urea to 8 M or 1 per cent SDS to denature and unfold the protein.

(ii) Disulphide bonds are reduced by adding dithiothreitol or dithioerythritol to 0·02 M and heating at 100 °C for 4 min.

(iii) $I^{14}CH_2CONH_2$ is added to 0·045 M and the mixture heated at 100 °C for 2 min.

(iv) The labelled protein is separated from low-molecular-weight contaminants by precipitation, dialysis, gel filtration, or ion-exchange chromatography etc.

Comments

The method can also be used for tritium labelling by substituting $IC^3H_2CONH_2$ for the ^{14}C-iodoacetamide used in step (iii). If necessary, urea or SDS can be left out in step (i) and/or step (iii) performed at a lower temperature for a longer time (e.g. 30–60 min at 37 °C), but alkylation is much less efficient if the protein is not completely denatured first.

(b) *Reductive methylation with borohydride (Rice and Means) 1971)*

This method is performed as described above for the tritiation of proteins, except that unlabelled KBH_4 or $NaBH_4$ is used in step (iv) but $H^{14}CHO$ is used in place of the unlabelled formaldehyde. Thus typically 0·1 mg of protein is dissolved in 0·1 ml 0·2 M sodium borate buffer pH 9·0, the solution is cooled in ice, and 10 μl of 0·04 M $H^{14}CHO$ is added. After 30 s four 2 μl portions of $NaBH_4$ (5 mg ml^{-1}) are added sequentially and a further 10 μl after 1 min. Low-molecular-weight components of the reaction mixture are removed by dialysis or gel filtration. Typically, 20 per cent of the protein amino groups are labelled by this method. The method is applicable to whole virus particles and some enzymes, but $NaBH_4$ also reduces disulphide bonds, may cleave peptide bonds, and can cause denaturation or inactivation. In addition the reaction is strongly pH dependent, and much of the formaldehyde is reduced directly to methanol in a side reaction so that the efficiency of methylation of amino groups is low.

(c) *Reductive methylation with cyanoborohydride (Jentoft and Dearborn 1979)*

(i) The reaction mixture contained protein (1 mg ml^{-1}), 2 mM $H^{14}CHO$ (10^5 dpm μmol^{-1}), 20 mM $NaCNBH_3$, and 0·1 M 4-(2-hydroxyethyl)-1-piperazine-ethanesulphonic acid (Hepes) buffer pH 7·5.

(ii) The mixture was incubated at 22 °C for 2–24 h.

(iii) Labelled protein was precipitated by the addition of trichloroacetic acid to 10 per cent and collected by centrifuging.

Comments

This is probably the method of choice at the present time since $NaCNBH_3$ is a much milder reducing agent than $NaBH_4$ and does not reduce aldehydes or ketones at neutral pH. The method overcomes most of the disadvantages of the $NaBH_4$ method, and at formaldehyde concentrations approximately equal to the concentration of the lysyl groups 50–70 per cent of the formaldehyde is covalently bound. When a sixfold excess of formaldehyde is used 80–90 per cent of the lysyl residues are converted into dimethyl derivatives, so under optimal conditions the extent of labelling is two to four times greater than by the $NaBH_4$ method. The reaction is not very sensitive to pH with the longer incubation times and at least the pH range 5–10 can be used, but if incubation times are kept short (e.g. 10 min) the pH range 8–10 is best. Many buffers are compatible with the reaction (e.g. phosphate, borate, EDTA, and tertiary amines such as Hepes), but primary and secondary amines (including tris) must be avoided. Urea, guanidine HCl, SDS, dithiothreitol, and 2-mercaptoethanol do not seriously interfere with the reaction.

A1.1.4. ^{32}P, ^{35}S, ^{14}C, ^3H, etc. by incubation methods

When bacteria or other cell cultures are actively proliferating *in vitro* the proteins can be readily labelled by adding the appropriate radioactive constituent to the culture medium. Similar *in vivo* experiments can be performed by injecting experimental animals with a suitable radioactive material or its metabolic precursor.

The proteins in cell cultures are usually labelled by incubation in the presence of radioactive amino acids. For example, Peterson and McConkey (1976) grew HeLa cells for 3 days in Eagles minimal essential medium containing penicillin-streptomycin, 10 per cent calf serum, and half the normal concentration of unlabelled methionine (7·5 μg ml^{-1}) but supplemented with 5 μCi ml^{-1} of ^{35}S-methionine. O'Farrell (1975) used much shorter labelling times for *Escherichia coli* proteins, and incubated cultures containing (1–3) x 10^8 cells ml^{-1} with 5–20 μCi ml^{-1} of ^{14}C-amino acids for only 5–30 min before adding a large excess of unlabelled amino acids, cooling in ice, and harvesting the cells by centrifuging at 10 000 g for 5 min. Alternatively, he labelled the proteins with ^{35}S by incubating cultures of 10^9 cells ml^{-1} for 6 h in a sulphate-free medium containing 200 μCi of carrier-free ^{35}SO$_4^{2-}$. A rather similar method was used by Lelay, Brunel, and Jeanteur (1978) for ^{32}P labelling of nuclear ribonucleoproteins of HeLa cells in which cultures containing 3 x 10^5 cells ml^{-1} were incubated for 36 h in a medium containing 0·05 mM unlabelled phosphate and 20 mCi of carrier-free H$_3$^{32}PO$_4$. These workers also labelled the ribonucleoproteins after isolation from cells by mixing 80 μg of the protein in 0·2 ml of 0·01 M tris–HCl buffer pH 8·3 containing 0·01 M MgCl$_2$ and 0·01 M 2-mercaptoethanol with 10 μl [γ-^{32}P] ATP (195 μM, 3·8 x 10^5 cpm μl^{-1}). After 10 min incubation at 30 °C the labelled protein was precipitated with 10 per cent trichloroacetic acid. ^{32}P-labelling of serine and threonine residues in membrane proteins with [γ-^{32}P] ATP in a similar way has recently been reported by Treiman, Pødenphant, Saermark, and Bock (1979), but in this case commercial preparations of [γ-^{32}P] ATP (Radiochemical Centre, Amersham) were used. Histones and other cellular proteins have been labelled using ^{32}P inorganic phosphate added to the culture medium in another similar procedure (Sanders, Groppi, and Browning 1980).

1.2. Nucleic acids

A1.2.1.   Incubation methods

In vitro incubation methods, in which radioactive precursors are added to the cell culture medium, are the usual way of incorporating ^{3}H, ^{14}C, or ^{32}P into nucleic acids (e.g. Friesen 1968). Addition of radioactive thymine or thymidine is a convenient way of labelling DNA and radioactive uridine is used for RNA labelling. If radioactive adenine, guanine, or cytosine is used all nucleic acids are labelled (Gall and Pardue 1971).

General labelling of nucleic acids can also be achieved in growing cell cultures by the simple addition of radioactive inorganic phosphate to the culture medium. For example, in a typical experiment to label phage RNA, Studier (1973) added 0·1 ml of *Escherichia coli* cells to 10 μl of chloramphenicol solution (4 mg ml^{-1}) plus 1–2 μl of carrier-free ^{32}PO$_4$ solution to give a final level of ^{32}P between 5 and 20 μCi ml^{-1}.

After about 5 s 0·2–0·5 μl of purified phage in CsCl solution was added, giving about 10 phage particles per cell. The tube was mixed (not shaken) briefly and incubated at 30 °C. After 6 min 1 ml of ice-cold medium was added and the tube placed on ice until the cells were harvested by centrifuging. For efficient labelling of RNA the cells should be grown in a low-phosphate medium before treatment. Proteins present are of course also ^{32}P-labelled at the same time.

A1.2.2. Chemical labelling with ^{125}I

The procedure of Commerford (1971) has been improved by Bean, Sriram, and Webster (1980) and can be used for labelling DNA and RNA. The procedure is as follows:

   (i)   4 μg nucleic acid in 8 μl H_2O is mixed with 15 μl 0·2 M sodium acetate buffer pH 5·0 containing 200 μg ml^{-1} sodium heparin.
   (ii) 5 μl 0·5 mM KI, 2μl^{125}I solution (carrier free, 100 mCi ml^{-1}, pH 8–9), and 1 μl thallic trichloride solution (6 mg ml^{-1}) are added.
   (iii) The mixture is incubated at 45 °C for 15 min.
   (iv) The mixture is chilled in ice and the iodination reaction is stopped by the addition of 1 μl 2-mercaptoethanol.
   (v)   The mixture is diluted with 0·6 ml of 0·1 M sodium acetate buffer pH 5·0 containing 10 per cent glycerol and 10 μg of unlabelled carrier nucleic acid.
   (vi) Residual iodine and other reactants are removed by gel filtration through a small (e.g. 9 ml) column of Sephadex G-25.
   (vii) Nucleic-acid-containing fractions are pooled, extracted once with phenol: chloroform, and precipitated with ethanol.

Comments

At the concentration used in step (ii) thallic trichloride solution is stable and if desired can be stored in the frozen state. After termination of the iodination reaction in step (iv), the working-up procedure described in steps (v)–(vii) can be modified to suit individual requirements. The efficiency of extraction and recovery of small amounts of RNA is improved by adding a carrier after step (iv), and synthetic polyuridylic acid is particularly suitable for this as there is no danger of it becoming labelled also, but with larger amounts of nucleic acid addition of carrier is unnecessary. The conditions used throughout this procedure are mild and were specifically developed to avoid fragmentation, denaturation, or any other disruption of the secondary structure of the nucleic acid being studied and to avoid changes in electrophoretic mobility, all of which tended to be difficulties encountered in earlier methods. The procedure specifically iodinates cytosine bases to 5-iodocytosine and reacts preferentially with single-stranded cytosines.

A1.2.3. Enzymic labelling with ^{32}P

A nucleic acid or restriction fragment can be labelled at either the 5' or 3' end, or both, by treatment with alkaline phosphatase to remove terminal phosphate groups, followed by labelling with ^{32}P by transfer from [γ-^{32}P] ATP with polynucleotide kinase or terminal transferase.

(a) Labelling 5' ends

In the method of Maxam and Gilbert (1977) the phosphorylation reaction was preceded by a heat denaturation in spermidine which increased the labelling 15-fold in the case of flush-ended restriction fragments. The procedure is as follows:

(i)   Dissolve dephosphorylated DNA in 75 μl of 10 mM glycine–NaOH
        buffer pH 9·5 containing 1 mM spermidine and 0·1 mM EDTA.
(ii)   Heat at 100 °C for 3 min and cool in ice water.
(iii)   Add 10 μl of 0·5 M glycine–NaOH buffer pH 9·5 containing 0·1 M MgCl$_2$
        and 0·05 M dithiothreitol.
(iv)   Add 10 μl of [γ-^{32}P] ATP (100 pmol, 1000 Ci mmol^{-1}).
(v)   Add several units of polynucleotide kinase to a final volume of 100 μl.
(vi)   Incubate at 37 °C for 30 min.
(vii)   Add 100μl of 4 M ammonium acetate, 20 μg of carrier tRNA and 600 μl
        of ethanol.
(viii) Mix well, cool to –70 °C, and centrifuge at 12 000 g.
(ix)   Rinse the pellet with ethanol.
(x)   Dry under vacuum.

(b) *Labelling 3' ends*

The 3' end is labelled in a similar way with [γ-^{32}P] ATP and terminal transferase
treatment (Maxam and Gilbert 1977).

(i)   Dissolve DNA in 70 μl of 10 mM tris–HCl buffer pH 7·5 containing 0·1
        mM EDTA.
(ii)   Heat at 100 °C for 3 min and cool in ice water.
(iii)   Add 10 μl of 1·0 M sodium cacodylate (pH 6·9).
(iv)   Add 2 μl of 0·05 M CoCl$_2$.
(v)   Mix and add 2 μl of 5 mM dithiothreitol.
(vi)   Add 10 μl of [γ-^{32}P] ATP (500 pmol, 100 Ci mmol^{-1}).
(vii)   Add several units of terminal transferase to a final volume of 100 μl.
(viii) Incubate at 37 °C for several hours.
(ix)   Perform steps (vii)–(x) of the 5' end labelling method as above.

(c) *Alternative labelling techniques*

3'-termini of RNA can also be labelled using T4 RNA ligase and [5'-^{32}P]pCp
(England, Bruce, and Uhlenbeck 1980), but if the RNA contains a 5'phosphate this
must be removed by a preliminary dephosphorylation with alkaline phosphatase or
it will compete with the labelling reaction.

   For most purposes the above methods give an adequate incorporation of
radiolabel, but for some procedures (e.g. blot transfers, blot hybridizations, colony
screening, *in situ* hybridizations etc.) DNA probes labelled to high specific
activities are required. This can be done (Feinberg and Vogelstein 1983) with the
large fragment of DNA polymerase I, using random oligonucleotides as primers.
This enables as much as 70 per cent of the precursor triphosphate to be
incorporated into complementary DNA and specific activities of over 10^9 dpm μg^{-1}
of DNA can be obtained. The resulting 'oligolabelled' DNA fragments can act as
highly efficient probes for hybridization experiments.

   High specific activity labelled RNA probes can be prepared in a rather similar
way with the aid of SP6 RNA polymerase (Zinn, Di Miae, and Mariatis 1983;
Church and Gilbert 1984).

APPENDIX 2

ADDRESSES OF SOME SUPPLIERS OF ELECTROPHORESIS EQUIPMENT, CHEMICALS, AND RELATED REAGENTS

Amersham/Searle Corp., 2636 S. Clearbrook Drive, Arlington Heights, Ill. 60005, U.S.A.

J.T. Baker Chemical Co., Phillipsburg, N.J. 08865, U.S.A.

Beckman Instruments Inc., 2500 Harbor Boulevard, Fullerton, Calif. 92634, U.S.A.

Bethesda Research Laboratories Inc., 411 North Stonestreet Avenue, Rockville, Md. 20850, U.S.A.

Bio-Rad Laboratories, 220 Wright Avenue, Richmond, Calif. 94804, U.S.A.

Birchcover Instruments Ltd., The Spirella Building, Letchworth, Herts SG6 4ET, Gt. Britain.

Brinkmann Instruments Inc., Cantiague Road, Westbury, N.Y. 11590, U.S.A.

British Drug Houses Chemicals Ltd., Broom Road, Poole, Dorset BN12 4NN, Gt. Britain.

Buchler Instruments, 1327 Sixteenth Street, Fort Lee, N.J. 07024, U.S.A.

Camag, Homburgerstrasse 24, CH-4132 Muttenz, Switzerland.

Camlab Ltd., Nuffield Road, Cambridge CB4 1TH, Gt. Britain.

Canalco—now part of Miles Laboratories.

Carl Zeiss, D–7082 Oberkochen, Federal Republic of Germany.

Chemetron, 20129 Milan, Via G. Modena 24, Italy.

Dakopatts A/S, Strandagervej 14, DK-2900 Hellerup, Denmark.

C. Desaga GmbH, P.O. Box 101969, D-6900 Heidelberg, 1, Federal Republic of Germany.

Eastman–Kodak Co., 1187 Ridge Road West, Rochester, N.Y. 14650, U.S.A.

Fuji Photo Film Co. Ltd., Chemical Products Dept. 26–30, Nishiazabu, 2–chrome, Minato-ku, Tokyo 106, Japan.

Gilford Instrument Laboratories Inc., Oberlin, Ohio 44074, U.S.A.

Hamilton Company, P.O. Box 10030, Reno, Nev. 89510, U.S.A.

Helena Laboratories, P.O. Box 752, Beaumont, Texas 77704, U.S.A.

Hellma Cells Inc., P.O. Box 544, Borough Hall Station, Jamaica, N.Y. 11424, U.S.A.

Hoechst, D-6230 Frankfurt (Main) 80, Federal Republic of Germany.

Hoffman-La Roche Inc., 340 Kingsland Street, Nutley, N.J. 07110, U.S.A.

Instrumentation Specialities Co (ISCO), P.O. Box 5347, Lincoln, Nebr. 68505, U.S.A.

LKB-Produkter AB, S-161 25 Bromma-1, Sweden.

Macherey-Nagel & Co., P.O. Box 307, D-5160 Düren, Federal Republic of Germany.

Marine Colloids Division, FMC Corporation, Bio Products, P.O. Box 308, Rockland, Maine 04841, U.S.A.

E. Merck, Frankfurter Strasse 250, D-6100 Darmstadt 1, Federal Republic of Germany.

Miles Laboratories Inc., P.O. Box 70, Elkhart, Ind. 46515, U.S.A.

Millipore Corp., Bedford, Mass. 01730, U.S.A.

New England Nuclear Corp, 549 Albany Street, Boston, Mass. 02118, U.S.A.
Nuclear Chicago Corp, 333 East Howard Avenue, Des Plaines, Ill. 60018, U.S.A.
Olympus Corp. of America, 4 Nevada Drive, New Hyde Park, N.Y. 11042, U.S.A.
Pharmacia Fine Chemicals AB, Box 175, S-751 04 Uppsala 1, Sweden.
Pierce Chemical Co., P.O. Box 117, Rockford, Ill. 61105, U.S.A.
Radiochemical Centre, White Lion Road, Amersham, Bucks. HP7 9LL, Gt.
Britain.
Radiometer A/S, 72 Emdrupvej, DK-2400 Copenhagen NV, Denmark.
RIA Products Inc., P.O. Box 914, Waltham, Mass. 02154, U.S.A.
Sartorius-Werke, Göttingen, Federal Republic of Germany.
Schleicher & Schüll GmbH, D-3354 Dassel, Federal Republic of Germany.
Serva Feinbiochemica GmbH, P.O. Box 105260, Carl-Benz-Strasse 7, D-6900
Heidelberg, Federal Republic of Germany.
Shandon Southern Products Ltd., 95 Chadwick Road, Astmoor Industrial Estate,
Runcorn, Cheshire WA7 1PR, Gt. Britain.
Sigma Chemical Co., P.O. Box 14508, Saint Louis, MO. 63178, U.S.A.
Technicon Instruments Corp, Tarrytown, N.Y. 10591, U.S.A.
Universal Scientific Ltd., 231 Pleshet Road, London E13 0QU, Gt. Britain.
Wellcome Reagents Ltd., 303 Hither Green Lane, London SE13 6TL, Gt. Britain.
Whatman Ltd., Springfield Mill, Maidstone, Kent ME14 2LE, Gt. Britain.

REFERENCES

AASTED, B. (1980). *Biochem. J.* **189**, 183.

ADAIR, W. S., JURIVICH, D., and GOODENOUGH, U. W. (1978). *J. cell Biol.*, **79**, 281.

AEBERSOLD, R., LEDERMANN, F., BRAUN, D.G., and CHANG, J-Y. (1984). *Anal. Biochem.* **136**, 465.

AIR, G. M., COULSON, A. R., FIDDES, J. C., FRIEDMANN, T., HUTCHISON, C. A., III, SANGER, F., SLOCOMBE, P. M., and SMITH, A. J. H. (1978). *J. molec. Biol.* **125**, 247.

—— SANGER, F., and COULSON, A. R. (1976). *J. molec. Biol.* **108**, 519.

ALBANESE, E. and GOODMAN, D. (1977). *Anal. Biochem.* **80**, 60.

ALDER, S. P., PURICH,P., and STADTMAN, E. R. (1975). *J. biol. Chem.* **250**, 6264.

ALEXANDER, A., CULLEN, B., EMIGHOLZ, K., NORGARD, M. V., and MONAHAN, J. J. (1980). *Anal. Biochem.* **103**, 176.

ALFAGEME, C. R., ZWEIDLER, A., MAHOWALD, A., and COHEN, L. H. (1974). *J. biol. Chem.* **249**, 3729.

ALHANATY, E., TAUBER-FINKELSTEIN, M., and SHALTIEL, S. (1981). *FEBS Lett.* **125**, 151.

ALICHANIDIS, E. and ANDREWS, A. T. (1977). *Biochim. biophys. Acta* **485**, 424.

ALLAN, B. J., WHITE, T. T., KIRK, J., and SCHILLING, J. J. (1981). *Anal. Biochem.* **113**, 1.

ALLEN, R. C. (1978). *J. Chromatogr.* **146**, 1.

—— (1980). *Electrophoresis* **1**, 32.

—— HARLEY, R. A., and TALAMO, R. C. (1974). *Am. J. Clin. Pathol.* **63**, 732.

—— MASAK, K. C., and McALLISTER, P. K. (1980). *Anal. Biochem.* **104**, 494.

ALLINGTON, W. B., CORDRY, A. L., McCULLOCH, G. A., MITCHELL, D. E. and NELSON, J. W. (1978). *Anal. Biochem.* **85**, 188.

ALLORE, R. J. and BARBER, B. H. (1984). *Anal. Biochem.* **137**, 523.

ALTEN, R. C., OULLA, P. M., ARNAUD, P., and BAUMSTARK, J. .S. (1977). In *Electrofocusing and isotachophoresis* (eds. B. J. Radola and D, Graesslin), p. 255. Walter de Gruyter, Berlin, New York.

ALTLAND, K. and KAEMPFER, M. (1980). *Electrophoresis* **1**, 57.

ALWINE, J. C., KEMP, D. J., and STARK, G. R. (1977). *Proc. Natl. Acad. Sci. U.S.A.* **74**, 5350.

AMALRIC, F., MERKEL, C., GELFAND, R., and ATTARDI, G. (1978). *J. molec. Biol.* **118**, 1.

AMBLER, J. (1978*a*). *Clin. Chim. Acta* **85**, 183.

—— (1978*b*). *Clin. Chim. Acta* **88**, 63.

—— (1978*c*). *Clin. Chim. Acta* **89**, 511.

—— (1983). In *Electrophretic techniques* (eds. C. F. Simpson and M. Whittaker), p. 81. Academic Press, New York.

—— Rogers, M. (1980). *Clin. Chem.* **26**, 1221.

—— and WALKER, G. (1979). *Clin. Chem.* **25**, 1320.

AMES, G. F.-L. (1974). *J. biol. Chem.* **249**, 634.

—— and NIKAIDO, K. (1976). *Biochemistry* **15**, 616.

AN DER LAN, B. and CHRAMBACH, A. (1980). *Electrophoresis* **1**, 23.

——, HORUK, R., SULLIVAN, J. V., and CHRAMBACH, A. (1983). *Electrophoresis* **4**, 335.

ANDERSON, B. L., BERRY, R. W., and TELSER, A. (1983). *Anal. Biochem.* **132**, 365.

ANDERSON, M. and ANDREWS, A. T. (1977). *J. dairy Res.* **44**, 223.

——, CAWSTON, T. E., and CHEESEMAN, G. C. (1974). *Biochem. J.* **139**, 653.

ANDERSON, N. G. and ANDERSON, N. L. (1978*a*). *Anal. Biochem.* **85**, 331.

——, ——, and TOLLAKSEN, S. L. (1979). *Clin. Chem.* **25**, 1199.

——, and POWERS, M. T., and TOLLAKSEN, S. L. (1982). *Clin. Chem.* **28**, 1045.

ANDERSON, N. L. and ANDERSON, N. G. (1977). *Proc. Natl. Acad. Sci. U.S.A.* **74**, 5421.

——, —— (1978*b*). *Anal. Biochem.* **85**, 341.

——, —— (1979). *Biochem. Biophys. Res. Commun.* **88**, 258.

——, NANCE, S. L., PEARSON, T. W., and ANDERSON, N. G. (1982). *Electrophoresis*, **3**, 135.

——, TAYLOR, J., SCANDORA, A. E., COULTER, B. P., and ANDERSON, N. G. (1981). *Clin. Chem.* **27**, 1807.

——, WILTSIE, J. C., LI, C. Y., WILLARD-GALLO, K. E., TRACEY, R. P., YOUNG, D. S., POWERS, M. T., and ANDERSON, N. G. (1983). *Clin. Chem.* **29**, 762.

ANDERSSON, L.-O., BORG, H., and MIKAELSSON, M. (1972). *FEBS Lett* **20**, 199.

ANDREWS, A. T. (1975). *J. dairy Res.* **42**, 89.

—— (1978). *Eur. J. Biochem.* **90**, 59.

—— (1984). In *Methods of enzymatic analysis* (ed. H. Bergmeyer, J. Bergmeyer and D. Grassl), Vol. V. p. 277. Verlag Chemie, Weinheim.

—— and PALLAVICINI, C. (1973). *Biochim. biophys. Acta* **321**, 197.

—— and REITHEL, F. J. (1970). *Arch. Biochem. Biophys.* **141**, 538.

ANKER, H. S. (1970). *FEBS Lett.* **7**, 293.

ARAI, K. and WATANABE, S. (1968). *J. Biol. Chem.* **243**, 2670.

ARNAUD, P., CHAPUIS-CELLIER, C., WILSON, G. B., KOISTININ, J., ALLEN, R. C., and FUDENBERG, H. H. (1977). In *Electrofocusing and isotachophresis* (eds. B. J. Radola and D. Graesslin), p. 265. Walter de Gruyter, Berlin, New York.

ARONSON, J. N. and BORRIS, D. P. (1967). *Anal. Biochem.* **18**, 27.

ARTONI, G., GIANAZZA, E., ZANONI, M., GELFI, C., TANZI, M. C., BAROZZI, C., FERRUTI, P., and RIGHETTI, P. G. (1984). *Anal. Biochem.* **137**, 420.

AXELSEN, N. H., KRØLL, J., and WEEKE, B. (1973). *A manual of quantitative immunoeletrophoresis.* Universitetsforlaget, Oslo.

BACHRACH, H. L. (1981). *Anal. Biochem.* **110**, 349.

BADER, J. P., RAY, D. A., and STECK, T. L. (1972). *Biochim. biophys. Acta* **264**, 73.

BAGLEY, E. A., LONBERG-HOLM, K., PANDYA, B. V., and BUDZYNSKI, A. Z. (1983). *Electrophoresis* **4**, 238.

BAILEY, J. M. and DAVIDSON, N. (1976). *Anal. Biochem.* **70**, 75.

BALDENSTEN, A. (1980). *Science Tools* **27**, 2.

BALLARD, C. M., ROBERTS, M. W. H., and DICKINSON, J. P. (1979). *Biochim. biophys. Acta* **582**, 102.

BANKER, G. A. and COTMAN, C. W. (1972). *J. biol. Chem.* **247**, 5856.

BARGER, B. O., WHITE, F. C., PACE, J. L., KEMPER, D. L., and RAGLAND, W. L. (1976). *Anal. Biochem.* **70**, 327.

BARKSDALE, A. D. and ROSENBERG, A. (1978). *Meth. Enzymol.* **48**, 321.

BASSET, P., BRACONNIER, F., AND ROSA, J. (1982). *J. Chromatogr.* **227**, 267.

——, FROISSART, C., VINCENDON, G., and MASSARELLI, R. (1980). *Electrophoresis* **1**, 168.

——, ZWILLER, J., REVEL, M.-O., and VINCENDON, G. (1983). *Electrophoresis* **4**, 399.

BASU, S. P., RAO, S, N., and HARTSUCK, J. A. (1978). *Biochim. biophys. Acta* **533**, 66.

BATES, D. L., PERHAM, R. N., and COGGINS, J. R. (1975). *Anal. Biochem.* **68**, 175.

BATTERSBY, R. V. and HOLLOWAY, C. J. (1982). *Electrophoresis* **3**, 275.

BAUMANN, G. and CHRAMBACH, A. (1975). *Anal. Biochem.* **64**, 530.

——, —— (1976a). *Anal Biochem.* **70**, 32.

——, —— (1976b). *Proc. Natl. Acad. Sci. U.S.A.* **73**, 732.

BAUMANN, H., CAO, K., and HOWALD, H. (1984). *Anal. Biochem.* **137**, 517.

BEAN, W. J., SRIRAM, J., and WEBSTER, R. G. (1980). *Anal. Biochem.* **102**, 228.

BECKER, M. M. and WANG, J. C. (1984). *Nature* (Lond.) **309**, 682.

BEIDLER, J. L., HILLIARD, P. R., and HILL, R. L. *Anal. Biochem.* **126**, 374.

BEISIEGEL, U. and UTERMANN, G. (1979). *Eur. J. Biochem.* **93**, 601.

BELL, L. and BYERS, B. (1983). *Anal. Biochem.* **130**, 527.

BELL, M. L. and ENGVALL, E. (1982). *Anal. Biochem.* **123**, 329.

BENNETT, J. C. (1967). *Meth. Enzymol.* **11**, 330.

BENSON, J. R. and HARE, P. E. (1975). *Proc. Natl. Acad. Sci. U.S.A.* **72**, 619.

BERRY, M. J. and SAMUEL, C. E. (1982). *Anal. Biochem.* **124**, 180.

BERSON, G. (1983). *Anal. Biochem.* **134**, 230.

BERTOLINI, M. J., TANKERSLEY, D. L., and SCHROEDER, D. D. (1976). *Anal. Biochem.* **71**, 6.

BERTOLOTTO, A. and MAGRASSI, M. L. (1984). *Electrophoresis* **5**, 97.

BROWN, A. S., MOLE, J. E., HUNTER, F., and BENNET, J. C. (1980). *Anal. Biochem.* **103**, 184.

BIER, M., CUDDERBACK, R. M., and KOPWILLEM, A. (1977). *J. Chromatogr.* **132**, 437.

—— and KOPWILLEM, A. (1977). In *Electrofocusing and isotachophoresis* (eds. B. J. Radola and D. Graesslin), p. 567. Walter de Gruyter, Berlin, New York.

BIGELIS, R. and BURRIDGE, K. (1978). *Bichem. Biophys. Res. Commun.* **82**, 322.

BIGGIN, M. D., GIBSON, T. J., and HONG, G. F. (1983). *Proc. Natl. Acad. Sci. U.S.A.* **80**, 3963.

BINION, S. B., RODKEY, L. S., EGEN, N. B., and BIER, M. (1982). *Electrophoresis* **3**, 284.

BISPINK, G. and NEUHOFF, V. (1977). In *Electrofocusing and isotachophoresis* (eds. B. J. Radola and D. Graesslin), p. 135. Walter de Gruyter, Berlin, New York.

BITTLE, J. L., HOUGHTEN, R. A., ALEXANDER, H., SHINNICK, T. M., SUTCLIFFE, J. G., LERNER, R. A., ROWLANDS, D. J., and BROWN, F. (1982). *Nature* (*Lond.*) **298**, 30.

BITTNER, M., KUPFERER, P., and MORRIS, C. F. (1980). *Anal. Biochem.* **102**, 459.

BJELLQVIST, B., EK, K., RIGHETTI, P. G., GIANAZZA, E., GÖRG, A., POSTEL, W., and WESTERMEIER, R. (1982). *J. Biochem. Biophys. Meth.* **6**, 317.

BJERRUM, O. J. (1977). *Biochim. biophys. Acta* **472**, 135.

—— (1978). *Anal. Biochem.* **90**, 331.

—— and BHAKDI, S. (1977). *FEBS Lett.* **81**, 151.

——, ——, BØG-HANSEN, T. C., KNÜFERMANN, H., and WALLACH, D. F. H. (1975). *Biochim. biophys. Acta* **406**, 489.

—— and BØG-HANSEN, T. C. (1976). *Biochim. biophys. Acta* **455**, 66.

——, GERLACH, J. H., BØG-HANSEN, T. C., and HERTZ, J. B. (1982). *Electrophoresis* **3**, 89.

——, INGILD, A., LØWENSTEIN, H., and WEEKE, B. (1973). In *A manual of quantitative immunoeletrophoresis* (eds. N. H. Axelsen, J. Krøll, and B. Weeke), p. 145. Universitetsforlagert, Oslo.

—— and LUNDAHL, P. (1973). In *A manual of quantitative immunoelectrophoresis* (eds. N. H. Axelsen, J. Krøll, and B. Weeke), p. 139. Universitetsforlaget, Oslo.

BLACKBURN, S. (1965). *Meth. Biochem. Anal.* **13** 1.

BLAD-HOLMBERG, D. (1979). *Biochim. biophys. Acta* **553**, 25.

BLAGROVE, R. J. and FRENKEL, M. J. (1978). *Anal. Biochem.* **87**, 287.

BLAKE, M. S., JOHNSTON, K. H., RUSSEL-JONES, G. J., and GOTSCHLICH, E. C. (1984). *Anal. Biochem.* **136**, 175.

BLAKESLEY, R. W. and BOEZI, J. A. (1977). *Anal. Biochem.* **82**, 580.

BLANK, A., SILBER, J. R., THELEN, M. P., and DEKKER, C. A. (1983). *Anal. Biochem.* **135**, 423.

——, SUGIYAMA, R. H., and DEKKER, C.A. (1982). *Anal. Biochem.* **120**, 267.

BLIN, N., GABAIN, A. V., and BUJARD, H. (1975). *FEBS Lett.* **53**, 84.

BLOOMSTER, T. G. and WATSON, D. W. (1981). *Anal. Biochem.* **113**, 79.

BLUME, P., MALLEY, A., KNOX, R. J., and SEAMAN, G. V. F. (1978). *Clin. Chem.* **24**, 1300.

BOBB, D. and HOFSTEE, B. H. J. (1971). *Anal. Biochem.* **40**, 209.

BOCCACI, M., MASSA, A. and TENTORI, L. (1981). *Clin. Chim. Acta.* **116**, 137.

BØG-HANSEN, T. C., BJERRUM, O. J., and BROGREN, C. H. (1977). *Anal. Biochem.* **81**, 78.

BOLTON, A. E. and HUNTER, W. M. (1973). *Biochem. J.* **133**, 529.

BOLTZ, R. C., TODD, P., STREIBEL, M. J., and LOUIE, M. K. (1973). *Prep. Biochem.* **3**, 383.

BONCINELLI, E., SIMEONE, A., DE FALCO, A., FIDANZA, V. and LA VOLPE, A. (1983). *Anal. Biochem.* **134**, 40.

BONNER, W. M. and LASKEY, R. A. (1974). *Eur. J. Biochem.* **46**, 83.

—— and STEDMAN, J. D. (1978). *Anal. Biochem.* **89**, 247.

—— WEST, M. H. P., and STEDMAN, J. D. (1980). *Eur. J. Biochem.* **109**, 17.

BOREJDO, J. and FLYNN, C. (1984). *Anal. Biochem.* **140**, 84.

BOSISIO, A. B., LOEHERLEIN, C., SNYDER, R. S., and RIGHETTI, P. G. (1980). *J. Chromatogr.* **189**, 317.

——, SNYDER, R. S., and RIGHETTI, P. G. (1981). *J. Chromatogr.* **209**, 265.

BOSSHARD, H. F. and DATYNER, A. (1977). *Anal. Biochem.* **82**, 327.

BOSSINGER, J., MILLER, M. J., VO, K.-P., GEIDUSCHEK, E. P., and XUONG, N.-H. (1979). *J. biol. Chem.* **254**, 7986.

BOTT, R. R., NAVIA, M. A., and SMITH, J. L. (1982). *J. biol. Chem.* **257**, 9883.

BOURS, J. (1973). *Sci. Tools* **20**, 29.

BOWEN, B., STEINBERG, J., LAEMMLI, U. K., and WEINTRAUB, H. (1980). *Nucl. Acid. Res.* **8**, 1.

BOYD, J. B. and MITCHELL, H. K. (1965). *Anal. Biochem.* **13**, 28.

BRADBURY, W. C., MILLS, S. D., PRESTON, M. A., BARTON, L. J., and PENNER, J. L. (1984). *Anal. Biochem.* **137**, 129.

BRANDT, P. and ERNST, D. (1982). *Electrophoresis* **3**, 174.

BRAY, D. and BROWNLEE, S. M. (1973). *Anal. Biochem.* **55**, 213.

BREWER, J. M. (1967). *Science* **156**, 256.

BRIDGEN, J. (1976). *Biochemistry* **15**, 3600.

BROGREN, C. H. (1977). In *Electrofocusing and isotachophoresis* (eds. B. J.

Radola and D. Graesslin), p. 549. Walter de Gruyter, Berlin, New York.
—— and PELTRE, G. (1977). In *Electrofocusing and isotachophoresis* (eds. B. J. Radola and D. Graesslin), p. 587. Walter de Gruyter, Berlin, New York.

BROWN, J. R. and HARTLEY, B. S. (1966). *Biochem. J.* **101**, 214.

BROWN, R. K., CASPERS, M. L., and VINOGRADOV, S. N. (1977). In *Electrofocusing and isotachophoresis* (eds. B. J. Radola and D. Graesslin), p. 87. Walter de Gruyter, Berlin, New York.

BROWN, W. E. and HOWARD, G. C. (1980). *Anal. Biochem.* **101**, 294.

BROWNLEE, G. G. and CARTWRIGHT, E. .M. (1977). *J. molec. Biol.* **114**, 93.

BROWNSTONE, A. D. (1969). *Anal. Biochem.* **27**, 25.

—— (1976). In *Chromatographic and electrophoretic techniques*, Vol. 2, *Zone electrophoresis* (ed. I. Smith) (4th edn), p. 387. Heinemann, London.

BRYAN, J. K. (1977). *Anal. Biochem.* **78**, 513.

BRYANT, M. L., NALEWAIK, R. P., TIBBS, V. L., and TODARO, G. J. (1979). *Anal. Biochem.* **96**, 84.

BRYCE, C. F. A., MAGNUSSON, C. G. M., and CRICHTON, R. R. (1978). *FEBS Lett.* **96**, 257.

BUDOWLE, B. (1984). *Electrophoresis* **5**, 165.

BURGHES, A. H. M., DUNN, M. J., and DUBOWITZ, V. (1982). *Electrophoresis* **3**, 354.

BURLINA, A. (1979). *Clin. Biochem.* **12**, 71.

BURNETT, D. and CHANDY, K. G. (1983). *Anal. Biochem.* **128**, 317.

BURNETTE, W. N. (1981). *Anal. Biochem.* **112**, 195.

BUZÁS, Z., HJELMELAND, L. M., and CHRAMBACH, A. (1983). *Electrophoresis* **4**, 27.

BYGREN, P., WIESLANDER, J., and HEINEGÅRD, D. (1979). *Biochim. biophys. Acta* **553**, 255.

CAIN, D. F. and PITNEY, R. E. (1968). *Anal. Biochem.* **22**, 11.

CALDWELL, R. C. and PIGMAN, W. (1965). *Arch Biochem. Biophys.* **110**, 91.

CAMPBELL, W. P., WRIGLEY, C. W., and MARGOLIS, J. (1983). *Anal. Biochem.* **129**, 31.

CANN, J. R. (1979). *Meth. Enzymol.* **61**, 142.

CANTAROW, W. D., SARAVIS, C. A., IVES, D. V., and ZAMCHECK, N. (1982). *Electrophoresis* **3**, 85.

CANTZ, M., CHRAMBACH, A., BACH, G., and NEUFELD, E. F. (1972). *J. biol. Chem.* **247**, 5456.

CAPEL, M., REDMAN, B., and BOURQUE, D. P. (1979). *Anal. Biochem.* **97**, 210.

CARBON, P., EHRESMANN, C., EHRESMANN, B., and ERBEL, J.-P. (1979). *Eur. J. Biochem.* **100**, 399.

CARDIN, A. D., WITT, K. R. and JACKSON, R. L. (1984). *Anal. Biochem.* **137**, 368.

CARMICHAEL, G. G. (1980). *Electrophoresis* **1**, 78.

CARON, M., FAURE, A., R. G. and CORNILLOT, P. (1977). *Biochim. biophys. Acta* **491**, 558.

CARREIRA, L. H., CARLTON, B. C., BOBBIO, S. M., NAGAO, R. T., and MEAGHER, R. B. (1980). *Anal. Biochem.* **106**, 455.

CASEY, B. R., WONG, S. T., MASON, A. J., LEE, R. and FORD, H. C. (1981). *Clin. Chim. Acta* **114**, 187.

CASSARA, G., GIANAZZA, E., RIGHETTI, P. G., POMA, S., VINCENTINI, L. and SCORTECCI, V. (1980). *J. Chromatogr.* **221**, 279.

CASTLE, S. L. and BOARD, P. G. (1983). *Electrophoresis* **4**, 277.

CATON, J. E. and GOLDSTEIN, G. (1971). *Anal. Biochem.* **42**, 14.

CATSIMPOOLAS, N. (1980). *Electrophoresis* **1**, 73.
—— and GRIFFITH, A. L. (1977). *Anal. Biochem.* **80**, 555.
CHAO, L. and CHAO, T. (1981). *Electrophoresis* **2**, 60.
CHAMBERLAIN, J.P. (1979). *Anal. Biochem.* **98**, 132.
CHAPUIS-CELLIER, C. and ARNAUD, P. (1981). *Anal. Biochem.* **113**, 325.
CHARLIONET, R., MARTIN, J.P., SESBOUE, R., MADEC, P.J., and LEFEBVRE, F. (1979). *J. Chromatogr.* **176**, 89.
CHEN, J-L. and MORAWETZ, H. (1981). *J. biol. Chem.* **256**, 9221.
CHENG, C. C., BROWNLEE, G. G., CAREY, N. H., DOEL, M. T., GILLAM, S., and SMITH, M. (1976). *J. molec. Biol.* **107**, 527.
CHOO, K. H., COTTON, R. G. H., and DANKS, D. M. (1980). *Anal. Biochem.* **103**, 33.
CHOULES, G.L. and ZIMM, B. H. (1965). *Anal. Biochem.* **13**, 336.
CHRAMBACH, A., AN DER LAN, B., MOHRMANN, H., and FELGENHAUER, K. (1981). *Electrophoresis* **2**, 279.
——, DOERR, P., FINLAYSON, G. R., MILES, L. E. M., SHERINS, R., and RODBARD, D. (1973). *Ann. N.Y. Acad. Sci.* **209**, 44.
—— and NGUYEN, N. Y. (1977) In *Electrofocusing and isotachophoresis* (eds. B. J. Radola and D. Graesslin), p. 51. Walter de Gruyter, Berlin, New York.
——, REISFELD, R. A., WYCKOFF, M., and ZACCARI, J. (1967). *Anal. Biochem.* **20**, 150.
—— and RODBARD, D. (1971). *Science* **172**, 440.
CHRISTOPHER, A. R., NAGPAL, M. L., CARROLL, A. R., and BROWN, J. C. (1978). *Anal. Biochem.* **85**, 404.
CHURCH, G. M. and GILBERT, W. (1984). *Proc. Natl. Acad. Sci. U.S.A.* **81**, 1991.
CINTRON, C., PECZON, B. D., and KUBLIN, C. L. (1978). *Anal. Biochem.* **87**, 622.
CLARK, P. M. S., KRICKA, L. J., and WHITEHEAD, T. P. (1980). *Clin Chem.* **26**, 201.
CLARKE, H. M. G. and FREEMAN, T. (1968). *Clin. Sci.* **35**, 403.
CLARKE, L., HITZEMANN, R., and CARBON, J. (1979). *Meth. Enzymol.* **68**, 436.
CLARKE, P., LIN, H-C., and WILCOX, G. (1982). *Anal. Biochem.* **124**, 88.
CLARKE, S. (1975). *J. biol Chem.* **250**, 5459.
—— (1981). *Biochim. biophys. Acta* **670**, 195.
CLEEVE, H. J. W. and TUA, D. C. (1984). *Clin. Chim. Acta* **137**, 333.
CLEGG, J. C. S. (1982). *Anal. Biochem.* **127**, 389.
CLEVE, H., PATUTSCHNICK, W., POSTEL, W., WESER, J., and GÖRG, A. (1982). *Electrophoresis* **3**, 342.
CLEVELAND, D. W., FISCHER, S. G., KIRSCHNER, M. W., and LAEMMLI, U. K. (1977). *J. biol. Chem.* **252**, 1102.
CLINE, L. J. and CROWLE, A. J. (1979). *Clin. Chem.* **25**, 1749.
COMINGS, D. E. and COHEN, L. W. (1979). *Biochim. biophys. Acta* **578**, 61.
COMMERFORD, S. L. (1971). *Biochemistry*, **10**, 1993.
CONDEELIS, J. S. (1977). *Anal. Biochem.* **77**, 195.
CONFAVREUX, C., GIANAZZA, E., CHAZOT, G., LASNE, Y., and ARNAUD, P. (1982). *Electrophoresis* **3**, 206.
COPPER, P. C. and BURGESS, A. W. (1982). *Anal. Biochem.* **126**, 301.
CORZO, J., RIOL-CIMAS, J. M., and MELÉNDEZ-HEVIA, E. (1984). *Electrophoresis* **5**, 168.
COSSU, G., MANCA, M., PIRASTRU, M. G., BULLITA, R., BOSISIO, A. B., and RIGHETTI, P. G. (1984). *J. Chromatogr.* **307**, 103.
COTRUFO, R., MONSURRO, M. R., DELFINO, G., and GERACI, G. (1983). *Anal. Biochem.* **134**, 313.

CREIGHTON, T. E. (1979). *J. molec. Biol.* **129**, 235.

CREMER, TH., DAMES, W. and NEUHOFF, V. (1972). *Hoppe-Seyler's Z. physiol. Chem.* **353**, 1317.

CULLIFORD, B. J. (1964). *Nature (Lond.)* **201**, 1092.

CUONO, C. B. and CHAPO, G. A. (1982). *Electrophoresis* **3**, 65.

CUTTING, J. A. and ROTH, T. F. (1973). *Anal. Biochem.* **54**, 386.

DABAN, J-R. and ARAGAY, A. M. (1984). *Anal. Biochem.* **138**, 223.

DAHLBERG, A, E., DINGMAN, C. W., and PEACOCK, A. C. (1969). *J. molec. Biol.* **41**, 139.

D'ALESSIO, J. M. (1982). In *Gel electrophoresis of nucleic acids: A practical approach* (eds. D. Rickwood and B. D. Hames), p. 173. I.R.L. Press Ltd., Oxford.

DAMES, W., MAURER, H. R., and NEUHOFF, V. (1972). *Hoppe-Seyler's Z. physiol. Chem.* **353**, 554.

——, —— (1974). In *Electrophoresis and isoelectric focusing in polyacrylamide gel* (eds. R. C. Allen and H. R. Maurer), p. 221. Walter de Gruyter, Berlin, New York.

DANNER, D. B. (1982). *Anal. Biochem.* **125**, 139.

DAS, M., MIYAKAWA, T., FOX, C. F., PRUSS, R. M., AHARONOV, A., and HERSCHMAN, H. R. (1977). *Proc. Natl. Acad. Sci. U.S.A.* **74**, 2790.

DAUSSANT, J., RENARD, H. A., and SKAKOUN, A. (1982). *Electrophoresis* **3**, 99.

DAVID, G. S. (1972). *Biochem. Biophys. Res. Commun.* **48**, 464.

—— and REISFELD, R. A. (1974). *Biochemistry* **13**, 1014.

DAVIE, J. R. (1982). *Anal. Biochem.* **120**, 276.

DAVIES, H. M. and MIFLIN, B. J. (1978). *J. Chromatogr.* **153**, 284.

DAVIES, R. W. (1982). In *Gel electrophoresis of nucleic acids: A practical approach* (eds. D. Rickwood and B. D. Hames), p. 117. I. R. L. Press Ltd., Oxford.

DAVIS, B. J. (1964). *Ann N.Y. Acad. Sci.* **121**, 404.

DEAN, H. R. and WEBB, R. A. (1926). *J. Pathol. Bacteriol.* **29**, 473.

DEININGER, P. L. (1983). *Anal. Biochem.* **135**, 247.

DELINCÉE, H. and RADOLA, B. J. (1977). In *Electrofocusing and isotachophoresis* (eds. B. J. Radola and D. Graesslin), p. 181. Walter de Gruyter, Berlin, New York.

——, —— (1978). *Anal. Biochem.* **90**, 609.

DELMOTTE, P. (1977). *Sci. Tools* **24**, 33.

—— (1979). *J. Chromatogr.* **165**, 87.

DEMACKER, P. N. M., VOS-JANSSEN, H. E., VAN'T LAAR, A., and JANSEN, A. P. (1978). *Clin. Chem.* **24**, 1439.

DEMUS, H. and MEHL, E. (1970). *Biochim. biophys. Acta* **211**, 148.

DERMER, G. B. and EDWARDS, J. J. (1983). *Electrophoresis* **4**, 212.

DE WACHTER, R. and FIERS, W. (1972). *Anal. Biochem.* **49**, 184.

——, —— (1982). In *Gel electrophoresis of nucleic acids: A practical approach* (eds. D. Rickwood and B. D. Hames), p. 77. I. R. L. Press Ltd., Oxford.

DI MARI, S. J., CUMMING, M. A., HASH, J. H., and ROBINSON, J. P. (1982). *Arch. Biochem. Biophys.* **214**, 342.

DOLJA, V. V., NEGRUK, V. I., NOVIKOV, V. K., and ATABEKOV, J. C. (1977). *Anal. Biochem.* **80**, 502.

DOLPHIN, P. J. (1980) *FEBS Lett.* **117**, 252.

DRAWERT, F. and BEDNAR, J. (1979). *J. Agric. Food Chem.* **27**, 3.

DRESCHER, D. G. and LEE, K. S. (1978). *Anal. Biochem.* **84**, 559.

DRETZEN, G., BELLARD, M., SASSONE-CORSI, P., and CHAMBON, P. (1981). *Anal. Biochem.* **112**, 295.

DRYSDALE, J. W. (1977). In *Electrofocusing and isotachophoresis* (eds. B. J. Radola and D. Graesslin), p. 241. Walter de Gruyter, Berlin, New York.

DUBRAY, G. and BEZARD, G. (1982). *Anal. Biochem.* **119**, 325.

DUDOV, K. P., DABEVA, M. D., and HADJIOLOV, A. A. (1976). *Anal. Biochem.* **76**, 250.

DUESBERG, P. H. and RUECKERT, R. R. (1965). *Anal. Biochem.* **11**, 342.

DUNCAN, R, and HESHEY, J. W. B. (1984). *Anal. Biochem.* **138**, 144.

DUNKER, A. K. and KENYON, A. J. (1976). *Biochem.* **153**, 191.

—— and RUECKERT, R. R. (1969). *J. biol. Chem.* **244**, 5074.

DUNN, M. J. and BURGHES, A. H. M. (1983). *Electrophoresis* **4**, 173.

DUPUIS, G. and DOUCET, J-P. (1981). *Biochim. biophys. Acta* **669**, 171.

DYKES, D., NELSON, M., and POLESKY, H. (1983). *Electrophoresis* **4**, 417.

EASTON, D. M., LIPNER, H., HINES, J., and LEIF, R. C. (1971). *Anal. Biochem.* **39**, 478.

ECKERZALL, P. D. and BEELEY, J. A. (1980). *Electrophoresis* **1**, 62.

——, and CONNER, J. G. (1984). *Anal. Biochem.* **138**, 52.

ECKHARDT, A. E., HAYES, C. E., and GOLDSTEIN, I. J. (1976). *Anal. Biochem.* **73**, 192.

EDSTRÖM, J.-E. (1956). *Biochim. biophys. Acta* **22**, 378.

EDWARDS, J. J., TOLLAKSEN, S. L., and ANDERSON, N. G. (1981). *Clin. Chem.* **27**, 1335.

EINARSSON, R., KARLSSON, B., and ÅKERBLOM, E. (1984). *J. Chromatogr.* **284**, 143.

——, and MOBERG, V. (1981). *J. Chromatogr.* **209**, 121.

EISENBACH, L. and EISENBACH, M. (1979). *Anal. Biochem.* **92**, 228.

EK, K. and RIGHETTI, P. G. (1980). *Electrophoresis* **1**, 137.

ELDER, J. K., AMOS, A., SOUTHERN, E. M., and SHIPPY, G. A. (1983). *Anal. Biochem.* **128**, 223.

——, and SOUTHERN, E. M. (1983). *Anal. Biochem.* **128**, 227.

ELEY, M. H., BURNS, P. C., KANNAPELL, C. C., and CAMPBELL, P. S. (1979), *Anal. Biochem.* **92**, 411.

ELSON, E. and JOVIN, T. M. (1969). *Anal. Biochem.* **27**, 193.

ENG, P. R. and PARKES, C. O. (1974). *Anal. Biochem.* **59**, 323.

ENGLAND, T. E., BRUCE, A. G., and UHLENBECK, O. C. (1980). *Meth. Enzymol.* **65**, 65.

ERICKSON, R. H. and KIM, Y. S. (1980). *Biochim. biophys. Acta* **614**, 210.

EVANS, R. W. and WILLIAMS, J. (1980). *Bichem. J.* **189**, 541.

EVENSON, M. A. and DEUTSCH, H. F. (1978). *Clin. Chem. Acta* **89**, 341.

EVERAERTS, F. M., BECKER, J. L., and VERHEGGEN, T. P. E. M. (1973). *Ann. N.Y. Acad. Sci.* **209**, 419.

——, ——, —— (1976). Isotachphoresis – theory, instrumentation and application. *Journal of Chromatography Library*, Vol. 6. Elsevier, Amsterdam.

——, and VERHEGGEN, T. P. E. M. (1983). In *Electrophoretic techniques* (eds. C. F. Simpson and M. Whittaker), p. 149. Academic Press, New York.

FAHEY, R. C., NEWTON, G. L., DORIAN, R., and KOSOWER, E. M. (1980). *Anal. Biochem.* **107**, 1.

FAIRBANKS, G., LEVINTHAL, C., and REEDER, R. H. (1965). *Bichem. Biophys. Res. Commun.* **20**, 393.

FARRELL, D. F., MACMARTIN, M. P., and CLARK, A. F. (1978). *Clin. Chim. Acta* **89**, 145.

FARWELL, D. C. and DION, A. S. (1979). *Anal. Biochem.* **95**, 533.

——, —— (1981). *Anal. Biochem.* **113**, 423.

FASOLI, A., SOMMARIVA, D., SCOTTI, L., MASSAROLI, C., and NEGRATI, M. (1978). *Clin. Chim. Acta*. **87**, 341.

FAULKNER, R. D., CARRAWAY, R., and BHATNAGAR, Y. M. (1982). *Biochim. biophys. Acta* **708**, 245.

FAWCETT, J. S. (1977). In *Electrofocusing and isotachophoresis* (eds. B. J. Radola and D. Graesslin), p. 59. Walter de Gruyter, Berlin, New York.

—— (1983) In *Electrophoretic techniques* (eds. C. F. Simpson and M. Whittaker), p. 57. Academic Press, New York.

—— and MORRIS, C. J. O. R. (1966). *Sep. Stud.* **1**, 9.

FEINBERG, A. P. and VOGELSTEIN, B. (1983). *Anal. Biochem.* **132**, 6.

FEINSTEIN, D. L. and MOUDRIANAKIS, E. N. (1984). *Anal. Biochem.* **136**, 362.

FELGENHAUER, K. (1974). In *Electrophoresis and isoelectric focusing in poly-acrylamide gel* (eds. R. C. Allen and H. R. Maurer), p. 283. Walter de Gruyter, Berlin, New York.

—— (1979). *J. Chromatogr.* **173**, 299.

—— and HAGEDORN, D. (1980). *Clin. Chim. Acta* **100**, 121.

FENNER, C., TRAUT, R. R., MASON, D. T., and COFFELT-WIKMAN, J. (1975). *Anal. Biochem.* **63**, 595.

FENTON, J. C. B. and SHINE, B. S. F. (1979). *Clin. Chim. Acta* **92**, 289.

FERGUSON, K. A. (1964). *Metabolism* **13**, 985.

FERNANDES, P. B., NARDI, R. V., and FRANKLIN, S. G. (1978). *Anal. Biochem.* **91**, 101.

FERNANDES-POL, J. A. (1982). *FEBS Letts.* **143**, 86.

FIERS, W., CONTRERAS, R., DUERINCK, F., HAEGEMAN, G., ISERENTANT, D., MERREGAERT, J., MIN JOU, W., MOLEMANS, F., RAEYMAEKERS, A., VAN DEN BERGHE, A., VOLCKAERT, G., and YSEBAERT, M. (1976). *Nature* (*Lond.*) **260**, 500.

FINLAYSON, G. R. and CHRAMBACH, A. (1971). *Anal. Biochem.* **40**, 292.

FISCHER, S. G. and LERMAN, L. S. (1980). *Meth. Enzymol.* **68**, 183.

FISH, W. W., REYNOLDS, J. A., and TANFORD, C. (1970). *J. biol. Chem.* **245**, 5166.

FISHBEIN, J. C., PLACE, A. R., ROPSON, I. J., POWERS, D. A., and SOFER, W. (1980). *Anal. Biochem.* **108**, 193.

FISHBEIN, W. N. (1972). *Anal. Biochem.* **46**, 388.

FISHER, P. M. and DINGMAN, C. W. (1971). *Biochemistry* **10**, 1895.

FORD, P. J. and MATHIESON, T. (1978). *Eur. J. Biochem.* **87**, 199.

FREDRIKSSON, S. (1977). In *Electrofocusing and isotachophoresis* (eds. B. J. Radola and D. Graesslin), p. 71 Walter de Gruyter, Berlin, New York.

—— (1978). *J. Chromatogr.* **151**, 347.

FREY, M. D. and RADOLA, B. J. (1982*a*) *Electrophoresis* **3**, 27.

——, —— (1982*b*). *Electrophoresis* **3**, 216.

FRIEDEL, K. and HOLLOWAY, C. J. (1981). *Electrophoresis* **2**, 116.

FRIES, E. (1976). *Anal. Biochem.* **70**, 124.

—— and HJERTÉN, S. (1975). *Anal. Biochem.* **64**, 466.

FRIESEN, J. D. (1968). *Meth. Enzymol.* **12B**, 625.

FUJITA, T., TODA, T., and OHASHI, M. (1984). *Anal. Biochem.* **139**, 463.

FUNDERBURGH, J. L. and CHANDLER, J. W. (1978). *Anal. Biochem.* **91**, 462.

FURLAN, M., PERRET, B. A. and BECK, E. A. (1979). *Anal. Biochem.* **96**, 208.

FURTH, A. J. (1980). *Anal. Biochem.* **109**, 207.

GALL, J. G. and PARDUE, M. L. (1971). *Meth. Enzymol.* **21**, 470.

GANNON, F., O'HARE, K., PERRIN, F., LE PENNEC, J. P., BENOIST, C., COCHET,

M., BREATHNACH, R., ROYAL, A., GARAPIN, A., CAMI, B., and CHAMBON, P. (1979). *Nature (Lond.)* **278**, 428.

GAROFF, H. and ANSORG, W. (1981). *Anal. Biochem.* **115**, 450.

——, FRISCHAUF, A. M., SIMONS, K., LEHRACH, H., and DELIUS, H. (1980). *Nature* (Lond). **288**, 236.

GARRELS, J. J. (1979). *J. biol. Chem.* **254**, 7961.

GELFI, C. and RIGHETTI, P. G. (1981*a*) *Electrophoresis* **2**, 213.

——, —— (1981*b*). *Electrophoresis* **2**, 220.

GELSEMA, W. J., DE LIGNY, C. L., and VAN DER VEEN, N. G. (1979*a*). *J. Chromatogr.* **171**, 171.

——, ——, —— (1979*b*). *J. Chromatogr.* **173**, 33.

GEORGE, H. J., MISRA, L., FIELD, D. J., and LEE, J. C. (1981). *Bichemistry* **20**, 2402.

GERSHONI, J. M. and PALADE, G. (1982). *Anal. Biochem.* **124**, 396.

——, —— (1983). *Anal. Biochem.* **131**, 1.

GEYSEN, J., DE LOOF, A., and VANDESANDE, F. (1984). *Electrophoresis* **5**, 129.

GHADGE, G. D., BODHE, A. M., MODAK, S. R., and VARTAK, H. G. (1983). *Anal. Biochem.* **128**, 468.

GIANAZZA, E., and AROSIO, P. (1980). *Biochim. biophys. Acta* **625**, 310.

——, CELENTANO, F., DOSSI, G., BJELLQVIST, B., and RIGHETTI, P. G. (1984). *Electrophoresis* **5**, 88.

——, CHILLEMI, F., DURANTI, M., and RIGHETTI, P. G. (1983). *J. Biochem. Biophys. Meth.* **8**, 339.

——, FRIGERIO, A., TABLIABUE, A., and RIGHETTI, P. G. (1984). *Electrophoresis* **5**, 209.

GIANNOVARIO, J. A., GRIFFIN, R. N., and GRAY, E. L. (1978). *J. Chromatogr.* **153**, 329.

GIBSON, D. R. and GRACY, R. W. (1979). *Anal. Biochem.* **96**, 352.

GIBSON, W. (1981). *Anal. Biochem.* **118**, 1.

GIOMETTI, C. A., ANDERSON, N. G., and ANDERSON, N. L. (1979). *Clin. Chem.* **25**, 1877.

——, BARANY, M., DANON, M. J., and ANDERSON, N. G. (1980). *Clin. Chem.* **26**, 1152.

GIULIANI, A., MARINUCCI, M., CAPPABIANCA, M. P., MAFFI, D., and TENTORI, L. (1978). *Clin. Chem. Acta* **90**, 19.

GLASS, W. F., BRIGGS, R. C., and HNILICA, L. S. (1981). *Anal. Biochem.* **115**, 219.

GLOSSMANN, H. and NEVILLE, D. M. (1971). *J. biol. Chem.* **246**, 6339.

GOLDBERG, R. L. and FULLER, G. C. (1978). *Anal. Biochem.* **90**, 69.

GOLDENBERG, D. P. and CREIGHTON, T. E. (1984). *Anal. Biochem.* **138**, 1.

GOLDMAN, D. and MERRIL, C. R. (1982). *Electrophoresis* **3**, 24.

GOLDMAN, E. and HATFIELD, G. W. (1979). *Meth. Enzymol.* **59**, 292.

——, HOLMES, W. M., and HATFIELD, G. W. (1979). *J. molec. Biol.* **129**, 567.

GONENNE, A. and LEBOWITZ, J. (1975). *Anal. Biochem.* **64**, 414.

GORDON, A. H. (1975). *Laboratory Techniques in Biochemistry and Molecular Biology*, Vol. 1, *Electrophoresis of Proteins in Polyacrylamide and Starch Gels* (e.g. T. S. Work and E. Work) (2nd edn). North-Holland, Amsterdam, London.

GORELIK, F. S., FREEDMAN, S. D., DELAHUNT, N. G., GERSHONI, J. M., and JAMIESON, J. D. (1982). *J. Cell. Biol.* **93**, 401a.

GÖRG, A., POSTEL, W., WESER, J., WEIDINGER, S., PATUTSCHNICK, W., and CLEVE, H. (1983). *Electrophoresis* **4**, 153.

——, ——, and WESTERMEIER, R. (1978). *Anal. Biochem.* **89**, 60.

——, ——, ——, BJELLQVIST, B., EK, K., GIANAZZA, E., and RIGHETTI, P. G. (1983). In *Electrophoresis '82* (ed. D. Stathakos), Walter de Gruyter, Berlin.

GOULIANOS, K., SMITH, K. K., and WHITE, S. N. (1980). *Anal. Biochem.* **103**, 64.

GOWER, D. C. and WOLEDGE, R. C. (1977). *Sci. Tools* **24**, 17.

GRABAR, P. and WILLIAMS, C. A. (1953). *Biochim. biophys. Acta* **10**, 193.

GRACY, R. W. (1977). *Meth. Enzymol.* **47**, 195.

GRAESSLIN, D., WEISE, H. C., and RICK, M. (1976). *Anal. Biochem.* **71**, 492.

GRATZER, W. B. and BEAVEN, G. H. (1961). *J. Chromatogr.* **5**, 315.

GREEN, G. R., POCCIA, D., and HERLANDS, L. (1982). *Anal. Biochem.* **123**, 66.

GREEN, M. R., PASTEWKA, J. V., and PEACOCK, A. C. (1973). *Anal. Biochem.* **56**, 43.

GRIFFITH, A., CATSIMPOOLAS, N., and KENNEY, J. (1973). *Ann. N.Y. Acad. Sci.* **209**, 457.

GRIFFITH, I. P. (1972). *Anal. Biochem.* **46**, 402.

GRIFFITHS, G. R. and HUANG, P. C. (1979). *J. biol. Chem.* **254**, 8057.

GROSSBACH, U. (1974). In *Electrophoresis and isoelectric focusing in poly-acrylamide gel* (ed. R. C. Allen and H. R. Maurer), p.207. Walter de Gruyter, Berlin, New York.

GUILLEMETTE, J. G. and LEWIS, P. N. (1983). *Electrophoresis* **4**, 92.

GUILLEUX, F., HAYER, M., THOMAS, N., and DE BORNIER, B. M. (1978). *Clin. Chim. Acta* **87**, 383.

GUINET, R. F. M. (1983). *Electrophoresis* **4**, 224.

GURD, F. R. N. (1967). *Meth. Enzymol.* **11**, 532.

HAFF, L. A., LASKY, M., and MANRIQUE, A. (1979). *J. biochem. biophys. Meth.* **1**, 275.

HAGEN, F. S. (1979). *Anal. Biochem.* **93**, 299.

HAGEN, I., BJERRUM, O., and SOLUM, N. O. (1979). *Eur. J. Biochem.* **99**, 9.

HAGER, D. A. and BURGESS, R. R. (1980). *Anal. Biochem.* **109**, 76.

HAGLUND, H. (1967). *Sci. Tools* **14**, 17.

—— (1970). *Sci. Tools* **17**, 2.

—— (1971). *Meth. Biochem. Anal.* **19**, 1.

HALLINAN, F. U. (1983). *Electrophoresis* **4**, 265.

HANNIG, K. (1978). *J. Chromatogr.* **159**, 183.

—— (1982). *Electrophoresis* **3**, 235.

——, WIRTH, H., MEYER, B.-H., and ZEILLER, K. (1975). *Hoppe-Seyler's Z. physiol. Chem.* **356**, 1209.

HANSEN, J.-E. S., LIHME, A., and BØGHANSEN, T. C. (1984). *Electrophoresis* **5**, 196.

HANSEN, J. N. (1976). *Anal. Biochem.* **76**, 37.

—— (1981). *Anal. Biochem.* **116**, 146.

——, PHEIFFER, B. H., and BOEHNERT, J. A. (1980). *Anal. Biochem.* **105**, 192.

HARADA, S. (1975). *Clin. Chim. Acta* **63**, 275.

HARBOE, N. and INGILD, A. (1973). In *A manual of quantitative immunoelectro-phoresis* (eds. N. H. Axelsen, J. Krøll, and B. Weeke), p. 161. Universitets-forlaget, Oslo.

HARDY, B., HOFFMAN, J., and OSSIMI, Z. (1984). *Biochem. Biophys. Res. Commun.* **120**, 325.

HARE, D. L., STIMPSON, D. I., and CANN, J. R. (1978). *Arch. Biochem. Biophys.* **187**, 274.

HARELL, D. and MORRISON, M. (1979). *Arch. Biochem. Biophys.* **193**, 158.

HARRINGTON, M. G., MERRIL, C. R., GOLDMAN, D., XU, X., and MCFARLIN, D. E. (1984). *Electrophoresis* **5**, 236.

HARTLEY, B. S. (1970). *Biochem. J.* **119**, 805.

HARTMANN, B. K. and UDENFRIEND, S. (1969). *Anal. Biochem.* **30**, 391.

HAUZER, K., TICHÁ, M., HOŘEJŠÍ, V., and KOCOUREK, J. (1979). *Biochim. biophys. Acta* **583**, 103.

HEARING, V. J., KLINGLER, W. G., EKEL, T. M., and MONTAGUE, P. M. (1976). *Anal. Biochem.* **72**, 113.

HEDRICK, J. L. and SMITH, A. J. (1968). *Arch. Biochem. Biophys.* **126**, 155.

——, SMITH, A. J., and BRUENING, G. E. (1969). *Biochemistry*, **8**, 4012.

HEEGAARD, N. H. H., HEBSGAARD, K. P., and BJERRUM, O. J. (1984). *Electrophoresis* **5**, 263.

HEILMANN, J., BARROLLIER, J., and WATZKE, E. (1957). *Z. physiol. Chem.* **309**, 219.

HELENIUS, A. and SIMONS, K. (1977). *Proc. Natl. Acad. Sci. U.S.A.* **74**, 529.

HENDERSON, L. E., OROSZLAN, S., and KONIGSBERG, W. (1979). *Anal. Biochem.* **93**, 153.

HENTSCHEL, C., IRMINGER, J.-C., BUCHER, P., and BIRNSTIEL, M. L. (1980). *Nature* (London) **285**, 147.

HERBERT, W. J. (1973). In *Handbook of experimental immunology*, Vol. 3 (ed. D. M. Weir), (2nd Edn) Appendixes 2 and 3. Blackwell, Oxford.

—— (1978). In *Handbook of Experimental Immunology*, Vol. 3 (ed. D. M. Weir) (3rd Edn) Appendixes 3 and 4. Blackwell, Oxford.

HILLIER, R. M. (1976). *J. dairy Res.* **43**, 259.

HINDLEY, J. (1983). *Laboratory Techniques in Biochemistry and Molecular Biology*, Vol. 10, DNA Sequencing. Elsevier Biomedical Press, New York.

HIROKAWA, T., TAKEMI, H., KISO, Y., TAKIYAMA, R., MORIO, M., FUJII, K., and KIKUCHI, H. (1984). *J. Chromatogr.* **305**, 429.

HJELMELAND, L. M. and CHRAMBACH, A. (1981). *Electrophoresis* **2**. 1.

——, —— (1983). *Electrophoresis* **4**, 20.

HJERTÉN, S. (1962). *Arch. Biochem. Biophys.* Suppl. 1, 147.

—— (1970). *Meth. Biochem. Anal.* **18**, 55.

—— (1983a). *J. Chromatogr.* **270**, 1.

—— (1983b). *Biochim. biophys. Acta* **736**, 130.

——, JERSTEDT, S., and TISELIUS, A. (1965). *Anal. Biochem.* **11**, 219.

——, ÖFVERSTEDT, L.-G., and JOHANSSON, G. (1980). *J. Chromatogr.* **194**, 1.

HO, N. W. Y. (1979). *Anal. Biochem.* **97**, 51.

—— (1983). *Electrophoresis* **4**, 168.

HOCH, S. O. (1982). *Biochem. Biophys. Res. Commun.* **106**, 1353.

HOEG, J. M., PAPADOPOULOS, N. M., GREGG, R. E., and BREWER, H. B. (1983). *Clin. Chem.* **29**, 1459.

HOFFMAN, W. L. and DOWBEN, R. M. (1978a). *Anal. Biochem.* **89**, 414.

——, —— (1978b). *Anal. Biochem.* **89**, 540.

HOFFMEISTER, H. (1974). In *Electrophoresis and isoelectric focusing in polyacrylamide gel* (eds. R. C. Allen and H. R. Maurer), p. 266. Walter de Gruyter, Berlin, New York.

HOH, J. F. Y., YEOH, G. P. S., THOMAS, M. A. W., and HIGGINBOTTOM, L. (1979). *FEBS Lett.* **97**, 330.

HOLLOWAY, C. J. and LÜSTORFF, J. (1980). *Electrophoresis* **1**, 129.

——, and PINGOUD, V. (1981). *Electrophoresis* **2**, 127.

HOLTLUND, J. and KRISTENSEN, T. (1981). *Anal. Biochem.* **87**, 425.

HONDA, S., WAKASA, H., TERAO, M., and KAKEHI, K. (1979). *J. Chromatogr.* **177**, 109.

HOPWOOD, J. J. and HARRISON, J. R. (1982). *Anal. Biochem.* **119**, 120.

HOŘEJŠÍ, V. (1979). J. Chromatogr. **178**, 1.

—— (1981). Anal. Biochem. **112**, 1.

—— and TICHÁ, M. (1981a). Anal. Biochem. **116**, 22.

——, —— (1981b). J. Chromatogr. **216**, 43.

——, —— and KOCOUREK, J. (1977a). Biochim. biophys. Acta **499**, 290.

——, ——, —— (1977b). Biochim. biophys. Acta **499**, 301.

——, ——, TICHÝ, P., and HOLÝ, A. (1982). Anal. Biochem. **125**, 358.

HORIGOME, T. and SUGANO, H. (1983). Anal. Biochem. **130**, 393.

HORST, M. N., BASHA, M. M., BAUMBACH, G. A., MANSFIELD, E. H., and ROBERTS, R. M. (1980). Anal. Biochem. **102**, 399.

——, BAUMBACH, G., and ROBERTS, R. M. (1979). FEBS Lett. **100**, 385.

HOUGHTON, M., EATON, M. A. W., STEWART, A. G., SMITH, J. C., DOEL, S. M., CATLIN, G. H., LEWIS, H. M., PATEL, T. P., EMTAGE, J. S., CAREY, N. H., and PORTER, A. G. (1980). Nucleic Acids Res. **8**, 2885.

HOUSTON, L. L. (1971). Anal. Biochem. **44**, 81.

HOWARD, L., SHULMAN, S., SADANANDAN, S., and KARPATKIN, S. (1982). J. biol. Chem. **257**, 8331.

IGLOI, G. L. (1983). Anal. Biochem. **134**, 184.

INOUYE, M. (1971). J. biol. Chem. **246**, 4834.

JACKSON, J. H. and RUSSELL, P. J. (1984). Anal. Biochem. **137**, 41.

JACKSON, P. and THOMPSON, R. J. (1984). Electrophoresis **5**, 35.

JACOBSON, K. B. and VAUGHAN, C. (1977). Anal. Biochem. **78**, 295.

JALKONEN, M., SALMI, A., PELTONEN, J., and KOURI, T. (1983). J. Chromatogr. **268**, 79.

JAMES, R. and BRADSHAW, R. A. (1984). Anal. Biochem. **140**, 456.

JANSSEN, P. T. and VAN BIJSTERVELD, O. P. (1981). Clin. Chim. Acta **114**, 207.

JANSSON, P. A., GRIM, L. B., ELIAS, J. G., BAGLEY, E. A., and LONBERG-HOLM, K. K. (1983). Electrophoresis **4**, 82.

JEFFERIS, R. (1975). Sci. Tools **22**, 33.

—— and BUTWELL, A. J. (1975). Sci. Tools **22**, 1.

JENTOFT, N. and DEARBORN, D. G. (1979). J. biol. Chem. **254**, 4359.

JEPPESEN, P. G. N. (1974). Anal. Biochem. **58**, 195.

—— (1980). Meth. Enzymol. **65**, 305.

JEPPSSON, J. O. (1977a). In Isoelectric focusing and isotachophoresis (eds. B. J. Radola and D. Graesslin), p. 273. Walter de Gruyter, Berlin, New York.

—— (1977b). Application Note 307. LKB Produkter AB.

——, FRANZÉN, B., and NILSSON, K. O. (1978). Sci. Tools **25**, 69.

JOHANSSON, B. G. and HJERTÉN, S. (1974). Anal. Biochem. **59**, 200.

JOHNSON, A. M. (1982) Clin. Chem. **28**, 1797.

JOHNSON, B. F. (1982). Anal. Biochem. **127**, 235.

JOHNSON, B. L. (1967). Nature (Lond.) **216**, 859; Science **158**, 131.

JOHNSON, M. H., WALKER, R. W. H., KEIR, G., and THOMPSON, E. J. (1982). Biochim. biophys. Acta **718**, 121.

JOHNSON, P. H. and GROSSMAN, L. I. (1977). Biochemistry **16**, 4217.

JOHNSON, S. J., METCALF, E. C., and DEAN, P. D. G. (1980). Anal. Biochem. **109**, 63.

JOHNSON, T. K., YUEN, K. C. L., DENELL, R. E., and CONSIGLI, R. A. (1983). Anal. Biochem. **133**, 126.

JONES, G. D., WILSON, M. T., and DARLEY-USMAR, V. M. (1981). Biochem. J. **193**, 1013.

JONES, M. J. and SPRAGG, S. P. (1983). Electrophoresis **4**, 291.

JONES, W. T., BROADHURST, R. B., and GURNSEY, M. P. (1982). *Biochim. biophys. Acta* **701**, 382.

JONSSON, M., FREDRISKSSON, S., JONTELL, M., and LINDE, A. (1978). *J. Chromatogr.* **157**, 235.

—— and RILBE, H. (1980). *Electrophoresis* **1**, 3.

JORDAN, E. M. and RAYMOND, S. (1969). *Anal. Biochem.* **27**, 205.

JORGENSON, J. W. and LUKACS, K. DE A. (1981). *Clin. Chem.* **27**, 1551.

JOVIN, T. M. (1973). *Biochemistry* **12**, 871, 879, 890; *Ann. N.Y. Acad. Sci.* **209**, 477.

——, CHRAMBACH, A., and NAUGHTON, M. A. (1964). *Anal. Biochem.* **9**, 351.

——, DANTE, M. L., and CHRAMBACH, A. (1970). *Multiphasic Buffer Systems Output, PB 196085–196092, 203016.* National Technical Information Service, Springfield, Va.

JUANG, R-H., CHANG, Y.-D., SUNG, H-Y., and SU, J-C. (1984). *Anal. Biochem.* **141**, 348.

JUDD, W. (1979). *Anal. Biochem.* **93**, 373.

KALTSCHMIDT, E. and WITTMANN, H. G. (1970). *Anal. Biochem.* **36**, 401.

KAPADIA, G., CHRAMBACH, A., and RODBARD, D. (1974). In *Electrophoresis and isoelectric focusing in polyacrylamide gel* (eds. R. C. Allen and H. R. Maurer), p. 115. Walter de Gruyter, Berlin, New York.

——, VAITUKAITIS, J. L., and CHRAMBACH, A. (1981). *Prep. Biochem.* **11**, 1.

KAPP, O. H. and VINOGRADOV, S. N. (1978). *Anal. Biochem.* **91**, 230.

KARLSSON, C., DAVIES, H., ÖHMAN, J., and ANDERSSON, U.-B. (1973). *Application Note No. 75.* LKB Produkter AB.

KARLSTAM, B. (1981). *J. Chromatogr.* **211**, 233.

KARPETSKY, T., BROWN, G. E., MCFARLAND, E., BRADY, S. T., ROTH, W., RAHMAN, A., and JEWETT, P. (1984). *Biochem. J.* **219**, 553.

KELLY, C., TOTTY, N. F., WATERFIELD, M. D., and CRUMPTON, M. J. (1983). *Biochem. Internat.* **6**, 535.

KEMPNER, D. H., SMOLKA, A. J. K., and REMBAUM, A. (1982). *Electrophoresis* **3**, 109.

KERCKAERT, J.-P. (1978). *Anal. Biochem.* **84**, 354.

——, BAYARD, B., and BISERTE, G. (1979). *Biochim. biophys. Acta* **576**, 99.

KESSLER, R. E. (1981). *Anal. Biochem.* **116**, 129.

KIM, Y. K., YAGUCHI, M., and ROSE, D. (1969). *J. dairy Sci.* **52**, 316.

KINDMARK, C. O. and THORELL, J. I. (1972). *Scand. J. clin. lab. Invest.* **29** (Suppl. 124), 49.

KING, E. E. (1970). *J. Chromatogr.* **53**, 559.

KINGSBURY, N. and MASTERS, C. J. (1970). *Anal. Biochem.* **36**, 144.

KINZKOFER, A. and RADOLA, B. J. (1981). *Electrophoresis* **2**, 174.

KIRKPATRICK, F. H. and ROSE, D. J. (1978). *Anal. Biochem.* **89**, 130.

KITAMURA, N., SEMLER, B. L., ROTHBERG, P. G., LARSEN, G. R., ADLER, C. J., DORNER, A. J., EMINI, E. A., HANECAK, R., LEE, J. J., VAN DER WERF, S., ANDERSON, C. W., and WIMMER, E. (1981). *Nature (Lond.).* **291**, 547.

KITAZOE, Y., MIYAHARA, M., HIRAOKA, N., VETA, H., and UTSUMI, K. (1983). *Anal. Biochem.* **134**, 295.

KITTLER, J. M., MEISLER, N. T., VICEPS-MADORE, D., CIDLOWSKI, J. A., and THANASSI, J. W. (1984). *Anal. Biochem.* **137**, 210.

KJELLIN, K. G., MOBERG, U., and HALLANDER, L. (1977). *Sci. Tools* **22**, 3.

KLASEN, E. C. and RIGUTTI, A. (1982). *Electrophoresis* **3**, 168.

KLEINE, T. O. and ENDERS, C. (1979). *J. clin. Chem. clin. Biochem.* **17**, 509.

KLOSTERMEYER, H. and OFFT, S. (1978). *Z. Lebensml.-Unters.-Forsch.* **167**, 158.

KLUGE-WILM, R. (1978). *Fleischwirtschaft* **58**, 1068, 1103.

KNECHT, D. A. and DIMOND, R. L. (1984). *Anal. Biochem.* **136**, 180.

KOCH, G. L. E. and SMITH, M. J. (1982). *Eur. J. Biochem.* **128**, 107.

KODAMA, H. and UASA, S. (1979). *J. Chromatogr.* **163**, 300.

KOHLRAUSCH, F. (1897). *Annln Phys. Chem.* **62**, 209.

KOLIN, A. (1977). In *Electrofocusing and isotachophoresis* (eds. B. J. Radola and D. Graesslin), p. 3. Walter de Gruyter, Berlin, New York.

KOLODNY, G. M. (1984). *Anal. Biochem.* **138**, 66.

KOPWILLEM, A. and LUNDIN, H. (1974). *LKB Application Note 183.*

KORANT, B. D. and LONBERG-HOLM, K. (1974). *Anal. Biochem.* **59**, 75.

KOSSMANN, K. T. (1984). *J. Clin. Chem. Clin. Biochem.* **22**, 253.

KOZIARZ, J. J., KÖHLER, H., and STECK, T. L. (1978). *Anal. Biochem.* **86**, 78.

KRISHNAMOORTHY, R., BOSISIO, A. B., LABIE, D., and RIGHETTI, P. G. (1978). *FEBS Letts.* **94**, 319.

KRØLL, J. (1973a). In *A manual of quantitative immunoelectrophoresis* (eds. N. H. Axelsen, J. Krøll, and B. Weeke), p. 61. Universitetsforlaget, Oslo.

—— (1973b). In *A manual of quantitative immunoelectrophoresis* (eds. N. H. Axelsen, J. Krøll, and B. Weeke), p. 79. Universitetsforlaget, Oslo.

—— (1973c). In *A manual of quantitative immunoelectrophoresis* (eds. N. H. Axelsen, J. Krøll, and B. Weeke), p. 83. Universitetsforlaget, Oslo.

KRUSKI, A. W. and NARAYAN, K. A. (1974). *Anal. Biochem.* **60**, 431.

KUBA, K., LIPPEL, K., and FRANTZ, I. D. (1979). *Clin. Chem.* **25**, 1471.

KUBAK, B. M. and YOTIS, W. W. (1982). *Biochim. biophys. Acta* **687**, 238.

KUBICZ, A. and WOLANSKA, L. (1977). In *Electrofocusing and isotachophoresis* (eds. B. J. Radola and D. Graesslin), p. 233. Walter de Gruyter, Berlin, New York.

KUMARASAMY, R. and SYMONS, R. H. (1979). *Anal. Biochem.* **95**, 359.

KURNICK, N. B. (1950). *Exp. cell Res.* **1**, 151.

LAAN, H. W. and DIAZ, D. P. (1978). *Clin. Chim. Acta.* **88**, 163.

LÅÅS, T. and OLSSON, I. (1981a). *Anal. Biochem.* **114**, 167.

——, —— (1981b). *Electrophoresis* **2**, 235.

——, ——, and SÖDERBERG, L. (1980). *Anal. Biochem.* **101**, 449.

LACKS, S. A. and SPRINGHORN, S. S. (1980). *J. biol. Chem.* **255**, 7467.

——, ——, and ROSENTHAL, A. L. (1979). *Anal. Biochem.* **100**, 357.

LAEMMLI, U. K. (1970). *Nature (Lond.)* **227**, 680.

LAI, C. Y. (1977). *Meth. Enzymol.* **47**, 236.

LAM, K. S. and KASPER, C. B. (1980). *Anal. Biochem.* **108**, 220.

LAMAS, L. (1979). *Eur. J. Biochem.* **96**, 93.

LAMBIN, P. (1978). *Anal. Biochem.* **85**, 114.

LANCASTER, M. V. and SPROUSE, R. F. (1977). *Anal. Biochem.* **77**, 158.

LANE, J. D., ZIMMER, H. G., and NEUHOFF, V. (1979). *Hoppe-Seyler's Z. physiol. Chem.* **360**, 1405.

LASKEY, R. A. (1980). *Meth. Enzymol.* **65**, 363.

LASKEY, R. A. and MILLS, A. D. (1975). *Eur. J. Biochem.* **56**, 335.

LASKY, M. and MANRIQUE, A. (1980). *Electrophoresis* **1**, 119.

LASNE, F., BENZERARA, O., and LASNE, Y. (1982). *Biochim. biophys. Acta* **703**, 49.

LASTICK, S. M. and MCCONKEY, E. H. (1976). *J. biol. Chem.* **251**, 2867.

LATNER, A. L. (1975). *Adv. clin. Chem.* **17**, 193.

LATTER, G. I., METZ, E., BURBECK, S., and LEAVITT, J. (1983). *Electrophoresis* **4**, 122.

LAURELL, C. B. (1965). *Anal. Biochem.* **10**, 358.
—— (1966). *Anal. Biochem.* **15**, 45.
LAW, B. A., ANDREWS, A. T., and SHARPE, M. E. (1977). *J. dairy Res.* **44**, 145.
LEABACK, D. H. (1977). In *Electrofocusing and isotachophoresis* (eds. B. J. Radola and D. Graesslin), p. 155. Walter de Gruyter, Berlin, New York.
LEACH, B. S., COLLAWN, J. F., and FISH, W. W. (1980). *Biochemistry* **19**, 5734.
LEE, G. T. Y. and ENGELHARDT, D. L. (1979). *J. molec. Biol.* **129**, 221.
LEE, Y. F.and FOWLKS, E. R. (1982). *Anal. Biochem.* **119**, 224.
LEGOCKI, R. P. and VERMA, D. P. S. (1981). *Anal. Biochem.* **111**, 385.
LEHRACH, H., DIAMOND, D., WOZNEY, J. M., and BOEDTKER, H. (1977). *Biochemistry* **16**, 4743.
LEIBOWITZ, M. J. and WANG, R. W. (1984). *Anal. Biochem.* **137**, 161.
LELAY, M.-N., BRUNEL, C., and JEANTEUR, P. (1978). *FEBS Lett.* **90**, 54.
LEMKIN, P. F., LIPKIN, L. E., and LESTER, E. P. (1982). *Clin. Chem.* **28**, 840.
LESTER, E. P., LEMKIN, P. F., LOWRY, J. F., and LIPKIN, L. E. (1982). *Electrophoresis* **3**, 364.
——, MILLER, J. B., and YACHNIN, S. (1978). *Biochim. biophys. Acta* **536**, 165.
LEVY, A. L. and CHUNG, D. (1953). *Anal. Chem.* **25**, 396.
LEWICKI, P. P. and SINSKEY, A. J. (1970). *Anal. Biochem.* **33**, 273.
LI, Y.-T. and LI, S.-C. (1973). *Ann. N.Y. Acad. Sci.* **209**, 187.
LINDEBERG, E. G. G. (1976). *J. Chromatogr.* **117**, 439.
LING, V. (1972). *J. molec. Biol.* **64**, 87.
LINK, H. and KOSTULAS, V. (1983). *Clin. Chem.* **29**, 810.
LISCHWE, M. A. and OCHS, D. (1982). *Anal. Biochem.* **127**, 453.
LITTLE, P. F. R. (1984). *Nature (Lond.)* **310**, 369.
——, ANNISON, G., DARLING, S., WILLIAMSON, R., CAMBA, L., and MODELL, B. (1980). *Nature (Lond.)* **285**, 144.
LIVELY, M. O., MORAN, T. F., and POWERS, J. C. (1979). *J. biol. Chem.* **254**, 262.
LOENING, U. E. (1967). *Biochem. J.* **102**, 251.
—— (1969). *Biochem. J.* **113**, 131.
LONBERG-HOLM, K., BAGLEY, E. A., NUSBACHER, J., and HEAL, J. (1982). *Clin. Chem.* **28**, 962.
LONSDALE-ECCLES, J. D., LYNLEY, A. M., and DALE, B. A. (1981). *Biochem. J.* **197**, 591.
LORENTZ, K. (1976). *Anal. Biochem.* **76**, 214.
McCONKEY, E. H. (1979). *Anal. Biochem.* **96**, 39.
—— and ANDERSON, C. (1984a). *Electrophoresis* **5**, 230.
——, —— (1984b). *Electrophoresis* **5**, 233.
McCOY, J. P., VARANI, J., and GOLDSTEIN, I. J. (1983). *Anal. Biochem.* **130**, 437.
McEWEN, B. S. (1968). *Anal. Biochem.* **25**, 172.
McFARLANE, A. S., (1958). *Nature (Lond.)* **182**, 53.
McGREGOR, J. L., CLEMETSON, K. J., JAMES, E., CAPITANIO, A., GREENLAND, T., LÜSCHER, E. F., and DECHAVANNE, M. (1981). *Eur. J. Biochem.* **116**, 379.
McGUIRE, J. K., MILLER, T. Y., TIPPS, R. W., SNYDER, R. S., and RIGHETTI, P. G. (1980). *J. Chromatogr.* **194**, 323.
McILWAIN, H. and BUDDLE, H. L. (1953). *Biochem. J.* **53**, 412.
McKENZIE, D. and MILLER, R. C. (1983). *Clin. Chem.* **29**, 189.
McPHIE, P., HOUNSELL, J., and GRATZER, W. B. (1966). *Biochemistry* **5**, 986.
MAHADIK, S. P., KORENOVSKY, A., and RAPPORT, M. M. (1976). *Anal. Biochem.* **76**, 615.
MAIZEL, J. V. (1971). *Meth. Virol.* **5**, 179.
MALAMUD, D. and DRYSDALE, J. W. (1978). *Anal. Biochem.* **86**, 620.

MALHOTRA, L. C., MURTHY, M. R. V., and CHAUDARY, K. D. (1978). *Anal. Biochem.* **86**, 363.

MANABE, T., KOJIMA, K., JITZUKAWA, S., HOSHINO, T., and OKUYAMA, T. (1982). *Clin. Chem.* **28**, 819.

——, ODA, O., and OKUYAMA, T. (1982). *J. Chromatogr.* **241**, 361.

MANCINI, G., CARBONARA, A. O., and HEREMANS, J. F. (1965). *Immuno-chemistry*, **2**, 235.

MANIATIS, T. and EFSTRATIADIS, A. (1980). *Meth. Enzymol.* **65**, 209.

MANROW, R. E . and DOTTIN, R. P. (1980). *Proc. Natl. Acad. Sci. U.S.A.* **77**, 730.

MANSKE, W., BOHN, B., and BROSSMER, R. (1977). In *Electrofocusing and isotachophoresis* (eds. B. J. Radola and D. Graesslin), p. 495. Walter de Gruyter, Berlin, New York.

MANWELL, C. (1977). *Biochem. J.* **165**, 487.

MARCEL, Y. L., BERGSETH, M., and NESTRUCK, A. C. (1979). *Biochim. biophys. Acta* **573**, 175.

MARCH, S. C., PARIKH, I., and CUATRECASAS, P. (1974). *Anal. Biochem.* **60**, 149.

MARDIAN, J. K. W. and ISENBERG, I. (1978). *Anal. Biochem.* **91**, 1.

MARGOLIS, J. (1973). *Lab. Pract.* **22**, 107.

—— and KENRICK, K. G. (1968). *Anal. Biochem.* **25**, 347.

—— and WRIGLEY, C. W. (1975). *J. Chromatogr.* **106**, 204.

MARGULIES, M. M. and TIFFANY, H. L. (1984). *Anal. Biochem.* **136**, 309.

MARIASH, C. N., SEELIG, S., and OPPENHEIMER, J. H. (1982). *Anal. Biochem.* **121**, 388.

MARINKA, K. (1972). *Anal. Biochem.* **50**, 304.

MARJANEN, L. A. and RYRIE, I. J. (1974). *Biochim. biophys. Acta* **371**, 442.

MARKWELL, M. A. K. (1982). *Anal. Biochem.* **125**, 427.

MARSHALL, T. (1984a). *Electrophoresis* **5**, 245.

—— (1984b). *Anal. Biochem.* **139**, 506.

—— and WILLIAMS, K. M. (1984). *Anal. Biochem.* **139**, 502.

MATIOLI, G. T. and NIEWISCH, H. B. (1965). *Science* **150**, 1824.

MATOUŠEK, V. and HOŘEJŠÍ, V. (1982). *J. Chromatogr.* **245**, 271.

MATSUDAIRA, P. T. and BURGESS, D. R. (1978). *Anal. Biochem.* **87**, 386.

MATTICK, J. S., TSUKAMOTO, Y., NICKLESS, J. and WAKIL, S. J. (1983). *J. biol. Chem.* **258**, 15291.

MAURER, H. R. (1968). *Hoppe-Seyler's Z. physiol. Chem.* **349**, 115.

—— (1971). *Disc electrophoresis and related techniques of polyacrylamide gel electrophoresis* (2nd edn). Walter de Gruyter, Berlin, New York.

—— and ALLEN, R. C. (1972). *Clin. Chim. Acta* **40**, 359.

—— and DATI, F. A. (1972). *Anal. Biochem.* **46**, 19.

MAXAM, A. M. and GILBERT, W. (1977). *Proc. Natl. Acad. Sci. U.S.A.* **74**, 560.

——, —— (1980). *Meth. Enzymol.* **65**, 499.

MEHTA, P. D., MEHTA, S. P. and PATRICK, B. A. (1984). *Clin. Chem.* **30**, 735.

MEINKOTH, J. and WAHL, G. (1984). *Anal. Biochem.* **138**, 267.

MELGAR, E. and GOLDTHWAIT, D. A. (1968). *J. biol. Chem.* **243**, 4409.

MÉNDEZ, E. (1982). *Anal. Biochem.* **126**, 403.

—— and LAI, C. Y. (1975). *Anal. Biochem.* **65**, 281.

MERRIL, C. R., DUNAU, M. L. and GOLDMAN, D. (1981). *Anal. Biochem.* **110**, 201.

——, GOLDMAN, D. and VAN KEUREN, M. L. (1982). *Electrophoresis* **3**, 17.

MENDEZ, E. and LAI, C. Y. (1975). *Anal. Biochem.* **65**, 281.

MIKKERS, F. E. P., EVERAERTS, F. M., and PEEK, J. A. F. (1979a). *J. Chromatogr.* **168**, 293.

——, ——, —— (1979*b*). *J. Chromatogr.* **168**, 317.
——, —— and VERHEGGEN, T. P. E. M. (1979). *J. Chromatogr.* **169**, 11.
——, RINGOIR, S., and DE SMET, R. (1979). *J. Chromatogr.* **162**, 341.
——, VERHEGGEN, T., EVERAERTS, F., HULSMAN, J., and MEIJERS, C. (1980). *J. Chromatogr.* **182**, 496.
MILLE, M. J., VO, P. K., NIELSEN, C., GEIDUSCHER, E. P., and XUONG, N. H. (1982). *Clin. Chem.* **28**, 867.
MILLS, E. N. C. and FREEDMAN, R. B. (1983). *Biochim. biophys. Acta* **734**, 160.
MOLD, D. E., WEINGERT, J., ASSARAF, J., LUBAHN, D. B., KELNER, D. N., SHAW, B. R., and McCARTY, K. S. (1983). *Anal. Biochem.* **135**, 44.
MONAHAN, J. J., WOO, S. L. C., LIARAKOS, C. D., and O'MALLEY, B. W. (1977). *J. biol. Chem.* **252**, 4722.
MORGAN, M. R. A., BROWN, P. J., LEYLAND, M. J., and DEAN, P. D. G. (1978). *FEBS Lett.* **87**, 239.
——, GEORGE, E., and DEAN, P. D. G. (1980). *Anal. Biochem.* **105**, 1.
——, SLATER, N. A., and DEAN, P. D. G. (1979). *Anal. Biochem.* **92**, 144.
MOROI, M. and JUNG, S. M. (1984). *Biochim. biophys. Acta* **798**, 295.
MORRIS, J. E., CANOY, D. W., and RYND, L. S. (1981). *J. Chromatogr.* **224**, 407.
MORRISSEY, J. H. (1981). *Anal. Biochem.* **117**, 307.
MOSS, B. A. and BROWNLEE, G. G. (1981). *Nucleic Acid Res.* **9**, 1941.
MOULIN, S., FRUCHART, J. S., DEWAILLY, P., and SEZILLE, G. (1979). *Clin. Chim. Acta* **91**, 159.
MUKERJEE, H. (1978). *J. Chromatogr.* **153**, 247.
MUNIZ, N. (1977). *Clin. Chem.* **23**, 1826.
MURACH, K.-F., BELTLE, W., and BOPP, M. (1982). *Electrophoresis* **3**, 337.
MYEROWITZ, R. L., CHRAMBACH, A., RODBARD, D., and ROBBINS, J. B. (1972). *Anal. Biochem.* **48**, 394.
NAKAMURA, K., KUWAHARA, A., and TAKEO, K. (1979). *J. Chromatogr.* **171**, 89.
NAKANISHI, S., INOUE, A., KITA, T., NAKAMURA, M., CHANG, A. C. Y., COHEN, S. N., and NUMA, S. (1979). *Nature (Lond.)* **278**, 423.
NAKASHIMA, H., NAKAGAWA, Y., and MAKINO, S. (1981). *Biochim. biophys. Acta* **643**, 509.
NARAYANAN, K. R. and RAJ, A. S. (1977). In *Electrofocusing and isotachophoresis* (eds. B. J. Radola and D. Graesslin), p. 221. Walter de Gruyter, Berlin, New York.
NATHANS, D. and SMITH, H. O. (1975). *Annu. Rev. Biochem.* **44**, 273.
NELLES, L. P. and BAMBURG, J. R. (1976). *Anal. Biochem.* **73**, 522.
NEUHOFF, V., SCHILL, W.-B., and STERNBACH, H. (1969). *Hoppe-Seyler's Z. physiol. Chem.* **350**, 335.
——, ——, —— (1970). *Biochem. J.* **117**, 623.
NEVILLE, D. M. (1971). *J. biol. Chem.* **246**, 6328.
—— and GLOSSMANN, H. (1971). *J. biol. Chem.* **246**, 6335.
NEWBY, A. C. and CHRAMBACH, A. (1979). *Biochem. J.* **177**, 623.
—— RODWELL, M., and CHRAMBACH, A. (1978). *Arch. Biochem. Biophys.* **190**, 109.
NGUYEN, N. Y., BAUMANN, G., ARBEGAST, E., GRINDLELAND, R. E., and CHRAMBACH, A. (1981). *Prep. Biochem.* **11**, 139.
—— and CHRAMBACH, A. (1976). *Anal. Biochem.* **74**, 145.
——, —— (1977). *Anal. Biochem.* **82**, 54.
——, —— (1978). *Anal. Biochem.* **87**, 576.
——, —— (1979). *Anal. Biochem.* **94**, 202.
——, —— (1980). *Electrophoresis* **1**, 14.

——, RODBARD, D., SVENDSEN, P. J., and CHRAMBACH, A. (1977). *Annal. Bichem.* **77**, 39.

NIEDIECK, B. (1978). *Immunochemistry* **15**, 11.

NIELSEN, B. L. and BROWN, L. R. (1984). *Anal. Biochem.* **141**, 311.

NIELSEN, C. S. and BJERRUM, O. J. (1975). *Scand. J. Immunol.* **4** (suppl. 2), 73.

——, —— (1977). *Biochim. biophys. Acta* **466**, 496.

NIELSEN, P. J., MANCHESTER, K. L., TOWBIN, H., GORDON, J., and THOMAS, G. (1982). *J. biol. Chem.* **257**, 12316.

NILSSON, K. and OLSSON, J. E. (1978). *Clin. Chem.* **24**, 1134.

NORRILD, B., BJERRUM, O. J., and VESTERGAARD, B. F. (1977). *Anal. Biochem.* **81**, 432.

OAKLEY, B. R., KIRSCH, D. R., and MORRIS, N. R. (1980). *Anal. Biochem.* **105**, 361.

OAKLEY, C. L. and FULTHORPE, A. J. (1953). *J. Pathol. Bacteriol.* **65**, 49.

OCHS, D. C., McCONKEY, E. H., and SAMMONS, D. W. (1981). *Electrophoresis* **2**, 304.

O'CONNELL, P. B. H. and BRADY, C. J. (1976). *Anal. Biochem.* **76**, 63.

OERLEMANS, F., DE BRUYN, C., MIKKERS, F., VERHEGGEN, T., and EVERAERTS, F. (1981). *J. Chromatogr.* **225**, 369.

O'FARRELL, P. H. (1975). *J. biol. Chem.* **250**, 4007.

O'FARRELL, P. Z., GOODMAN, H. M., and O'FARRELL, P. H. (1977). *Cell,* **12**, 1133.

ÖFVERSTEDT, L. G., HAMMERSTRÖM, K., BALGOBIN, N., HJERTÉN, S., PETTERSSON, U., and CHATTOPADHYAYA, J. (1984). *Biochim. biophys. Acta* **782**, 120.

——, SUNDELIN, J., and JOHANSSON, G. (1983). *Anal. Biochem.* **134**, 361.

OGSTON, A. G. (1958). *Trans. Faraday Soc.* **54**, 1754.

OHSAWA, K. and EBATA, N. (1983). *Anal. Biochem.* **135**, 409.

OKA, S., HIROTSUNE, M., and SHIGETA, S. (1979). *Anal. Biochem.* **98**, 417.

OLDEN, K. and YAMADA, K. M. (1977). *Anal. Biochem.* **78**, 483.

OLMSTED, J. B. (1981). *J. biol. Chem.* **256**, 11955.

OMENYI, S. N. and SNYDER, R. S. (1983). *Prep. Biochem.* **13**, 437.

OPPENHEIM, J. D., NACHBAR, M. S., and BLANK, M. (1983). *Electrophoresis* **4**, 53.

ORNSTEIN, L. (1964). *Ann. N.Y. Acad. Sci.* **121**, 321.

OTAVSKY, W. I. and DRYDALE, J. W. (1976). *Anal. Biochem.* **65**, 533.

OTTO, M. and ŠNEJDÁRKOVÁ, M. (1981). *Anal. Biochem.* **111**, 111.

OUCHTERLONY, Ö. (1958). In *Progress in allergy*, (ed. P. Kallós), Vol. 5, p. 1. Karger, Basel, New York.

—— and NILSSON, L.-Å. (1978). In *Handbook of experimental immunology* (ed. D. M. Weir) (3rd edn.), Vol. 1, p. 190. Blackwell, Oxford.

OUDIN, J. (1952). *Methods in medical research* (ed. A. C. Corcoran), Vol. 5, p. 335. Year Book Publishers, Chicago.

OWEN, P., OPPENHEIM, J. D., NACHBAR, M. S. and KESSLER, R. E. (1977). *Anal. Biochem.* **80**, 446.

—— and SALTON, M. R. J. (1976). *Anal. Biochem.* **73**, 20.

PANYIM, S. and CHALKLEY, R. (1969). *Arch. Biochem. Biophys.* **130**, 377.

——, THITIPONGPANICH, R., and SUPATIMUSRO, D. (1977). *Anal. Biochem.* **81**, 320.

PARK, C. M. (1973). *Ann. N.Y. Acad. Sci.* **209**, 237.

PARKER, R. C., WATSON, R. M., and VINOGRAD, J. (1977). *Proc. Natl. Acad. Sci. U.S.A.* **74**, 851.

PATTERSON, M. S. and GREEN, R. C. (1965). *Anal. Chem.* **37**, 854.

PAYNE, J. W. (1973). *Biochem. J.* **135**, 867.

PEACOCK, A. C., BUNTING, S. L., and NISHINAGA, K. (1977). *Biochim. biophys. Acta* **475**, 352.

—— and DINGMAN, C. W. (1968). *Bichemistry,* **7**, 668.

PEATS, S. (1984). *Anal. Biochem.* **140**, 178.

PEFEROEN, M., HUYBRECHTS R., and DE LOOF, A. (1982). *FEBS Lett.* **145**, 369.

PERDEW, G. H., SCHAUP, H. W., and SELIVONCHICK, D. P. (1983). *Anal. Biochem.* **135**, 453.

PETER, R., WOLFRUM, D. I., and NEUHOFF, V. (1976). *Comp. Biochem. Physiol. B.* **55**, 583.

PETERSON, J. L. and MCCONKEY, E. H. (1976). *J. biol. Chem.* **251**, 548.

PETERSON, R. F. and PETERSON, R. F. (1969). *Abstr. 158th Mtg Amer. Chem. Soc.,* B-151. Am. Chem. Soc., New York.

PETROPAKIS, H. J., ANGELMEIR, A. F., and MONTGOMERY, M. W. (1972). *Anal. Biochem.* **46**, 594.

PFLUG, W., DE LA VIGNE, V., and BRUDER, W. (1981). *Electrophoresis* **2**, 327.

PHILLIPS, D. R. and MORRISON, M. (1970). *Biochem. Biophys. Res. Commun.* **40**, 284.

PHILLIPS, G. R. (1974). In *Electrophoresis and isoelectric focusing in polyacrylamide gel* (eds. R. C. Allen and H. R. Maurer), p. 255. Walter de Gruyter, Berlin, New York.

PINO, R. M. and HART, T. K. (1984). *Anal. Biochem.* **139**, 77.

PITT-RIVERS, R. and IMPIOMBATO, F. S. A. (1968). *Biochem. J.* **109**, 825.

PLANK, L., HYMER, W. L., KUNZE, M. E., MARKS, G. M., LANHAM, J. W., and TODD, P. (1983). *J. Biochem. Biophys. Meth.* **8**, 275.

PLAZA, S. (1981). *J. Chromatogr.* **207**, 407.

PLUMLEY, F. G. and SCHMIDT, G. W. (1983). *Anal. Biochem.* **134**, 86.

PODUSLO, J. F. (1981). *Anal. Biochem.* **114**, 131.

——, and RODBARD, D. (1980). *Anal. Biochem.* **101**, 394.

POEHLING, H.-M. and NEUHOFF, V. (1980). *Electrophoresis* **1**, 90.

——, WYSS, U., and NEUHOFF, V. (1980). *Electrophoresis* **1**, 198.

POLSKY, F., EDGELL, M. H., SEIDMAN, J. G., and LEDER, P. (1978). *Anal. Biochem.* **87**, 397.

POPESCU, O. (1983). *Electrophoresis* **4**, 432.

PORETZ, R. D. and PIECZENIK, G. (1981). *Anal. Biochem.* **115**, 170.

PORTER, A. G., BARKER, C., CAREY, N. H., HALLEWELL, R. A., THRELFALL, G., and EMTAGE, J. S. (1979). *Nature (Lond.)* **282**, 471.

POSNER, 1. (1976). *Anal. Biochem.* **70**, 187.

PRAT, J. P., LAMY, J. N., and WEILL, J. D. (1969). *Bull. Soc. Chim. biol.* **51**, 1367.

PROUDFOOT, N. J. (1976). *J. molec. Biol.* **107**, 491.

PRUNELL, A. and BERNARDI, G. (1977). *J. molec. Biol.* **110**, 53.

——, KOPECKA, H., STRAUSS, F., and BERNARDI, G. (1977). *J. molec. Biol.* **110**, 17.

——, STRAUSS, F., and LEBLANC, B. (1977). *Anal. Biochem.* **78**, 57.

PRUSINIER, S. B., GROTH, D. F., BILDSTEIN, C., MASIARZ, F. R., MCKINLEY, M. P., and COCHRAN, S. P. (1980). *Proc. Natl. Acad. Sci. U.S.A.* **77**, 2984.

PUSZKIN, S., SCHOOK, W., MAIMON, J., and PUSZKIN, E. (1977). *Biochim. biophys. Acta* **494**, 144.

PULLEYBLANK, D. E., SHURE, M., TANG, D., VINOGRAD, J., and VOSBERG, H.-P. (1975). *Proc. Natl. Acad. Sci. U.S.A.* **72**, 4280.

PURRELLO, M. and BALAZS, I. (1983). *Anal. Biochem.* **128**, 393.

PUTNAM, F. W. (ed.) (1975). *The plasma proteins* (2nd edn.). Academic Press, New York.

QUEEN, C. and KORN, L. (1980). *Meth. Enzymol.* **65**, 595.

QUITSCHKE, W. and SCHECHTER, N. (1982). *Anal. Biochem.* **124**, 231.

RADOLA, B. J. (1975). *Biochim. biophys. Acta* **386**, 181.

—— (1980). *Electrophoresis* **1**, 43.

RAMON, G. (1922). *C.r. Seances Soc. biol. Paris* **86**, 661.

RAYMOND, S. (1964). *Ann. N.Y. Acad. Sci.* **121**, 350.

—— and WANG, V. J. (1960). *Anal. Biochem.* **1**, 39.

REID, M. S. and BIELESKI, R. L. (1968). *Anal. Biochem.* **22**, 374.

REIJENGA, J. C., VERHEGGEN, T. P. E. M., and EVERAERTS, F. M. (1984). *J. Chromatogr.* **283**, 99.

REISER, J. and WARDALE, J. (1981). *Eur. J. Biochem.* **114**, 569.

REISFELD, R. A., LEWIS, U. J., and WILLIAMS, D. E. (1962). *Nature (London)* **195**, 281.

RENART, J., REISER, J., and STARK, G. R. (1979). *Proc. Natl. Acad. Sci. U.S.A.* **76**, 3116.

REPÁŠOVÁ, Ľ., POLONSKÝ, J., KOŠÍK, M., and VODNÝ, Š. (1984). *J. Chromatogr.* **286**, 347.

RERABEK, J. E. (1977). *Clin. Chem.* **23**, 186.

RESSLER, N., SPRINGGATE, R., and KAUFMAN, J. (1961). *J. Chromatogr.* **6**, 409.

REYNOLDS, J. A. and TANFORD, C. (1970a). *Proc. Natl. Acad. Sci. U.S.A.* **66**, 1002.

——, —— (1970b). *J. biol. Chem.* **245**, 5161.

RICE, R. H. and MEANS, G. E. (1971). *J. biol. Chem.* **246**, 831.

RICH, S. A., ZIEGLER, F. D., and GRIMLEY, P. M. (1979). *Clin. Chim. Acta* **96**, 133.

RICHARDS, E. G., COLL, J. A., and GRATZER, W. B. (1965). *Anal. Biochem.* **12**, 452.

—— and LECANIDOU, R. (1971). *Anal. Biochem.* **40**, 43.

——, —— (1974a). In *Electrophoresis and isoelectric focusing in polyacrylamide gel* (eds. R. C. Allen and H. R. Maurer), p. 16. Walter de Gruyter, Berlin, New York.

——, —— (1974b). In *Electrophoresis and isoelectric focusing in polyacrylamide gel* (eds. R. C. Allen and H. R. Maurer), p. 245. Walter de Gruyter, Berlin, New York.

RIGHETTI, P. G. (1984). *J. Chromatogr.* **300**, 165.

——, BROST, B. C. W., and SNYDER, R. S. (1981). *J. Biochem. Biophys. Meth.* **4**, 347.

——, BROWN, R. P., and STONE, A. L. (1978). *Biochim. biophys. Acta* **542**, 232.

—— and CARAVAGGIO, T. (1976). *J. Chromatogr.* **127**, 1.

—— and CHILLEMI, F. (1978). *J. Chromatogr.* **157**, 243.

——, DELPECH, M., MOISAND, F., KRUH, J., and LABIE, D. (1983). *Electrophoresis* **4**, 393.

——, EK, K., and BJELLQVIST, B. (1984). *J. Chromatogr.* **291**, 31.

—— and GIANAZZA, E., (1980). In *Electrophoresis '79* (ed. B. J. Radola), pp. 23–38. Walter de Gruyter, Berlin.

——, GACON, G., GIANAZZA, E., LOSTANLEN, D., and KAPLAN, J. C. (1978b). *Biochem. Biophys. Res. Commun.* **85**, 1575.

——, GELFI, C., and BOSISIO, A. B. (1981). *Electrophoresis* **2**, 291.

—— and GIANAZZA, E. (1980). In *Electrophoresis '79* (ed. B.J. Radola), pp. 23–38. Walter de Gruyter, Berlin.

——, GIANAZZA, E., BJELLQVIST, B., EK, K., GÖRG, A., and WESTERMEIER, R. (1983). In *Electrophoresis '82* (ed. D. Stathakos). Walter de Gruyter, Berlin.

——, ——, and EK, K. (1980). *J. Chromatogr.* **184**, 415.

——, ——, SALAMINI, F., GALANTE, E., VIOTTI, A., and SOAVE, C. (1977). In *Electrofocusing and isotachophoresis* (eds. B. J. Radola and D. Graesslin), p. 199. Walter de Gruyter, Berlin, New York.

——, KRISHNAMOORTHY, R., GIANAZZA, E., and LABIE, D. (1978*a*). *J. Chromatogr.* **166**, 455.

——, ——, LAPOUMEROULIE, C., and LABIE, D. (1979). *J. Chromatogr.* **177**, 219.

——, TUDOR, G., and EK, K. (1981). *J. Chromatogr.* **220**, 115.

——, VAN OSS, C. J., and VANDERHOFF, J. W. (1979). *Electrokinetic separation methods.* Elsevier/North Holland Biomedical Press, Amsterdam.

RILBE, H. (1973). *Ann. N.Y. Acad. Sci.* **209**, 11.

—— (1977). In *Electrofocusing and isotachophoresis* (eds. B. J. Radola and D. Graesslin), p. 35. Walter de Gruyter, Berlin, New York.

RILEY, R. F., and COLEMAN, M. K. (1968). *J. lab. clin. Med.* **72**, 714.

RITTENHOUSE, J. and MARCUS, F. (1984). *Anal. Biochem.* **138**, 442.

RITZMAN, S. R. and DANIELS, S. C. (1975). *Serum protein abnormalities: diagnostic and clinical aspects.* Little & Brown, Boston, Mass.

ROBERTS, R. M. and JONES, J. S. (1972). *Anal. Biochem.* **49**, 592.

ROCHETTE, J., RIGHETTI, P. G., BOSISIO, A. B., VERTONGEN, F., SCHNECK, G., BOISSEL, J. P., LABIE, D., and WAJCMAN, H. (1984). *J. Chromatogr.* **285**, 143.

RODBARD, D. and CHRAMBACH, A. (1970). *Proc. Natl. Acad. Sci. U.S.A.* **65**, 970.

——, —— (1971). *Anal. Biochem.* **40**, 95.

——, —— (1974). In *Electrophoresis and isoelectric focusing in polyacrylamide gel* (eds. R. C. Allen and H. R. Maurer), p. 28. Walter de Gruyter, Berlin, New York.

——, —— and WEISS, G. H. (1974). In *Electrophoresis and isoelectric focusing in polyacrylamide gel* (eds. R. C. Allen and H. R. Maurer), p. 62. Walter de Gruyter, Berlin, New York.

——, KAPADIA, G., and CHRAMBACH, A. (1971). *Anal. Biochem.* **40**, 135.

ROSENBLUM, B. B., NEEL, J. V., and HANASH, S. M. (1983). *Proc. Natl. Acad. Sci. U.S.A.* **80**, 5002.

ROSENGREN, A., BJELLQVIST, B., and GASPARIC, V. (1977). In *Electrofocusing and isotachophoresis* (eds. B. J. Radola and D. Graesslin), p. 165. Walter de Gruyter, Berlin, New York.

ROSSI, R. (1978). *Anal. Biochem.* **85**, 291.

ROTHE, G. M. and PURKHANBABA, H. (1982). *Electrophoresis* **3**, 43.

ROUTS, R. (1973). *Ann. N.Y. Acad. Sci.* **209**, 445.

ROVERA, G., MAGARIAN, C., and BORUN, T. W. (1978). *Anal. Biochem.* **85**, 506.

ROWE, D. S. (1969). *Bull. W.H.O.* **40**, 613.

RUBIN, R. W. and MILIKOWSKI, C. (1978). *Biochim. biophys. Acta* **509**, 100.

RÜCHEL, R. (1977). *J. Chromatogr.* **132**, 451.

RUDDELL, A. and JACOBS-LORENA, M. (1982). *Anal. Biochem.* **122**, 248.

SAGA, M. and KANO, S. (1979). *Clin. Chim. Acta* **95**, 521.

SAKAI, Y., ITAKURA, K., KANADA, T., EBATA, N., SUGA, K., AIKAWA, H., NAKAMURA, K., and SATA, T. (1984). *Anal. Biochem.* **137**, 1.

SALMON, J. E., NUDEL, U., SCHILIRO, G., NATTA, C. L., and BANK, A. (1978). *Anal. Biochem.* **91**, 146.

SAMMONS, D. W., ADAMS, L. D., and NISHIZAWA, E. E. (1981). *Electrophoresis* **2**, 135.

SANDERS, M. M., GROPPI, V. E., and BROWNING, E. T. (1980). *Anal. Biochem.* **103**, 157.

SANGER, F., AIR, G. M., BARRELL, B. G., BROWN, N. L., COULSON, A. R., FIDDES, J. C., HUTCHISON III, C. A., SLOCOMBE, P. M., and SMITH, M. (1977). *Nature (Lond.)* **265**, 687. .

—— and BROWNLEE, G. G. (1967). *Meth. Enzymol.* **12A**, 361.

——, ——, and BARRELL, B. G. (1965). *J. molec. Biol.* **13**, 373.

—— and COULSON, A. R. (1975). *J. molec. Biol.* **94**, 441.

——, —— (1978). *FEBS Lett.* **87**, 107.

——, ——, BARRELL, B. G., SMITH, A. J. H., and ROE, B. A. (1980) *J. Mol. Biol.* **143**, 161.

——, ——, FRIEDMANN, T., AIR, G. M., BARRELL, B. G., BROWN, N. L., FIDDES, J. C., HUTCHISON III, C. A., SLOCOMBE, P. M., and SMITH, M. (1978). *J. molec. Biol.* **125**, 225.

——, ——, HONG, G. F., HILL, D. F., and PETERSEN, G. B. (1982). *J. Mol. Biol.* **162**, 729.

——, DONELSON, J. E., COULSON, A. R., KÖSSEL, H., and FISCHER, D. (1973) *Proc. Natl. Acad. Sci. U.S.A.* **70**, 1209.

——, ——, ——, ——, ——, (1974). *J. molec. Biol.* **90**, 315.

——, NICKLEN, S. and COULSON, A. R. (1977). *Proc. Natl. Acad. Sci. U.S.A.* **74**, 5463.

SANO, K., NAKAO, M., SHIBA, A., and KOBAYASHI, K. (1984). *Clin. Chim. Acta* **137**, 115.

SARAVIS, C. A. (1984). *Electrophoresis* **5**, 54.

——, O'BRIEN, M., and ZAMCHECK, N. (1979). *J. Immunol. Meth.* **29**, 97.

SARGENT, J. R. and GEORGE, S. G. (1975). *Methods in zone electrophoresis*, (3rd edn.), p. 23. B.D.H. Chemicals, Poole, Gt. Britain.

SARIS, C. J. M., VAN EENBERGEN, J., JENKS, B. G., and BLOEMERS, H. P. J. (1983). *Anal. Biochem.* **132**, 54.

SCHAFFER, H. E. and SEDEROFF, R. R. (1981). *Anal. Biochem.* **115**, 113.

SCHANBACHER, F. L. and SMITH, K. L. (1974). *Anal. Biochem.* **59**, 235.

SCHAUP, H. W., BEST, J. B., and GOODMAN, A. B. (1969). *Nature (Lond.)* **221**, 864.

SCHEELE, G., PASH, J., and BIEGER, W. (1981). *Anal. Biochem.* **112**, 304.

SCHEIDEGGER, J. J. (1955). *Int. Arch. Allergy* **7**, 103.

SCHICK, M. (1975). *Anal. Biochem.* **63**, 345.

SCHMIDT, B. (1979). *J. clin. Chem. clin. Biochem.* **17**, 508.

SCHMIDT, B. L. (1980). *Clin. Chim. Acta* **102**, 253.

SCHNEIDER, W. and KLOSE, J. (1983). *Electrophoresis* **4**, 284.

SCHUCHMAN, E. H. and DESMICK, R. J. (1981). *Anal. Biochem.* **117**, 419.

SCHUERCH, A. R., MITCHELL, W. R., and JOKLIK, W. K. (1975). *Anal. Biochem.* **65**, 331.

SCHUMACHER, J., RANDLES, J. W., and RIESNER, D. (1983). *Anal. Biochem.* **135**, 288.

SCHWINGHAMER, M. W. and SHEPHERD, R. J. (1980). *Anal. Biochem.* **103**, 426.

SEAMAN, G. V. F. (1975). In *The red blood cell* (ed. D. MacN. Surgenor), Vol. 2, p. 1135. Academic Press, New York.

SERWER, P. (1981). *Anal. Biochem.* **112**, 351.

—— (1983). *Electrophoresis* **4**, 375.

—— and ALLEN, J. L. (1984). *Biochemistry* **23**, 922.

SESBOÜÉ, R., MARTIN, J. P., and LEBRETON, J. P. (1981). *Clin. Chem. Acta* **109**, 229.

SHAFRITZ, D. A. and DRYSDALE, J. W. (1975). *Biochemistry* **14**, 61.

SHAPIRO, A. L., VIÑUELA, E., and MAIZEL, J. V. (1967). *Biochem. Biophys. Res. Commun.* **28**, 815.

SHAPSHAK, P., TOURTELLOTTE, W. W., STAUGAITIS, S., COWAN, T., INGRAM, T., WEIL, M. L., BLISS, D., and TOURTELLOTTE, W. G. (1983). *Anal. Biochem.* **132**, 305.

SHARP, P. A., SUGDEN, B., and SAMBROOK, J. (1973). *Biochemistry* **12**, 3055.

SHEPHERD, G. R. and GURLEY, L. R. (1966). *Anal. Biochem.* **14**, 356.

SHULMAN, G. (1979). *Clin. Biochem.* **12**, 93.

SHULMAN, S. and KARPATKIN, S. (1980). *J. biol. Chem.* **255**, 4320.

SIMON, M. and CUAN, J. (1982). *Clin. Chem.* **28**, 9.

SINGH, J. and WASSERMAN, A. R. (1970). *Biochim. biophys. Acta* **221**, 379.

SJÖDAHL, J. and HJALMARSSON, S. G. (1978). *FEBS Lett.* **92**, 22.

SLATER, G. G. (1969). *Anal. Chem.* **41**, 1039.

SMITH, A. J. H. (1980). *Meth. Enzymol.* **65**, 560.

SMITH, B. A. and WARE, B. R. (1978). In *Contemporary topics in analytical and clinical chemistry* (eds. D. M. Hercules, G. M. Heiftje, L. R. Snyder, and M. A. Evenson), Vol. 2, p. 29. Plenum Press, New York.

SMITH, I. (ed.). (1968). *Chromatographic and electrophoretic techniques*, Vol. 2, *Zone electrophoresis.* Heinemann, London.

SMITH, M. and KHORANA, H. G. (1963). *Meth Enzymol.* **6**, 645.

SMITH, M. R., DEVINE, C. S., COHN, S. M., and LIEBERMAN, M. W. (1984). *Anal. Biochem.* **137**, 120.

SMITHIES, O. (1955). *Biochem. J.* **61**, 629.

—— (1959). *Adv. protein Chem.* **14**, 65.

SMOLKA, A., KEMPNER, D., and REMBAUM, A. (1982). *Electrophoresis* **3**, 300.

——, MARGEL, S., NERREN, B. H., and REMBAUM, A. (1980). *Biochim. biophys. Acta* **588**, 246.

SONDEREGGER, P., JAUSSI, R., GEHRING, H., BRAUNSCHWEILER, K., and CHRISTEN, P. (1982). *Anal. Biochem.* **122**, 298.

SOUTHERN, E. M. (1975). *J. Mol. Biol.* **98**, 503.

—— (1979). *Anal. Biochem.* **100**, 319.

—— and MITCHELL, A. (1971). *Biochem. J.* **123**, 613.

SPIKER, S. (1980). *Anal. Biochem.* **108**, 263.

SREEKRISHNA, K., JONES, C. E., GUETZOW, K. A., PRASAD, M. R., and JOSHI, V. C. (1980). *Anal. Biochem.* **103**, 55.

STANLEY, J. and VASSILENKO, S. (1978). *Nature (London)* **274**, 87.

STAPRANS, I., FELTS, J. M., and BUTTS, R. J. (1983). *Anal. Biochem.* **134**, 240.

STECK, G., LEUTHARD, P., and BÜRK, R. R. (1980). *Anal. Biochem.* **107**, 21.

STEGMANN, H. (1967). *Hoppe-Seyler's Z. physiol. Chem.* **348**, 951.

—— (1978). In *Proc. 11th FEBS Meeting, Copenhagen, 1977*, Vol. 44, *Symp. A3: Biochemical Aspects of New Protein foods* (eds. J. Adler-Nielsen, B. O. Eggum, L. Munk, and H. S. Olsen). p. 11. Pergamon Press, Oxford, New York.

—— and LOESCHSKE, V. (1976). *Index of European potato varieties (Electrophoretic spectra, National registers, Appraisal of characteristics, Genetic data).* Verlag Paul Parey, Berlin.

STEIN, S., CHANG, C. H., BÖHLEN, P., IMAI, K., and UDENFRIEND, S. (1974). *Anal. Biochem.* **60**, 272.

STEINBERG, R. A. (1984). *Anal. Biochem.* **141**, 220.

STELLWAGEN, N. C. (1983). *Biochemistry,* **22**, 6180, 6186.

STEPHENS, R. E. (1975). *Anal. Biochem.* **65**, 369.
STEWARD, M. L. and CROUCH, R. J. (1981). *Anal. Biochem.* **111**, 203.
STINSON, R. A. (1977). *Biochem. J.* **167**, 65.
STORRING, P.-L. and TIPLADY, R. J. (1978). *Biochem. J.* **171**, 79.
STROTTMANN, J. M., ROBINSON, J. B., and STELLWAGEN, N. C. (1983). *Anal. Biochem.* **132**, 334.
STUDIER, F. W. (1972). *Science* **176**, 367.
—— (1973). *J. molec. Biol.* **79**, 237.
SUGDEN, B., DE TROY, B., ROBERTS, R. J. and SAMBROOK, J. (1975). *Anal. Biochem.* **68**, 36.
SUISSA, M. (1983). *Anal. Biochem.* **133**, 511.
SUTHERLAND, J. C., MONTELEONE, D. C., TRUNK, J., and COLENO, A. (1984). *Anal. Biochem.* **139**, 390.
SVENDSEN, P. J. (1973). In *A manual of quantitative immunoelectrophoresis* (eds. N. H. Axelsen, J. Krøll, and B. Weeke), p. 69. Universitetsforlaget, Oslo.
—— and ROSE, C. (1970). *Sci Tools* **17**, 13.
SVENSSON, H. (1961). *Acta chem. Scand.* **15**, 325.
SWANK, R. T. and MUNKRES, K. D. (1971). *Anal. Biochem.* **39**, 462.
SWANSTROM, R. and SHANK, P. R. (1978). *Anal. Biochem.* **86**, 184.
SWITZER, R. C., MERRIL, C. R., and SHIFRIN, S. (1979). *Anal. Biochem.* **98**, 231.
SYKES, S. and DENNIS, P. M. (1977). *Clin. Chim. Acta* **79**, 309.
SZEWCZYK, B. (1983). *Anal. Biochem.* **130**, 60.
TAKEISHI, K. and KANEDA, S. (1981). *Anal. Biochem.* **113**, 212.
TAKEO, K. (1984). *Electrophoresis* **5**, 187.
—— and NAKAMURA, S. (1972). *Arch. Biochem. Biophys.* **153**, 1.
TAKETA, K. (1983). *Electrophoresis* **4**, 371.
TALBOT, D. N. and YPHANTIS, D. A. (1971). *Anal. Biochem.* **44**, 246.
TANAKA, Y., DYER, T. A., and BROWNLEE, G. G. (1980). *Nucleic Acids Res.* **8**, 1259.
TATA, S. J. and MOIR, G. F. J. (1976). *Anal. Biochem.* **70**, 495.
TAUTZ, D. and RENZ, M. (1983). *Anal. Biochem.* **132**, 14.
TEDESCO, T. A., BONOW, R., and MELLMAN, W. J. (1972). *Anal. Biochem.* **46**, 173.
TEREBUS-KEKISH, O., BARCLAY, M., and STOCK, C. C. (1978). *Clin. Chim. Acta* **88**, 9.
THATCHER, D. R. and HODSON, B. (1981). *Biochem. J.* **197**, 105.
THE, T. H. and FELTKAMP, T. E. W. (1970). *Immunology* **18**, 865.
THOMPSON, B. J., BURGHES, A. H. M., DUNN, M. J., and DUBOWITZ, V. (1981). *Electrophoresis* **2**, 251.
——, DUNN, M. J., BURGHES, A. H. M., and DUBOWITZ, V. (1982). *Electrophoresis* **3**, 307.
THOMSON, A. R. (1983). In *Electrophetic techniques* (eds. C. F. Simpson and M. Whittaker), p. 253. Academic Press, London & New York.
THORUN, W. and MAURER, H. R. (1971). In *Disc electrophoresis and related techniques of polyacrylamide gel electrophoresis* (ed. H. R. Maurer) (2nd edn), p. 4. Walter de Gruyter, Berlin, New York.
THURING, R. W. J., SANDERS, J. P. M., and BORST, P. (1975). *Anal. Biochem.* **66**, 213.
TICHÁ, M., HOŘEJŠI, V., and BARTHOVÁ, J. (1978). *Biochim. biophys. Acta* **534**, 58.
TIJSSEN, P. and KURSTAK, E. (1983). *Anal. Biochem.* **128**, 26.

TISHLER, P. V. and EPSTEIN, C. J. (1968). *Anal. Biochem.* **22**, 89.

TIVOL, W. F. and BENISEK, W. F. (1977). *Anal. Biochem.* **78**, 93.

TODA, T., FUJITA, T., and OHASHI, M. (1984). *Electrophoresis* **5**, 42.

TODD, R. D. and GARRARD, W. T. (1977). *J. biol. Chem.* **252**, 4729.

TOWBIN, H., STAEHELIN, T., and GORDON, J. (1979). *Proc. Natl. Acad. Sci. U.S.A.* **76**, 4350.

TRACY, R. P. and CHAN, S. H. P. (1979). *Biochim. biophys. Acta* **576**, 109.

——, CURRIE, R. M., and YOUNG, D. S. (1982a). *Clin. Chem.* **28**, 890.

——, ——, —— (1982b). *Clin. Chem.* **28**, 908.

TRAH, T. J. and SCHLEYER, M. (1982). *Anal. Biochem.* **127**, 326.

TREIMAN, M., PØDENPHANT, J., SEARMARK, T., and BOCK, E. (1979). *FEBS Lett.* **97**, 147.

TRIVEDI, S. M., FRONDOZA, C. G., and HUMPHREY, R. L. (1983). *Clin. Chem.* **29**, 836.

TSAI, C. M. and FRASCH, C. E. (1982). *Anal. Biochem.* **119**, 115.

TSCHÖPE, W. and RITZ, E. (1980). *J. Chromatogr.* **221**, 59.

TSHABALALA, M. A., SCHRAM, S. B., GERBERICH, F. G., LOWMAN, D. W., and ROGERS, L. B. (1981). *J. Chromatogr.* **207**, 353.

TSUGITA, A., SASADA, S., VAN DEN BROEK, R., and SCHEFFLER, J. J. (1982). *Eur. J. Biochem.* **124**, 171.

TUSZYNSKI, G. P., BUCK, C. A., and WARREN, L. (1979). *Anal. Biochem.* **93**, 329.

——, DAMSKY, C. H., FUHRER, J. P., and WARREN, L. (1977). *Anal. Biochem.* **83**, 119.

——, KNIGHT, L. C., KORNECKI, E., and SRIVASTAVA, S. (1983). *Anal. Biochem.* **130**, 166.

TZENG, M. C., (1983). *Anal. Biochem.* **128**, 412.

UI, N. (1971). *Biochim. biophys. Acta* **229**, 567.

—— (1973). *Ann. N.Y. Acad. Sci.* **209**, 198.

URIEL, J. and BERGES, J. (1966). *C.r. Acad. Sci. Paris* **262**, 164.

——, —— (1974). In *Electrophoresis and isoelectric focusing in polyacrylamide gel* (eds. R. C. Allen and H. R. Maurer), p. 235. Walter de Gruyter, Berlin, New York.

UTERMANN, G. and BEISIEGEL, U. (1979). *FEBS Lett.* **97**, 245.

UYTTENDAELE, K., DE GROOTE, M., BLATON, V., PEETERS, H., and ALEXANDER, F. (1977). *J. Chromatogr.* **132**, 261.

VALKONEN, K. H. and PIHA, R. S. (1980). *Anal. Biochem.* **104**, 499.

VAN DEN HOEK, A. K. and ZAIL, S. S. (1977). *Clin. Chim. Acta* **79**, 7.

VANDERHOFF, J. W. and VAN OSS, C. J. (1979). In *Electrokinetic separation methods* (eds. P. G. Righetti, C. J. Van Oss and J. W. Vanderhoff), p. 257, Elsevier/North Holland Biomedical Press, Amsterdam.

VAN DER SLUIS, P. J., BOER, G. J. and POOL, C. W. (1981). *Anal. Biochem.* **133**, 226.

VANEIJK, H. G., VANNOORT, W. L., KROOS, M. J., and VANDERHEUL, C. (1978). *J. clin. Chem. clin. Biochem.* **16**, 557.

VAN LENTE, F. and GALEN, R. S. (1978). *Clin. Chim. Acta* **87**, 211.

VAN ORMONDT, H. and HATTMAN, S. (1976). *Anal. Biochem.* **74**, 207.

VAN TOL, R. G. L. and VAN VLOTEN-DOTING, L. (1979). *Eur. J. Biochem.* **93**, 461.

VECCHIO, G., RIGHETTI, P. G., ZANONI, M., ARTONI, G., and GIANAZZA, E. (1984). *Anal. Biochem.* **137**, 410.

VERBRUGGEN, R. (1975). *Clin. Chem.* **21**, 5.

VERHEGGEN, T., MIKKERS, F., EVERAERTS, F., OERLEMANS, F., and DE BRUYN, C. (1980). *J. Chromatogr.* **182**, 317.

VERSEY, J. (1976). In *Chromatographic and electrophoretic techniques* (ed. I. Smith), Vol. 2, *Zone electrophoresis* (4th edn.), p. 347. Heinemann Medical Books, London.

VESTERBERG, O. (1971). *Meth. Enzymol.* **22**, 389.

—— and GRAMSTRUP-CHRISTENSEN, B. (1984). *Electrophoresis* **5**, 282.

—— and HANSEN, L. (1978). *Biochim. biophys. Acta* **534**, 369.

—— and SVENSSON, H. (1966). *Acta chem. Scand.* **20**, 820.

VO, K. P., MILLER, M. J., GEIDUSCHEK, E. P., NIELSEN, C., OLSON, A., and XUONG, N-H. (1981). *Anal. Biochem.* **112**, 258.

VOORDOUW, G. and ROCHE, R. S. (1975). *Biochemistry* **14**, 4667.

WACHSLICHT, H. and CHRAMBACH, A. (1978). *Anal. Biochem.* **84**, 533.

WAHRMANN, J. P., GROS, F., PIAU, J. P., and SCHAPIRA, G. (1980). *Biochim. biophys. Acta* **612**, 421.

WALLACE, R. W. and DIECKERT, J. W. (1976). *Anal. Biochem.* **75**, 498.

——, YU, P. H., DIECKERT, J. P., and DIECKERT, J. W. (1974). *Anal. Biochem.* **61**, 86.

WALTERS, J. A. L. I. and BONT, W. S. (1979). *Anal. Biochem.* **93**, 41.

WALTON, K. E., STYER, D., and GRUERNSTEIN, E. I. (1979). *J. biol. Chem.* **254**. 7951.

WARD, L. D. and WINZOR, D. J. (1981). *Arch. Biochem. Biophys.* **209**, 650.

WARDI, A. H. and ALLEN, W. S. (1972). *Anal. Biochem.* **48**, 621.

WARNICK, G. R., MAYFIELD, C., ALBERS, J. J., and HAZZARD, W. R. (1979). *Clin. Chem.* **25**, 279.

WARREN, D. F., NAUGHTON, M. A., and FINK, M. A. (1982). *Anal. Biochem.* **121**, 331.

WATTS, J. W., KING, J. M., and SAKAI, F. (1977). *Anal. Biochem.* **79**, 579.

WEBB, K. S., MICKEY, D. D., STONE, K. R., and PAULSON, D. F. (1977). *J. Immunol. Meth.* **14**, 343.

WEBER, K. and KUTER, D. J. (1971). *J. biol. Chem.* **246**, 4505.

—— and OSBORN, M. (1969). *J. biol. Chem.* **244**, 4406.

—— —— (1975). In *The proteins* (eds. H. Neurath, R. L. Hill, and C.-L. Boeder) (3rd edn), Vol. 1. Academic Press, New York, San Francisco, London.

——, PRINGLE, J. R., and OSBORN, M. (1972). *Meth. Enzymol.* **26**, 3.

WEEKE, B. (1973a). In *A manual of quantitative immunoelectrophoresis, methods and applications* (eds. N. H. Axelsen, J. Krøll, and B. Weeke), p. 15. Universitetsforlagert, Oslo.

—— (1973b). In *A manual of quantitative immunoelectrophoresis, methods and applications* (eds. N. H. Axelsen, J. Krøll, and B. Weeke), p. 37. Universitetsforlagert, Oslo.

WEIDERKAMM, E., WALLACH, D. F. H., and FLÜCKIGER, R. (1973). *Anal. Biochem.* **54**, 102.

WEIDINGER, S. and CLEVE, H. (1984). *Electrophoresis* **5**, 223.

WEIGELE, M., DE BERNARDO, S., LEIMGRUBER, W., CLEELAND, R., and GRUNBERG, E. (1973). *Biochem. Biophys. Res. Commun.* **54**, 899.

WEINER, A. M., PLATT, T., and WEBER, K. (1972). *J. biol. Chem.* **247**, 3242.

WEISS, P. M., DEL VECCHIO, V. G., and BURNETT, J. B. (1978). *Anal. Biochem.* **84**, 512.

WEITZMAN, S., SCOTT, V., and KEEGSTRA, K. (1979). *Anal. Biochem.* **97**, 438.

WELLNER, D. and HAYES, M. B. (1973). *Ann. N.Y. Acad. Sci.* **209**, 34.

WEST, D. W. and TOWERS, G. E. (1976). *Anal. Biochem.* **75**, 58.

WESTERMEIER, R., POSTEL, W., WESER, J., and GÖRG, A. (1983). *J. Biochem. Biophys. Meth.* **8**, 321.

WHEELER, F. C., FISHEL, R. A., and WARNER, R. C. (1977). *Anal. Biochem.* **78**, 260.

WHITE, M. D. and RALSTON, G. B. (1981). *Electrophoresis* **2**, 240.

WHITEHEAD, J. S., KAY, E., LEW, J. Y., and SHANNON, L. M. (1971). *Anal. Biochem.* **40**, 287.

WHITNEY, J. B., COPLAND, G. T., SKOW, L. C., and RUSSELL, E. C. (1979). *Proc. Natl. Acad. Sci. U.S.A.* **76**, 867.

WIENAND, U., SCHWARZ, Z. and FEIX, G. (1979). *FEBS Lett.* **98**, 319.

WIESLANDER, L. (1979). *Anal. Biochem.* **98**, 305.

WILBUR, W. J. and LIPMAN, D. J. (1983). *Proc. Natl. Acad. Sci. U.S.A.* **80**, 726.

WILLARD, K. E. and ANDERSON, N. G. (1981). *Clin. Chem.* **27**, 1327.

——, GIOMETTI, C. S., ANDERSON, N. L., O'CONNOR, T. E., and ANDERSON, N. G. (1980). *Anal. Biochem.* **100**, 289.

WILLEMS, M., WIERINGA, B., MULDER, J., AB, G., and GRUBER, M. (1979). *Eur. J. Biochem.* **93**, 469.

WILLIAMS, D. E. and REISFELD, R. A. (1964). *Ann. N.Y. Acad. Sci.* **121**, 373.

WILLIAMS, J. G. and GRATZER, W. B. (1971). *J. Chromatogr.* **57**, 121.

WILLOUGHBY, E. W. and LAMBERT, A. (1983). *Anal. Biochem.* **130**, 353.

WILSON, C. M. (1979). *Anal. Biochem.* **96**, 263.

WILSON, J. T., WILSON, L. B., REDDY, V. B., CAVALLESCO, C., GHOSH, P. K., DE RIEL, J. K., FORGET, B. G., and WEISSMAN, S. M. (1980). *J. biol. Chem.* **255**, 2807.

WINTER, A. (1977). In *Electrofocusing and isotachophoresis* (eds. B. J. Radola and D. Graesslin), p. 433. Walter de Gruyter, Berlin, New York.

—— and KARLSSON, C. (1976). *Application Note 219*. LKB-Produkter AB.

——, PERLMUTTER, H., and DAVIES, H. (1975). *Application Note 198*. LKB-Produkter AB.

WOODS, E. F. (1967). *J. biol. Chem.* **242**, 2859.

WOTJKOWIAK, Z., BRIGGS, R. C., and HNILICA, L. S. (1983). *Anal. Biochem.* **129**, 486.

WRAY, W., BOULIKAS, T., WRAY, V. P., and HANCOCK, R. (1981). *Anal. Biochem.* **118**, 197.

WRIGLEY, C. W. (1971). *Meth. Enzymol.* **22**, 559.

WU, R. S., STEDMAN, J. D., WEST, M. H. P., PANTAZIS, P., and BONNER, W. M. (1982). *Anal. Biochem.* **124**, 264.

WYCKOFF, M., RODBARD, D., and CHRAMBACH, A. (1977). *Anal. Biochem.* **78**, 459.

YAKIN, H. M., KRONBERG, H., ZIMMER, H-G., and NEUHOFF, V. (1982). *Electrophoresis* **3**, 244.

YOUNG, D. S. and TRACY, R. P. (1983). *Electrophoresis* **4**, 117.

YOUNG, O. A. and DAVEY, C. L. (1981). *Biochem. J.* **195**, 317.

YOUNG, R. W. and FULHORST, H. W. (1965). *Anal. Biochem.* **11**, 389.

YUEN, K. C. L., JOHNSON, T. K., DENELL, R. E., and CONSIGLI, R. A. (1982). *Anal Biochem.* **126**, 398.

ZACHARIUS, R. M., ZELL, T. E., MORRISON, J. H. and WOODLOCK, J. J. (1969). *Anal. Biochem.* **30**, 148.

ZANNIS, V. I. and BRESLOW, J. L. (1980). *J. biol. Chem.* **255**, 1759.

ZEILLER, K., LÖSER, R., PASCHER, G., and HANNIG, K. (1975). *Hoppe-Seyler's Z. physiol. Chem.* **356**, 1225.

ZELENSKÝ, I., ZELENSKÁ, V., KANIANSKY, D., HAVAŠI, P., and LEDNÁROVÁ, V. (1984). *J. Chromatogr.* **294**, 317.

ZIEGLER, A. and KÖHLER, G. (1976). *FEBS Lett.* **71**, 142.

ZINN, K., DI MIAO, D., and MANIATIS, T. (1983). *Cell* **34**, 865.

ZIOLA, B. R. and SCRABA, D. G. (1976). *Anal. Biochem.* **72**, 366.

ZWEIDLER, A. and COHEN, L. H. (1972). *Fed. Proc. Fed. Am. Soc. exp. Biol.* **31**, 926.

GLOSSARY

ACTH	adrenocorticotropin
ADP	adenosine diphosphate
AMP	adenosine monophosphate
c-AMP	cyclic adenosine monophosphate
ANS	anilinonaphthalene sulphonate
ATP	adenosine triphosphate
Bis	N,N'-methylenebisacrylamide
C (per cent)	proportion of cross-linking agent (e.g. Bis etc.) in total gel monomer mixture (acrylamide + Bis) in grams per 100 g of mixture
CMP	cytidine monophosphate
CPC	N-cetylpyridinium chloride
CSF	cerebrospinal fluid
CTAB	cetyltrimethylammonium bromide
DATD	N,N'diallyltartardiamide
DEAE	diethylaminoethyl
DHEBA	N,N'-(1,2-dihydroxyethylene)bisacrylamide
DMAPN	dimethylaminopropionitrile
DMSO	dimethyl sulphoxide
DNA	deoxyribonucleic acid
DNP	dinitrophenol
DNS	dansyl
DOC	deoxycholate
EDIA	ethylenediacrylate
EDTA	ethylenediaminetetraacetic acid
FITC	fluorescein isothiocyanate
GMP	guanosine monophosphate
HDL	high density lipoprotein
IEF	iso-electric focusing
IMP	inosine monophosphate
ITP	isotachophoresis
KDS	potassium dodecyl sulphate
LDL	low density lipoprotein
MDPF	2-methoxy-2, 4-diphenyl-3(2H)-furanone
MW	molecular weight
NAD	nicotinamide adenine dinucleotide (oxidised)
NADH	nicotinamide adenine dinucleotide (reduced)
NCS	Nuclear Chicago solubiliser
NEPHGE	non-equilibrium pH gradient electrophoresis
OPA	o-phthaldialdehyde
PAGE	polyacrylamide gel electrophoresis
PAGIF	polyacrylamide gel iso-electric focusing
PAS	periodic acid–Schiff
pI	iso-electric point
POPOP	1,3-bis-2-(5-phenyloxazole)

PPO	2,5-diphenyloxazole
RNA	ribonucleic acid
SDS	sodium dodecyl sulphate
SDS–PAGE	polyacrylamide gel electrophoresis performed in the presence of sodium dodecyl sulphate
T (per cent)	total concentration of monomer (acrylamide + Bis) in grams per 100 ml used in polyacrylamide gel preparation
TCA	trichloroacetic acid
TEMED	N,N,N′,N′-tetramethylethylenediamine
TLC	thin-layer chromatography
TLE	thin-layer electrophoresis
TMP	thymidine monophosphate
TPCK	L-1-tosylamide-2-phenylethylchloromethyl ketone
tris	2-amino-2-(hydroxymethyl)propane-1,3-diol
UMP	uridine monophosphate
VLDL	very low density lipoprotein

INDEX